Nanoelectronic Circuit Design

Niraj K. Jha • Deming Chen

Editors

Nanoelectronic Circuit Design

 Springer

Editors
Niraj K. Jha
Department of Electrical Engineering
Princeton University
NJ, USA
jha@princeton.edu

Deming Chen
Department of Electrical and Computer
Engineering
University of Illinois at
Urbana-Champaign
IL, USA
dchen@illinois.edu

ISBN 978-1-4419-7444-0   e-ISBN 978-1-4419-7609-3
DOI 10.1007/978-1-4419-7609-3
Springer New York Dordrecht Heidelberg London

Printed on acid-free paper

Springer is part of Springer Science+Business Media (www.springer.com)

Preface

After enjoying a three-decade ride as the top semiconductor technology for implementing integrated circuits (ICs), the era of single-gate complementary metal-oxide semiconductors (CMOS) is coming to an end. Waiting in the wings as replacements are various interesting new nanotechnologies. In the past decade, a contender for replacing bulk CMOS technology has been double-gate field-effect transistor technology, most manufacturable of which are FinFETs. There are also other contenders, such as nanowires, carbon nanotubes, graphene nanoribbons, resonant tunneling diodes, quantum cellular automata, etc. We may also soon see hybrid nano/CMOS designs in which the memory has been implemented in a new technology, while the processing elements have been implemented in CMOS.

Although these nanotechnologies have attracted significant attention over the last decade, the emphasis of interest, as expected, has been more on their physics, chemistry, and fabrication aspects. However, interesting new nanoelectronic circuit designs are beginning to emerge that herald an exciting new era of IC design. These designs deal with both logic as well as interconnect.

Our aim in this book is to introduce readers to the emerging design paradigms in various nanotechnologies, and to bridge the existing gap between nanodevice research and nanosystems design. The book focuses on state-of-the-art research activities, yet, at the same time, covers the fundamental principles behind the nanotechnology developments. The ultimate goal is to expose the great potential of nanoelectronic technology and the unique challenges it poses along the device-to-system spectrum. A rich set of references is included at the end of each chapter to give pointers to readers who want to dig deeper. In addition, some exercises are also included in each chapter to allow the use of the book for a first-year graduate-level course on nanoelectronic circuit design.

The book is organized by grouping together chapters on each nanotechnology. Chapter 1 introduces various nanotechnologies. Chapters 2 and 3 deal with FinFET logic and memory design. Chapter 4 describes a nano/CMOS dynamically reconfigurable architecture. Chapters 5–7 discuss nanowire-based ICs and architectures. Chapters 8 and 9 describe how reliable logic circuits and field-programmable gate arrays can be built using nanotubes. Chapter 10 deals with circuit design based on graphene nanoribbon transistors. Chapter 11 compares copper, nanotube, graphene, and optics for implementing interconnects on chips. Chapters 12 and 13 discuss circuit design with resonant tunneling diodes and quantum cellular automata, respectively. The chapters are fairly independent of each other. Thus, any subset can be chosen for a one-semester course.

Last, but not the least, Niraj would like to thank his father, Dr. Chintamani Jha, his wife, Shubha, and his son, Ravi, for their encouragement and understanding. Deming would like to thank his wife, Li, and his sons, Jeffrey and Austin, for their love and understanding.

Niraj K. Jha
Deming Chen

Contents

Contributors

Csaba Andras Moritz
Department of Electrical and Computer Engineering,
University of Massachusetts, Amherst, MA, USA
andras@ecs.umass.edu

Deming Chen
Department of Electrical and Computer Engineering, University of Illinois
at Urbana-Champaign, IL, USA
dchen@illinois.edu

Scott Chilstedt
Department of Electrical Engineering, University of Illinois
at Urbana-Champaign, Urbana, IL, USA

Chi On Chui
Department of Electrical Engineering, The University of California,
Los Angeles, CA, USA

Giovanni De Micheli
Institute of Electrical Engineering, EPFL, Lausanne, Switzerland
giovanni.demicheli@epfl.ch

Chen Dong
Department of Electrical and Computer Engineering, University of Illinois,
at Urbana-Champaign, IL, USA
cdong3@illinois.edu

Soumya Eachempati
Department of Computer Science and Engineering, The Pennsylvania
State University, University Park, PA, USA

Aman Gayasen
Department of Computer Science and Engineering, The Pennsylvania
State University, University Park, PA, USA

Pallav Gupta
Core CAD Technologies, Intel Corporation, Folsom, CA 95630, USA
pallav.gupta@intel.com

M. Haykel Ben Jamaa
Commissariat à l'Energie Atomique, DRT-LETI-DACLE-LISAN,
Grenoble Cedex, France

Niraj K. Jha
Department of Electrical Engineering, Princeton University, NJ, USA
jha@princeton.edu

Rajiv Joshi
IBM, Thomas J. Watson Research Center, Yorktown Heights, NY, USA
rvjoshi@us.ibm.com

Rouwaida Kanj
IBM Austin Research Laboratory, Austin, TX, USA

Keunwoo Kim
IBM, Thomas J. Watson Research Center, Yorktown Heights, NY, USA

Kyung-Hoae Koo
Department of Electrical Engineering, Stanford University, Stanford, CA, USA

Albert Lin
Department of Electrical Engineering, Stanford University, Stanford, CA, USA

Prateek Mishra
Department of Electrical Engineering, Princeton University, Princeton, NJ, USA

Subhasish Mitra
Department of Electrical Engineering, Stanford University, Stanford, CA, USA
subh@stanford.edu

Kartik Mohanram
Electrical and Computer Engineering, Rice University, Houston, TX, USA
kmram@rice.edu

Anish Muttreja
nVidia, Santa Clara, CA, USA

Pritish Narayanan
Department of Electrical and Computer Engineering,
University of Massachusetts, Amherst, MA, USA

Nishant Patil
Department of Electrical Engineering, Stanford University, Stanford, CA, USA
nppatil@stanford.edu

H.-S. Philip Wong
Department of Electrical Engineering, Stanford University, Stanford, CA, USA

Krishna C. Saraswat
Department of Electrical Engineering, Stanford University, Stanford, CA, USA
saraswat@stanford.edu

Li Shang
Department of Electrical, Computer, and Energy Engineering,
University of Colorado, Boulder, CO, USA

N. Vijaykrishnan
Department of Computer Science and Engineering, The Pennsylvania
State University, University Park, PA, USA
vijay@cse.psu.edu

Hai Wei
Department of Electrical Engineering, Stanford University, Stanford, CA, USA

Xuebei Yang
Department of Electrical and Computer Engineering, Rice University,
Houston, TX, USA

Jie Zhang
Department of Electrical Engineering, Stanford University, Stanford, CA, USA

Wei Zhang
School of Computer Engineering, Nanyang Technological University, Singapore

Introduction to Nanotechnology

Deming Chen and Niraj K. Jha

The semiconductor industry has showcased a spectacular exponential growth in the number of transistors per integrated circuit for several decades, as predicted by Moore's law. Figure 1 shows the future technology trend predicted by ITRS (International Technology Roadmap for Semiconductors) [1]. By 2023, the physical gate length would scale down to 4.5 nm. Actually, according to a study [2], future devices could theoretically scale down to 1.5 nm with 0.04 ps switching speed and 0.017 eV energy consumption. However, maintaining such an exponential growth rate is a major challenge. Physical dimensions and electrostatic limitations faced by conventional process and fabrication technologies will likely thwart the dimensional scaling of complementary metal-oxide-semiconductor (CMOS) devices within the next decade. Figure 2 from ITRS shows that after 2016, the manufacturable solutions are unknown (the shaded area).

To enable future technology scaling, new device structures for next-generation technology have been proposed. The most promising ones so far include carbon nanotube field-effect transistors (CNFETs), FinFETs, nanowire FETs, III/V compound-based devices, graphene nanoribbon devices, resonant tunneling diodes, and quantum dot devices. Many of these devices have been shown to have favorable device properties and new device characteristics, and require new fabrication techniques. These nanoscale devices have a significant potential to revolutionize the fabrication and integration of electronic systems and scale beyond the perceived scaling limitations of traditional CMOS.

While innovations in nanotechnology originate at the individual device level, realizing the true impact of electronic systems demands that we translate these device-level capabilities into system-level benefits. Existing books on nanotechnology mainly focus on nanodevices, and typically do not deal with nanoscale circuits and their design issues. This book strives to bridge the existing gap between nanodevice research and nanosystem design. The book is organized by grouping together

D. Chen (✉)
Department of Electrical and Computer Engineering, University of Illinois
at Urbana-Champaign, IL, USA
e-mail: dchen@illinois.edu

N.K. Jha and D. Chen (eds.), *Nanoelectronic Circuit Design*,
DOI 10.1007/978-1-4419-7609-3_1, © Springer Science+Business Media, LLC 2011

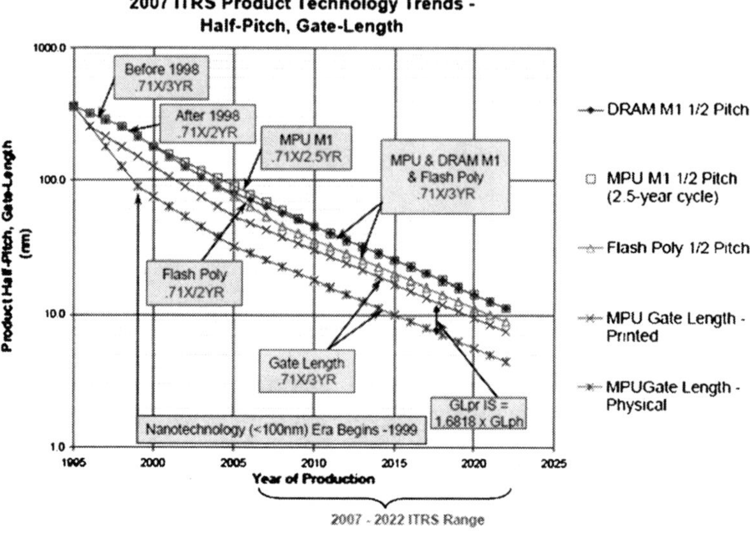

Fig. 1 Technology trend predicted by ITRS [1]

Year of Production	2016	2017	2018	2019	2020	2021	2022
DRAM ½ pitch (nm) (contacted)	22	20	18	16	14	13	11
DRAM and Flash							
DRAM ½ pitch (nm)	23	20	18	16	14	13	11
Flash ½ pitch (nm) (un-contacted poly)	18	16	14	13	11	10	9
Contact in resist (nm)	25	22	20	18	16	14	12
Contact after etch (nm)	23	20	18	16	14	13	11
Overlay [A] (3 sigma) (nm)	4.5	4.0	3.6	3.2	2.8	2.5	2.3
CD control (3 sigma) (nm) [B]	1.9	1.7	1.5	1.3	1.2	1.0	0.9

Fig. 2 Manufacturable solutions unknown after 2016 [1]

chapters on each nanotechnology. The sequence in which the nanotechnologies are covered is FinFETs, nano/CMOS, nanowires, carbon nanotubes (CNTs), graphene, resonant tunneling diodes, and quantum cellular automata.

In this chapter, we familiarize readers with the fundamentals of nanomaterials and nanodevices. We focus on device-level characteristics and their unique features. With regard to using these devices to build next-generation logic and circuits, we refer readers to the individual book chapters. Chapters 2 and 3 deal with FinFET logic and memory design. Chapter 4 describes a nano/CMOS dynamically reconfigurable architecture. Chapters 5–7 discuss nanowire-based integrated circuits (ICs) and architectures. Chapters 8 and 9 describe how reliable logic circuits and field-programmable gate arrays can be built using nanotubes. Chapter 10 deals with circuit design based on graphene nanoribbon transistors. Chapter 11 compares copper, nanotube, graphene, and optics for implementing interconnects on chips. Chapters 12 and 13 discuss circuit design with resonant tunneling diodes and quantum cellular automata, respectively.

1 FinFETs

Bulk CMOS technology has led to a steady miniaturization of transistors with each new generation, yielding continuous performance improvements. However, according to the ITRS, the scaling of bulk CMOS to sub-22-nm gate lengths faces several significant challenges: short-channel effects, high sub-threshold leakage, device-to-device variations, etc. To the rescue are likely to come double-gate field-effect transistors (DG-FETs) [3]. Such FETs provide better control of short-channel effects, lower leakage, and better yield in aggressively scaled CMOS processes.

Figure 3 shows a planar DG-FET. In it, two gates, the top and bottom ones, control the channel. However, because of the difficulty in aligning the top and bottom gates, a particular type of DG-FET, called FinFET, is the most popular DG-FET implementation because FinFETs are relatively easy to manufacture and are free of alignment problems that plague many other DG-FET geometries. Also, the use of a lightly doped channel in FinFETs makes them resistant to random dopant variations.

A FinFET is shown in Fig. 4. A FinFET is a FET in which the channel is turned on its edge and made to stand up. The two gates of a FinFET can be made independently controllable by etching away its top part, as shown in Fig. 5. Such a FinFET is called an independent-gate (IG) FinFET, whereas when the two gates of the FinFET are shorted (as in Fig. 4), the FinFET is called a shorted-gate (SG) FinFET, or sometimes tied-gate FinFET.

SG FinFETs provide improved drive strength and control of the channel. Thus, logic gates based on SG FinFETs are the fastest. On the other hand, the voltage bias in an IG FinFET can be used to linearly modulate the threshold voltage of the front gate. This phenomenon can be used to reduce the leakage power of the logic gate based on IG FinFETs by one to two orders of magnitude by reverse-biasing their back gates. This, however, comes at the price of increased logic gate delay.

Inverters based on SG and IG FinFETs, respectively, are shown in Fig. 6a, b. In Fig. 6b, V_{hi} and V_{low} depict the back-gate bias for the pFinFET and nFinFET, respectively. If the bias V_{hi} (V_{low}) is above the supply voltage (below ground), the

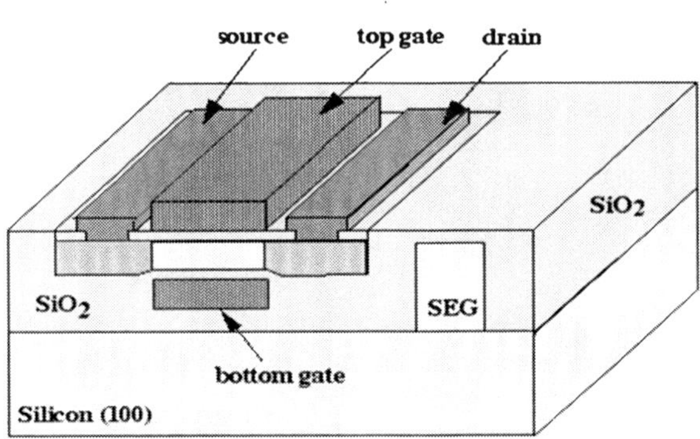

Fig. 3 A planar DG-FET

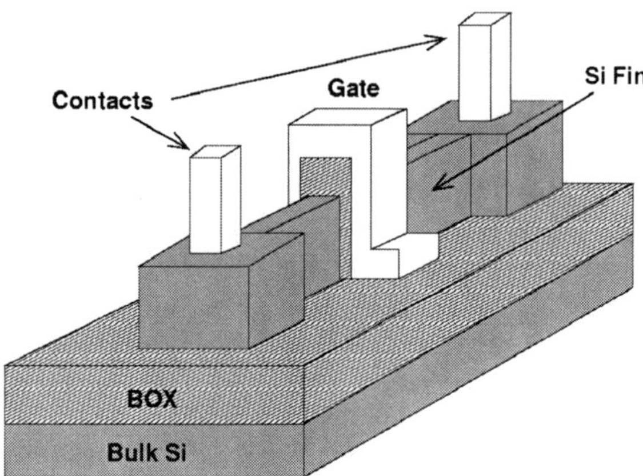

Fig. 4 A FinFET

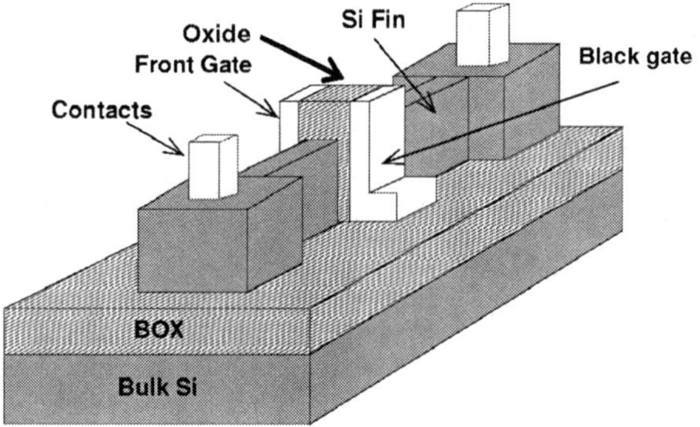

Fig. 5 An independent-gate FinFET

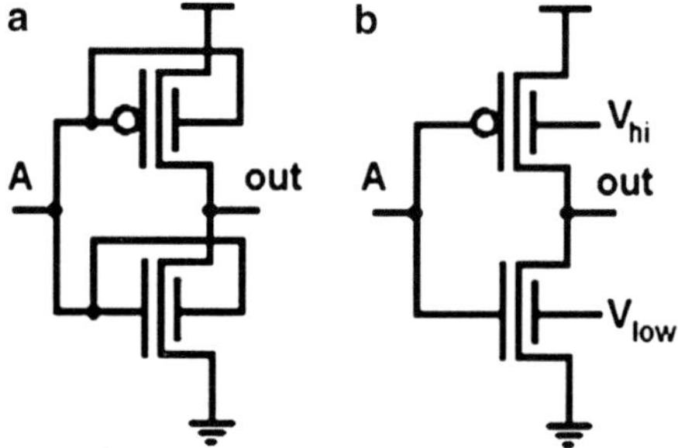

Fig. 6 Inverters based on SG
and IG FinFETs

corresponding transistors are reverse-biased. This reduces leakage at the expense of increased delay. On the other hand, if the bias V_{hi} (V_{low}) is below the supply voltage (above ground), the corresponding transistors are forward-biased. This reduces delay at the expense of leakage.

Use of IG FinFETs also enables novel circuit modules [4–6]. In digital logic design, the ability to independently control the two gates of a FinFET can be exploited in several ways: (1) merging of pairs of parallel transistors to reduce circuit area and capacitance, (2) use of a back-gate voltage bias to modulate transistor threshold voltage, etc. A parallel transistor pair consists of two transistors with their source and drain terminals tied together. Merging of transistors reduces parasitic capacitances. We will study these techniques in more detail in Chaps. 2 and 3.

Because a transistor's threshold voltage affects both its leakage power and delay, it is important to consider the assignment of FinFET back-gate bias during technology mapping or along with a gate-sizing step for a technology-mapped circuit. Simultaneous assignment of gates sizes and back-gate bias voltages for FinFETs was studied in [7], where sized and biased FinFET circuits were compared to bulk CMOS circuits at the same channel length and shown to have lower leakage as well as better area and delay characteristics.

The above characteristics of FinFETs, chiefly their double-gate structure, open up the possibility for novel circuit design and very fine-grain run-time power-delay trade-offs (at the granularity of a few clock cycles) [8] that were hitherto not possible.

2 Carbon Nanotube Devices

Carbon is a Group 14 element that resides above silicon in the Periodic Table. Like silicon and germanium, carbon has four electrons in its valence shell. In its most common state, it is an amorphous nonmetal, like coal or soot. In a crystalline, tetrahedrally bonded state, carbon becomes diamond, an insulator with relatively large band gap. However, when carbon atoms are arranged in crystalline structures composed of hexagonal benzene-like rings, they form a number of allotropes that offer exceptional electrical properties. For the replacement of silicon in future transistor channels, the two most promising of these allotropes are CNTs [9] and graphene.

In their semiconducting forms, these carbon nanomaterials exhibit room-temperature mobilities over ten times greater than silicon. This translates to devices with significant improvements in performance and power savings, allowing higher integration at the same power density. In addition, they can be scaled to smaller feature sizes than silicon while maintaining their electrical properties. It is for these reasons that the Emerging Research Devices and Emerging Research Materials working groups of the ITRS have selected carbon-based nanoelectronics as their recommended "beyond CMOS" technology [1]. In this section, we focus on CNTs and introduce graphene nanoribbons in the next section.

CNTs can be categorized into two groups: single-walled carbon nanotubes (SWCNTs) and multiwalled carbon nanotubes (MWCNTs) (Fig. 7). An SWCNT is a hollow cylinder with a diameter of roughly 1–4 nm, and can be thought of as a rolled-up sheet of monolayer graphene. An MWCNT is composed of a number of SWCNTs nested inside one another in order of diameter, and can be thought of as a rolled-up sheet of multilayer graphene. MWCNTs have dimensions greater than SWCNTs and are typically from four to several tens of nanometers in diameter. Carbon nanotubes vary in length and have been produced in lengths of up to 1 mm. With diameters of less than 10 nm, this allows for exceptionally high aspect ratios, making nanotubes essentially a one-dimensional material.

Due to their cylindrical symmetry, there is a discrete set of directions in which a graphene sheet can be rolled to form an SWCNT. To characterize each direction, two atoms in the graphene sheet are chosen, one of which serves the role of the origin. The sheet is rolled until the two atoms coincide. The vector pointing from the first atom to the second is called the chiral vector, and its length is equal to the circumference of the nanotube. The direction of the nanotube axis is perpendicular to the chiral vector.

Depending on the rolling method, SWCNTs can be either metallic or semiconducting. The mean free path of SWCNTs is in the micrometer range. Within this length, ballistic transport is observed in SWCNTs. Thus, its resistance is a constant without scattering effects. Metallic SWCNTs have high electron mobility and robustness and can carry a current density $\sim$1,000$\times$ larger than copper (Cu), making them attractive for use in nanoscale interconnects and nano-electromechanical systems (NEMS).

Semiconducting SWCNTs, on the other hand, have ideal characteristics for use in FETs. It is reported that single CNFET devices can be 13$\times$ and 6$\times$ faster than pMOS and nMOS transistors with a similar gate length, based on the intrinsic CV/I gate delay metric, when local interconnect capacitances and CNT imperfections are not considered [10, 11]. In the past, it was difficult to mass-produce CNT-based circuits because of the inability to accurately control nanotube growth and position.

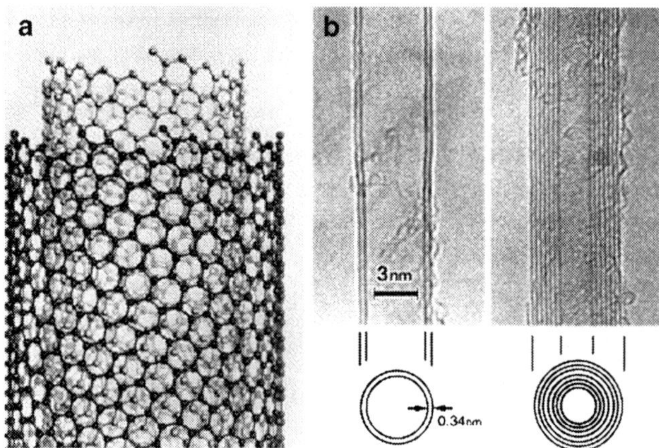

Fig. 7 Multiwalled carbon nanotubes discovered in 1991: (**a**) schematic; (**b**) TEM image [9]

However, recent research has demonstrated the fabrication of dense, perfectly aligned arrays of linear SWCNTs [12], and wafer-scale CNT-based logic devices [13, 14].

The first reports of operational room-temperature CNFETs came in 1998 from groups at IBM [15] and the Delft University of Technology [16]. The structures of two CNFETs are shown in Fig. 8. These designs have similar architectures: a single nanotube (either single-walled or multiwalled) behaves as the channel region and rests on metal source/drain electrodes. Both designs use SiO_2 dielectric on top of a silicon back-gate. Au or Pt contacts are used as source/drain (S/D) electrodes because their work functions are close to the CNT work function to improve transistor speed.

In 2001, the group from Delft University of Technology enhanced its previous CNFET design by using aluminum local gates to control individual transistors, as shown in Fig. 9a, b [17]. This design uses a narrow Al gate and is insulated by thin native Al_2O_3. The Al gate is defined by e-beam lithography on silicon oxide, and the gate insulator is grown by exposing the Al gate to air. SWCNTs are then deposited onto the wafer on top of the predefined gates. Finally, Au source and drain contacts are created by e-beam lithography. The device I-V characteristic shows that this new CNFET works as an enhancement-mode p-type device. In order to build logic functions, it is necessary to have both n-type and p-type transistors available. However, all of the aforementioned CNFETs are of p-type. n-Type CNFETs can be obtained by annealing of a p-type device or through doping.

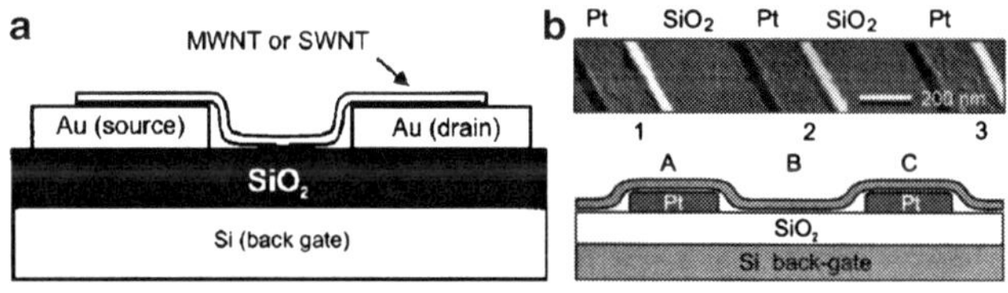

Fig. 8 Si back-gated CNFET: (**a**) with Au S/D contacts [15]; (**b**) with Pt S/D contacts [16]

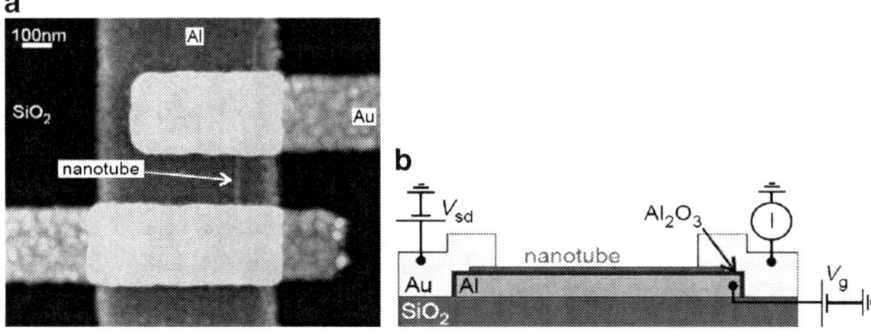

Fig. 9 (**a**) Atomic force microscopy (AFM) image of a single-nanotube transistor; (**b**) CNFET with individual Al back gate [17]

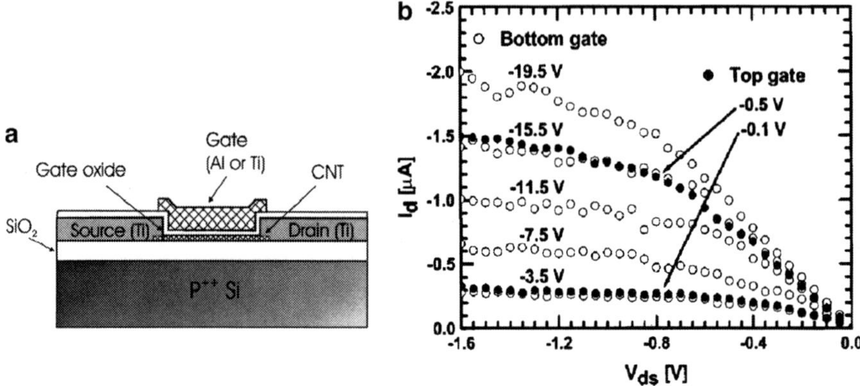

Fig. 10 Top-gated CNFET: (**a**) device structure; (**b**) I-V comparison to back-gated device [18]

A major improvement in CNT devices was made in 2002 through the creation of an MOS structure with a gate electrode on top of the nanotube channel and separated by a thin layer of SiO$_2$ dielectric. The structure of this top-gated design resembles common silicon MOSFET designs and is shown in Fig. 10 [18]. Top-gated designs offer several important improvements over back-gated devices, including operation at lower voltage, high-frequency operation, and higher reliability.

Single-nanotube devices provide great performance improvements over existing solutions. However, the integration of a single nanotube into existing integrated circuits is still a great challenge. Due to limited fabrication control of nanotube properties, a single-CNT device is susceptible to large performance fluctuations. One feasible solution is to use densely packed, horizontal arrays of nonoverlapping SWCNTs in the channel. This creates parallel conducting paths that can provide larger current than a lone CNT.

Having multiple CNTs in the channel also statistically averages the device-to-device variation and offers increased reliability against a single tube failure. However, how to create large-scale, high-density, and perfectly aligned nanotube arrays for mass production is still a challenge. The second challenge is that intrinsically, one-third of the fabricated carbon nanotubes are metallic. The metallic CNTs cannot be controlled by gate voltage and are always conducting, which deteriorates the transistor on/off ratio. Metallic nanotubes can be removed by techniques such as electrical breakdown.

Initial efforts to solve these problems have shown great promise [19–21]. For example, in [19], researchers used photolithography-defined parallel patterns on a quartz surface and grew CNTs with chemical vapor deposition (CVD) along these predefined patterns. Using this technique, nanotube arrays can be successfully fabricated with average diameters of ∼1 nm and lengths over 300 µm, with 99.9% alignment (Fig. 11). The nanotubes can then be transferred to the desired substrate, such as silicon or even a flexible plastic.

The resistivity of currently used Cu interconnects increases with downscaling dimensions due to electron-surface scattering and grain-boundary scattering. In the

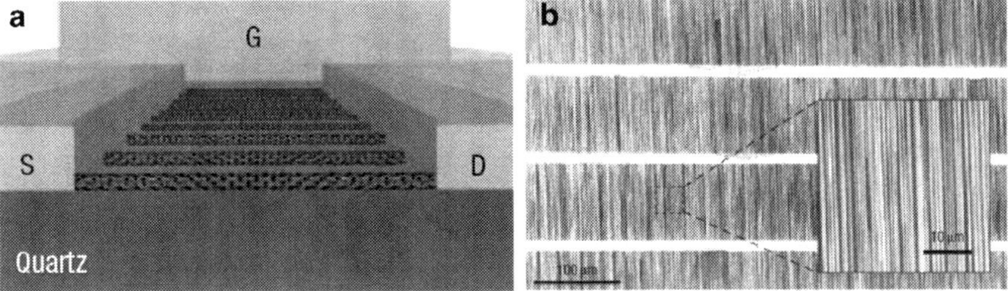

Fig. 11 (a) Cross-section of a CNFET with a channel of multiple parallel nanotubes; (b) dense arrays of SWCNTs [19]

meantime, current density is expected to become larger in future IC technologies [1]. These requirements motivate intensive studies on new solutions for nanoscale interconnect materials and structures. A CNT bundle is typically a bundle of SWCNTs, in which the tubes line up parallel to one another. A rope or bundle of SWCNTs conducts current in parallel and significantly reduces the resistance value. Thus, the SWCNT bundle interconnect can outperform Cu interconnect in propagation delay, especially for intermediate and long interconnects [22, 23]. In addition, SWCNT bundle vias offer high performance and high thermal conductivity (more than fifteen times higher than Cu [24]). This thermal property of a SWCNT bundle is specifically useful for 3D ICs to combat thermal problems.

3 Graphene Nanoribbon Devices

Like CNTs, graphene can exist in a number of forms. Monolayer graphene is made from a sheet of carbon exactly one atom thick, making it a pure two-dimensional material. Because graphene does not wrap around and connect back to itself like a CNT, its edges are free to bond with other atoms. Because unbonded edges are unstable, the edges are usually found to be passivated by absorbents such as hydrogen.

Bulk planar graphene can be patterned by lithography to create narrow strips called graphene nanoribbons (GNRs). Such ribbons can also be created through techniques such as chemical synthesis [25] and the "unzipping" of CNTs [26]. The smaller the width of the nanoribbon, the more impact the edge structure has on its properties. The crystallographic orientation of the edges is especially important. Figure 12 demonstrates two possible edge-state orientations, known as *armchair* and *zigzag*, and common width designations for each. In Fig. 12a, an armchair-GNR is shown with a size of $N = 10$, where N is the number of carbon atoms in width. In Fig. 12b, a zigzag GNR is shown with a size of $N = 5$, where N is the number of zigzag chains in width.

The intrinsic physical and electrical properties of graphene make it desirable for a large number of potential applications, ranging from biosensors to flexible

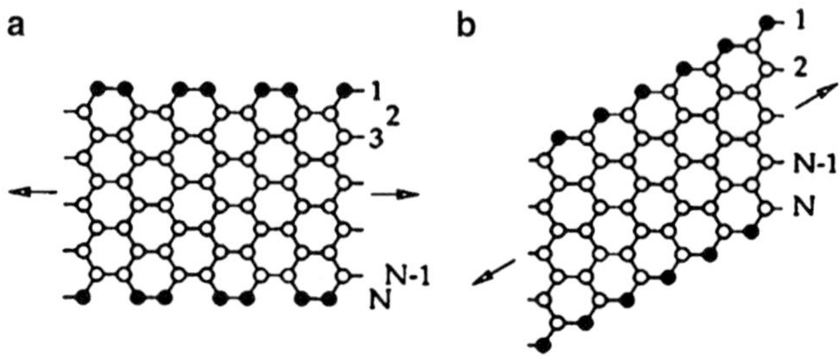

Fig. 12 Graphene nanoribbon classification: (**a**) armchair; (**b**) zigzag [27]

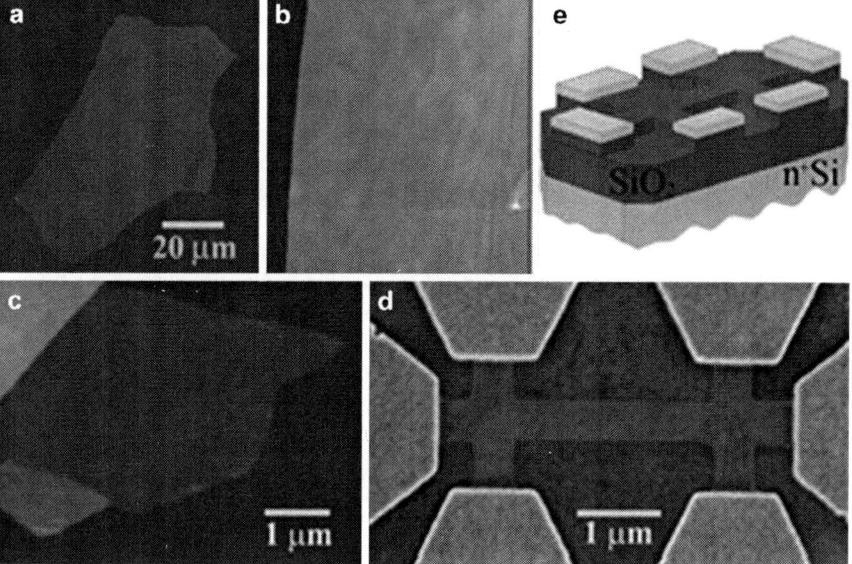

Fig. 13 (**a**)–(**c**) First images of few-layer graphene; (**d**)–(**e**) test structure [30]

electronics to solar cell electrodes. GNRs offer a mean free path $25\times$ longer than Cu, and carrier mobilities over $10\times$ higher than silicon [28, 29]. Furthermore, its intrinsic mobility limit is 2×10^5 cm^2/Vs [28]. Thus, one of the most exciting applications is the use of graphene channels in future high-performance transistors. To that end, physicists and materials scientists have been characterizing and experimenting with graphene to understand how it might be used to make such devices a reality.

Few-layer graphene was initially discovered at the University of Manchester in 2004 (Fig. 13a–c) [30]. The first graphene transistor-like device was created shortly thereafter. Also, a simple test structure was constructed to measure graphene's field effect behavior (Fig. 13d–e).

The major challenge posed by graphene is that in its native state as a large sheet, it behaves like a zero-band gap semiconductor, or semi-metal, meaning that it conducts electrons freely. This is desirable if graphene is to be used as an interconnect, and of course many researchers are exploring this possibility. However, in order to be effective for use in transistors, graphene needs to be made semiconducting and demonstrate a high on/off ratio. To obtain the necessary off state, a band gap must be introduced.

One way to open a band gap is to pattern graphene into a narrow ribbon to laterally confine the charge carriers in a quasi-one-dimensional system, somewhat analogous to a CNT. This idea was experimentally demonstrated in [31], where researchers at Columbia University used e-beam lithography to define two dozen graphene nanoribbons with widths ranging from 10 to 100 nm and varying crystallographic orientations. Conductance of these GNRs was measured at both 300 and 1.6 K. The measurements show that for a given crystallographic direction, the energy gap depends strongly on the width of the GNR. As the ribbons are made smaller, less conductance is observed, which indicates a stronger semiconducting behavior. A similar experiment in [32] leads to a similar conclusion. In both cases, researchers also note an apparent dependence of the electrical behavior on the edge states of the nanoribbons.

One of the first works to demonstrate sub-10-nm width GNRFETs was presented in [33]. The authors were able to achieve such dimensions because instead of patterning GNRs from a planar sheet with e-beam lithography, they started with GNRs that had been chemically derived at smaller dimensions. In this process, exfoliated graphene is dispersed into a chemical solution by sonification, creating very small fragments. The solution is then applied to a substrate and dried, and GNRs are identified with AFM. These GNRs ranged from monolayer to trilayer, and were deposited on a SiO_2 dielectric over a highly doped silicon back-gate, and contacted with Pd source/drain electrodes. A schematic of this design is shown in Fig. 14a. A number of devices were created using the bilayer GNRs, including both wide (10–60 nm) and small (<10 nm) GNRFETs. When tested, all of the large GNRs demonstrated metallic behavior due to vanishingly small band gaps, while all of the sub-10-nm GNRFETs were found to be semiconducting. AFM images of two example devices are shown in Fig. 14b, c.

The importance of GNR edge states was predicted by first-principles physics calculations [34, 35]. A recent experimental work used scanning tunneling

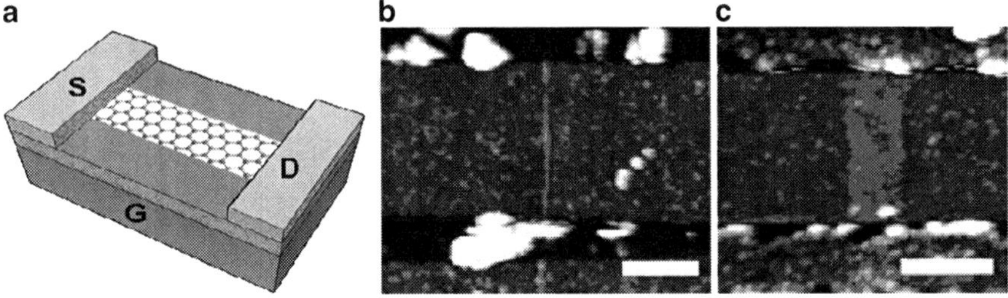

Fig. 14 Back-gated GNRFET: (**a**) schematic; (**b**) AFM image with $w \sim 2 \pm 0.5$ nm; (**c**) AFM image of GNR with $w \sim 60 \pm 5$ nm [33]

microscopy to verify this prediction, confirming that the crystallographic orientation of the edges significantly influences the electronic properties of nanometer-sized graphene [36]. By measuring the band gap of graphene samples and noting their predominant edge chirality, they observed that nanoribbons with predominantly zigzag edges are metallic, while predominantly armchair edges are semiconducting. For GNR transistors, all-semiconducting armchair edges are the most desirable. However, in the ribbons produced so far, the edges are not always atomically smooth and often contain a mixture of segment types, as seen in Fig. 15. In these cases, the semiconducting properties weaken, and the band gap becomes dependent on the ratio of armchair segments to zigzag segments [36].

In addition to logic applications, GNRFETs are well-suited for realizing individual ultra-high-frequency analog transistors [37]. Top-gated graphene transistors of various gate lengths have been fabricated with peak cutoff frequencies up to 26 GHz for a 150-nm gate [38]. Results also indicate that if the high mobility of graphene can be preserved during the device fabrication process, a cutoff frequency approaching terahertz may be achieved for graphene FETs with a gate length of just 50 nm and carrier mobility of 2,000 cm^2/Vs [37, 38].

Recent developments in the use of growth on copper foils and films have produced promising results. In ref. [39], a technique was demonstrated that produces 1-cm single layer graphene on copper films, and patterns GNRFETs on this layer. This allows direct fabrication of uniform transistor arrays using known thin film technology, without the need for delicate transfer processes (Fig. 16). This represents a potential advantage over the CNFET-based fabrication process where CNT transfer is required from the growth substrate to device substrate. However, in order to obtain uniform performance and achieve the scaling advantage of GNRFETs, advanced patterning techniques will be needed to define GNRs with known width, edge smoothness, and chirality. If atomically smooth edges are desired, sub-nanometer resolution patterning will be needed. Such a high-resolution patterning technique on a large scale is not available today. After all, differences

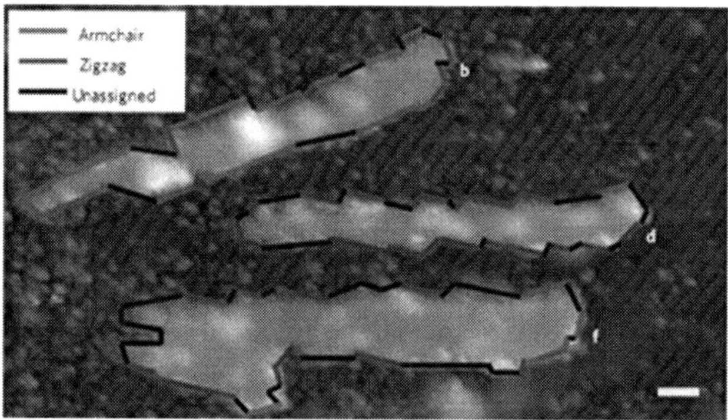

Fig. 15 GNR with different dominant edge chiralities. *Top*: Armchair GNR with 0.38-eV band gap. *Middle*: Zigzag GNR with 0.14-eV band gap. *Bottom*: Zigzag GNR with 0.12-eV band gap [36]

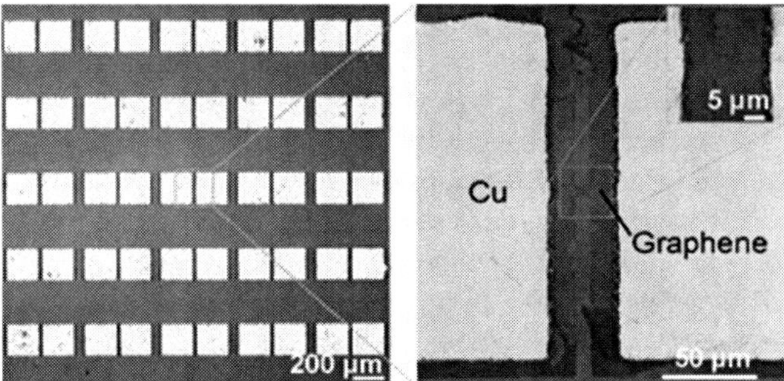

Fig. 16 Array of FETS fabricated on monolayer graphene [39]

in edge doping, lattice defects, oxide thicknesses, and ripples in the graphene sheets will still contribute to device variation. Therefore, variation-aware and fault-tolerant design techniques will need to be heavily employed.

A major opportunity lies in the development of advanced modeling and computer-aided design (CAD) techniques. With accurate GNRFET models, fast simulators, and GNR-centric CAD flows, high-level circuit and architecture design spaces can be more quickly explored. Such results can be used to determine the most promising directions for future development and help guide research on device fabrication.

4 Nanowire Devices

A nanowire is a nanostructure with the diameter constrained to tens of nanometers or less and an unconstrained length. Many different types of nanowires exist, including metallic (such as Ni or Au nanowires), semiconducting (such as silicon nanowires), and insulating (such as SiO_2 nanowires).

Silicon nanowire field-effect transistors (SiNW FETs) represent a promising alternative to conventional CMOS devices at the end of the ITRS roadmap because of the improved electrostatic control of the channel via the gate voltage and the consequent suppression of short-channel effects. To create active electronic elements, the first step is to chemically dope a semiconductor nanowire. This has already been done to individual nanowires to create p-type and n-type semiconductors. The next step is to find a way to create a p-n junction. This can be achieved in two ways. The first way is to physically cross a p-type wire over an n-type wire. The second way is to chemically dope a single wire with different dopants along the length. This method creates a p-n junction with only one wire.

Initial transport studies of SiNW FETs showed relatively low transconductance and carrier mobility. This was mainly due to poor contacts between the SiNWs and source-drain electrodes. Later on, researchers explored the limits of SiNW FETs by examining the influence of source-drain contact annealing and surface passivation on transistor properties. Thermal annealing and passivation of oxide defects by chemical modification were found to increase the average transconductance from 45 to 800 nS and average mobility from 30 to 560 cm^2/Vs with peak values from 2,000 nS and 1,350 cm^2/Vs, respectively [40].

The silicon nanowire transistor can suppress off-leakage because its thin wire-shaped silicon channel (nanowire channel) is effectively controlled by the surrounding gate (gate-all-around technology). The most recent advance was reported by Toshiba, which developed a 16-nm silicon nanowire transistor using multiple nanowires in the channel. It optimized gate fabrication and reduced the thickness of the gate sidewall to improve on-current [41]. Figure 17 shows a schematic of the device.

Recently, a junctionless nanowire transistor has been developed [42]. Figure 18 shows a schematic of an *n*-channel nanowire transistor. The underlying insulator layer (buried oxide) is not shown. The doping concentration in the channel is identical to that in the source and drain. In a classical trigate device, the source and drain are heavily doped *n*-type and the channel region under the gate is lightly doped *p*-type. In the junctionless gated resistor, the silicon nanowire is uniformly doped *n*-type and the gate material is *p*-type polysilicon.

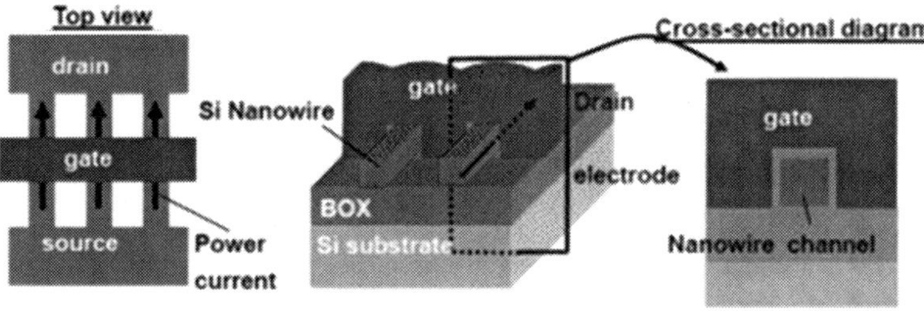

Fig. 17 A 3-dimensional structure of a silicon nanowire transistor [41]

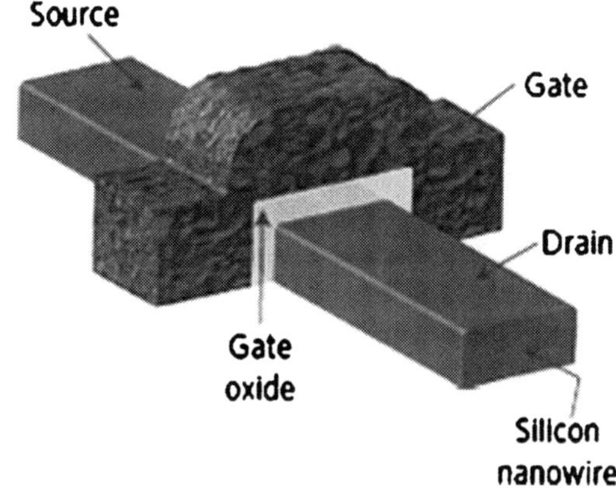

Fig. 18 Schematic of an *n*-channel junctionless nanowire transistor [42]

Opposite dopant polarities are used for p-channel devices. The key to fabricating a junctionless gated resistor is the formation of a semiconductor layer that is thin and narrow enough to allow for full depletion of carriers when the device is turned off. The semiconductor also needs to be heavily doped to allow for a reasonable amount of current flow when the device is turned on. Putting these two constraints together imposes the use of nanoscale dimensions and high doping concentrations, thus motivating the use of nanowires. The transistor exhibits near-ideal subthreshold slope, extremely low leakage currents, and less degradation of mobility than classical transistors when the gate voltage is increased [42].

There are two basic approaches to fabricating a group of nanowires: top-down and a bottom-up approach. In the top-down approach, a large piece of material is cut down to nanowires. A popular method is nanoimprint, a type of lithography that creates patterns by mechanical deformation of imprint resist through a mold, which has predefined topological patterns. The imprint resist is typically a monomer or polymer formulation that is cured by heat or ultraviolet light during the imprinting process. Using nanoimprint lithography, parallel 2-dimensional nanowires of 5 nm width and 14 nm pitch have been fabricated [43]. Figure 19 shows a scanning electron microscope (SEM) image of the patterned polymer. Such patterns can then be used to produce high-density parallel metal nanowires through etching and deposition stages.

Nanoimprint can be used to fabricate crossbar memories or routing structures for VLSI circuits. HP and several other research groups have fabricated and tested crossbar memories using metal nanowires and organic molecular switches [44]. The crossbar circuits, shown schematically in Fig. 20a, were fabricated by nanoimprint lithography. A monolayer of the [2] rotaxane molecules was sandwiched between bottom and top nanowires. The basic element in the circuit is the Pt/rotaxane/Ti junction formed at each crosspoint, which acts as a reversible and nonvolatile

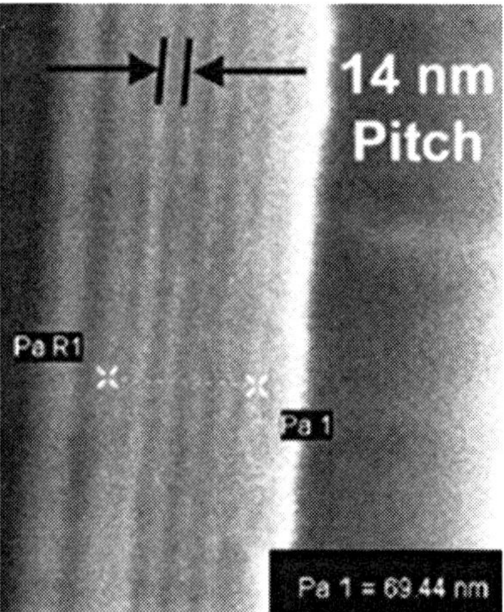

Fig. 19 Demonstration of ultrahigh density photocurable nanoimprint lithography with an SEM image of polymer patterned by a mold with a 14 nm pitch grating with 7 nm linewidth [43]

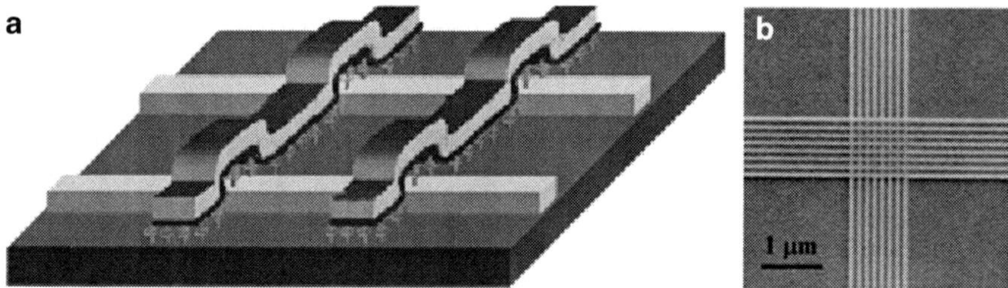

Fig. 20 (**a**) Schematic representation of the crossbar circuit structure. A monolayer of the [2] rotaxane is sandwiched between an array of Pt/Ti nanowires on the bottom and an array of Pt/Ti nanowires on the top. (**b**) An SEM image showing that the two sets of nanowires cross each other in the central area [44]

switch, and 64 such switches are connected to form an 8×8 crossbar circuit within a 1 μm^2 area. Figure 20b shows the SEM image of the crossbar memory.

In the bottom-up approach, the nanowires are grown from seed catalysts that define their diameter [45]. These nanowires are then assembled into aligned arrays using flow alignment and/or Langmuir–Blodgett techniques [46]. In ref. [46], a general and efficient solution-based method for controlling organization and hierarchy of nano-wire structures over large areas has been developed. Nanowires were aligned using the Langmuir–Blodgett technique and transferred to planar substrates in a layer-by-layer process to form parallel and crossed nanowire structures. The parallel and crossed nanowire structures were patterned efficiently into repeating arrays of con-trolled dimensions and pitch using photolithography to yield hierarchical structures with order defined from the nanometer through centimeter length scales.

A major challenge for nanowire-based devices or circuit structures is the higher defect rate compared to the conventional CMOS (this is also an issue for CNFET- and GNRFET-based circuits). For example, for the crossbar memory, it was reported [44] that only 50% of the cross-point switches could switch. Thus, new fabrication techniques for improving yield and fault-tolerant circuit design meth-odologies are active research activities now.

5 Resonant Tunneling Diodes and Quantum Cellular Automata

In this section, we discuss resonant tunneling diodes and quantum cellular automata.

5.1 Resonant Tunneling Diodes

A tunnel diode is a type of semiconductor diode capable of very fast operation through the use of quantum-mechanical effects. These diodes have a heavily doped *p-n* junction with a narrow depletion region (<10 nm wide). Under normal forward bias operation, as voltage begins to increase, electrons in the *n*-region conduction band are energetically aligned to the holes in the valence band of the *p*-region. Tunneling occurs and forward current is produced. As voltage increases further, electron and hole states become more misaligned and the current drops. This is called *negative resistance* because current decreases with increasing voltage.

As voltage increases even further, the diode begins to operate as a normal diode, where electrons travel by conduction across the *p-n* junction, and no longer by tunneling through the *p-n* junction barrier. Resonant tunneling diodes (RTDs) require that electrons must have a certain minimum energy above the energy level of the quantized states in the quantum well in order for tunneling to occur. Once the bias voltage is large enough to provide enough energy, RTDs looks like a normal tunneling diode. Resonant tunneling diodes have received a great deal of attention following the pioneering work by Esaki and Tsu [47] in 1974.

Figure 21a illustrates the I-V characteristics of an RTD [48]. As the voltage applied across the RTD terminals is increased from zero, the current increases until V_p, the peak voltage of the RTD. The corresponding current is called the peak current, I_p. As the voltage across the RTD is increased beyond V_p, the current through the device drops abruptly until the voltage reaches V_v, the valley voltage. The current at this voltage is the valley current, I_v. Beyond V_v, the current starts increasing again. For a current in $[I_v, I_p]$, there are two possible stable voltages; $V_1 < V_p$, or $V_3 > V_v$. In order for RTDs to be useful in high-performance circuits, it is necessary for them to exhibit a steep and narrow valley region (i.e., a high peak-to-valley current ratio). This allows for picosecond switching and sharp transitions.

Usually, RTDs are formed as a single quantum well structure surrounded by very thin layer barriers. This structure is called a *double-barrier structure*. Carriers such as electrons and holes can only have discrete energy values inside the quantum well. Figure 22 illustrates the concept. RTDs can be fabricated using many different types of materials (such as III–V, IV, II–VI semiconductor) and different types of resonant tunneling structures. RTDs are typically realized in III–V compound

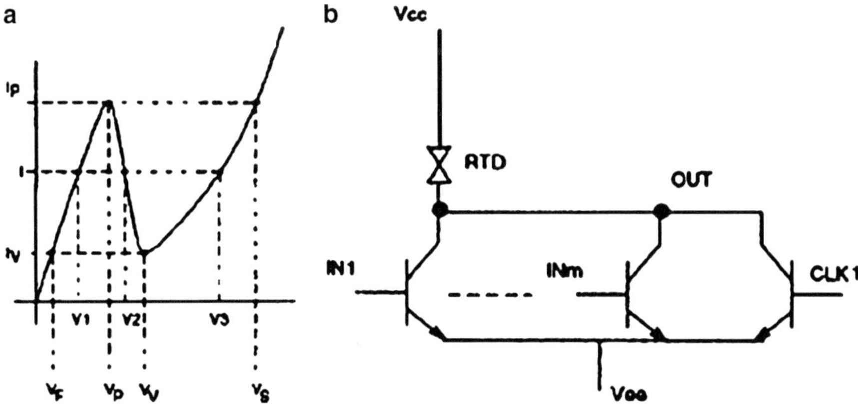

Fig. 21 (**a**) RTD I-V characteristics; (**b**) the operation of a bistable RTD + HBT logic gate [48]

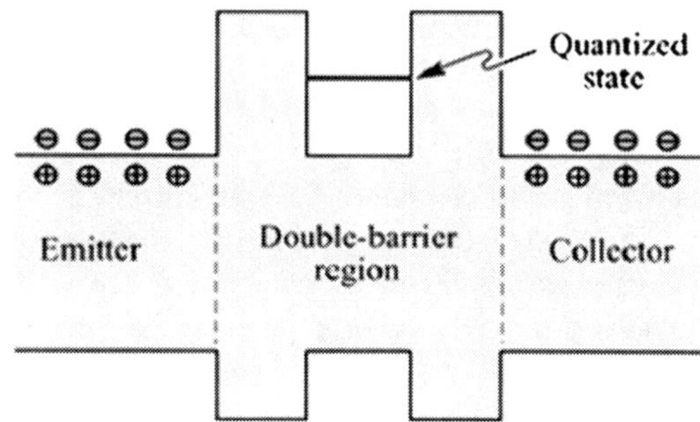

Fig. 22 Schematic of the double-barrier structure for RTD

material systems, where heterojunctions made of various III–V compound semi-conductors are used to create the double or multiple potential barriers in the conduction band or valence band. They can also be realized using the Si/SiGe materials system.

Researchers have developed digital circuits using RTDs combined with HBTs (heterojunction bipolar transistors), MODFETs (modulation-doped field-effect transistors), and other technologies. Figure 21b shows an example, where RTD and HBT together can implement nonweighted threshold logic function [48].

The terahertz (THz) frequency range has been receiving considerable attention recently because of its many applications, such as ultrahigh-speed wireless communications and imaging. RTDs have the potential for use as compact and coherent THz sources. For example, reference [49] describes sub-THz and THz oscillators with RTDs integrated on planar circuits. The RTD has a GaInAs/AlAs double-barrier structure on a semi-insulating InP substrate. Fundamental oscillation up to 0.65 THz and harmonic oscillation up to 1.02 THz were obtained at room temperature. This represents one of the best frequencies achieved through electron devices.

The main challenge for RTDs is its integration into large-scale circuits in useful numbers and the device density. Other challenges include the variation in the current–voltage characteristic of the RTDs across a wafer and from wafer to wafer, fabrication-dependent parasitic impedances, and edge effects [50]. Researchers are actively improving RTD designs and dealing with these issues. They also continue to develop RTDs on a silicon substrate. For example, reference [51] reports the use of an oxidation process for the fabrication of fluoride RTDs on Si substrates, which achieves low leakage and very large peak-to-valley current ratio. Such efforts will increase the potential for large-scale technology implementation of RTDs in the future.

5.2 Quantum Cellular Automata

Quantum cellular automata (QCA) refers to the model of quantum computing, which have been developed in contrast to conventional models of cellular automata introduced by von Neumann. QCA is a nanotechnology that has been recognized as

one of the top emerging technologies with potential applications in future computers [52, 53]. Several studies have reported that QCA can be used to design general-
purpose computational and memory circuits [54, 55]. First proposed in 1993 by
Lent et al. and experimentally verified in 1997, QCA is expected to achieve high
device density, high speed, and low power consumption.

Quantum dots are nanostructures modeled as quantum wells. Once electrons are
trapped inside the dot, it requires higher energy for electrons to escape. QCA
technology is based on the interaction of bi-stable QCA cells constructed from
four quantum dots. The cell is charged with two free electrons, which are able to
tunnel between adjacent dots. Due to their mutual electrostatic repulsion, the two
electrons tend to occupy diagonal sites. Thus, there exist two equivalent arrangements of the two electrons in the QCA cell with minimal energy. These are shown
in Fig. 23 [53]. These two arrangements are denoted as cell polarization $P = +1$ and
$P = -1$. By using cell polarization $P = +1$ to represent logic "1" and $P = -1$ to
represent logic "0," binary information is encoded in the charge configuration of the
QCA cell. The fundamental QCA logic primitives include QCA wire, QCA
inverter, and QCA majority gate [56]. Based on these components, QCA can be
used to construct universal quantum logic gates [56–58].

There exist semiconductor QCA (or metal-dot QCA) and molecular QCA. The
advantages of semiconductor QCA are robust semiconductor fabrication and use of
existing tools, techniques, and infrastructure for QCA development. Furthermore, it
is easy to apply inputs and observe outputs. On the other hand, creating QCA cells
from molecules allows us to scale beyond what photolithography can achieve and
obtain high logic density (up to 1,013 devices/cm^2). Also, QCA cells can be made
homogeneous using chemical synthesis and self-assembly techniques. This is
preferable over photolithography, which introduces variations in device characteristics at the nanometer scale. Finally, molecular QCA has a large potential to
provide significant speedups over the semiconductor QCA. The unique challenge
for molecular QCA, however, is whether hundreds or thousands of atoms can be
precisely synthesized and oriented to make QCA computation feasible.

While QCA is an exciting research area, it faces significant challenges before
widespread adoption. These challenges include development of new manufacturing
techniques for large-scale QCA circuit fabrication, investigation of defect-free and

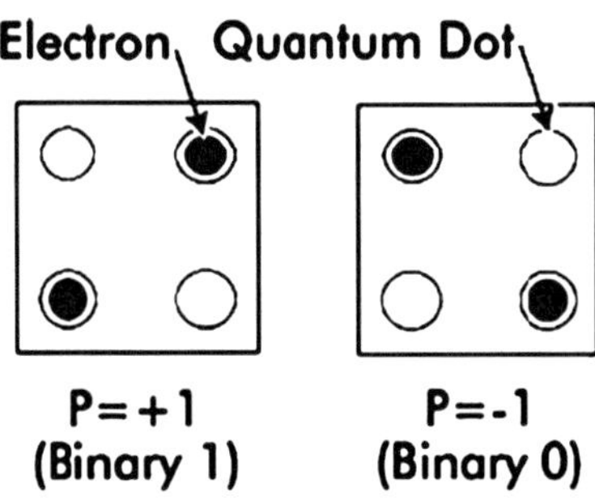

Fig. 23 QCA cells showing
how binary information is
encoded in the two fully
polarized diagonals of the
cell [53]

fault-tolerant circuits and architectures, effective communication with the external environment, and QCA-centric CAD tool development.

6 Conclusions

Single-gate CMOS technology has been the mainstay of the semiconductor industry for the last three decades. However, the CMOS era is coming to an end. This has led to very active investigation of alternative nanotechnologies. While the previous two decades saw the development of many exciting new nanodevices, the coming decade will see burgeoning interest in nanocircuit research. Since it typically takes a decade before academic research comes to fruition in actual products, the time has already arrived to start exploring novel nanocircuits and nanoarchitectures. This book takes a step in this direction. Since no single nanotechnology is expected to dominate all market segments, the book presents some of the state-of-the-art nanosystems that have been developed so far using various nanotechnologies.

References

1. International Technology Roadmap for Semiconductors, http://www.itrs.net/
2. V.V. Zhirnov, R.K. Cavin, J.A. Hutchby, and G.I. Bourianoff, "Limits to binary logic switch scaling: A Gedanken model," *Proc. IEEE* **91**(11):1934–1939, 2003.
3. T.-J. King, "FinFETs for nanoscale CMOS digital integrated circuits," *Proc. IEEE/ACM Int. Conf. Computer-Aided Design*, San Diego, CA, pp. 207–210, 2005.
4. P. Beckett, "A fine-grained reconfigurable logic array based on double-gate transistors," *Proc. IEEE Int. Field-Programmable Technol. Conf.*, December 16–18, pp. 260–267, 2002.
5. K. Yuen, T. Man, and A.C.K. Chan, "A 2-bit MONOS nonvolatile memory cell based on double-gate asymmetric MOSFET structure," *IEEE Electron. Device Lett.* **24**:518–520, 2003.
6. A. Bhoj and N.K. Jha, "Pragmatic design of gated-diode FinFET DRAMs," *Proc. IEEE Int. Conf. Computer Design*, Lake Tahoe, CA, October 4–7, 2009.
7. B. Swahn and S. Hassoun, "Gate sizing: FinFETs vs. 32-nm bulk MOSFETs," *Proc. Design Automation Conf.*, Lake Tahoe, CA, July 24–28, pp. 528–531, 2006.
8. C.-Y. Lee and N.K. Jha, "FinFET-based dynamic power management of on-chip interconnection networks through adaptive back-gate biasing," *Proc. IEEE Int. Conf. Computer Design*, Lake Tahoe, CA, October 4–7, 2009.
9. S. Iijima, "Carbon nanotubes: Past, present, and future," *Phys. B* **323**(1–4):1–5, 2002.
10. J. Deng, N. Patil, K. Ryu, A. Badmaev, C. Zhou, S. Mitra, and H.-S. Wong, "Carbon nanotube transistor circuits: Circuit-level performance benchmarking and design options for living with imperfections," *Proc. IEEE Int. Solid-State Circuits Conf. (ISSCC)*, San Francisco, CA, pp. 586–588, 2007.
11. J. Guo, S. Datta, and M. Lundstrom, "Assessment of silicon MOS and carbon nanotube FET performance limits using a general theory of ballistic transistors," *Proc. IEEE Int. Electron Devices Meeting*, pp. 711–714, 2002.
12. S.J. Kang et al., "High-performance electronics using dense, perfectly aligned arrays of single-walled carbon nanotubes," *Nat. Nanotechnol.* **2**(4):230–236, 2007.

13. N. Patil, A. Lin, E. Myers, H.S.-P. Wong, and S. Mitra, "Integrated wafer-scale growth and transfer of directional carbon nanotubes and misaligned-carbon-nanotube-immune logic structures," *Proc. Symp. VLSI Technology*, Honolulu, HE, pp. 205–206, 2008.
14. W. Zhou, C. Rutherglen, and P. Burke, "Wafer-scale synthesis of dense aligned arrays of single-walled carbon nanotubes," *Nano Res.* **1**:158–165, 2008.
15. R. Martel, T. Schmidt, H.R. Shea, T. Hertel, and P. Avouris, "Single- and multi-wall carbon nanotube field-effect transistors," *Appl. Phys. Lett.* **73**(17):2447, 1998.
16. S.J. Tans, A.R.M. Verschueren, and C. Dekker, "Room-temperature transistor based on a single carbon nanotube," *Nature* **393**(6680):49–52, 1998.
17. A. Bachtold, P. Hadley, T. Nakanishi, and C. Dekker, "Logic circuits with carbon nanotube transistors," *Science* **294**(5545):1317–1320, 2001.
18. S.J. Wind, J. Appenzeller, R. Martel, V. Derycke, and P. Avouris, "Vertical scaling of carbon nanotube field-effect transistors using top gate electrodes," *Appl. Phys. Lett.* **80** (20):3817–3819, 2002.
19. S.J. Kang, C. Kocabas, T. Ozel, M. Shim, N. Pimparkar, M.A. Alam, S.V. Rotkin, and J.A. Rogers, "High-performance electronics using dense, perfectly aligned arrays of single-walled carbon nanotubes," *Nat. Nanotechnol.* **2**(4):230–236, 2007.
20. P.G. Collins, M.S. Arnold, and P. Avouris, "Engineering carbon nanotubes and nanotube circuits using electrical breakdown," *Science* **292**(5517):706–709, 2001.
21. N. Patil, A. Lin, J. Zhang, H. Wei, K. Anderson, H.-S.P. Wong, and S. Mitra, "VMR: VLSI-compatible metallic carbon nanotube removal for imperfection-immune cascaded multi-stage digital logic circuits using carbon nanotube FETs," *Proc. IEEE Int. Electron Devices Meeting*, pp. 573–576, 2009.
22. A. Naeemi, R. Sarvari, and J.D. Meindl, "Performance comparison between carbon nanotube and copper interconnects for gigascale integration (GSI)," *IEEE Electron. Device Lett.* **26**:84–86, 2005.
23. N. Srivastava and K. Banerjee, "Performance analysis of carbon nanotube interconnects for VLSI applications," *Proc. IEEE Int. Conf. Computer-Aided Design*, pp. 383–390, 2005.
24. A. Kawabata et al., "Carbon nanotube vias for future LSI interconnects," *Proc. IEEE Int. Interconnect Tech. Conf.*, pp. 251–253, June 2004.
25. X. Li et al., "Chemically derived, ultrasmooth graphene nanoribbon semiconductors," *Science* **319**(5867):1229–1232, 2008.
26. L. Jiao, L. Zhang, X. Wang, G. Diankov, and H. Dai, "Narrow graphene nanoribbons from carbon nanotubes," *Nature* **458**(7240):877–880, 2009.
27. K. Nakada and M. Fujita, "Edge state in graphene ribbons: Nanometer size effect and edge shape dependence," *Phys. Rev. B* 54:17954–17961, 1996.
28. J.H. Chen, C. Jang, S. Xiao, M. Ishigami, and M.S. Fuhrer, "Intrinsic and extrinsic performance limits of graphene devices on SiO_2," *Nat. Nanotechnol.* **3**(4):206–209, 2008.
29. H. Li, C. Xu, N. Srivastava, and K. Banerjee, "Carbon nanomaterials for next-generation interconnects and passives: Physics, status, and prospects," *IEEE Trans. Electron Devices: Special Issue on Compact Interconnect Models for Gigascale Integration* **56**(9):1799–1821, 2009.
30. K.S. Novoselov et al., "Electric field effect in atomically thin carbon films," *Science* **306** (5696):666–669, 2004.
31. M.Y. Han, B. Ozyilmaz, Y. Zhang, and P. Kim, "Energy band-gap engineering of graphene nanoribbons," *Phys. Rev. Lett.* **98**(20):206805, 2007.
32. Z. Chen, Y.-M. Lin, M.J. Rooks, and P. Avouris, "Graphene nano-ribbon electronics," *Physica E* **40**(2):228–232, 2007.
33. X. Wang, Y. Ouyang, X. Li, H. Wang, J. Guo, and H. Dai, "Room-temperature all-semiconducting sub-10-nm graphene nanoribbon field-effect transistors," *Phys. Rev. Lett.* **100** (20):206803–206907, 2008.
34. K. Nakada and M. Fujita,"Edge state in graphene ribbons: Nanometer size effect and edge shape dependence," *Phys. Rev. B* **54**:17954–17961, 1996.

35. Y.-W. Son, M.L. Cohen, and S.G. Louie, "Energy gaps in graphene nanoribbons," *Phys. Rev. Lett.* **97**(21):216803, 2006.
36. K.A. Ritter and J.W. Lyding, "The influence of edge structure on the electronic properties of graphene quantum dots and nanoribbons," *Nat. Mater.* **8**(3):235–242, 2009.
37. A.K. Geim, "Graphene: Status and prospects," *Science* **324**(5934):1530–1534, 2009.
38. Y. Lin et al., "Operation of graphene transistors at gigahertz frequencies," *Nano Lett.* **9**(1):422–426, 2009.
39. M.P. Levendorf, C.S. Ruiz-Vargas, S. Garg, and J. Park, "Transfer-free batch fabrication of single layer graphene transistors," *Nano Lett.* **9**(12):4479–4483, 2009.
40. Y. Cui, Z. Zhong, D. Wang, W. Wang, and C.M. Lieber, "High-performance silicon nanowire field-effect transistors," *Nano Lett.* **3**(2):149–152, 2003.
41. Toshiba Corporation, "Toshiba develops silicon nanowire transistor for 16-nm generation and beyond," press release, June 15, 2010. http://www.physorg.com/news195834466.html
42. J.P. Colinge, C.W. Lee, A. Afzalian, N.D. Akhavan, R. Yan, I. Ferain, P. Razavi, B. O'Neill, A. Blake, M. White, A.M. Kelleher, B. McCarthy, and R. Murphy, "Nanowire transistors without junctions," *Nat. Nanotechnol.* **5**:225–229, 2010.
43. M.D. Austin et al., "Fabrication of 5-nm linewidth and 14-nm pitch features by nanoimprint lithography," *Appl. Phys. Lett.* **84**(26):5299–5301, 2004.
44. Y. Chen et al., "Nanoscale molecular-switch crossbar circuits," *Nanotechnology* **14**:462–468, 2003.
45. Y. Cui, L.J. Lauhon, M.S. Gudiksen, J. Wang, and C.M. Lieber, "Diameter-controlled synthesis of single crystal silicon nanowires," *Appl. Phys. Lett.* **78**(15):2214–2216, 2001.
46. D. Whang, S. Jin, Y. Wu, and C.M. Lieber, "Large-scale hierarchical organization of nanowire arrays for integrated nanosystems," *Nano lett.* **3**(9):1255–1259, 2003.
47. L.L. Chang, L. Esaki, and R. Tsu, "Resonant tunneling in semiconductor double barriers," *Appl. Phys. Lett.* **24**:593–595, 1974.
48. P. Mazumder, S. Kulkarni, M. Bhattacharya, and A. Gonzalez, "Circuit design using resonant tunneling diodes," *Proc. Int. Conf. VLSI Design: VLSI for Signal Processing*, 1998.
49. M. Asada, S. Suzuki, and N. Kishimoto, "Resonant tunneling diodes for sub-terahertz and terahertz oscillators," *Jpn. J. Appl. Phys.* **47**(6):4375–4384, 2008.
50. J.P. Sun et al., "Resonant tunneling diodes: Models and properties," *Proc. IEEE* **86**(4):641–661, 1998.
51. S. Watanabe, M. Maeda, T. Sugisaki, and K. Tsutsui, "Fluoride resonant tunneling diodes on Si substrates improved by additional thermal oxidation process," *Jpn. J. Appl. Phys.* **44**(4B):2637–2641, 2005.
52. M. Wilson et al., *Nanotechnology: Basic Science and Emerging Technologies*, London: Chapman & Hall, 2002.
53. R. Zhang, K. Walus, W. Wang, and G.A. Jullien, "A method of majority logic reduction for quantum cellular automata," *IEEE Trans. Nanotechnol.* **3**(4):443–450, 2004.
54. A. Vetteth et al., "RAM design using quantum-dot cellular automata," *Proc. Nanotechnol. Conf. Tradeshow* **2**:160–163, 2003.
55. A. Vetteh et al., "Quantum dot cellular automata carry-look-ahead adder and barrel shifter," *Proc. IEEE Emerging Telecommun. Technol. Conf.*, pp. 1–5, 2002.
56. I. Amlani et al., "Digital logic gate using quantum-dot cellular automata," *Science* **284**(5412):289–291, 1999.
57. E.N. Ganesh, L. Kishore, and M. Rangachar, "Implementation of quantum cellular automata combinational and sequential circuits using majority logic reduction method," *Int. J. Nanotechnol. Appl.* **2**(1):89–106, 2008.
58. X.S. Hu, M. Crocker, M. Niemier, M. Yan, and G. Bernstein, "PLAs in quantum-dot cellular automata," *Proc. Emerging VLSI Technologies and Architectures Conf.*, 2006.

FinFET Circuit Design

Prateek Mishra, Anish Muttreja, and Niraj K. Jha

Abstract Fin-type field-effect transistors (FinFETs) are promising substitutes for bulk CMOS at the nanoscale. FinFETs are double-gate devices. The two gates of a FinFET can either be shorted for higher perfomance or independently controlled for lower leakage or reduced transistor count. This gives rise to a rich design space. This chapter provides an introduction to various interesting FinFET logic design styles, novel circuit designs, and layout considerations.

Keywords Circuit design · FinFETs · Layout · Leakage power · Power optimization

1 Introduction

As nanometer process technologies have advanced, chip density and operating frequency have increased, making power consumption in battery-operated portable devices a major concern. Even for nonportable devices, power consumption is important because of the increased packaging and cooling costs as well as potential reliability problems. Thus, the main design goal for VLSI (very-large-scale integration) designers is to meet performance requirements within a power budget. Therefore, power efficiency has assumed increased importance. This chapter explores how circuits based on FinFETs (fin-type field-effect transistors), an emerging transistor technology that is likely to supplement or supplant bulk CMOS (complementary metal-oxide-semiconductor) at 22-nm and beyond, offer interesting delay–power tradeoffs.

N.K. Jha (✉)

Department of Electrical Engineering, Princeton University, NJ, USA

e-mail: jha@princeton.edu

N.K. Jha and D. Chen (eds.), *Nanoelectronic Circuit Design*,
DOI 10.1007/978-1-4419-7609-3_2, © Springer Science+Business Media, LLC 2011

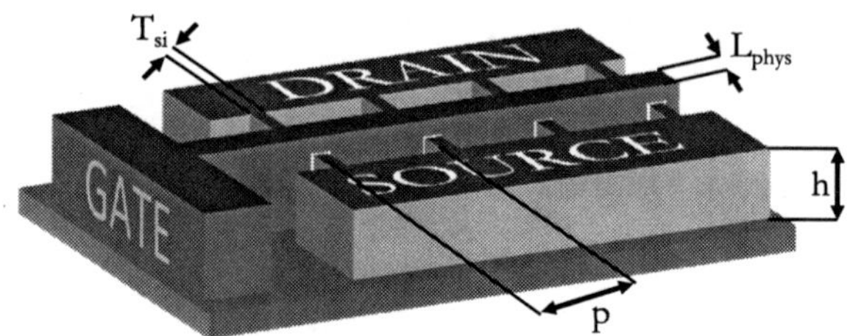

Fig. 1 Multi-fin FinFET [17]

The steady miniaturization of metal-oxide-semiconductor field-effect transistors (MOSFETs) with each new generation of CMOS technology has provided us with improved circuit performance and cost per function over several decades. However, continued transistor scaling will not be straightforward in the sub-22 nm regime because of fundamental material and process technology limits [1]. The main challenges in this regime are twofold: (a) minimization of leakage current (subthrehsold + gate leakage), and (b) reduction in the device-to-device variability to increase yield [2]. FinFETs have been proposed as a promising alternative for addressing the challenges posed by continued scaling. Fabrication of FinFETs is compatible with that of conventional CMOS, thus making possible very rapid deployment to manufacturing.

Figure 1 shows the structure of a multi-fin FinFET. The FinFET device consists of a thin silicon body, the thickness of which is denoted by T_{Si}, wrapped by gate electrodes. The current flows parallel to the wafer plane, whereas the channel is formed perpendicular to the plane of the wafer. Due to this reason, the device is termed *quasi-planar*. The independent control of the front and back gates of the FinFET is achieved by etching away the gate electrode at the top of the channel. The effective gate width of a FinFET is $2nh$, where n is the number of fins and h is the fin height. Thus, wider transistors with higher on-currents are obtained by using multiple fins. The fin pitch (p) is the minimum pitch between adjacent fins allowed by lithography at a particular technology node. Using spacer lithography, p can be made as small as half of the lithography pitch [3].

1.1 Shorted-Gate and Independent-Gate FinFETs

FinFET devices come in many flavors. In shorted-gate (SG) FinFETs, the two gates are connected together, leading to a three-terminal device. This can serve as a direct replacement for the conventional bulk-CMOS devices. In independent-gate (IG) FinFETs, the top part of the gate is etched out, giving way to two independent gates. Because the two independent gates can be controlled separately, IG-mode FinFETs offer more design options (Fig. 2).

Fig. 2 (**a**) SG-mode FinFET;
(**b**) IG-mode FinFET

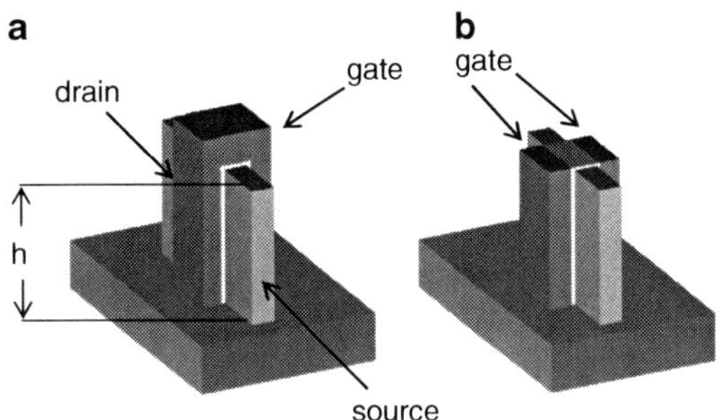

Fig. 2 (**a**) SG-mode FinFET; (**b**) IG-mode FinFET

2 Logic Design Using SG/IG-Mode FinFETs

The performance and power characteristics of FinFET logic gates using transistors in various connected configurations are explored next. Some guidelines for "back of the envelope" logic design using FinFETs are also presented. In general, three modes of FinFET logic gates are logically obvious: (1) SG-mode, in which FinFET gates are tied together; (2) low-power (LP)-mode, in which the back-gate bias is tied to a reverse-bias voltage to reduce subthreshold leakage [4]; and (3) IG-mode, in which independent signals drive the two device gates. Figure 3 shows the implementation of a two-input NAND gate in each of the above modes. A hybrid IG/LP-mode NAND gate, which employs a combination of LP and IG modes is also presented. Similarly, other Boolean functions can be implemented in CMOS styles in each of the above-mentioned modes. (V_{hi} and V_{low} are the corresponding reverse-bias voltages.)

Before proceeding with a discussion of FinFET NAND gates, let us consider some characteristics of the FinFET device that have a bearing on digital design. SPICE-simulated DC transfer characteristics, that is, I_{ds} vs. V_{gfs}, for a 32-nm n-type FinFET are shown in Fig. 4. Here V_{gfs} denotes the potential difference between the front gate (gf) and source terminals. The transistor's source terminal was tied to ground for these simulations and the drain was tied to the power supply. Transfer characteristics are presented for various back-gate voltages (V_{gbs}). The University of Florida double-gate (UFDG) SPICE model [5] was used for 32-nm FinFET simulations. The power supply was fixed at 1 V. Curves corresponding to SG, LP, and IG modes of operation are indicated. Similar results can be obtained for p-type FinFET.

The operating temperature was fixed at 70°C for all simulations. FinFET structures suffer from considerable self-heating. Their thermal simulations are shown to yield a temperature close to 70°C [6] if the switching activity is assumed to be 0.1.

The variation in on- and off-state FinFET currents, I_{on} and I_{off}, across the three modes of FinFET operations is notable. FinFETs offer the best drive strength in SG-mode. I_{on} reduces by about 60% in the IG and LP modes. Application of a

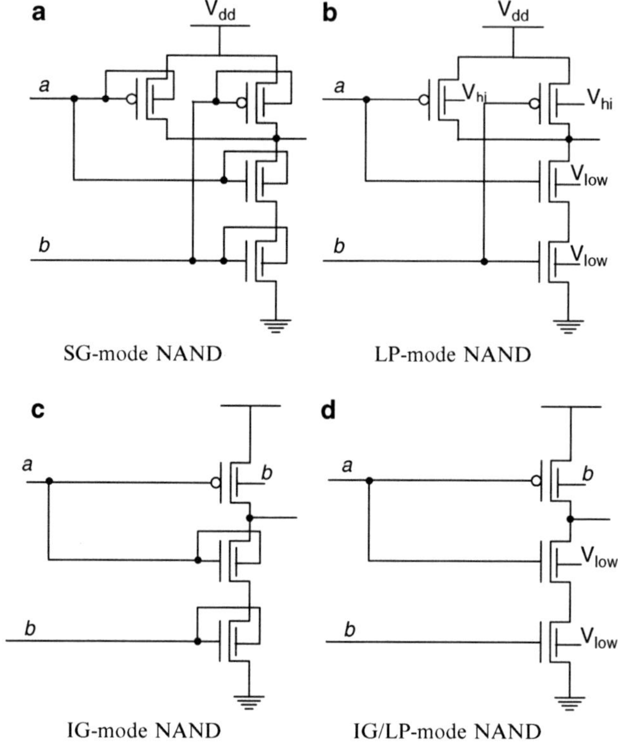

Fig. 3 Different FinFET-based NAND gate designs [4]

reverse bias on the back gate in the LP-mode leads to a further reduction in I_{on}, albeit at a smaller rate. However, a FinFET with one gate fed by logic 0, as in the pull-up p-type FinFET of an IG-mode NAND gate, is not a significantly better driver than a FinFET with a reverse-biased back-gate. On the other hand, I_{off} decreases much more rapidly with increasing reverse bias. A strong reverse-bias reduces I_{off} by more than an order of magnitude compared to the SG-mode, which displays the highest I_{off}.

It is useful to consider the implications of the above device characteristics on the design of an LP-mode FinFET inverter. Figure 5 plots the variation in average delay and leakage power against change in the back-gate bias voltage for a minimum-sized LP-mode inverter, driving a load four times its size and driven by a slope of 5 ps.

Both pull-up and pull-down were driven by a back-gate bias of equal strength in this experiment. For instance, if the strength of the back-gate bias was 0.2 V, a voltage of −0.2 V was used for the back-gate bias of the pull-down FinFET and a voltage of 1.20 V was used to bias the pull-up FinFET. The figure also depicts delay and leakage for an SG-mode inverter. It can be seen that inverter delay degrades sharply in going from the SG-mode to zero reverse-bias LP-mode, and more slowly with increasing reverse bias. The leakage current, however, strongly varies with

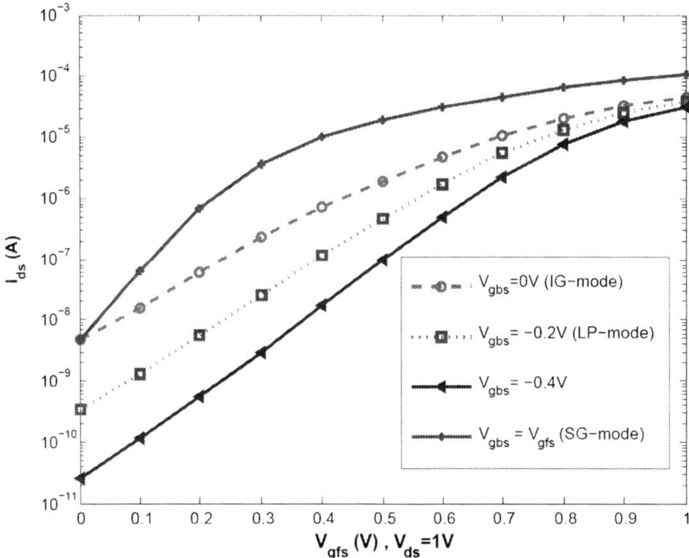

Fig. 4 Simulated I_{ds} vs. V_{gfs} characteristics for a 32-nm n-type FinFET

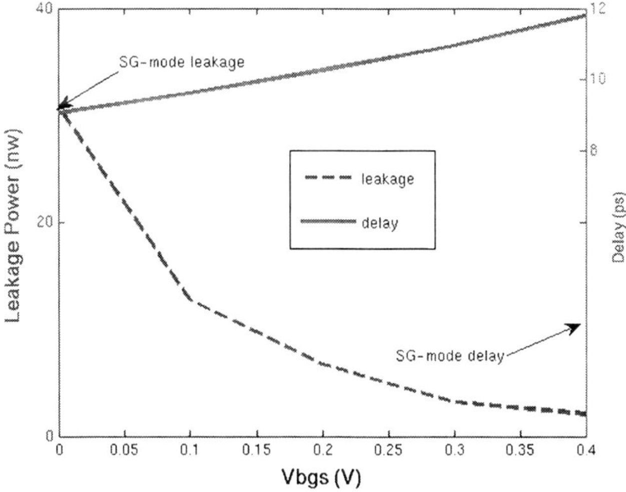

Fig. 5 LP-mode FinFET inverter delay and leakage power variation

back-gate bias. The leakage curve shows an initial sharp decline, but flattens out at back-gate bias voltages exceeding 0.26 V. Further increase in the bias can only lead to delay overheads without much corresponding savings in leakage. With this in mind, a suitable value for the back-gate bias for n-type FinFETs can be set to -0.26 V. For p-type FinFETs, the back-gate voltage was adjusted to 1.18 V to equalize rise and fall delays.

2.1 Design of Logic Gates

Let us revisit the NAND gate designs shown in Fig. 3. Let us consider transistor sizing for each gate. All logic gates are designed to have the minimum possible size that the available FinFET models allowed. All FinFETs in the pull-up and pull-down blocks, respectively, were sized equally. Let the ratio of the widths of a pull-up (W_p) and a pull-down FinFET (W_n) be denoted as β. To obtain β, we use the following considerations:

1. Electron mobility exceeds hole mobility by $2\times$.
2. The electrical width of a FinFET is quantized based on the number of fins in it. It is assumed that wider transistors can only be obtained by increasing the number of fins. The height of each fin is assumed to be fixed.
3. As we observed earlier, using a FinFET in the LP or IG modes reduces its drive strength by almost 60%.

Values for β, and estimates for input capacitance, average off-state current consumption under different input vectors, and delay (average of rise and fall delays) for each gate are shown in Table 1. Assuming a ratio of 2:1 between the electron and hole mobility, a matched CMOS NAND gate may be designed with $\beta = 1$. The SG-mode NAND gate can be obtained by directly translating the CMOS NAND design to FinFETs, while retaining the same sizing. Table 1 reports delay measurements obtained using HSPICE, under three load conditions: unloaded and with loads of 4 (FO4) and 20 (FO20) minimum-sized SG-mode FinFET inverters, respectively, for each design mode. An input slope of 5 ps was used to drive the gates.

In the LP-mode gate, the drive strength of every FinFET is reduced equally. Thus, we can continue to use $\beta = 1$. As expected, the average delay of the LP-mode gate is almost twice that of the SG-mode gate. On the other hand, the input capacitance of an LP-mode gate is only half that of an SG-mode gate, because only one FinFET gate is driven by the input signal. More significantly, leakage power, averaged over all input combinations, is reduced by over 90% because of threshold voltage control.

The IG-mode gate was designed to have asymmetric rise and fall delays. Only one transistor gate is used for pull-up in the IG-mode NAND gate. To achieve balanced rise and fall delays, the pull-up would need to be scaled up. However, using equally sized pull-up and pull-down, i.e., $\beta = 1$, yields savings in area, input capacitance, and

Table 1 Comparison of delay and leakage current for SG-, LP-, IG-, and IG/LP-mode NAND gates

Design mode	β	C_{in} (aF)	I_{off} (nA)	Unloaded delay (ps)	FO4 delay (ps)	FO20 delay (ps)
SG	1	340	4.44	1.31	5.05	13.94
LP	1	170	0.22	5.64	19.95	61.84
IG	1	255	3.41	2.89	10.21	20.12
IG/LP	1	170	0.42	3.44	28.82	60.32

diffusion capacitance at the gate output. As a result, under unloaded conditions, the IG-mode NAND gate has an average delay comparable to, or even better than, the SG-mode NAND gate, but consumes less area and power. Unfortunately, the asymmetry in the pull-up and pull-down drive strengths of an IG-mode gate can lead to large disparities in the rise and fall delays under conditions of greater load. If both transitions through a gate are critical, an IG-mode gate may not be suitable.

As an alternative, consider the IG/LP-mode design. In the fashion of an IG-mode gate, in the IG/LP-mode, parallel transistors, that is, the pull-up for a NAND and pull-down for a NOR gate, are merged. However, unlike the IG-mode design, delays are balanced by reducing the strength of the complementary series structure. This can be seen by comparing the IG- and IG/LP-mode NAND gate results in Table 1.

Strength reduction is achieved by tying the back gates of FinFETs in series to a strong reverse bias. Essentially, the faster transition is slowed down to match the transition made slow by merging transistors, in exchange for significant savings in leakage. At first sight, this might seem to be a large loss in performance. However, often an IG/LP-mode gate has better worst-case rise and fall delays than its IG-mode counterpart. For instance, IG/LP-mode NAND gates actually have a worst-case (rise) delay that is smaller than or comparable to their IG-mode counterparts, under all load conditions, because of reduced competition from the pull-down network during a rising transition at the output. The same observation applies to the falling transition for corresponding NOR gates. This might make IG/LP gates more useful in situations where both rising and falling transitions through a gate are critical.

Also, the IG/LP-mode NAND gate shows savings in excess of 56% and 33% in leakage, averaged across input vectors, and switched capacitance, respectively, compared to the IG-mode design, while retaining the same transistor area. In the interest of brevity, this section presented data only for two-input NAND gates. However, the design techniques examined are generally applicable. NOR and AND-OR-INVERT (AOI) gates can be designed using the same principles and similar trade-offs will be observed for power and delay. Including varied implementations of each logic gate, as proposed in this section, in a cell library provides a level of flexibility that might be used to obtain useful tradeoffs in the power-delay design space.

The circuits could be synthesized using a combination of logic gates from the FinFET cell library and, thereafter, the power–delay spectrum can be generated through various power optimization tools under relaxed/severe delay constraints [4].

3   Threshold Voltage Control Through Multiple Supply Voltages for Power-Efficient FinFET Interconnects

In modern circuits, interconnect efficiency is a central determinant of circuit efficiency. Moreover, as the technology is scaled down, the importance of efficient interconnect design is increasing. FinFET interconnect design can provide several new promising interconnect synthesis schemes.

A mechanism for improving FinFET efficiency, called *threshold voltage control through multiple supply voltages* (TCMS) is described here. The scheme is significantly different from conventional supply voltage schemes. A circuit design for a FinFET buffer using TCMS is developed. It is shown in [7] that on average TCMS can provide power savings of 50.4% and device area savings of 9.2% as compared to the state-of-art dual-V_{dd} interconnect synthesis schemes.

It has been demonstrated that digital logic circuits using FinFETs can be significantly more power-efficient than their counterparts implemented in bulk CMOS at the same gate length. Thus, directly translating bulk CMOS interconnects to FinFETs may also be expected to provide corresponding power savings. However, beyond the obvious technology-driven efficiency benefits, circuits can take advantage of FinFET's double-gate structure to further optimize power and performance.

An important FinFET characteristic is threshold voltage (V_{th}) controllability. The V_{th} at each gate of a FinFET can be controlled through the application of a voltage at the other gate. Because the V_{th} governs both transistor power consumption and delay, V_{th} controllability is a powerful tool for circuit optimization. We describe an innovative synthesis style for interconnect circuits with multiple supply voltages, which takes advantage of V_{th} controllability in connected-gate FinFETs. Traditionally, in multiple supply voltage circuits, power is saved through the use of a lower supply voltage on off-critical paths. For instance, a conventional dual supply voltage (V_{dd}) circuit may use the nominal high-performance V_{dd} for the technology process at hand and, on off-critical paths, a lower V_{dd}, which is typically 60–70% in magnitude compared to the higher V_{dd}.

In a significant departure from convention, a new multiple-V_{dd} scheme, called TCMS, which does not employ a lower V_{dd}, is described. Three supply voltages are used in TCMS: a nominal supply voltage (V_{dd}^{L}), a slightly higher supply voltage (V_{dd}^{H}), and a slightly negative supply voltage (V_{ss}^{H}). The design is based on the principle that in an overdriven FinFET inverter (an inverter that is driven by an input voltage that is higher than its V_{dd}), both leakage and output-current drive (and, thereby, delay) can be controlled simultaneously.

Subthreshold leakage is reduced because of an increase in the V_{th} of the leaking transistor, while current drive is increased because of the larger gate drive experienced by the active transistor. This allows both subthreshold leakage and device width, and, thereby, dynamic power consumption, to be reduced. TCMS is also a voltage-level shifter-free style of circuit design, that is, inverters tied to high and low V_{dd} are allowed to freely alternate, without the need for dedicated level shifters. Instead, level-shifting is built into inverters that require it, through the use of higher-V_{th} FinFETs. FinFETs with higher V_{th} are used at the input of buffers that are connected to V_{dd}^{H}, but driven by buffers operating at V_{dd}^{L} in order to eliminate static leakage current. This allows frequent use of overdriven buffers in TCMS interconnects, thereby increasing the attendant power savings obtainable through the use of overdriven inverters.

3.1 The Principle of TCMS

The V_{th} at each FinFET gate cannot only be controlled statically through the control of process parameters, such as channel-dopant concentration or the value of the gate work function, but also dynamically through the application of a voltage to the other gate (gate–gate coupling). A generalized model for the relationship between the threshold voltage ($V_{th_{gf}}$) at the front gate (gf) of a FinFET and the voltage applied to its back gate (gb) is derived in [8]. However, the following approximate relationship suffices.

$$V_{th_{gf}} \approx \begin{cases} V_{th_{gf}}^0 -\delta(V_{gbs} - V_{th_{gb}}) & \textit{if } V_{gbs} < V_{th_{gb}}, \\ V_{th_{gf}}^0 & \textit{otherwise}, \end{cases}$$

where s denotes the source terminal of the FinFET, δ is a positive value determined by the ratio of gate and body capacitances, and $V_{th_{gf}}^0$ is the minimum observed $V_{th_{gf}}$. The above equation is given for an n-type FinFET, but may also be used for a p-type FinFET with the usual changes in sign. If the FinFET is operated with both its gates tied together, the threshold voltages of both gates respond simultaneously to change in voltage at the other gate. As the above equation predicts, gate–gate coupling is observed only in the weak inversion region of operation. In the region of strong inversion, the presence of inversion charge in the channel shields FinFET gates from each other and no coupling is observed.

As mentioned earlier, TCMS is based on the observation that in an overdriven inverter, both subthreshold leakage and current drive can be controlled simultaneously. We illustrate this concept using Fig. 6, where inverter (a) is a so-called high-V_{dd} inverter, that is, it is connected to V_{dd}^H and V_{ss}^H. Inverter (b) is a low-V_{dd} inverter, that is, it is connected to V_{dd}^L and V_{ss}^L. The values of V_{dd}^H, V_{ss}^H, and V_{dd}^L are

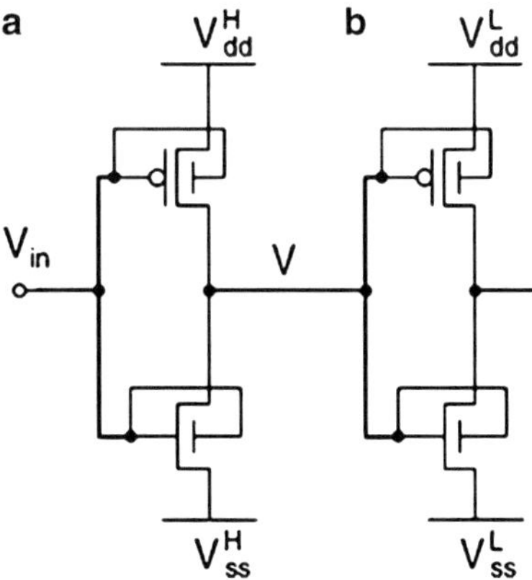

Fig. 6 Principle of TCMS [7]

1.08, −0.1 and 1.0 V, respectively. V_{ss}^{L} is assumed to be tied to a ground. Inverter (b) is, thus, overdriven.

From this point on, it is assumed that any inverter connected to V_{dd}^{H} is also connected to V_{ss}^{H} and similarly for the lower supply voltages. To illustrate the principle of TCMS, let V_{in} in Fig. 6 be held at the logic 0 voltage and, thus, $V = V_{dd}^{H}$. Then, subthreshold leakage through inverter (b) is determined by the leakage through the p-type FinFET, while the inverter's delay is determined, to a large extent, by the current through the n-type FinFET.

According to the above equation, the V_{th} of the p-type FinFET is increased due to the reverse-biased voltage of 0.08 V observed at its gates, thereby reducing subthreshold leakage. The V_{th} of the n-type FinFET, which is operating in the strong-inversion region, is not appreciably altered. Nevertheless, it experiences a forward-biased voltage of 1.08 V, which is higher than the normal gate drive of 1.0 V at the inverter input. This leads to a somewhat higher drive current. Similarly, the application of a logic 1 voltage at the circuit input leads to a reduction in the leakage of the n-type FinFET and improvement in the drive strength of the p-type FinFET.

SPICE-simulated DC transfer characteristics for an overdriven n-type FinFET is shown in Fig. 7. A PTM for 32 nm FinFETs, available from [9, 10], was used for this simulation. In the simulation, the drain terminal of the n-type FinFET was tied to V_{dd}^{L}, and the source terminal was tied to ground. The voltage at the gate terminals was swept from V_{ss}^{H} to V_{dd}^{H}. On-currents through the FinFET at normal drive ($V_{gs} = V_{dd}^{L}$) and overdrive ($V_{gs} = V_{dd}^{H}$) are indicated by I_{on}^{L} and I_{on}^{H} in the figure, respectively. Though it may not be evident on the logarithmic scale of Fig. 7, the value of I_{on}^{H} exceeds the value of I_{on}^{L} by about 3.4%. The relationship between the corresponding values of off-current I_{off}^{H} and I_{off}^{L}, respectively, is evident in Fig. 7. I_{off}^{H} is over 6× smaller in value than I_{off}^{L}.

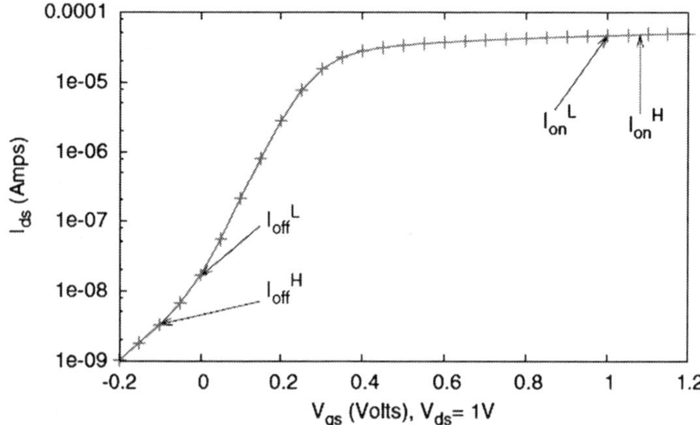

Fig. 7 Simulated I_{ds}–V_{gs} characteristics for an overdriven 32 nm n-type FinFET [7]

3.2   Circuit Design Considerations

Next, we look at TCMS circuits in more detail.

3.2.1   Power Consumption in TCMS Circuits

The mechanisms governing power consumption in TCMS are considerably different from conventional multiple-V_{dd} schemes. Conventionally, the use of a lower supply voltage leads to lower dynamic and leakage power consumption in logic gates tied to it. In TCMS, on the other hand, power savings are observed mainly in overdriven inverters. Three mechanisms that govern the power consumption are observed.

1. Subthreshold leakage power consumption in overdriven inverters is reduced due to V_{th} control.
2. The use of a higher V_{dd} leads to higher short-circuit and leakage power consumption in inverters tied to it, as well as higher switching power consumption in capacitances driven by these inverters.

The following guidelines balance the above opposing mechanisms and minimize power consumption. Values for V_{dd}^{H} and V_{ss}^{H} need to be chosen carefully. Higher values lead to larger savings in subthreshold leakage power consumption through TCMS, but also, higher dynamic power consumption. Thus, circuits should be designed such that the parasitic capacitances charged through V_{dd}^{H} and the widths of high-V_{dd} inverters are minimized. Further, the circuit topology should allow repeated use of overdriven inverters in order to maximize power savings.

3.2.2   Exploratory Buffer Design for TCMS

A circuit that provides an ideal theoretical substrate to explore TCMS is a long interconnect wire, driven by identical buffers, inserted at regular intervals. The source and sink nodes are also assumed to be driven by identical buffers. The buffers comprise a pair of inverters in series, with the input inverter being smaller in size than the output inverter. Figure 8 depicts a link in this wire, designed to use

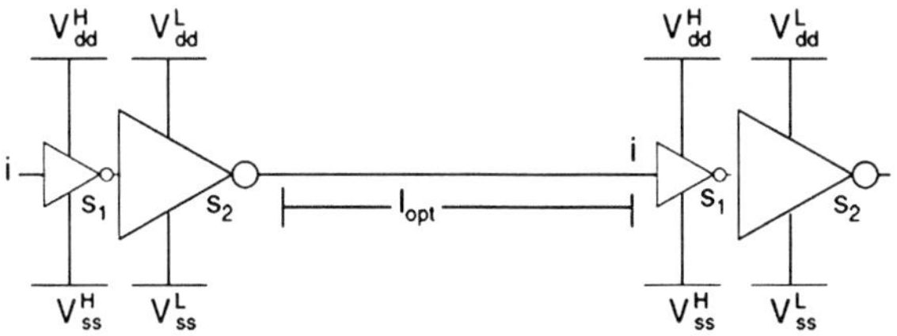

Fig. 8   A buffered link on an infinitely long identical interconnect wire [7]

TCMS. The size of each inverter (s_1 and s_2) is indicated below it in Fig. 8. The length of the link, l_{opt}, is also shown. The smaller inverter in each buffer is tied to V_{dd}^H and the larger buffer is overdriven. The design was chosen because of the following features that might help in maximizing power savings:

1. The size of the high-V_{dd} inverter is kept small.
2. The only parasitic capacitances charged through V_{dd}^H are the input capacitances of the larger inverters, and output capacitances of the smaller inverters. No wire capacitances, which are typically much larger in value than inverter parasitic capacitances, are charged through V_{dd}^H.
3. The larger inverters are overdriven and thus have a much higher I_{on}, presenting a further avenue for saving power: the size of the larger inverters may be reduced, while maintaining delay across the link. Shrinking the larger inverters also reduces the load on the high-V_{dd} inverters feeding them and allows their width to be reduced as well.
4. High- and low-V_{dd} inverters alternate, providing maximum opportunity for power savings. A significant challenge in the above design is the static leakage that would arise in all the high-V_{dd} buffers since they are driven by voltages lower than their power supply and will not switch off properly. To avoid the use of level-converters, high-V_{th} is used in high-V_{dd} buffers to eliminate the static leakage.

The TCMS buffer described above can now be used for low-power buffered interconnect synthesis. As suggested in [7], TCMS buffers outperform dual-V_{dd} buffer insertion schemes in terms of area (by 9% in fin-count) and power consumption (by $2\times$).

3.3 Logic Design Using TCMS

The concept of TCMS was illustrated through its application to a buffer chain in the previous section. However, TCMS can be extended to any logic gate based on SG-mode FinFET logic gates. This is explained next.

Consider the two-input NAND gate shown in Fig. 9. The power supply voltages for the NAND gates are V_{dd}^L and V_{ss}^L. Consider the two inputs a and b. They may be the outputs of a high-V_{dd} gate or a low-V_{dd} gate (a high-V_{dd} gate is connected to V_{dd}^H and V_{ss}^H, and a low-V_{dd} gate to V_{dd}^L and V_{ss}^L). During circuit synthesis, when this gate is embedded in a larger circuit, it might so happen that a is the output of a high-V_{dd} gate and b comes from a low-V_{dd} gate, or vice versa. Suppose the former is true. Thus, the FinFETs connected to input a follow the TCMS principle explained above. FinFETs connected to input b cannot employ the TCMS principle because there is no gate-to-source voltage difference to exploit.

On the other hand, if the power supply voltages for the NAND gate are V_{dd}^H and V_{ss}^H, then the FinFETs connected to input a will not be able to take advantage of the TCMS principle. Also, input b is from the output of a low-V_{dd} gate and is driving a

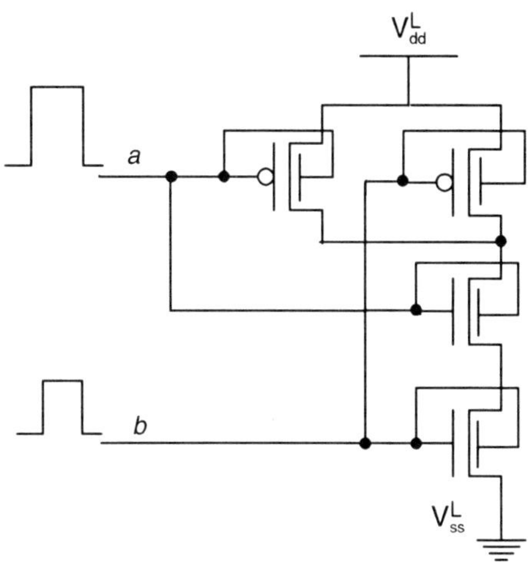

Fig. 9 TCMS principle employed on a NAND gate

high-V_{dd} gate. This results in an increased leakage current because the p-type FinFET is forward-biased by $V_{dd}^{H} - V_{dd}^{L}$. To avoid this problem, a level-converter may be used to restore the signal to V_{dd}^{H}. These level-converters may be combined with a flip-flop, as in the clustered voltage scaling (CVS) technique [11], to minimize the power for voltage-level restoration. In an "asynchronous" approach, such as extended clustered voltage scaling (ECVS) [12], level-converters may be inserted between logic gates connected to V_{dd}^{L} and V_{dd}^{H}. In such schemes, the power and delay overheads for the level-converters are large.

In the case of TCMS, using level-converters is not an attractive option, because power savings are obtained through the use of overdriven gates, the frequent use of which necessitates frequent level conversion. However, level conversion can be built into logic gates without requiring the use of level-converters [13], through the use of a high-V_{th} FinFET at the inputs of high-V_{dd} gates that need to be driven by a low-V_{dd} input voltage. FinFET V_{th} may be controlled through a number of mechanisms. For example, there are several process-related options to statically control the V_{th} of a FinFET, e.g., channel doping, gate work function engineering, or asymmetrical double gates [14].

The first step towards evaluating the utility of the TCMS principle for arbitrary logic circuits involves the design of technology libraries, consisting of high-V_{dd} cells, low-V_{dd} cells, low-V_{dd} cells that are being driven by high-V_{dd} cells, and high-V_{dd} cells that are being driven by low-V_{dd} cells. All these cells have to be characterized at both high-V_{th} and low-V_{th}. Thus, the design variables that need to be targeted are supply voltage, input gate voltage, and threshold voltage. Hence, for a two-input NAND gate of a given size, we have five design variables: supply voltage, gate input voltage for input a, gate input voltage for input b, V_{th} for FinFETs connected to input a, and V_{th} for FinFETs connected to input b.

If the V_{th} of a p-type FinFET connected to an input is high (low), then the corresponding n-type FinFET connected to the same input also has a high (low) V_{th}.

It can be easily seen that 32 two-input NAND gates of a particular size are possible, because of the five design variables. For example, one type of NAND gate may have a high supply voltage (V_{dd}^{H}, V_{ss}^{H}), a low-input a gate voltage, a high-input b gate voltage, a high V_{th} for FinFETs connected to input a, and a low V_{th} for FinFETs connected to input b.

Let 1 denote the case when either a high-supply voltage or a high-input gate voltage or a high V_{th} is used. Similarly, let 0 denote when either a low-supply voltage or a low-input gate voltage or a low V_{th} is used. Using this convention, the example NAND gate can be termed *nand10110*. The first 1 in nand10110 denotes a high-supply voltage; thereafter, 0 denotes a low-input a gate voltage, third 1 denotes a high-input b gate voltage, the fourth 1 represents the high V_{th} for input a, and the fifth 0 represents a low V_{th} for input b. Thus, 32 NAND gate modes are possible ranging from nand00000 to nand11111. However, certain combinations of design variables are not allowed: a logic gate with a high-supply voltage and low-input gate voltages cannot employ low-V_{th} transistors as this will lead to a large leakage current, as explained earlier. Thus, nand10000, nand10001, nand10010, nand10100, nand10101, nand11000, and nand11010 are not allowed. This leads to 25 NAND gate modes instead of 32.

Similarly, there are 25 NOR gate modes. Because the inverter is a one-input gate, it has three design variables: supply voltage, input gate voltage, and V_{th}. This leads to seven valid modes for inverters. For each NAND, NOR, and inverter mode, we include five sizes: X1, X2, X4, X8, and X16. These numbers denote the size relative to the minimum-sized logic gates. The library is characterized by simulating the delay, leakage, and short-circuit power consumption of each constituent cell in HSPICE. Transistor capacitance is also measured using HSPICE. To model interconnect delay and load, fanout and size-dependent wire load models were obtained by scaling the wire characteristics available as part of a 130-nm technology library, according to the method presented in [15].

A possible power minimization flow is shown in Fig. 10. It starts by mapping the logic netlist to low-V_{dd} gates with low-V_{th} and finding its delay-minimized configuration, using Synopsys Design Compiler. The library of low-V_{dd} gates with low-V_{th} is referred to as the SG library because of its use of SG FinFETs. Thereafter, in Phase I, the circuit is divided into alternate levels of high-V_{dd} and low-V_{dd} gates. The gates at odd levels are changed to high-V_{dd} gates with high-V_{th}. The gates at even levels are changed to other modes of low-V_{dd} gates to maintain circuit consistency, as will be clearer later.

Next, in Phase II, a linear programming-based algorithm is used to assign gate sizes and modes to the mapped circuit by selecting cells from the TCMS library. Cell selection is based on the algorithm presented in [16]. The linear programming formulation can be used for reducing both delay and power in the circuits. The iteration terminates when all the delay constraints are met and the change in power consumption between successive iterations is less than some prespecified percent [17]. The TCMS is shown to reduce both the overall power consumption and fincount of the benchmark circuits by a factor of $3\times$ under relaxed delay constraints. It also outperforms ECVS.

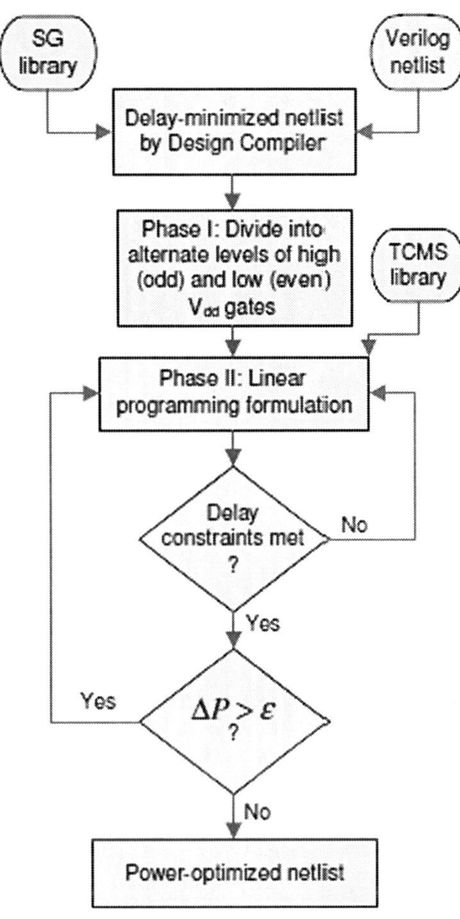

Fig. 10 A possible power-optimization flow

4 Schmitt Trigger

A Schmitt trigger is a high-performance circuit used to shape input pulses and reduce noise. It responds to a slow-changing input pulse with a fast-changing output transition. The voltage-transfer characteristics of the circuit show different switching thresholds for positive and negative going input slopes [18].

Consider the CMOS Schmitt trigger (CMOS-ST) shown in Fig. 11. It consists of an inverter followed by another inverter which is driven by positive feedback. The idea behind the circuit is that the switching threshold of the inverter depends upon the pull-up and pull-down ratio of the transistors. Increasing the ratio raises the switching threshold and vice versa. Adapting the ratio depending upon the direction of the transition results in a shift in threshold and the hysteresis effect [19].

Figure 12 shows the Schmitt trigger based on IG-mode FinFETs. The second inverter and the first inverter in the bulk CMOS case are clubbed together in a single IG-mode inverter. The independent control of the back gate reduces the number of

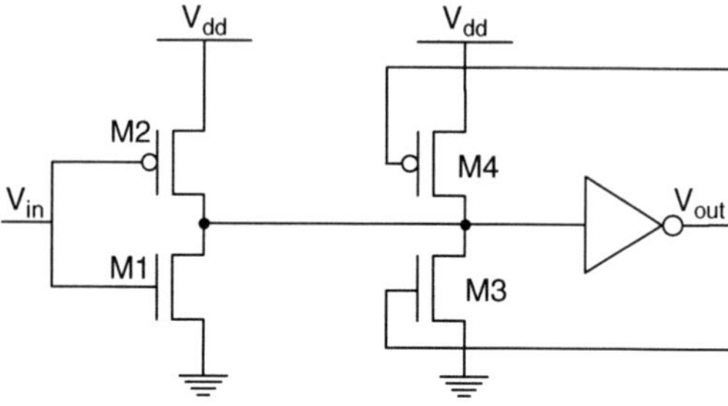

Fig. 11 CMOS Schmitt trigger

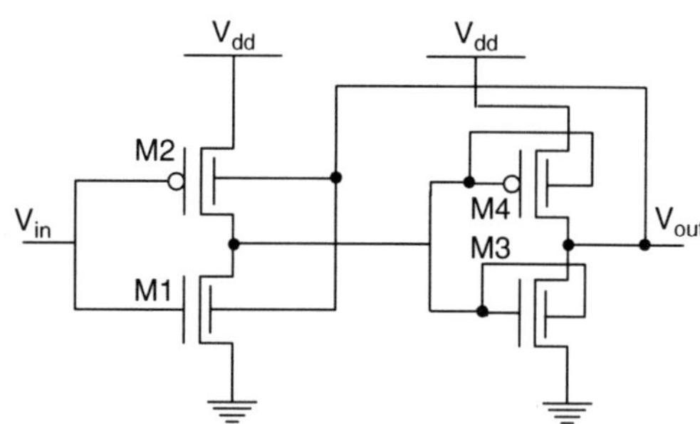

Fig. 12 IG-mode implementation of a FinFET Schmitt trigger

transistors in the circuit by two. Suppose that, initially, the input voltage $V_{in} = 0$. Then both the gates in the IG-mode p-type FinFET will be active because output voltage $V_{out} = 0$. This leads to a shift to the right of voltage-transfer characteristics of the inverter formed by FinFETs M1 and M2. On the other hand when $V_{in} = 1$, IG-mode n-FinFET M1 is active, leading to a shift to the left for the inverter. Figure 13 show the hysteresis loop created by this circuit.

The FinFET Schmitt trigger (FinFET-ST) created in this manner leads to a very large hysteresis window because of the contention between front and back gates of M1 and M2. The shape of the hysteresis window in FinFET-ST can be regulated using asymmetric FinFETs. In asymmetric FinFETs, the back-gate oxide thickness (t_{box}) is increased to approximately $10\times$ that of the front-gate oxide thickness (t_{fox}) to weaken its control on the gate channel, thereby reducing the contention between the front and back gates [20]. Let the ratio t_{box}/t_{fox} be labeled S.

Table 2 gives the percentage gate area reduction, delay reduction, and power savings obtained by switching from CMOS-ST to FinFET-ST at iso-noise margins, at two different loading conditions [20]. The reduction in the power–area metrics can be attributed to: (1) reduced number of transistors in FinFET-ST decreases all components of power dissipation, and (2) during switching, the contention between M2 (M1) and the feedback inverter is replaced by a weaker contention offered by the back gates of FinFETs M1 and M2. The gate delays of CMOS-ST and FinFET-ST remain roughly the same due to the weaker contention offered by the back gates of FinFETs M1 and M2.

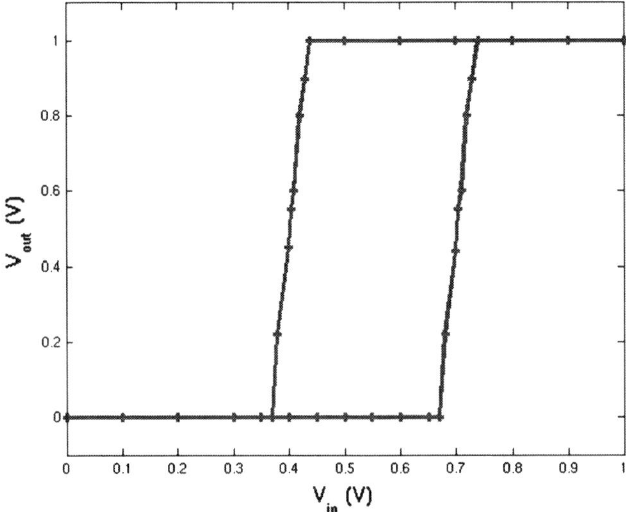

Fig. 13 DC transfer characteristics of IG-mode Schmitt trigger [20]

Table 2 Percent reduction in area, delay, and power of FinFET-ST w.r.t CMOS-ST at iso-noise margins under two different loading conditions [20]

$S = t_{\text{box}}/t_{\text{fox}}$	Gate area (%)	CL = 1 pF		CL = 1 fF	
		Power dissipation (%)	Delay (%)	Power dissipation (%)	Delay (%)
1	–	–	–	–	–
4	13	12	−7	20	−8
7	12	14	1	25	−0.4
10	10	11	1	24	−0.7

5 Latch Design Using SG/IG-Mode FinFETs

We next switch our attention to FinFET latch design.

5.1 SG-Mode Latch Design

Let us first consider the design of brute-force static FinFET latches. A brute-force latch consists of an inverter driving a transmission gate, which in turn drives a cross-coupled inverter. Careful sizing of the transistors needs to be done for the latch to become operational. The data transfer to the latch shown in Fig. 14 happens due to brute force. To be able to transfer data into the latch, the driver inverter I1 and transmission gate T1 must be stronger than feedback inverter I2. Figure 15 shows the circuit schematic of an SG-mode implementation of this latch. This

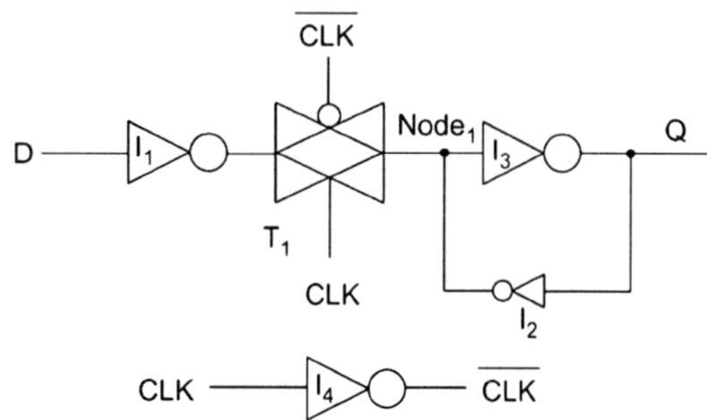

Fig. 14 Brute-force latch [21]

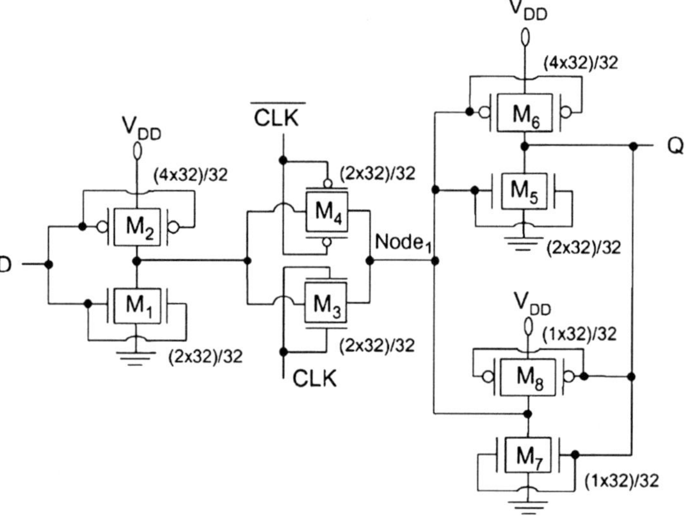

Fig. 15 Circuit schematic of the brute-force latch [21]

schematic has a direct one-to-one correspondence with the conventional bulk CMOS static latch. The driver inverter consisting of M1 and M2, and the transmission gate transistors (M3 and M4) must be sized stronger than the feedback inverter (M7 and M8). The corresponding sizes are shown in the figure. The size of each transistor is given as (number of fins $\times$ h/L_{phys}) [21].

5.2   IG-Mode FinFET Latch

Figure 16 shows the IG-mode implementation of a static FinFET latch [21]. The feedback inverter contains IG-mode transistors. The back gate of M8 is attached to V_{dd} and the back gate of M7 is reverse-biased to ground. This increases the threshold voltage of these transistors, leading to a weakened feedback inverter. Therefore, one does not need to size up the driver inverter and the transmission gate

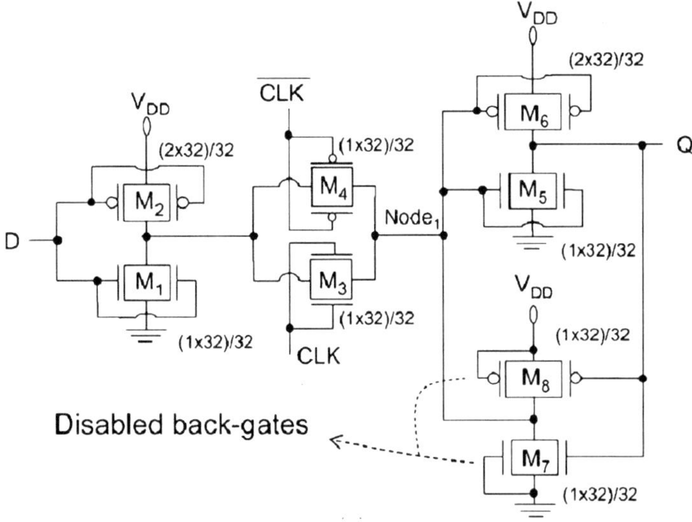

Fig. 16 Circuit schematic of an IG-mode brute-force latch [21]

in order to overpower the feedback inverter. This leads to smaller transistors in the latch. Consequently, we have significantly lower power consumption (switching + leakage) as compared to the SG-mode brute-force FinFET latch.

6 Precharge-Evaluate Logic Circuits

Next, we look at how dynamic logic circuits, such as domino, can exploit FinFETs.

6.1 *FinFET Domino Logic Circuits with SG-Mode FinFETs*

Domino logic circuits are preferentially used in high-performance architectures. However, they suffer from reliability problems due to the dynamic storage of charge. This effect worsens with technology scaling. Thus, noise immunity vs. speed and power tradeoffs need to be carefully investigated while designing dynamic logic circuits. SG/IG-mode FinFET domino logic provides excellent power–delay savings.

Figure 17 shows a standard footless domino logic circuit in SG-mode FinFET technology. During the precharge phase, the clock (CLK) signal is low, which charges the dynamic node. The output is low, which activates the keeper. During the precharge phase, the inputs must be low in order to avoid a direct short-circuit current path from the power supply to ground. This, in general, is achieved by intentionally delaying the clock signal from the preceding domino stages in a

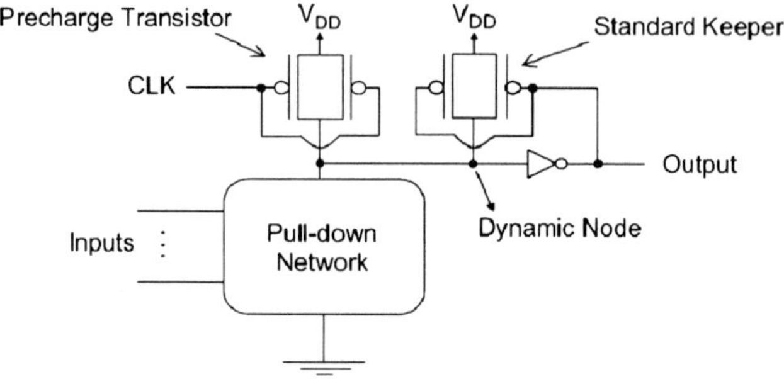

Fig. 17 SG-mode domino logic circuit [22]

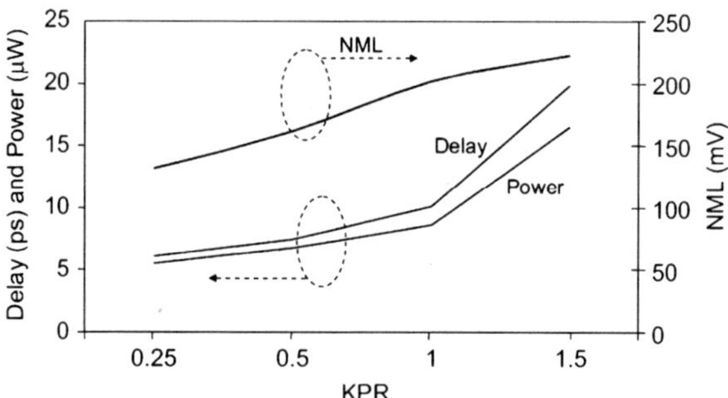

Fig. 18 Evaluation of power, delay, and noise margin of a 16-input OR gate for SG-mode domino in 32-nm technology [22]

multistage domino logic circuit. When the clock signal turns high, the circuit enters the evaluation phase. Provided the input combination provides a discharge path to ground, the dynamic node starts to discharge. However, the pull-down circuitry has to contend with the keeper transistors while it is trying to discharge the dynamic node [22].

Alternatively, if the dynamic node does not discharge during the evaluation phase, the keeper transistor prevents a change in the state of the dynamic node due to coupling noise, charge sharing and subthrehsold leakage.

Simulation results for a 16-input footless domino OR gate are shown in Fig. 18 for various keeper sizes (KPR). KPR is the ratio of keeper size to the size of one of the pull-down network transistors. NML represents the noise margin. NML is measured for the worst case assuming all the inputs of the OR gate are simultaneously excited by the same noise signal. As can be seen from the figure, NML is enhanced by almost 70% when KPR is increased from 0.25 to 1.5. However, this comes at the price of increased evaluation delay (by $3.3\times$) and power dissipation (by $3\times$). This problem can be alleviated in a FinFET design, as discussed next.

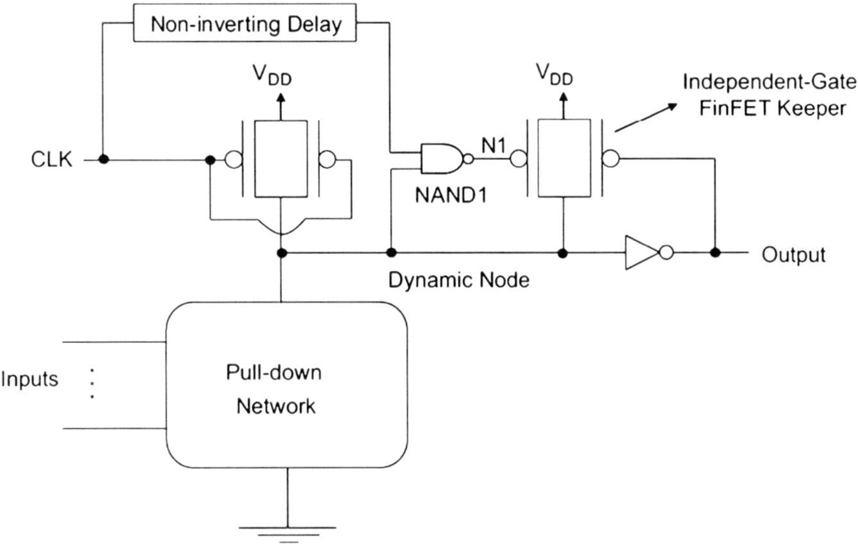

Fig. 19 Schematic of the IG-mode domino logic circuit [22]

6.2 *FinFET Domino Logic with IG-Mode FinFETs*

Figure 19 shows an innovative domino logic style made possible by IG-mode FinFETs. A variable-V_{th} keeper, based on the back-gate biasing technique, is used for simultaneous power and delay reduction at iso-noise margin as compared to the SG-mode FinFET domino circuit mentioned earlier [22].

The proposed circuit works as follows. In the precharge phase, the clock signal is low. The pull-down network is cut off. The keeper control signal N1 becomes high after a fixed delay. This biases one of the independent gates of the FinFET, thereby increasing the threshold voltage of the keeper FinFET. During the evaluation phase, when the clock changes to high and under a specific combination of input values, the dynamic node is discharged to ground. The evaluation speed is increased because of the lower contention current provided by the weakened keeper FinFET. The switching power is reduced because of the decreased contention current provided by the FinFET keeper despite the increased switched capacitance due to the inclusion of the noninverting delay element and NAND gate.

Alternatively, when the pull-down network is not activated, the dynamic node is at V_{dd}, and node N1 transitions to ground after some delay, thus shifting the bias of the IG-mode gate to ground. This essentially reduces the V_{th} of the keeper IG-FinFET. Hence, this FinFET operates with a low V_{th}, providing noise immunity that is similar to the standard SG-FinFET domino logic circuit with the same keeper transistor [22]. Thus, a variable-V_{th} keeper leads to shorter evaluation delay and reduced power consumption at almost iso-noise margins.

7 FinFET Layout

By now, we have some understanding of SG/IG-mode FinFET combinational/ sequential circuits, however, not much at the physical level of abstraction. Because the design of digital VLSI circuits is often based on the standard cell approach, an understanding of layout issues in SG/IG-mode standard cells is important.

Let the gate width of a FinFET with a single fin be W_{min}. As mentioned earlier, the gate width of a multi-fin FinFET is quantized in the number of fins. The higher values of widths are achieved by connecting a number of fins (N_{FIN}) in parallel. Figure 20 shows an SG-mode FinFET in which four fins have been connected in parallel. The width of this device is $4W_{min}$. The area occupied by this device is proportional to $(N_{FIN} - 1)/P_{FIN}$, where P_{FIN} is the fin pitch defined by the process technology.

Currently, there are two different kinds of technologies used to define P_{FIN}. In particular, for the lithography-defined FinFET technology, P_{FIN} is set by the lithographic resolution of the process. On the other hand, in the spacer-defined technology, P_{FIN} can be reduced to half, thus enabling the use of sublithographic control of fin pitch. The device area also consists of the source/drain diffusion and gate underlap. This area remains the same in both types of process technologies. Another method for reducing the layout area is based on increasing W_{min}, which, in turn, amounts to increasing the fin height [23]. However, to keep the short-channel effects under control, arbitrarily large values of fin height cannot be chosen. A certain ratios need to be maintained among fin height, channel length, and fin width for process feasibility and reduced short-channel effects.

Figure 21 shows the layout of an IG-mode FinFET. In such a FinFET, the front and back gates have a separate contact and, thus, the fin pitch needs to be increased to accommodate the back-gate contact. Consequently, the fin pitch in IG-mode FinFETs is greater than the fin pitch in SG-mode FinFETs. The fin pitch in IG-mode FinFETs is given by $T_{Si} + DR1 + 2(DR2 + DR3)$, where DR1, DR2, and DR3 are technology-specified design rules. Also, IG-mode FinFETs do not benefit from spacer lithography due to the increased fin pitch. Therefore, IG-mode devices are expected to have far inferior layout density as compared to that of SG-mode devices [23].

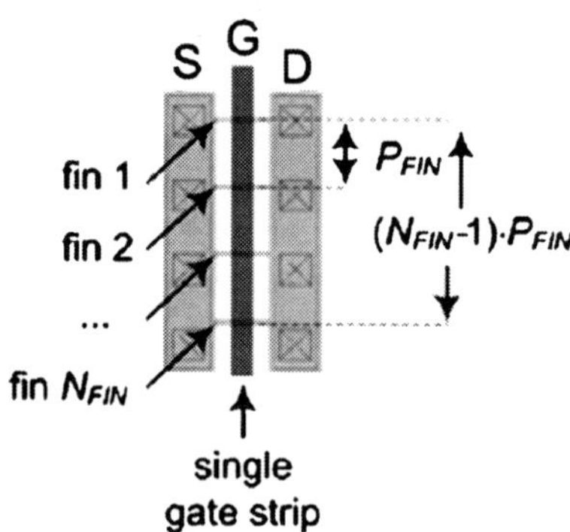

Fig. 20 SG-mode FinFET layout [23]

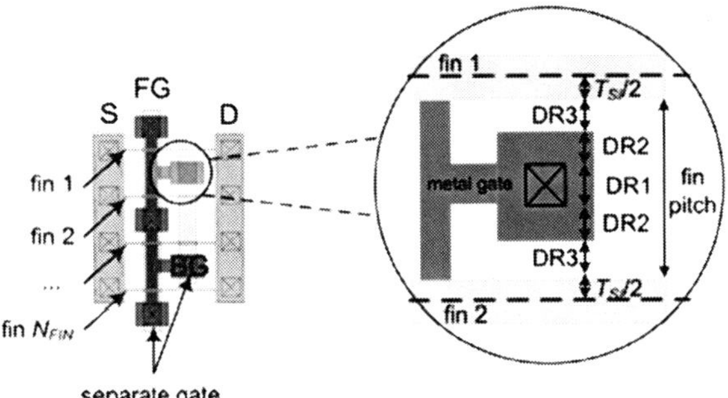

Fig. 21 IG-mode FinFET layout

Fig. 22 Layout of 2× NOR2 in CMOS and SG-mode FinFET [23]

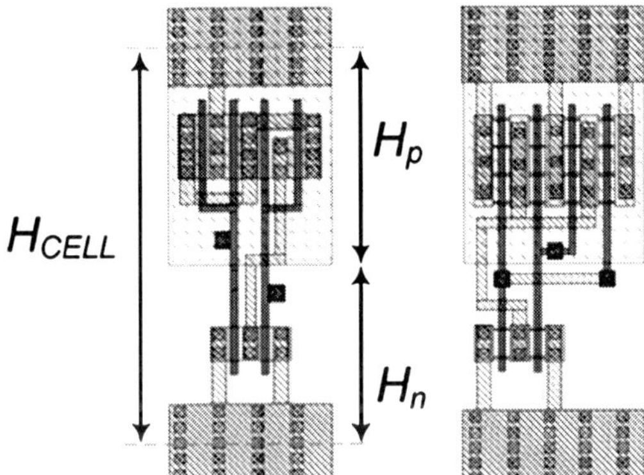

7.1 Layout Analysis of FinFET Standard Cells (SG/IG-Mode)

Figure 22 shows the layout of a 2× two-input NOR cell (referred to as NOR2) based on bulk CMOS and SG-mode FinFETs using lithography-defined technology. Figure 23 shows the layout of SG-mode FinFET in spacer technology. In Fig. 22, the standard cell height HCELL was nominally set to 14 tracks of metal 2, whereas the ground/supply rails were assumed to be 2.5 tracks tall.

In this layout, the height H_p corresponding to p-type FinFETs was set to 1.5 times the height H_n corresponding to n-type FinFETs. NAND, INV, AOI, and other kinds of standard library cells can be laid out in a similar fashion. Figure 24 shows the layout of a 2× NOR2 cell in IG-mode.

Table 3 gives the area results of FinFET 2× NOR2 cells (SG/IG-mode) normalized to the cell implemented in bulk CMOS using both spacer- and lithography-defined technologies. It can be seen from the table that the SG-mode NOR gate has superior layout density as compared to that of bulk CMOS when spacer-defined technology is used. On the other hand, SG-mode NOR and bulk CMOS NOR gates have comparable areas when lithography-defined technology is used. However, the

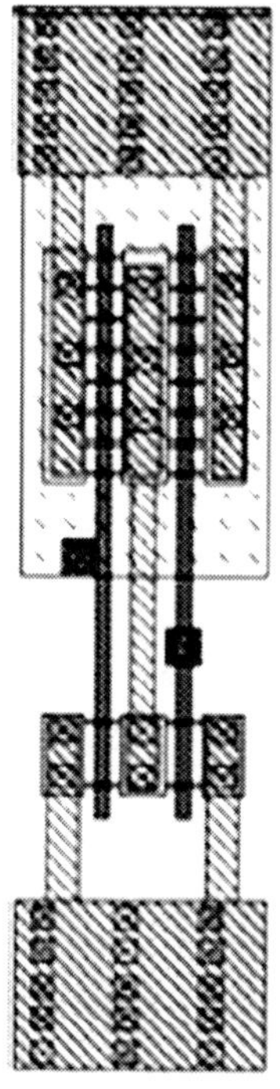

Fig. 23 SG-mode NOR2 layout in spacer technology [23]

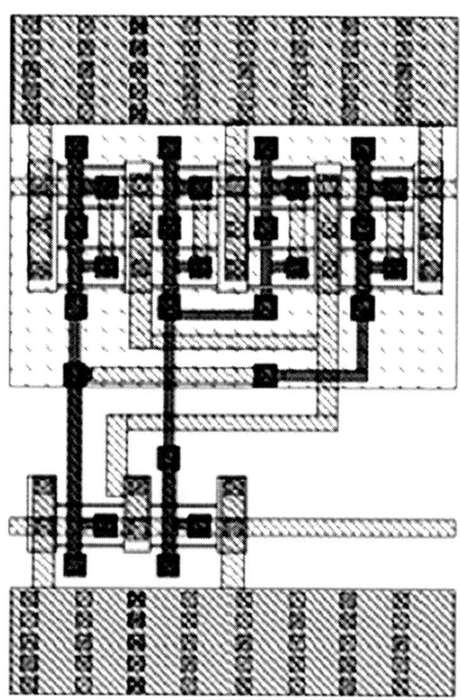

Fig. 24 IG-mode 2× NOR2 layout [23]

Table 3 Normalized area of 2× NOR2 cell w.r.t conventional CMOS layout [23]

	SG-mode	IG-mode
Lithography defined	1.1	2
Spacer defined	0.7	2

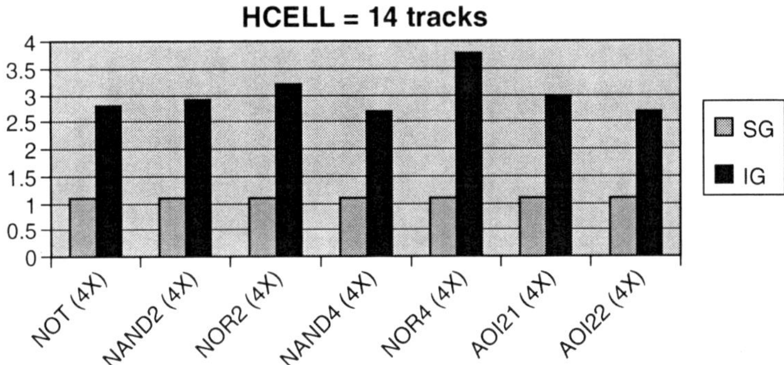

Fig. 25 Area of SG-mode and IG-mode cells using lithography-defined technology normalized w.r.t conventional CMOS technology [23]

IG-mode NOR gate occupies twice the area as compared to bulk CMOS. Similar trends are observed for other gates in the library implemented using lithography-defined technology, as shown in Fig. 25 [23]. Therefore, clever choices need to be made when area and leakage both are a major concern.

8 Oriented FinFETs

An added advantage of the FinFET is that it can be easily fabricated along different channel planes in a single die. Fabrication of conventional planar MOSFETs along any plane other than $\langle 100 \rangle$ is difficult due to increased process variations and interface traps. Furthermore, to obtain different transistor orientations, the devices have to be fabricated along trench sidewalls. With the advent of FinFETs, fabrication of transistors along the $\langle 110 \rangle$ plane has become feasible, leading to design of circuits using differently oriented transistors. Electron mobility is highest in the $\langle 100 \rangle$ plane and the hole mobility along the $\langle 110 \rangle$ plane. Thus, logic gates consisting of p-type FinFETs implemented in the $\langle 110 \rangle$ plane and n-type FinFETs in the $\langle 100 \rangle$ plane will be the fastest. This can be achieved by orienting the p-type FinFETs to be either perpendicular or parallel to the flat or notch of a $\langle 100 \rangle$ wafer and orienting the n-type FinFET fins to be rotated at a 45° angle . The non-Manhattan layout may pose certain yield issues [3].

To quantify the delay of the variously oriented transistors, simulations were performed using a double-gate HSPICE model, called BSIM [24], in conjunction with UFDG [5]. Figure 27 shows the variation in p-type (n-type) FinFET drain-to-

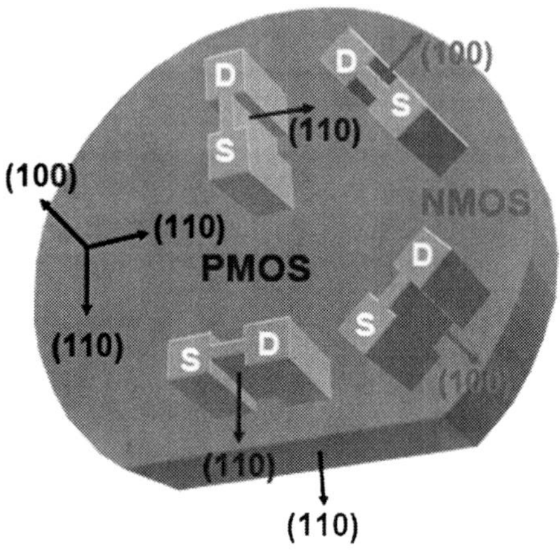

Fig. 26 Si wafer showing differently oriented transistors [3]

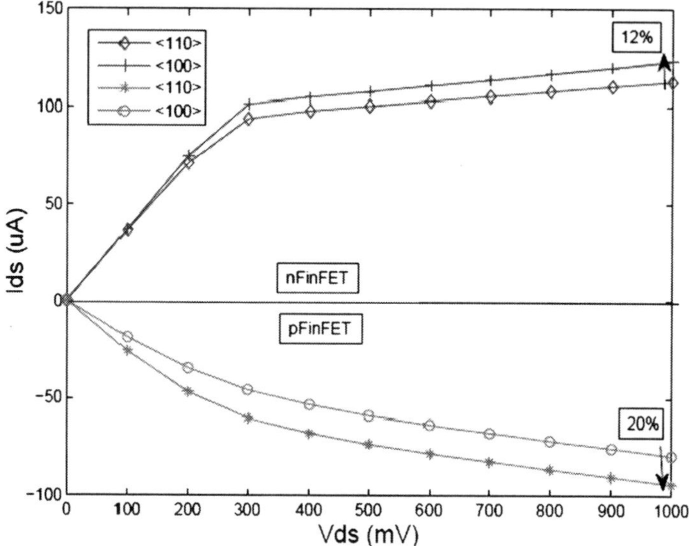

Fig. 27 I_{ds} vs. V_{ds} characteristics for *p*-type and *n*-type FinFETs in the $\langle 110 \rangle$ and $\langle 100 \rangle$ channel orientations

source current (I_{ds}) with drain-to-source voltage (V_{ds}) for different channel orientations at the 32-nm FinFET technology node. BSIM also takes into account the quantum-mechanical effects on carrier mobility for different channel orientations. As shown in Fig. 27, when the orientation changes from $\langle 100 \rangle$ to $\langle 110 \rangle$, the saturation current for *p*-type FinFETs increases by around 20%, whereas for *n*-type FinFETs, when the orientation changes from $\langle 110 \rangle$ to $\langle 100 \rangle$, it increases by 12%. There is a larger increase in the *p*-type FinFET current drive when the channel orientation changes because of the smaller dependence of the hole mobility on velocity saturation. The change in carrier mobility due to transistor orientation is diminished by the velocity saturation effect.

8.1 Library Design Using Oriented FinFETs

To exploit oriented FinFETs, two additional kinds of libraries can be constructed: the oriented shorted-gate (OSG) and the oriented low-power (OLP) gate. The OSG library contains logic gates with a pull-up network that consists of p-type FinFETs oriented along the $\langle 110 \rangle$ plane, whereas the pull-down network uses n-type Fin-FETs in the $\langle 100 \rangle$ plane. Thus, these gates are faster than SG gates, as explained earlier. However, OSG logic gates incur an area penalty because of the oriented p-type FinFETs used in the pull-up network. OLP logic gates have oriented p-type FinFETs and a reverse voltage bias applied to the back gates of all their FinFETs in order to increase the effective V_{th} of the front gate. This allows leakage–delay tradeoffs. One needs to quantitatively study the performance and power characteristics of FinFET logic gates in various channel orientations.

The bar graphs in Fig. 28 show the percent decrease in delay (w.r.t. to $\langle 100 \rangle$ oriented gates) for a minimum-sized INV, NAND, and NOR gates in the $\langle 110 \rangle$ orientation and in the optimized orientation where the pull-up network is in the $\langle 110 \rangle$ plane and the pull-down network is in the $\langle 100 \rangle$ plane. The delay is the average of the rising and falling delays. The logic gates have been sized to closely match the rising and falling delays. For the $\langle 110 \rangle$ case, all the FinFETs in the logic gates are in the $\langle 110 \rangle$ plane. The reduction in delay when the INV is switched from the $\langle 100 \rangle$ plane to the optimized orientation is 8%, whereas the reduction in delay

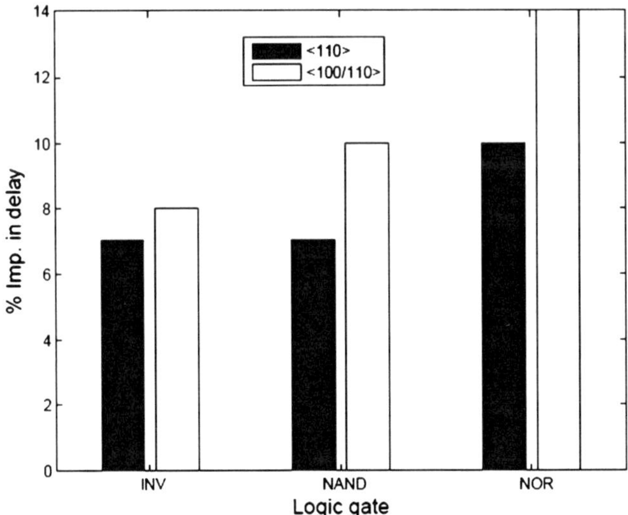

Fig. 28 Percent improvement in delay of various logic gates w.r.t $\langle 100 \rangle$ orientation

for the NAND (NOR) gate is 10% (14%). The reduction in delay for a NOR gate is maximum because the improvement in hole mobility has a maximal effect on stacked p-type FinFETs. There is also a reduction in delay when we move to the $\langle 110 \rangle$ orientation, despite degradation in electron mobility, because the increased hole mobility in this plane reduces the rising time delay. However, as expected, the delay reduction in the $\langle 110 \rangle$ orientation is smaller as compared to the optimal configuration because of the degraded electron mobility in the $\langle 110 \rangle$ plane. The delay reduction in the $\langle 110 \rangle$ plane for the INV is 7%, while the delay reduction for the NAND (NOR) gate is 7% (10%) [3].

Thus, a standard cell library consisting of OSG- and OLP-mode gates can be developed. This can further enrich the library comprising SG-, LP-, and IG-mode logic gates. Because the logic gates in the library offer power–delay tradeoffs, interesting delay-constrained power-optimized circuit netlists can be developed. Various gate-sizing techniques can be used to exploit the benefits of the unique standard cell libraries developed using FinFETs [3].

9 Conclusions

FinFETs are a promising substitute for bulk CMOS for meeting the challenges being posed by the scaling of conventional MOSFETs. Due to its double-gate structure, it offers innovative circuit design styles. In this chapter, we explored different kinds of cell libraries that are possible via SG-, LP-, IG-, and IG/LP-mode FinFET logic gates. We looked at power–delay tradeoffs made possible by these cell modes. We also discussed how delay-optimal FinFET gates can be obtained by fabricating p-type FinFETs in the $\langle 110 \rangle$ channel orientation and n-type FinFETs in the $\langle 100 \rangle$ orientation. This can further enrich the cell library. A power optimizer can be used to synthesize FinFET circuits using cells from these libraries under delay constraints. We also discussed an innovative circuit design technique called TCMS. It can be used for interconnect as well as logic synthesis. It offers power as well as area savings relative to conventional state-of-art dual-V_{dd} schemes.

FinFETs also enable new sequential circuit designs. We studied new FinFET designs of the Schmitt trigger, latch, and domino logic circuits. An IG-mode Schmitt trigger offers superior power, delay, and area numbers relative to SG-mode Schmitt trigger at iso-delay margins. New IG-mode latch and domino logic circuits provide superior power and area at iso-delay margins. We also discussed the physical implementation of FinFET circuits. We showed that layouts based on spacer lithography occupy smaller area relative to lithography-defined layouts. We also found that IG-mode layout density is inferior to that of SG-mode and bulk CMOS technology because of the large area occupied by the intermediate contacts.

Exercises

1. Why are FinFETs termed *quasi-planar* devices?
2. Consider a two-input NOR gate.

(a) Draw the SG-mode implementation of the NOR gate.
(b) Size the NOR gate such that the worst-case rising and falling delays are matched. Assume that hole mobility is half of electron mobility.
(c) Draw an LP-mode implementation of the NOR gate.
(d) Suppose that the driving strength of SG-mode FinFETs is 1.5 times that of LP-mode FinFETs. If the delay of an SG-mode NOR gate is t_p, what is the delay of the LP-mode NOR gate.
(e) If the dynamic power consumption of the unloaded SG-mode NOR gate is P, what is the dynamic power consumption of the unloaded LP-mode NOR gate?
(f) What is the worst-case leakage vector of a two-input NOR gate?
(g) If the worst-case leakage power of an SG-mode NOR gate is L, what is the worst-case leakage of an LP-mode NOR gate. (Assume that the leakage of an IG-mode FinFET is 10% that of an SG-mode FinFET.)
(h) If we double the number of fins in all IG-mode FinFETs, how are the delay, leakage, and dynamic power consumption of the LP-mode NOR gate affected?

3. Consider the figure shown below

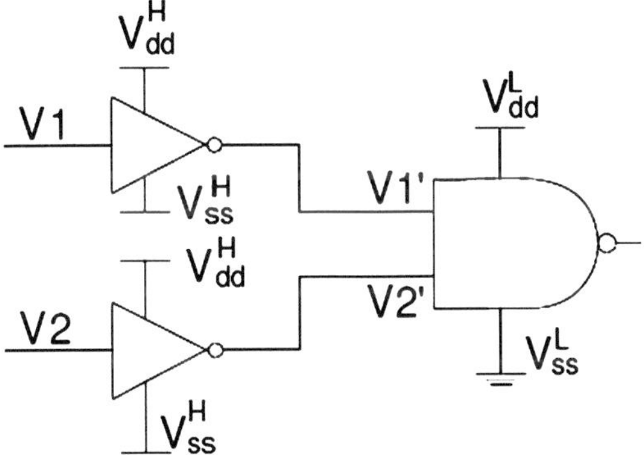

Let $V_{dd}^H = 1.08$ V, $V_{ss}^H = -0.8$ V, $V_{dd}^L = 1.0$ V, and $V_{ss}^L = 0$ V. Assume all the following leakage numbers for the different input vectors are in nW.

Leakage power for a high-V_{dd} inverter:	
1	385.22
0	150.00

Leakage power for a low-V_{dd} NAND gate driven by high-V_{dd} inputs:	
00	2.83
01	52.36
10	50.95
11	150.2

Also, consider an implementation of the circuit where all the gates are driven using V_{dd}^L and ground.

Leakage power for a low-V_{dd} inverter:

1	343.21
0	127.38

Leakage power for a low-V_{dd} NAND gate:

00	14.06
01	230.3
10	210.6
11	636.4

(a) Compare the leakage of the TCMS and low-V_{dd} circuits across all input vector combinations.

(b) Which circuit leaks less and why?

(c) Explain how the circuit shown above makes use of the TCMS concept?

(d) Assuming that the inverters see an input capacitance of 170 aF, what is the dynamic power consumption under two different scenarios: TCMS and low-V_{dd}. (Assume switching activity $= 0.1$ and operational frequency $= 1$ GHz.)

(e) Compare the total power consumption under the two scenarios?

4. Consider the circuit shown in the figure below:

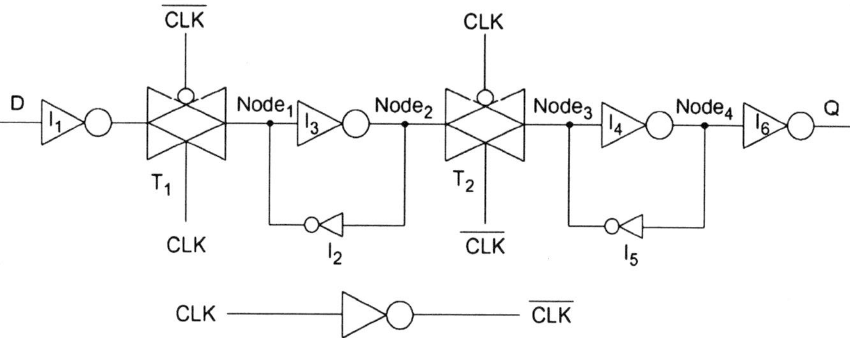

(a) What does the above circuit do?

(b) Which transistors need to be sized stronger and which weaker in order for the above circuit to operate?

(c) Draw an SG-mode circuit schematic of the above circuit.

(d) Taking inspiration from Sect. 5.2, draw an IG-mode implementation of the above circuit. (Bias p-type FinFETs to V_{dd} and n-type FinFETs to ground)

(e) Which transistors were changed to IG-mode FinFETs and why?

(f) Comment on the sizing of the IG-mode implementation of the circuit?

(g) Which implementation is expected to have better average propagation delay and why?

(h) Which implementation is expected to have better average leakage current and why?

(i) Which implementation is expected to have a lower clock power consumption and why?

5. Consider the circuit shown below

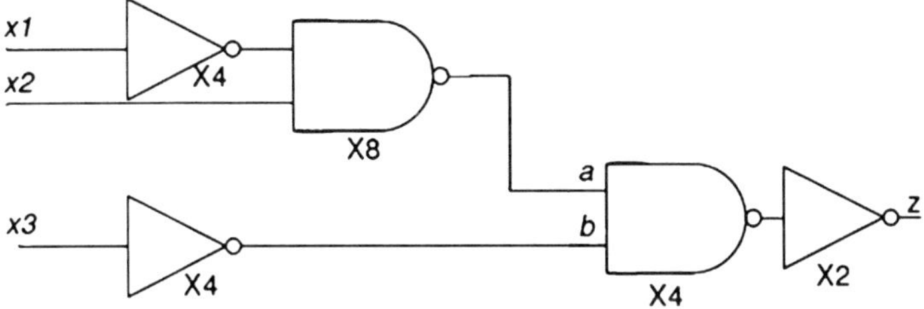

(a) Divide the circuit into levels.
(b) Divide the circuit into alternate levels of V_{dd}^H and V_{dd}^L such that a minimum number of level-converters are required. (Assume that the input level is the same as that of the V_{dd} of the input gate.)
(c) Which gates need to employ high-V_{th} in order to avoid the need for level-converters?

References

1. T.-J. King, "FinFETs for nanoscale CMOS digital integrated circuits," in *Proc. Int. Conf. Computer-Aided Design*, Nov. 2005, pp. 207–210.
2. K. Bernstein, C.-T. Chuang, R.V. Joshi, and R. Puri, "Design and CAD challenges in sub-90 nm CMOS technologies," in *Proc. Int. Conf. Computer-Aided Design*, Nov. 2003, pp. 129–136.
3. P. Mishra and N.K. Jha, "Low-power FinFET circuit synthesis using surface orientation optimization," in *Proc. Design Automation and Test in Europe*, Mar. 2010, pp. 311–314.
4. A. Muttreja, N. Agarwal, and N.K. Jha, "CMOS logic design with independent gate FinFETs," in *Proc. Int. Conf. Computer Design*, Oct. 2007, pp. 560–567.
5. J.G. Fossum et al., "A process/physics-based compact model for nanoclassical CMOS device and circuit design," *IEEE J. Solid-State Electron.*, **48**:919–926, June 2004.
6. J.H. Choi, J. Murthy, and K. Roy, "The effect of process variation on device temperatures in FinFET circuits," in *Proc. Int. Conf. Computer-Aided Design*, Nov. 2007, pp. 747–751.
7. A. Muttreja, P. Mishra, and N.K. Jha, "Threshold voltage control through multiple supply voltages for power-efficient interconnects," in *Proc. Int. Conf. VLSI Design*, Jan. 2008, pp. 220–227.
8. V.P. Trivedi, J.G. Fossum, and W. Zhang, "Threshold voltage and bulk inversion effects in nonclassical CMOS devices with undoped ultra thin devices," *Solid State Electron.*, 1(1):170–178, Dec. 2007.
9. W. Zhao and Y. Cao, "New generation of predictive technology model for sub-45 nm design exploration," in *Proc. Int. Symp. Quality of Electronic Design*, May 2006, pp. 585–590.
10. W. Zhao and Y. Cao, "Predictive technology model for nano-CMOS design exploration," *ACM J. Emerg. Technol. Comput. Syst.*, **3**(1):1–17, Apr. 2007.
11. K. Usami and M. Horowitz, "Clustered voltage scaling for low-power design," in *Proc. Int. Symp. on Low-Power Electronics and Design*, Aug. 1995, pp. 3–8.
12. K. Usami et al., "Automated low-power technique exploiting multiple supply voltages applied to media processor," *IEEE J. Solid-State Circuits*, 33(3):463–472, Mar. 1998.

13. A. Diril et al., "Level-shifter free design of low-power dual supply voltage CMOS circuits using dual threshold voltages," *IEEE Trans. VLSI Syst.*, **13**(3):1103–1107, Sep. 2005.
14. L. Chang et al., "Gate length scaling and threshold voltage control of double-gate MOSFETs," in *Proc. Int. Electronic Device Meeting*, Dec. 2000, pp. 719–722.
15. D. Sylvester and K. Keutzer, "Getting to the bottom of deep sub-micron," in *Proc. Int. Conf. on Computer-Aided Design*, Nov. 1998, pp. 203–211.
16. D. Chinnery and K. Keutzer, "Linear programming for sizing V_{dd} and V_{th} assignment," in *Proc. Int. Symp. on Low-Power Electronics and Design*, Aug. 2005, pp.149–154.
17. P. Mishra, A. Muttreja, and N.K. Jha, "Low-power FinFET circuit synthesis using multiple supply and threshold voltages," *ACM J. Emerg. Tech. Comput. Syst.*, **5**(2):1–23, Jul. 2009.
18. K. Roy et al., "Double-gate SOI devices for low-power and high-performance applications," in *Proc. Int. Conf. on Computer-Aided Design*, Nov. 2005, pp. 217–224.
19. J.M. Rabaey, *Digital Integrated Circuits: A Design Perspective*, Englewood Cliffs, NJ: Prentice-Hall, 1996.
20. T. Cakici, A. Bansal, and K. Roy, "A low-power four transistor Schmitt trigger for asymmetric double gate fully depleted SOI devices," in *Proc. IEEE Int. SOI Conf.*, Sep. 2003, pp. 21–22.
21 S.A. Tawfik and V. Kursun, "Low-power and compact sequential circuits with independent-gate FinFETs," *IEEE Trans. Electron Dev.*, **55**(1):60–70, Jan. 2008.
22. S.A. Tawfik and V. Kursun, "High-speed FinFET domino logic circuits with independent gate-biased double-gate keepers providing dynamically adjusted immunity to noise," in *Proc. Int. Conf. on Microelectronics*, Dec. 2007, pp. 175–178.
23. M. Alioto, "Comparative evaluation of layout density in 3T, 4T, and MT FinFET standard cells," *IEEE Trans. VLSI Syst.*, **18**(2):1–12, Feb. 2010.
24. M. Dunga et al. "BSIM-MG: a versatile multi-gate FET model for mixed signal design," in *Proc. Int. Symp. VLSI Technology*, Jun. 2007, pp. 60–61.

FinFET SRAM Design

Rajiv Joshi, Keunwoo Kim, and Rouwaida Kanj

Abstract We present a comprehensive review of finFET devices taking into consideration different levels of interest ranging from the physics of FinFET devices, design considertaions, and applications to memory design and statistics.

We start with fundamental equations that describe the advantages of finFETs in terms of leakage reduction and ON current improvement compared to planar devices. Following this, we look at variability aspects of finFETs such as quantization (L, W, Vt) from a memory design perspective. We then lay the foundation for SRAM yield optimization in terms of cell dynamic behaviour and study different cell designs that leverage the finFET device structure. We also apply statistical methodology to evaluate the finFET variability.

Keywords FinFET · SRAM · Statistical · Leakage · double gate · backgate · MOSFET

1 Introduction to Nonplanar SRAM

Continued CMOS scaling has been the main driving factor behind silicon technology advancement, including the improvements in MOSFET performance. However, the scaling process in sub-nanometer channel lengths will have to cope with unknown lithographic capabilities. Furthermore, whether continued improved performance gain is possible through such miniaturized channel lengths is unclear due to severe short-channel effects (SCEs) and fundamental limitations of silicon material characteristics that will prevail near the end of the ITRS roadmap [23].

In order to overcome these problems, new device structures for next-generation technology have been proposed, such as silicon-on-insulator (SOI) MOSFET,

R. Joshi (✉)
IBM, Thomas J. Watson Research Center, Yorktown Heights, NY, USA
e-mail: rvjoshi@us.ibm.com

N.K. Jha and D. Chen (eds.), *Nanoelectronic Circuit Design*,
DOI 10.1007/978-1-4419-7609-3_3, © Springer Science+Business Media, LLC 2011

double-gate (DG) MOSFET, SiGe MOSFET, carbon nanotube FET, low-temperature CMOS, and even quantum dot device [1]. Among them, SOI and DG MOSFETs are most interesting due to migratory process/fabrications from conventional CMOS technology along with ideal device characteristics under electrical coupling of the two gates [2].

SRAMs are the key component in microprocessor chips and applications, and the fraction of total chip area devoted to SRAM arrays is large for state-of-the-art designs. As the device is scaled down, process variation effects, including device random fluctuations, become a crucial factor in SRAM design. This is particularly true for SRAM cell devices; SRAM cells use the most scaled device widths/dimensions to reduce the area. The smaller cell area aggravates random statistical fluctuations, and hence stable SRAM design becomes more challenging and complex as the Si technology advances.

This chapter mainly focuses on design considerations of double-gate (DG) FinFET SRAM for high-performance and low-power integrated circuit (IC) applications using unique physical and theoretical concepts of DG-FET configuration. Optimal FinFET SRAM design is presented and compared with conventional planar technologies such as its bulk-Si or SOI counterparts. Although FinFETs have intrinsic advantages over their planar counterparts, we describe circuit design methods to further maximize FinFET circuit performance and minimize its power and leakage. Several novel FinFET SRAM circuit design techniques are analyzed.

2 Why FinFETs?

Figure 1 shows the schematic cross-section of a DG-nFET. The front and back gates are electrically coupled to better control SCEs by substantially lowering both drain-induced barrier lowering (DIBL) and sub-threshold slope S [1]. Therefore, DG

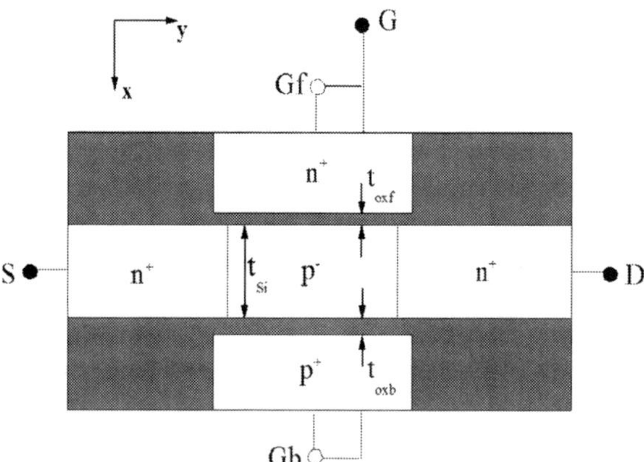

Fig. 1 DG-nFET structure: L_{eff} = effective channel length, t_{oxf} = front-gate insulator thickness, t_{oxb} = back-gate insulator thickness, and t_{Si} = Si-film/body thickness

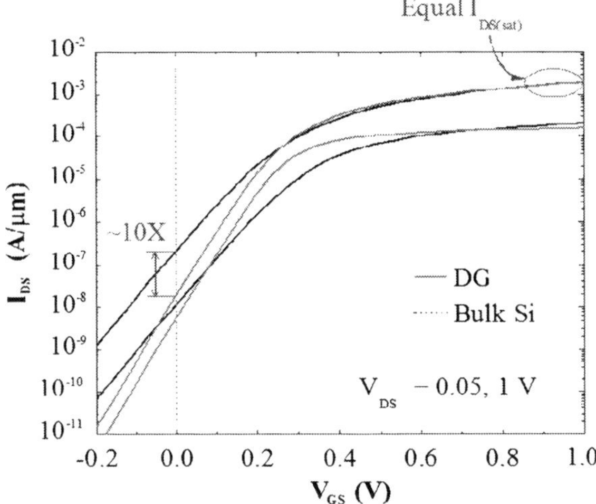

Fig. 2 MEDICI-predicted I_{DS}–V_{GS} characteristics at $V_{DS} = 0.05$, 1 V for 25-nm bulk-Si and DG-nFETs. DG-nFET is designed with $L_{eff} = 25$ nm, $t_{oxf} = t_{oxb} = 1.5$ nm, $t_{Si} = 5$ nm, and $N_{body} = N_A = 10^{15}$ cm^{-3}

devices are most suitable for low-power designs as they enable significant reduction in standby power while simultaneously providing increased performance. To illustrate, we choose a 25-nm device design. The analyzed device structure is shown in Fig. 1.

Figure 2 shows MEDICI-predicted I_{DS}–V_{GS} characteristics at $V_{DS} = 0.05$, 1 V for 25-nm bulk-Si and DG-nFETs. The DG-nFET is designed with $L_{eff} = 25$ nm, $t_{oxf} = t_{oxb} = 1.5$ nm, $t_{Si} = 5$ nm, and $N_{body} = N_A = 10^{15}$ cm^{-3}. Relatively thicker oxide for extremely short L_{eff} is used to avoid gate-oxide tunneling current (I_{Gate}). Thus, $I_{off} \gg I_{Gate}$ in this study. Note that the thicker oxide can be used in DG devices due to fewer SCEs. Thus, less scaling can be allowed. Due to the lightly doped body in a DG device, reverse/forward source/drain junction band-to-band and trap-assisted tunneling currents is negligibly small [3].

For possible DG device options with uniformly or halo-doped body, we neglect the junction tunneling effects. Quantum-mechanical (QM) confinements of carriers, which increase threshold voltage V_t and decreases effective gate capacitance, are considered in the device design by solving one-dimensional (1-D) self-consistent Schrödinger–Poisson equations [4].

Both devices are made with equal drive current around $V_{DD} = V_{GS} = V_{DS} = 1$ V and comparable V_t at $V_{DS} = 1$ V. Since a DG device has two identical channels, on-current I_{on} and total gate capacitance are divided by two to compare the performance for the same device width with that of the bulk-Si device. A DG-MOSFET shows far superior device characteristics to its bulk-Si counterparts with much lower sub-threshold slope S (65 mV/V vs. 90 mV/V) and much suppressed DIBL (35 mV/V vs. 105 mV/V), which offer over 10× reduced off-current I_{off}. The MOSFET I_{off} for $V_{DS} = V_{DD} \gg k_BT/q$ can be expressed as

$$I_{\text{off}} = I_{\text{o}} \times 10^{-V_{\text{t(sat)}}/S},\qquad(1)$$

where I_{o} is I_{DS} at $V_{\text{GS}} = V_{\text{t(lin)}}$, and $V_{\text{t(lin)}}$ and $V_{\text{t(sat)}}$ are V_{t} at $V_{\text{DS}} = 0.05$ V and V_{DD}, respectively. A lower S for a DG device yields a $11.7\times$ reduced I_{off} at equal $V_{\text{t(sat)}} \sim 0.25$ V, as seen from (1) [5].

I_{off} reduction due to a DG device is further increased for higher (equal) V_{t}, again as seen from (1), which indicates that a DG device may be more suitable for low-power designs. Note that the model parameters in 2-D device simulations are physically calibrated against Monte Carlo-predicted data using full-band structure of Si with consistently calculated scattering rate and Fermi–Dirac statistics [6]; therefore, MEDICI-predicted results would be representative of device performance and power.

On the other hand, the substantial reduction in either DIBL or S by bulk-Si device design is not possible in the sub-nanoscale arena because the reduced DIBL worsens S due to the decreased depletion width and the improved S degrades SCEs, including DIBL, due to the reduced channel-to-body coupling [2]. Furthermore, a PD/SOI device has much higher DIBL than a bulk-Si device and an FD/SOI device suffers from higher V_{t} fluctuations due to process variation [1], which can make the devices leakier. Reverse biasing for bulk-Si stacked devices helps reduce I_{off}, yet it can degrade the performance. Note that SOI structures have no reverse body-bias effect in stacked devices; the same holds for DG devices because the body is floating. To summarize, DG devices enable exponential reduction in I_{off} by lowering both S and DIBL and avoiding the detrimental reverse body-bias effect, which cannot be achieved with bulk-Si (or SOI) technology.

3 Physics, Theory, and Modeling of FinFET Devices for SRAM Applications

Because of their near-ideal intrinsic features, DG-MOSFETs, with very thin Si-film bodies, will quite possibly constitute the CMOS technology of the future as the lateral scaling limit is approached. In addition to the inherent suppression of SCEs and naturally steep sub-threshold slope, DG-MOSFETs offer high I_{on} and transconductance, generally attributed to the two-channel property of the symmetrical DG device [7].

More important, we believe, is the electrical coupling of the two gate structures through the charged Si film. This charge coupling underlies the noted features of the device, which translate to high $I_{\text{on}}/I_{\text{off}}$ ratios when the threshold voltage is properly controlled. Such control has been shown to be easily effected via asymmetrical gates of n^{+} and p^{+} polysilicon [2], which, however, would seem to undermine the current drive because the resulting device has only one predominant channel. Contrarily, the gate–gate coupling in the asymmetrical DG-MOSFET is more beneficial than in the symmetrical counterpart, resulting in superiority of the former device for more reasons than just the threshold-voltage control.

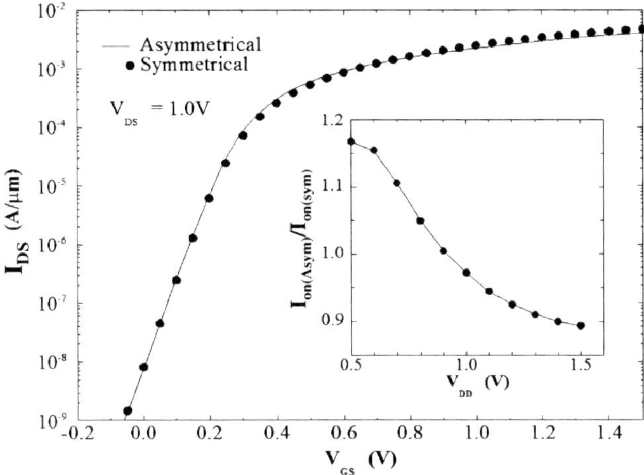

Fig. 3 MEDICI-predicted I–V characteristics of the modified symmetrical DG-FET, contrasted to that of the asymmetrical device having equal I_{off}. The predicted V_{DD} dependence of asymmetrical/symmetrical I_{on} ratio for the DG-nFETs is shown in the insert

Theoretically, a "near-mid-gap" gate material with gate work function $F_M = \chi_{(Si)} + 0.375\, E_{g(Si)}$, where $\chi_{(Si)}$ is the silicon–electron affinity and $E_{g(Si)}$ is the energy gap of silicon, will reduce I_{off} to the noted equality, as shown in Fig. 3, where the predicted I_{DS}–V_{GS} characteristic of the so-modified symmetrical-gate MOSFET at $V_{DS} = 1.0$ V is compared with that of the asymmetrical-gate device.

Interestingly, when the off-state currents are made equal, the on-state currents for both devices are comparable, even though the asymmetrical device has only one predominant channel (for low and moderate V_{GS}), as revealed by the plots of its current components in Fig. 4. The corresponding $I_{on(asym)}/I_{on(sym)}$ ratio, plotted versus V_{DD} in the insert of Fig. 3, is actually greater than unity for lower V_{DD}. We attribute this V_{DD} dependence to better suppression of DIBL [3] in this asymmetrical device. Simulations of longer metallurgical channel length $L_{met} = 0.5$ μm devices yield $I_{on(asym)}/I_{on(sym)} \cong 1$, independent of V_{DD}. For the longer devices, the predicted I_{DS}–V_{GS} characteristics for the two devices are demonstrated to have comparable I_{on} [2]. Thus, the near-equality of the currents in both devices is clearly evident.

The MEDICI simulations are based on semi-classical physics with analytical accounting for quantization effects [8]. To ascertain that the predicted symmetrical-versus-asymmetrical DG benchmarking results are not precluded by the effects of QM confinement of electrons in the thin Si film, we used SCHRED-2 [4], a 1-D (in x) self-consistent solver of the Poisson and Schrödinger equations, to check them.

SCHRED-predicted real electron charge density (Q_c, which correlates with the channel current) versus V_{GS} in both devices is plotted for $V_{DS} = 0$ in Fig. 5. Although the QM electron distributions across the Si film (with the electron density $n(x)$ forced to zero at the two Si–SiO$_2$ boundaries by the wave function conditions)

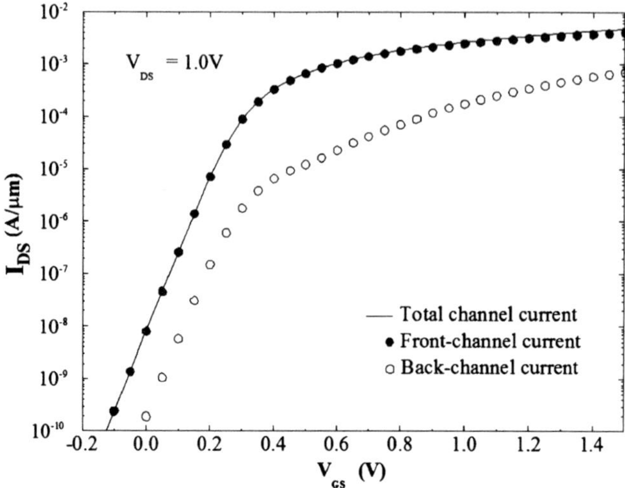

Fig. 4 MEDICI-predicted channel-current components (integrated over the front and back halves of the Si film) for the 50-nm asymmetrical and symmetrical DG-nFETs. The back-channel current (with p^+ gate) is not significant, but the front-channel current is enhanced by the gate–gate coupling and beneficial inversion-layer capacitance

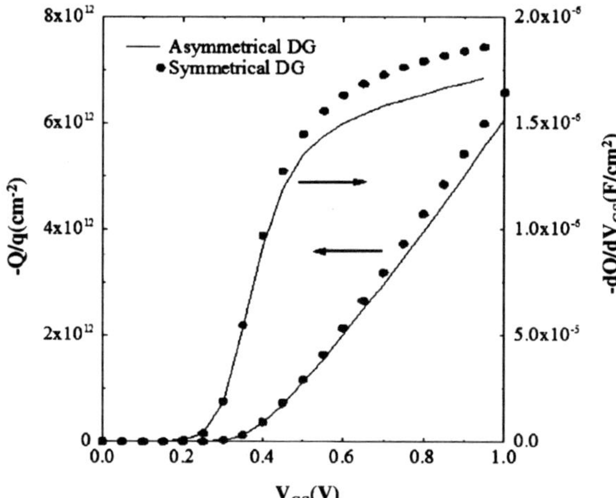

Fig. 5 SCHRED-predicted integrated electron charge density in the asymmetrical and symmetrical DG-nFET; $V_{DS} = 0$. Also shown are the corresponding predicted V_{GS}-derivatives of the charge densities

differ noticeably from the classical results, Q_c is nearly the same in both devices, implying nearly equal currents as predicted by MEDICI. We therefore are confident that the QM effects will not alter the main conclusions of our DG device benchmarking, which we now explain analytically.

3.1 First-Order Poisson Equations

To gain physical insight and explain the surprising benchmark results presented earlier, we begin with a first-order analytical solution of Poisson's equation (1-D in x) applied to the thin Si-film body. For a general DG-nMOSFET with long L_{met} at low V_{DS}, assuming inversion-charge sheets (inversion layer thickness $t_{\mathrm{i}} = 0$) at the front and back surfaces of the fully depleted Si film, we have [9]

$$V_{\mathrm{GfS}} = V_{\mathrm{FBf}} + \left(1 + \frac{C_{\mathrm{b}}}{C_{\mathrm{of}}}\right)\Psi_{\mathrm{sf}} - \frac{C_{\mathrm{b}}}{C_{\mathrm{of}}}\Psi_{\mathrm{sb}} - \frac{Q_{\mathrm{cf}}}{C_{\mathrm{of}}} - \frac{Q_{\mathrm{b}}}{2C_{\mathrm{of}}} \tag{2}$$

and

$$V_{\mathrm{GbS}} = V_{\mathrm{FBb}} - \frac{C_{\mathrm{b}}}{C_{\mathrm{ob}}}\Psi_{\mathrm{sf}} + \left(1 + \frac{C_{\mathrm{b}}}{C_{\mathrm{ob}}}\right)\Psi_{\mathrm{sb}} - \frac{Q_{\mathrm{cb}}}{C_{\mathrm{ob}}} - \frac{Q_{\mathrm{b}}}{2C_{\mathrm{ob}}}, \tag{3}$$

where V_{GfS} and V_{GbS} are the front and back gate-to-source voltages, V_{FBf} and V_{FBb} are the front- and back-gate flatband voltages, Ψ_{sf} and Ψ_{sb} are the front- and back-surface potentials, Q_{cf} and Q_{cb} are the front- and back-surface inversion charge densities, $Q_{\mathrm{b}} = -qN_{\mathrm{A}}t_{\mathrm{Si}}$ is the depletion charge density, $C_{\mathrm{of}} = \varepsilon_{\mathrm{ox}}/t_{\mathrm{oxf}}$ and $C_{\mathrm{ob}} = \varepsilon_{\mathrm{ox}}/t_{\mathrm{oxb}}$ are the front- and back-gate oxide capacitances, and $C_{\mathrm{b}} = \varepsilon_{\mathrm{Si}}/t_{\mathrm{Si}}$ is the depletion capacitance. By setting $V_{\mathrm{GS}} = V_{\mathrm{GfS}} = V_{\mathrm{GbS}}$ (implying no gate–gate resistance [3]) and eliminating the y_{sb} terms from (2) and (3), we derive the following expression for the DG structure.

$$V_{\mathrm{GS}} = \Psi_{\mathrm{sf}} + \frac{1}{1+r}\left[(V_{\mathrm{FBf}} + rV_{\mathrm{FBb}}) - \left(\frac{Q_{\mathrm{cf}}}{C_{\mathrm{of}}} + r\frac{Q_{\mathrm{cb}}}{C_{\mathrm{ob}}}\right) - \left(\frac{Q_{\mathrm{b}}}{2C_{\mathrm{of}}} + r\frac{Q_{\mathrm{b}}}{2C_{\mathrm{ob}}}\right)\right], \tag{4}$$

where r is a gate–gate coupling factor expressed as

$$r = \frac{C_{\mathrm{b}}C_{\mathrm{ob}}Q_{\mathrm{cf}}}{C_{\mathrm{of}}(C_{\mathrm{b}} + C_{\mathrm{ob}})} \cong \frac{3t_{\mathrm{oxf}}}{3t_{\mathrm{oxb}} + t_{\mathrm{Si}}}. \tag{5}$$

The approximation in (5) follows from $\varepsilon_{\mathrm{Si}}/\varepsilon_{\mathrm{ox}} \cong 3$. Note that r decreases with increasing t_{Si}.

3.2 Sub-threshold Slope Tracking

Since (4) applies to a general DG device structure, it implies that both the asymmetrical- and symmetrical-gate devices should have near-ideal sub-threshold slope, or gate swing:

$$S = \left(\frac{KT}{q} \ln(10)\right) \frac{dV_{GS}}{d\Psi_{sf}}, \tag{6}$$

because with Q_{cf} and Q_{cb} negligible for weak inversion, $dV_{GS}/d\psi_{sf} = 1$. The $L_{met} = 0.5$ μm simulation results are in accord with this result [2]. The 50-nm devices in Fig. 3 show $S = 65$ mV for both devices, greater than 60 mV (at $T = 300$ K) due to mild SCEs (which subside for thinner t_{Si}). The gate–gate coupling, implicit in (4), underlies the near-ideal S in the DG-MOSFET.

Hence, it is evident that the asymmetrical DG-MOSFET current tracks that of the symmetrical counterpart as V_{GS} is increased from the off-state condition where the currents are equal, in accord with (4)–(6). We now have to explain why this tracking continues, in essence, into the strong-inversion region, as illustrated in Fig. 5, and as implied by the electron charge densities $Q_c(V_{GS})$ in Fig. 6.

3.3 Strong-Inversion Region Charge Tracking

Effects of finite inversion-layer thickness (t_i) [3] on $Q_c(V_{GS})$ in DG devices are quite important. For the symmetrical DG (n-channel) device, the effective $t_{i(sym)}$ (for the front and back channels) is defined by integrating the electron density over half of the Si film [29]:

$$t_{i(sym)} = \frac{\int_0^{t_{si}/2} xn(x)dx}{\int_0^{t_{si}/2} n(x)} = -\frac{2q}{Q_{c(sym)}} \int_0^{t_{si}/2} xn(x)dx \tag{7}$$

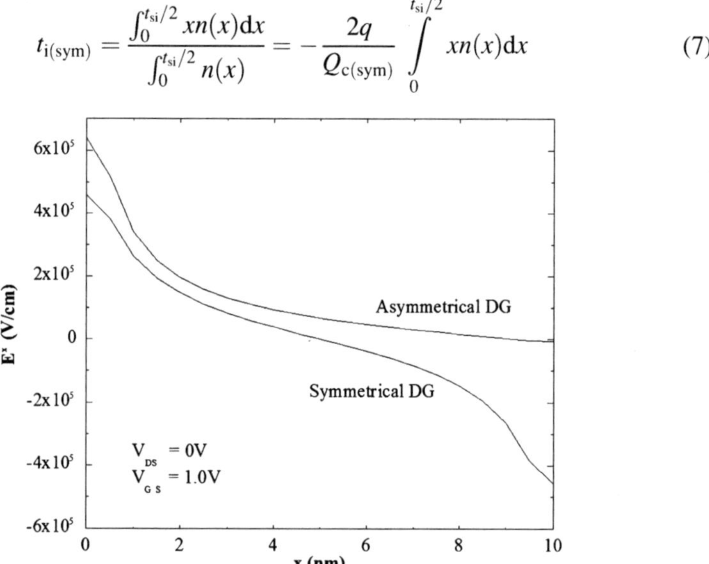

Fig. 6 MEDICI-predicted transverse electric field variations across the Si film ($t_{Si} = 10$ nm) of the asymmetrical and symmetrical DG-nFETs. Note that the field in the symmetrical device is always zero at the center of the film ($x = t_{Si}/2$)

and the total $Q_{c(sym)}$ is analytically expressed by combining Poisson's equation and Gauss's law [2]:

$$Q_{c(sym)} = -2C_{Gf(sym)}(V_{GS} - V_{Tf(sym)}), \tag{8}$$

$$C_{Gf(sym)} = \frac{C_{of}}{1 + \frac{C_{of}}{C_{i(sym)}}} \cong \frac{C_{of}}{1 + \frac{1}{3}\frac{t_{i(sym)}}{t_{oxf}}}, \tag{9}$$

where C_{of} is the total front-gate (or back-gate) capacitance, and $V_{Tf(sym)}$ is a nearly constant threshold voltage for strong-inversion conditions. In (9), $C_{i(sym)} = \varepsilon_{Si}/t_{i(sym)}$ represents the (front or back) inversion-layer capacitance, $-(dQ_{c(sym)}/d\Psi_{sf})/2$ [3]. The factor of 2 in (8) reflects two identical channels and gates. Relative to the single-gate device, for which C_{Gf} could differ from $C_{Gf(sym)}$ slightly, we thus infer an approximate doubling of the drive current, but exotic gate material is needed, as noted in Sect. 2.

For the asymmetrical DG (n-channel) device, we define the effective $t_{i(asym)}$ (for the predominant front channel) by integrating the electron density over the entire Si film:

$$t_{i(asym)} = \frac{\int_0^{t_{si}} xn(x)dx}{\int_0^{t_{si}} n(x)dx} = -\frac{q}{Q_{c(asym)}} \int_0^{t_{si}} xn(x)dx. \tag{10}$$

To derive the counterpart to (8), we first write the 1-D Poisson equation as

$$\frac{d}{dx}\left(x\frac{d\Psi}{dx}\right) \cong \frac{q}{\varepsilon_{si}}x(n(x) + N_A) + \frac{d\Psi}{dx}. \tag{11}$$

Integrating (11) across the entire Si film yields

$$\Psi_{sf} \cong \Psi_{sb} - \frac{Q_{c(asym)}}{C_{i(asym)}} - \frac{Q_b}{2C_b} + t_{Si}E_{sb}, \tag{12}$$

where

$$E_{sb} = -\left.\frac{d\Psi}{dx}\right|_{x=t_{Si}}$$

is the electric field at the back surface.

The gate–voltage (Faraday) relations for the DG-MOSFET structure are

$$V_{GfS} = \Psi_{sf} + \Psi_{of} + \phi_{Gfs} \tag{13}$$

and

$$V_{GbS} = \Psi_{sb} + \Psi_{ob} + \phi_{Gbs}, \tag{14}$$

where Ψ_{of} and Ψ_{ob} are the potential drops across the front and back oxides, and Ψ_{Gfs} and Ψ_{Gbs} are the front and back gate-body work-function differences. Applying Gauss's law across the entire Si film of the asymmetrical device, we get (for no interfacial charge)

$$\Psi_{of} = \frac{1}{C_{of}} \left(\varepsilon_{si} E_{sb} - Q_{c(asym)} - Q_b \right) \tag{15}$$

and applying Gauss's law to the back-surface, we get

$$\Psi_{ob} = -\frac{\varepsilon_{si} E_{sb}}{C_{ob}}. \tag{16}$$

Now, by setting $V_{GS} = V_{Gfs} = V_{Gbs}$ in (13) and (14) and eliminating Ψ_{sb} and E_{sb} from (12) and (14)–(16), we finally derive the integrated electron charge expression:

$$Q_{c(asym)} \cong -C_{Gf(asym)}(1 + r)(V_{GS} - V_{Tf(asym)}), \tag{17}$$

where

$$C_{Gf(asym)} = \frac{C_{of}}{1 - r\frac{C_{of}}{C_{i(aym)}}} \cong \frac{C_{of}}{1 - \frac{r}{3}\frac{t_{i(asym)}}{t_{oxf})}} \tag{18}$$

partly defines the total front-gate capacitance, and

$$V_{Tf(asym)} = \Psi_{sf} + \frac{1}{1 + r}\left[(\phi_{Gfs} + r\phi_{GbS}) - \left(\frac{Q_b}{C_{of}} - r\frac{Q_b}{2C_b} \right) \right] \tag{19}$$

is a nearly constant threshold voltage for strong-inversion conditions. In (17)–(19), $r > 0$ [defined in (5) by a charge-sheet analysis] reflects the benefit of the "dynamic threshold voltage" [2] of the asymmetrical DG-MOSFET due to the gate–gate charge coupling. This effect is preempted in the symmetrical DG device because of the (back-channel) inversion charge, which shields the electric-field penetration in the Si film and pins y_{sb}. For the particular asymmetrical device simulated, $r = 0.47$.

We note further that in (17) and (18), the dependence of $Q_{c(asym)}$ on $t_{i(asym)}$ is different from the $t_{i(sym)}$ dependence in (8) and (9), which reflects an additional benefit due to the $n(x)$ distribution in the Si-film channel. By comparing (18) with (9), which are illustrated by the SCHRED-predicted $-dQ_c/dV_{GS}$ plots included in

Fig. 6, we see that for finite $t_{i(asym)}$ and $t_{i(sym)}$, which are comparable (classically as well as quantum-mechanically), $C_{Gf(asym)} > C_{of}$ whereas $C_{Gf(sym)} < C_{of}$.

The latter inequality for the symmetrical device is the well-known effect of finite inversion-layer capacitance [3], as characterized in (9); an incremental increase in V_{GS} must support an incremental increase in the potential drop across the inversion layer, at the expense of an increase in $-Q_{c(sym)}$. (Note that the inversion-layer potential drop is zero when $t_i = 0$ and Q_c is a charge sheet.) The former inequality for the asymmetrical device, however, is unusual. It can be explained by referring to the predicted transverse electric-field variations ($E_x(x)$) across the Si film (channel) shown in Fig. 6 for the asymmetrical and symmetrical devices. The fact that $E_x(x = t_{Si}/2) = 0$ is always true in the symmetrical device underlies the noted, detrimental (regarding current and transconductance) inversion-layer capacitance effect.

However, in the asymmetrical device, typically $E_x(x) > 0$ everywhere, but an incremental increase in V_{GS} will, while increasing $E_x(x = 0)$, decreases $E_x(x = t_{Si})$ ($\equiv E_{sb}$ in (12)), ultimately forcing $E_{sb} < 0$, as in Fig. 6 where $V_{GS} = 1.0$ V. This field perturbation results in an incremental decrease in the potential drop across the Si film (inversion layer), and hence more increase in $\neq Q_{c(asym)}$, as reflected by (18). For the particular DG devices simulated, $C_{Gf(asym)}/C_{Gf(sym)} = 1.21$ at $V_{GS} = 1.0$ V, and this ratio is even larger for lower V_{GS}.

Quantitatively, the two noted benefits to $Q_{c(asym)}$ yield $Q_{c(asym)}/Q_{c(sym)} \cong 0.88$ at $V_{GS} = 1.0$ V, which is consistent with Figs. 4 and 5, when small differences in average electron mobility in the two devices are accounted for. We conclude then that the near-equality of the currents in the asymmetrical and symmetrical DG devices is due to:

1. the *gate–gate charge coupling* (characterized by r), which underlies near-ideal sub-threshold slope in (6) and the $(1 + r)$ enhancement of $Q_{c(asym)}$ in (17), and
2. the *beneficial inversion-layer capacitance effect* on $C_{Gf(asym)}$ in (18).

This is in contrast to the (common) detrimental effect on $C_{Gf(sym)}$ in (9). We thus infer for the asymmetrical DG-MOSFET a near-doubling of the drive current, and conventional polysilicon gates are adequate. Note further in (17)–(19) the possibility for structural design optimization of the asymmetrical device, which is not possible for the symmetrical device, as evident from (8) and (9).

We have discussed fundamental DG device physics. This insight can be applied to SRAM design in FinFET as well as Trigate technology [10]. We will investigate FinFET SRAM as an example, but notice that Trigate-based SRAM design can be obtained in a similar fashion. Figure 7 shows the description of a Trigate structure.

4 SRAM Design

Arrays constitute the majority of design area, have the highest density on the chip, and tend to suffer from technology issues first. As the technology scales down to the 22-nm regime, random dopant fluctuations, oxide thickness variation, line edge

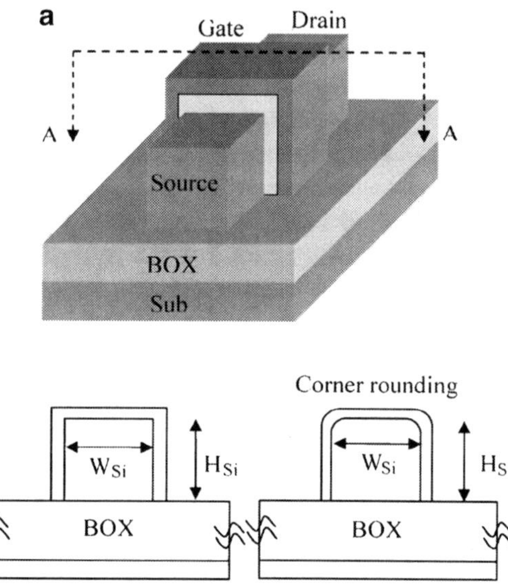

Fig. 7 (**a**) Three-dimensional trigate structure, and (**b**) two-dimensional cross-sections (A–A') of the triple-gate device with rectangular and rounded corners [10]

roughness, and SCEs result in a significant variation in the device threshold voltage and drive currents of planar devices [7].

Furthermore, due to high transverse electric fields and impurity scattering, carrier mobility is severely degraded. Designing large arrays requires design for five or more standard deviations. This is often challenged by an increase in process parameter variation with device scaling. Increasing variations make it difficult to design "near-minimum-sized" cells in embedded low-power large array designs, which in turn challenge the underlying high-density requirements of state-of-the-art memories [11].

Designed for improved scalability [12], the FinFET design is a good candidate for memory applications. First, we study conventional 6T SRAM cell metrics and design tradeoffs. FinFET-based SRAM cell designs are then presented.

4.1 SRAM Design Requirements and Functionality Metrics

Large array designs must be properly designed to address the tradeoffs among performance, power, area, and stability. To maintain good functionality and design margins, device strength ratio tradeoffs (pass-gate-to-pull-down and pass-gate-to-pFET) in terms of device size and threshold voltage are analyzed. Often higher-threshold voltages are set to address standby leakage power, and in some cases the device length is increased [12]. Increasing the device length to lower leakage can impact wordline and bitline loading and, hence, impact performance. Optimal design operating conditions, in terms of the design supply voltage, are also identified.

To minimize power dissipation, the supply voltage is often determined as the minimum supply voltage, V_{min}, that meets design functionality requirements. More recently, state-of-the-art processor designs [13] rely on separate logic and SRAM supplies to improve design stability in the presence of unacceptable leakage and process variation. This helps relax the requirements on V_{min} [13] compared to the single supply case.

Determining the minimum supply voltage for low-power operation requires the analysis of the SRAM cell functionality. Traditionally, static noise margins were used to study cell stability. However, it has been shown that, particularly for PD/SOI technologies, device-to-device fluctuations, in addition to hysteric V_t variation caused by the floating-body configuration can result in cell instability during Read/Write operations even if the static stability, in terms of the static noise margin of the cell, is high [13, 14].

Hence, SRAM performance metrics must be derived from the dynamic behavior of the cell. The recommended performance metrics are the dynamic "read stability" and "writability." These metrics are more comprehensive because they are derived from actual transient simulations. In addition to capturing SOI body effects, they account for the impact of cell placement and peripheral circuitry. We will next discuss dynamic stability and writability of the cell. Without loss of generality, we will assume an SRAM cell storing "0" on node L and "1" on node R (see Fig. 8).

4.1.1 Metric 1: Read Stability

During read "0," the pass-gate and pull-down transistors, PG and PD, respectively, act as a voltage divider circuit between the precharged bitline, *BL_L*, and node *L*. Usually, this induces noise at node *L* (Fig. 9a). However, it is possible that the cell can be weak due to process variation. In this case, the induced noise can be significant enough to flip the contents of the cell. This is known as a destructive read (Fig. 9b). For a cell to be stable, the maximum noise on node *L* must satisfy the "acceptance criteria" stated in (20).

$$V_{max}(L) < k^*V_{DD}; \text{ where } k < 1. \tag{20}$$

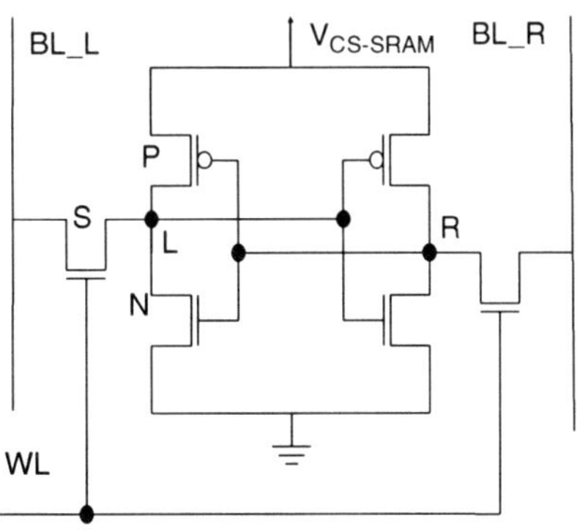

Fig. 8 An SRAM cell schematic

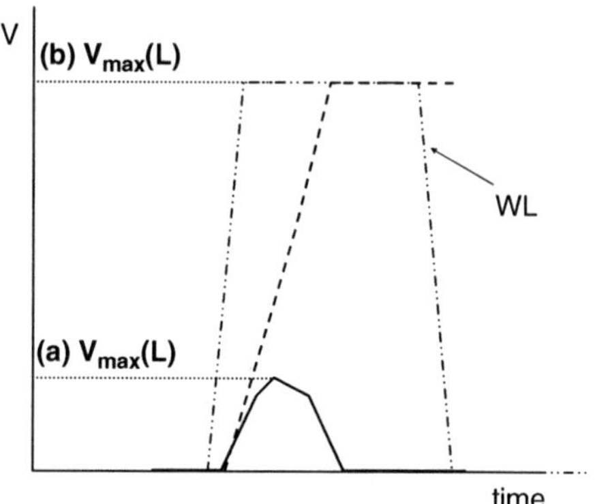

Fig. 9 Read "0" at node L. (a) The left pass-gate (PG) and pull-down (PD) transistors act as a voltage divider between L and BL_L, and $0 < V_{\max}(L) < k \times V_{\mathrm{DD}}$ during read. (b) Destructive read: It is possible that the cell is weak due to process variation, and $V_{\max}(L) = V_{\mathrm{DD}}$

To statistically analyze the read stability of the cell we check whether the cell violates this criterion at a given set of samples in the process parameter space [11]. Usually, we want to account for those samples that correspond to the cell flipping. However, we can be more conservative by setting $1/3 < k < 1/2$ (20).

4.1.2 Metric 2: Writability

A cell is writable if we are able to flip its contents (Fig. 10). A good measure of writability is the time it takes the cell to write a "0." This is equivalent to writing a "0" on node R for the cell of Fig. 8. Because the cell is a storage element with cross-coupled inverters, a conservative metric is to measure the write "0" delay from the time the wordline, WL, gets activated to the time node L charges up; i.e., write "0" on R is successful only if node L gets charged up.

Equation (21) describes the acceptance criterion for a write operation. Violating this criterion implies a write failure, and can occur due to write delay (time-to-write) being greater than the delay criterion, τ, or the cell not being able to flip its contents. Higher values of the parameter k (21) are intended to guarantee that we do not mistake a cell stuck at its bistable point for a cell that has successfully flipped.

$$T_{\mathrm{write}} = (t_2 - t_1) < \tau.$$

τ is the write delay criteria.

$$t_1 : V_{WL}(t_1) = \frac{V_{DD}}{2},$$

$$t_2 : V_L(t_2) = k^* V_{\mathrm{DD}}, \quad \frac{1}{2} < k < 1. \tag{21}$$

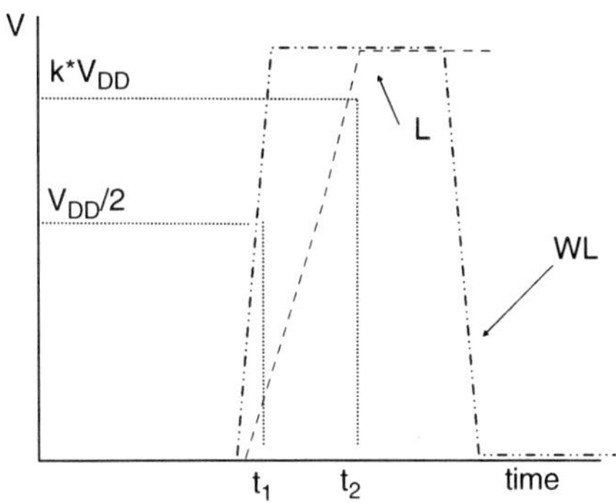

Fig. 10 $T_{\text{write}} = t_2 - t_1$ (21): time to write "0" on node R

5 FinFET Design for SRAM

The FinFET device is designed to overcome much of the sub-20-nm technology concerns in a typical planar device. This in turns relaxes much of the constraints on the SRAM devices. The thin body of the FinFET suppresses much of the SCEs, and, hence, the body channel can be lightly doped. This, in turn, reduces the threshold voltage variation [1]. It also helps to minimize the transverse electric field, hence the impurity scattering and carrier mobility. It also helps reduce the depletion charge and capacitance which, in turn, leads to a steep sub-threshold slope and, hence, reduced leakages. Furthermore, FinFET devices have lower capacitance. This helps reduce bitline loading and improves SRAM performance. In fact, FinFET devices are found to have increasing performance improvements over planar devices as technology scales.

State-of-the-art SRAM designs, in general, rely on the two types of FinFET operating modes (Fig. 11).

This is independent of the underlying physics (symmetrical vs. asymmetrical device). Theoretically, it is preferred that DG devices be built as asymmetrical devices.

Double-gate (DG) mode. The two gates are biased together to switch the FinFET on/off.

Back-gate (BG) mode. The two gates are biased independently. Front gate is used to switch the FinFET on/off and the back-gate is used to change the threshold voltage of the device. This mode offers tuning capabilities.

5.1 *Example of FinFET Simulation Models Used for SRAM Design*

Simulation is key to studying the functionality of the SRAM design. Because FinFET devices are relatively new, analytical models are still being refined [16]

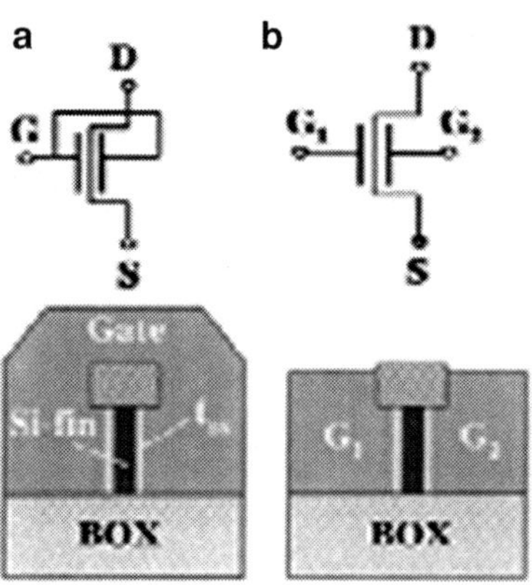

Fig. 11 (a) A tied-gate FinFET and (b) independent-gate controlled double-gate FinFET. Notice that for asymmetrical double-gate FinFET, "n^+-polysilicon" is used for front gate G_1, while "p^+-polysilicon" is used for back-gate G_2 [15]

and are not widely available in design environments like Cadence tools. Alternatives are lookup-table (LUT)-based models [16], mixed-mode simulations [8], or planar-based DG models [1].

LUT-based models are derived from TCAD [16] simulations in terms of $I\text{--}V$ tables. The LUT approach can be used for circuit simulation by using

$$I_X(V_G, V_D, V_S, t) = I_{\text{DCx}}(V_G, V_D, V_S, t) + \frac{\mathrm{d}Q_X}{\mathrm{d}t}, \qquad (22)$$

$$\frac{\mathrm{d}Q_X}{\mathrm{d}t} = \frac{\mathrm{d}Q_X}{\mathrm{d}V_G}\frac{\mathrm{d}V_G}{\mathrm{d}t} + \frac{\mathrm{d}Q_X}{\mathrm{d}V_D}\frac{\mathrm{d}V_D}{\mathrm{d}t} + \frac{\mathrm{d}Q_X}{\mathrm{d}V_S}\frac{\mathrm{d}V_S}{\mathrm{d}t}, \qquad (23)$$

where $X = G, D, S$. This equation is valid if the quasi-static approximation holds [16], which is generally true for short-channel devices; a general-purpose simulator can then be used to solve these equations.

On the other hand, mixed-mode device simulation can be used to simulate the DC transfer characteristics of SRAM cells under different biasing conditions [12]. Mixed-mode simulations rely on the device drift-diffusion models and density gradient models [8]. DC characteristics can underestimate the drain current. The trends between two different designs may be maintained. Performance characteristics require transient simulations, which can be expensive in mixed-mode simulation.

The fin-sidewall surface orientation can affect FinFET performance. In fact, the PMOS and NMOS FinFET mobility can be improved based on crystal surface orientation [1]. By optimizing the orientation of the devices, the SRAM cell margins can be improved.

Hence, for purposes of simulation, it is important to feed the mixed-mode simulator with appropriate carrier mobility [12]; the calibration can rely on experimental data for the different surface orientations (see Fig. 12).

Fig. 12 Description of possible FinFET orientations

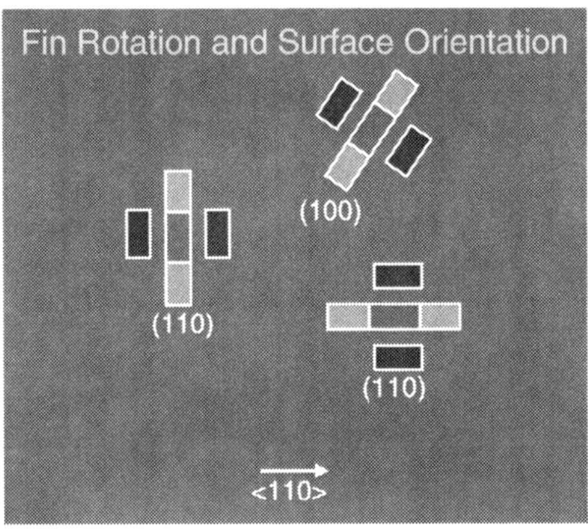

5.2 3-D Thermal Modeling of FinFET

The hotspot analysis in SRAM designs is very crucial. Hotspots degrade performance and impact reliability. In this section, we make an attempt to look at wordline driver and clock buffers, which generate the clock signal for memory design peripheral logic. Such an analysis is very helpful for pinpointing the thermal gradients from peripheral logic to SRAM cells. In general, portions of the design that have higher capacitance generate more power and, hence, suffer more heating effects.

This phenomenon is aggravated in FinFET designs. The increase of thermal resistance of the thin Si film constituting the channel due to sub-micron heat transport phenomena can impact the FinFET device drive. In [17], a detailed 3-D thermal model for multifinger FinFET devices was proposed to quantify the thermal behavior of the device. First, an in situ measurement and extraction methodology for thermal conductivity of thin Si film in real nanoscale CMOS devices was proposed. One can then perform 3-D thermal simulations of FinFET multifinger devices based on the measured/extracted thin-film thermal conductivity. The methodology was demonstrated at the 90, 65, and 45 nm technology nodes.

The experiments involved three-dimensional simulations employing Fourier's law of the temperature distribution of FinFET multifinger devices; since the fin height (device width) is fixed, we need multifinger devices to represent different SRAM cell device width [18]. The FinFET devices were arranged in multiple parallel arrays of six FinFETs situated one after the other (see Fig. 13) [17]. Each parallel array of FinFETs is equivalent to a finger in planar technology.

The values employed for the thermal conductivities of the materials are listed in Table 1. Note that in order to account for sub-micron heat transport phenomena in the channel region due to the actual dimensions of the FinFET, the thermal conductivity of silicon at the junction has been decreased to a value of 40 W/mK, instead of its bulk value, based on the experimentally measured/extracted data. This assumption will not alter though the temperature distribution in the rest of the domain [18].

Fig. 13 Cross-sectional temperature distribution at (**a**) the top surface and (**b**) its close-up of the shallow trench isolation of a ten-finger device with six FinFETs per finger according to the 90-nm technology node for a power dissipation of 0.2 mW/μm

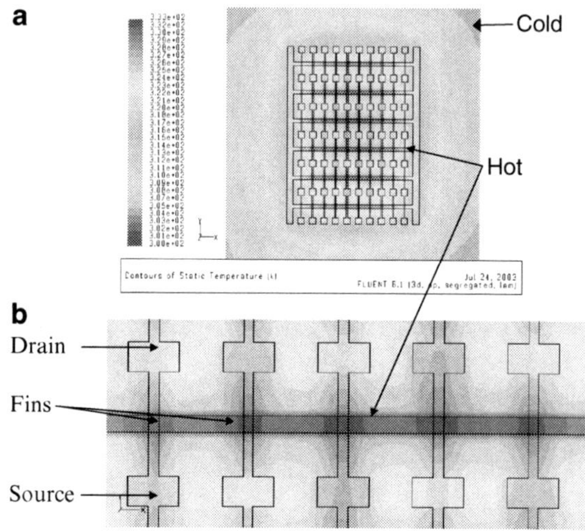

Table 1 Thermal conductivity values used in the simulation

	W/mK
Substrate (Si)	140
Shallow trench isolation (SiO$_2$)	1.4
Si at the junction (doped Si)	40
Gate (polycrystalline Si)	40
MCBAR (tungsten)	190
Metal interconnects and vias (Cu)	390

During the simulation, the thermal conductivities are considered to be temperature-independent since the device does not experience temperature changes that would vary the materials' thermal conductivities significantly.

The following analysis attempts to model the wordline driver of the memory design. The simulated devices have five levels of metal interconnects (M1–M5) connected by their respective vias (see Fig. 14). Joule heating due to current flow through the metal interconnects is also considered in order to more realistically represent the operating state of the device. For the simulations, adiabatic boundary conditions have been assumed on the four sides of the computational domain. These boundary conditions reflect the fact that practically no heat leaves through those walls and no temperature gradient is expected on those sides. For the top boundary, a convective boundary condition has been assumed to model a solder bump with a convective coefficient same as that of air. For the bottom surface, an isothermal boundary condition of 300 K has been imposed.

The maximum temperature rise occurs at the junction with a value of 32.6 K. Most of the generated heat, around 95% of the total, dissipates down the silicon substrate and practically all the temperature drop takes place in the shallow trench isolation (STI). Also, increasing the number of FinFETs would not yield a different result in terms of temperature rise both at the junction and/or in the entire domain.

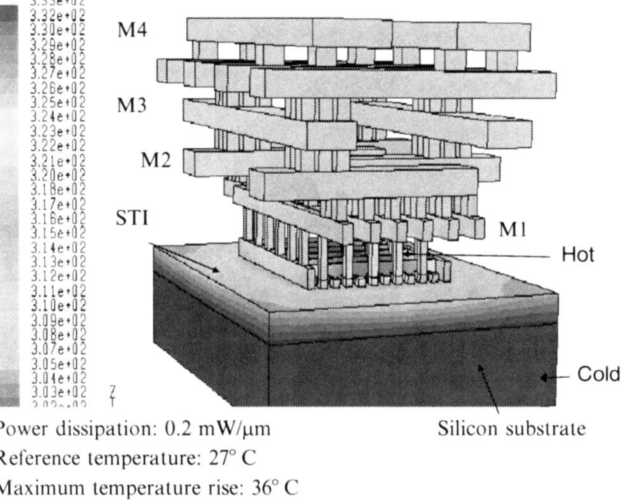

Power dissipation: 0.2 mW/μm
Reference temperature: 27° C
Maximum temperature rise: 36° C

Fig. 14 Three-dimensional temperature distribution of a ten-finger device with six FinFETs based on the 90-nm technology node

The temperature rise experienced by a FinFET is influenced by the temperature rise of the neighboring FinFETs, but it would only be affected by its closest neighbors within a characteristic distance determined by the thermal diffusion length in Si, and not by the entire array of FinFETs. Figure 13 clearly shows this certainty and one can see the smaller temperature rise at the junction for FinFETs situated at the end of the active area with respect to those situated in the center.

Figure 15a shows the temperature rises at the junction for the 90, 65, and the 45-nm technology nodes. As one can see, not only the actual value of the temperature rise, but the rate at which the temperature rises, increases with scaled technologies. This is expected, though, since the heat dissipation volume decreases as the device is scaled down in size.

However, when one compares the thermal resistance, defined as the change of the temperature rise at the junction per dissipated power and per unit device width:

$$[\text{Device width}] = [\text{Fin height}] \times [\text{Number of fins per finger}] \times [\text{Number of fingers}].$$

All three technologies exhibit roughly equal thermal resistance (see Fig. 15b). For a scaled technology, the gate length of the FinFET decreases and more FinFETs can be aligned together for a given device width. For a fixed total power in one finger, the power dissipated in each FinFET is, therefore, less in scaled technology. This reduced power dissipation for each FinFET compensates for the reduced heat dissipation volume, and hence maintains the thermal resistance constant as the technology scales. Notice then that the thermal resistance depends on the specific dimensions and geometry of the FinFET and, therefore, the thermal resistance is ought to be considered as a geometrical factor associated with the thermal behavior of the entire device.

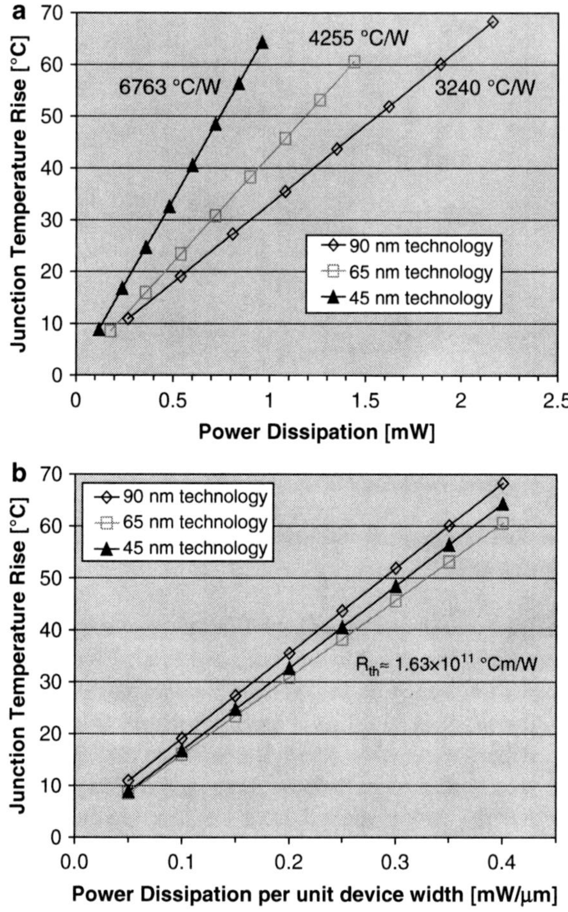

$R_{th} \approx 1.63 \times 10^{11}$ °Cm/W

Fig. 15 (**a**) Temperature rise at the junction for different power dissipations in FinFET multifinger devices. (**b**) Thermal resistance of FinFET multi-finger devices at the 90, 65, and 45-nm technology nodes

6 Low-Power, High-Performance 90-nm DG-FinFET SRAM Design

In this section, we investigate and compare the SRAM behavior of FinFET technology with 90-nm node planar PD/SOI technology. The goal is to demonstrate FinFET applicability to SRAM in a full cell design. Unique FinFET circuit behavior in SRAM applications, resulting from the near-ideal device characteristics, is demonstrated by full cell cross-section simulation in terms of high performance and low active and standby power.

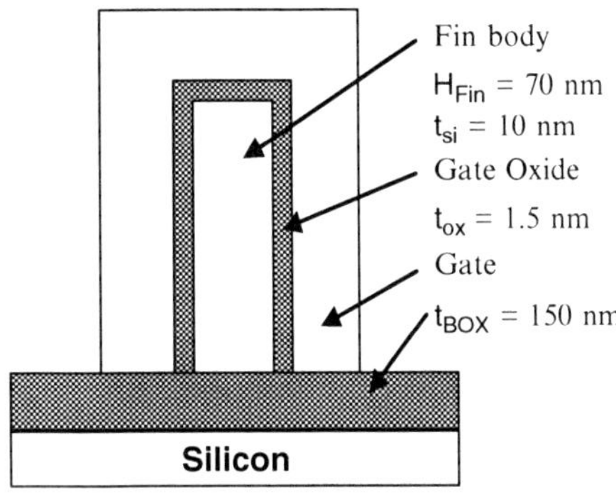

Fig. 16 Structural details of the N/P FinFETs (source/drain direction into the page)

6.1 Design Overview

The reference point for this study is a 90-nm node PD/SOI planar technology with extremely scaled oxide ($\sim$1 nm) and low supply voltage ($\sim$1 V) for SRAM analysis. The planar SOI technology is similar to one discussed in [1]. The FinFET technology in this study is designed with comparable channel length and digital logic performance. Figure 16 shows the FinFET channel design fabricated in a substrate with a buried oxide. The FinFET silicon is 10 nm wide (t_{Si}).

The DG technology advantage allows the gate oxide to be less aggressive ($\sim$1.5 nm), which results in lower gate-leakage power. Particularly, the FinFET devices are characterized by reduced I_{off}, $V_{\mathrm{t(lin)}}$, and DIBL effects.

The FinFET performance advantage derives from the reduced threshold (V_t) possible in a dual-gate device (enabled by better sub-threshold swing). The FinFET devices have lower $V_{\mathrm{t(lin)}}$ and DIBL because the wraparound FinFET gate exerts more control on the channel charge relative to the planar single gate [30] and due to the elimination of floating-body effects. This V_t reduction compensates performance for the difference in NFET I_{on}. Such attributes are ideal for SRAMs.

6.2 Device Models

The FinFET compact models used in this study are assembled using a modified version of planar PD/SOI compact model and considering specific FinFET DC/AC features [13]. This model is incorporated into a sub-circuit that includes the BSIMPD model [19] and current sources to model the gate-to-source and gate-to-drain leakage present in scaled SOI technologies. This approach captures important first-order FinFET features, such as low sub-threshold slope, reduced DIBL, reduced junction capacitance, and suppressed body factor, while automatically including SOI features, such as source/drain isolation from the substrate and self-heating.

Due to the thin fin thickness t_{Si}, the fin bodies are fully depleted under typical doping conditions. Full depletion implies that the mobile majority charge in the body will be zero, and to the first order, in strong inversion, the body potential is constant [14]. Therefore, the body charges including the body-to-diffusion junction capacitance terms are zeroed out through model parameter changes. Diode and impact ionization effects are also shut off due to full depletion together with the assumption that impact ionization effects will be low because of the scaled power supply voltage. For numerical considerations, the floating body is tied to the source node through a resistor R_{numer} since in this approximation it has little influence on the overall result. Note that because the body is fully depleted and the body does not move, the SOI history effect is negligible. The schematic of the modified model is shown in Fig. 17.

Finally, while the fin thickness is narrow enough to experience double-gate inversion charge control, it is assumed to be insufficiently narrow to significantly influence charge transport properties between the source and drain; the same mobility model formulation is used as in the planar PD/SOI technology. Using this mobility formulation, the calibrated FinFET models portray improved sub-threshold characteristics.

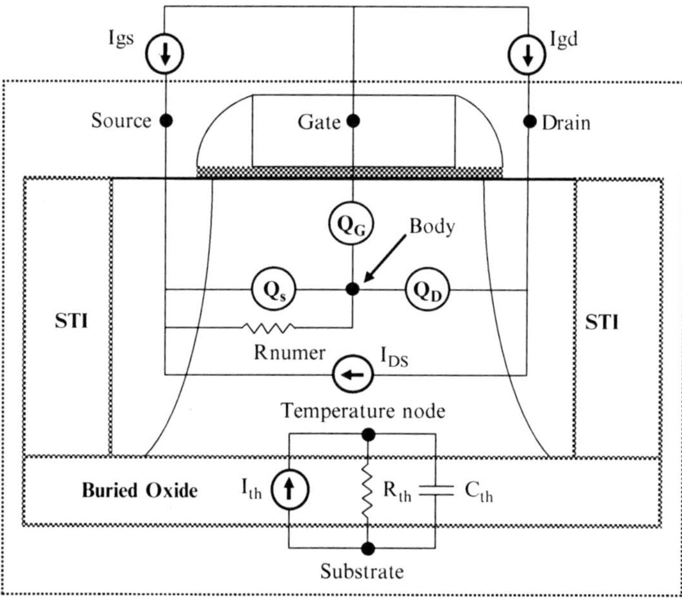

Fig. 17 FinFET sub-circuit compact model. The elements in the dotted box are from the BSIMPD Version 2.2.3 model [19]

Table 2 Quantization effect on the SRAM cell

FET type	Fin count	Change in width (%)
Pull-up	1	13.1
Pull-down	2	−3
Pass-gate	1	5

6.3   Width Quantization

The width of the FinFETs is quantized in multiples of $2 \times H_{\text{fin}}$ based on the process technology. For scaled technologies, the effect of width quantization can be large and the SRAM cell response can be very sensitive to it. In this study, the fin count was chosen to approximate the widths used in the comparison planar technology and also correspond to the fin packing density possible in the original PD/SOI planar cell design. An exact match in widths is not possible, and the percentage change is given in Table 2.

In fact, the authors in [31] also indicate that width quantization limits SRAM sizing choices. The quasi-planarity, on the other hand, allows increased cell current by increasing the fin height. Hence, the right combination of device structure can be achieved to improve yield and minimize leakage. Increasing V_t with fin height and body thickness improves stability, decreases variability, and decreases source–drain leakage exponentially. To control SCEs, this requires a small t_{ox} though; this degrades gate leakage. Increasing t_{ox} can control gate leakage; however, this worsens stability. Careful co-design of the device structure in terms of t_{ox}, fin height, V_t and V_{DD} are needed to achieve a good balance between leakage and yield of FinFET designs.

6.4   Layout and Surface Orientation: A Brief Overview

Conventional SRAM cells are designed with a small aspect ratio (BL height to BL width). This allows straight poly-Si gate lines and active regions. This enables accurate critical-dimension control, and reduces gate-length variations and corner-rounding issues [12]. Also, shorter cells and more relaxed metal pitch lead to a reduced BL capacitance. To reduce WL capacitance, WL segmentation is employed.

Figure 18 shows the corresponding case with individual back-gate biases with asymmetrical DG devices; although the device is 3-D, the layout is derived based on 2-D rule checks. The individual back-gate biases require two wiring channels, which increase the area of the cell. Each DG-FinFET is illustrated by the intersection of the solid silicon body and the square-dashed poly gate electrode/wiring. The larger rectangular solid areas at either end of the FinFET bodies are FinFET source/drain contacts. The lower two transistors are the pull-up PFETs. Directly above the two PFET devices are two NFETs, which have their front- and back-gates

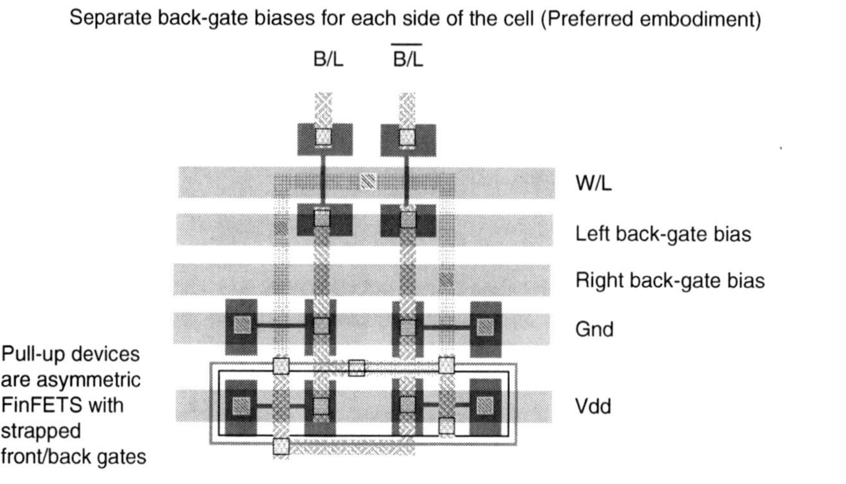

Fig. 18 Layout of an asymmetric DG device

wired to separate electrical nodes. These four devices form the cross-coupled pair The back-gates of the pass-gate NFETs and pull-down NFETs are wired together and brought out to a horizontal bus wire for the cell back-gate bias control. The front gates of the NFETs are wired as a conventional cross-coupled inverter pair using M1 and poly wiring. The top two DG-FinFETs are the transfer (pass-gate) devices. The front gates of the transfer devices are tied to the horizontal wordline (W/L) bus, while the back-gates of the transfer devices are connected to the back-gate bias control bus. The common back-gate bias control bus of all transfer devices and all pull-down devices allows their strengths to be modulated together. Overall, the cell area for the back-gate scheme is slightly increased compared to the DG (tied-gate) counterpart [20].

Furthermore, the ability to construct a rotated NFET pull-down layout [12] has been demonstrated. Such FinFET SRAM cells can be built by rotating the fins by 45°. This can be lithographically challenging, but helps improve the mobility of the devices.

6.5 Performance

The SRAM cell plan view is shown in Fig. 19. Since the fin height is quantized, it is important to study the quantization effect on performance. Two planar SOI cells with Regular V_t (RVT) and high threshold (HVT, V_t offset $\sim$80 mV) devices are compared with the quantized FinFET cell topology. To measure the read

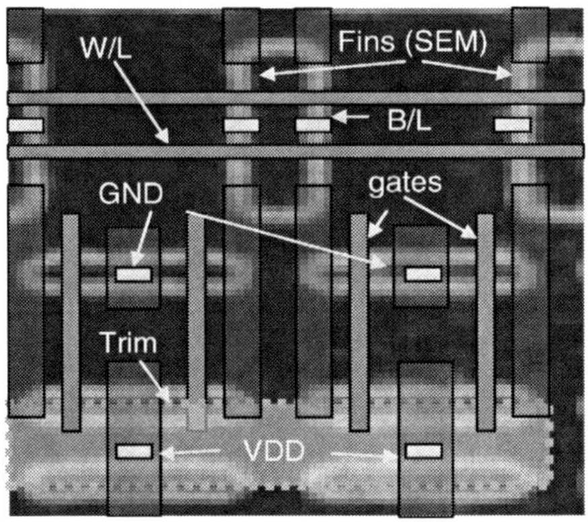

Fig. 19 SEM of the fin pattern in silicon is shown with gate, Trim, and via levels superimposed. Trim level is used to remove undesired fin sections. Interconnect for the cell is not shown

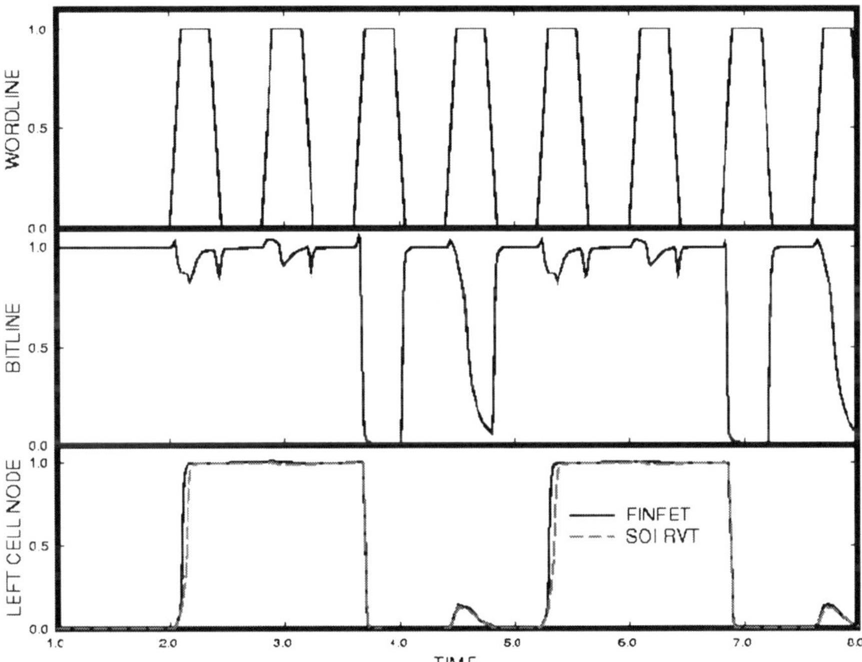

Fig. 20 Typical write operation of the SRAM cell showing the fast FinFET cell node rise time

performance, a sense amplifier is used (not shown here but is common to both analyses). The combined quantization effect on the cell and sense amplifier is analyzed using circuit delays (see Fig. 20). The read delays are measured from the wordline rise through the cell to the sense amplifier output.

The read delays improve significantly for the FinFET structure as the cell pass-gate strength increases due to quantization, thereby slightly changing the beta ratio of the cell. Since the FinFETs are fully depleted, the device junction capacitance is negligible, giving additional delay improvement. As a result, the delays improve substantially compared to both the RVT and HVT cells. This is particularly true at lower V_{DD}, where the gains were found to be 12–13%. The HVT cell creates a different beta ratio compared to the RVT cell, and resistance to pull-down "0" increases (due to the slowdown of pass-gate and pull-down devices).

For write delays, however, the comparison point changes. The write delays for FinFET are higher than those for the RVT SOI cell. The write delays are measured from the 50% bitline switch point to the 50% switch point of the flipping of the opposite cell node. Thus, this is a true representation of the cell delay. Due to width quantization, the PFET strength (the width of which is smaller to start with) increases, making the write ("0" to "1") operation more difficult. For the RVT cell, the pass-gate and pull-down NFETs become stronger, while for the HVT cell, the pass-gate and pull-down stacks get weaker. Thus, the strength of the pass-gate and pull-down transistors affects the write delays. FinFET delay was found to be 8% slower than RVT and 12% faster than HVT at low V_{DD} (0.8 V).

6.6 Power

Next, we discuss the power comparison for FinFET and PD/SOI cells for read and write operations. The active power shows a marked difference between the FinFET cell versus PD/SOI cell. The FinFET cell has much lower power: RVT power was found to be $\sim$4.5$\times$ that of FinFET. Since FinFET devices are fully depleted, the device junction capacitance is negligible compared to that of PD/SOI devices, and the active power is reduced.

The half-select power (when bit lines are held at V_{DD}) is slightly lower for the FinFET cell. As the FinFET is fully depleted, V_t changes due to floating-body effects are eliminated, compared to PD/SOI. This reduced V_t variation results in reduced leakage during the half-cell select operation.

A similar analysis is helpful in explaining the difference between the two for standby power. The slope for leakage power is much smaller for the FinFET cell while it is much sharper for PD/SOI. This is due to the lower FinFET I_{off} and DIBL. The RVT device power exponentially increases with V_{DD}. The FinFET device power shows weak linear dependence. At 0.8 V, the RVT power was 3$\times$ compared to that for FinFETs, and at 1 V, it was close to 6$\times$.

6.7 Stability

We now compare the read–disturb stability of the FinFET and PD/SOI cells. The stability depends on process-induced V_t fluctuations, and on length and width variation of the cell devices. As the magnitude of the V_t scatter increases, the maximum stable voltage (V_{max}, the voltage at which cell does not flip) needs to be lowered for "half-select or read stability." A similar analysis holds for writability. In addition, when the device drive strength increases, the NFET devices get stronger, which allows the cell to flip with bitlines precharged to V_{DD}. For the RVT SOI cell, a similar V_t scatter makes the cell more unstable. As a result, the maximum stable voltage decreases. We also study the minimum voltage (V_{min}) for stability as a function of scatter. The FinFET shows a lower value until the sensitivity of its lower threshold voltages is realized.

7 A High Performance, Low Leakage, and Stable SRAM Row-Based Back-Gate Biasing Scheme in FinFET Technology

In this section, we describe a back-gate biasing scheme using independent-gate controlled asymmetrical (n^+/p^+ polysilicon gates) FinFETs devices and its applications to 6T and 8T SRAM. Row-based above-V_{DD}/below-GND bias is applied to the back-gates of the pass-gate and pull-down cell NFETs to enhance the Read/Write performance, reduce standby leakage, and mitigate process (V_t) variability. The application of the technique to stacked Read transistors in 8T SRAM is also discussed.

7.1 Back-Gate Design Overview

As CMOS devices are scaled beyond 45 nm, dopant concentrations reach their practical limits, and random dopant fluctuation and V_t control become major concerns [19]. This affects the SRAM devices the most as the sizes are extremely small and the fluctuations are inversely proportional to the square root of length and width product. A known solution is to control the V_t by using body/well or back-gate biases. One serious problem with individual well/body or back-gate bias is the increased layout area and complexity, and therefore higher cost.

Individual back-gate biasing also requires more wiring resources. Furthermore, well/body bias in bulk CMOS or PD/SOI devices suffers from the following constraints: (1) a large reverse well/body bias causes an exponential increase in reverse junction leakage, including band-to-band tunneling current, while a forward well/body bias results in an exponential increase in the forward diode leakage; (2) the V_t modulation effect diminishes with device scaling due to the low body factor in scaled, low V_t transistor; and (3) the distributed RC for well/body contact limits the viable operating frequency.

In the following section, we study a dense row-based back-gate biasing scheme for independent-gate controlled asymmetrical (n^+/p^+ polysilicon gates) DG Fin-FET, as shown previously in Fig. 11 [1, 2]. Although, in the planar DG technology, the self-alignment of the top- and bottom-gate remains a technological challenge, the FinFET structure with selectively implanted n^+ and p^+ polysilicon gates on the sides of the fin can be more easily processed [1]. The device has a predominant front-channel with a significantly lower V_t and much larger current drive compared with the "weak" back-channel. The front-channel V_t (and current) can be modulated by back-gate biasing through gate-to-gate coupling.

The proposed scheme has the following advantages: (1) row-based back-gate biasing results in a very dense layout with minimum wiring resource; (2) this V_t modulation mechanism is significantly stronger than the well/body bias in bulk CMOS and PD/SOI devices; (3) the V_t modulation effect improves with device scaling due to stronger gate-to-gate coupling; (4) the frequency is only limited by the gate RC (same as the core logic); and (5) higher than V_{DD} (or lower than GND) back-gate bias can be applied to the back-gate, thus further enhancing the V_t modulation effect without causing excessive leakage (whereas a large forward/ reverse bias in not possible in well/body bias for bulk CMOS or PD/SOI).

Figure 21 shows the proposed DG FinFET SRAM structure. The selected and unselected rows with common back-gate biases are shown. All the cells in a row have a common back-gate bias. Row-based forward bias above V_{DD} is applied to the back-gates of the pass-gate NFET and pull-down NFET during the Read/Write operation to enhance performance. In the Read mode, the strengthened pass-gate and pull-down NFETs speed up the bitline signal development to improve the Read performance and margin [20]. In the Write mode, the strengthened pass-gate NFET helps the pull-down of cell storage node to improve the Write time (or time-to-write). During standby or sleep modes, a reverse bias below GND is applied to the back-gates of the pass-gate and pull-down NFETs to reduce the leakage.

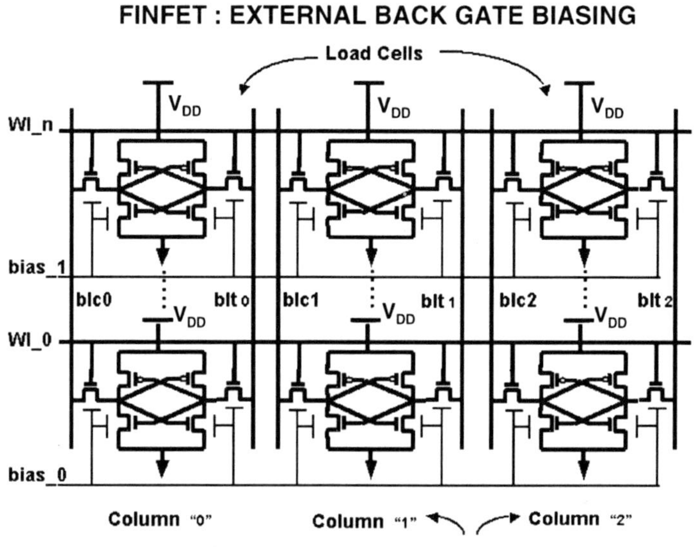

Fig. 21 Schematic diagram of row-based back-gate biasing scheme for DG FinFET SRAM

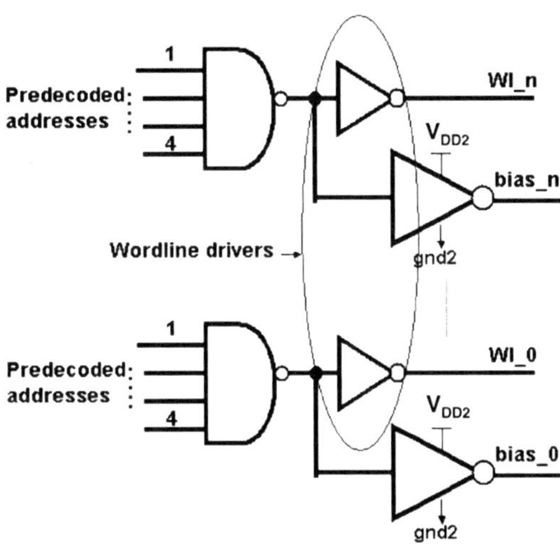

Fig. 22 Schematic diagram of the bias generator

7.2 Dynamic Bias Generator

The bias generator is shown in Fig. 22. The bias is tapped off from the earlier trigger point preceding the wordline driver. The bias generator creates "V_{DD2}" and "gnd2" levels through a simple inverter. When the wordline is activated, the bias generator is activated in a synchronized fashion to enhance the strength of NFETs in the SRAM cells during the active mode. For unselected cells, a back-gate bias of "gnd2" is applied. This below-GND bias reduces the leakages of the unselected cells.

The use of a mix-mode bias generator helps to improve performance and stability. Since the bias generator is simply an inverter, the cost of implementing it is very minimal (only 3–5% in area). However, the benefits are overwhelming. Even if the defect is present between the power and bias line, the cell will function, as the bias is only used to modulate the cell transistor strength depending on the operation and power saving mode (see Fig. 21). The row decoder is based on conventional design except that it also drives the bias generator.

7.3 Analysis of the FinFET SRAM

We analyze SRAM behavior based on 25-nm effective channel-length FinFET technology, comparing conventional and proposed schemes using mixed-mode simulations [8] for Read performance, leakage, and dynamic Read disturb. The device structure consists of an independent-gate controlled asymmetrical DG device. We assume fin height = 40 nm, and both conventioanl and proposed

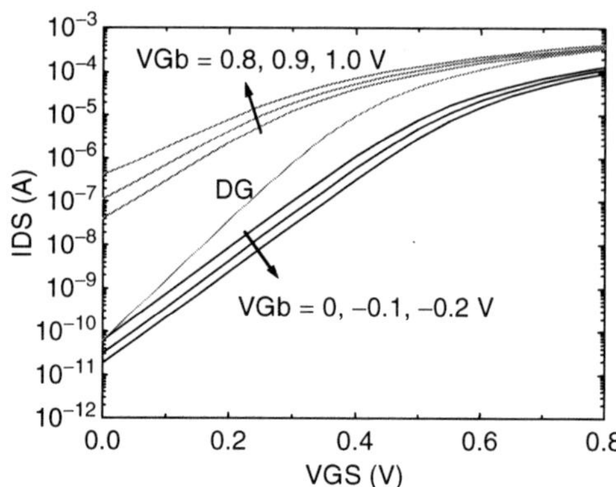

Fig. 23 MEDICI-predicted I_{DS} versus V_{GS} for the proposed scheme, compared with the conventional DG-mode (tied-gate) FinFET

schemes have one fin for the pull-up PFET and pass-gate NFET devices, and two fins for pull-down NFET devices.

Figure 23 shows MEDICI-predicted DC IV characteristics for the proposed scheme and conventional DG-mode (tied front/back gates) FinFET. Since larger-than-V_{DD} forward bias can be applied to the back-gate in the proposed scheme, I_{on} is significantly improved.

This larger I_{on} enhances the Read performance [17]. MEDICI was used to predict waveforms during the Read operation for the proposed bias generator. In the standby mode, (gnd2 < gnd) was applied to reduce the leakage current. In the Read mode, $V_{DD2} > V_{DD}$ was applied to improve the performance. The proposed scheme improved the Read performance by ∼20–40% by applying 0.1–0.2 V higher-than-V_{DD} back-gate bias to the selected row (compared to conventional tied-gate DG device.)

By providing lower-than-ground bias to the back-gates of the pass-gate and pull-down NFETs, V_t is increased, thereby substantially reducing the sub-threshold leakage current, according to analysis of the DC characteristics. MEDICI-predicted sub-V_t leakage current (I_{off}) decreases significantly as a function of back-gate voltage (V_{GB}) for the proposed scheme, compared with the conventional tied-gate FinFET, where the back-gate voltage (0.0 V) is the same as the front-gate voltage (0.0 V). For a reverse back-gate bias of −0.2 V, the leakage is reduced by more than 3×, compared with the conventional tied-gate FinFET scheme.

Figure 24 shows MEDICI-predicted Read noise voltage at the cell logic "0" storage node for conventional (tied-gate) and proposed schemes in FinFET technologies. The proposed scheme offers better stability for larger V_t mismatch since the NFET devices (especially the pull-down NFET) become stronger. The slightly worse stability for a smaller V_t mismatch is due to the lowered V_t.

By strengthening the pass-gate NFET during the Write operation, the proposed scheme speeds up the pull-down of the cell logic "1" storage node to improve the Write performance. MEDICI simulations also indicate that the time-to-write is improved by ∼15% compared with the conventional tied-gate scheme.

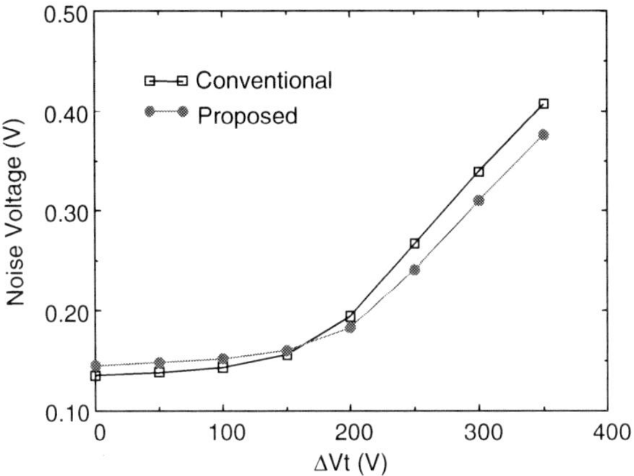

Fig. 24 MEDICI-predicted read noise voltage for the proposed scheme, compared with the conventional tied-gate scheme

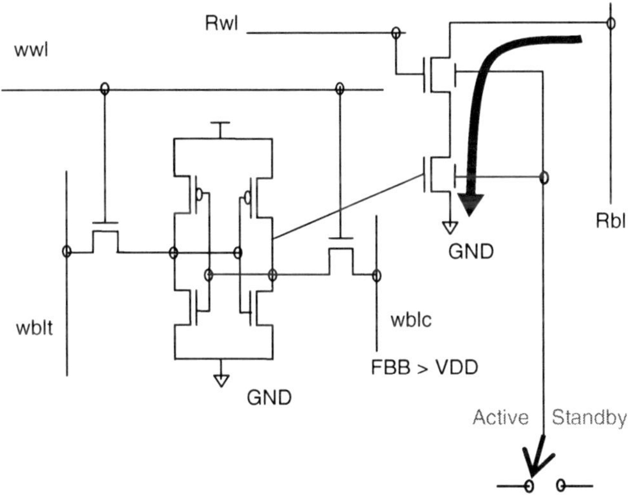

Fig. 25 Proposed multiport FinFET 8T SRAM cell

7.4 8T SRAM Cell

The proposed scheme can also be applied to multiport SRAMs [32]. Figure 25 shows the proposed multiport 8T SRAM cell. By controlling the back-gate bias for the stacked Read devices, performance enhancement can be achieved without increasing leakage power.

Since the cell nodes are not disturbed by the Read current in the 8T cell, Read stability is the same as Hold stability [12]. In the proposed 8T scheme, Read/Write operations are the same as in the conventional 8T scheme. Therefore, Read stability and Write stability can be the same. However, in the Read mode, the drain current increases with an increasing back-gate voltage. Thus, the Read performance is improved. The MEDICI-predicted Read delay for the conventional and proposed schemes again shows a significant improvement for the proposed scheme. Note that the back-gate voltage is switched to 1.0 V for 0.8 V of cell supply voltage during the Read operation for the proposed scheme. In the standby mode, the back-gate voltage is switched to ground or negative voltage to reduce the leakage current. This scheme can also be applied to register files as well.

8 Low-Power and Stable FinFET SRAM

In this section, we study a low-power FinFET SRAM design that relies on a programmable FinFET footer for improved performance and design margins [21].

8.1 Design Overview

It is very desirable to have a variable V_t diode to reduce leakage current in the standby/sleep mode and to compensate for the process variations and/or V_t fluctuations, especially in SRAMs. The conventional SRAM cell scheme uses a (footer) sleep transistor and a MOS clamping diode to reduce the leakage current in standby mode. However, it degrades the Read margin and performance due to the fixed V_{GND} (or V_{diode}) and the weakened pull-down devices in the Read operation.

In contrast, the scheme under study dynamically modulates V_{GND}, as shown in Fig. 26a. In the standby mode, the use of low V_{BG} increases V_t, and consequently, increases V_{diode} and V_{GND} to reduce the leakage current. In the Read mode, the use of high V_{BG} decreases V_t, and consequently, decreases V_{diode} and, hence, V_{GND}, to improve the Read margin and performance. Figure 26b shows MEDICI-predicted V_{GND} versus V_{BG} for the proposed scheme. V_{GND} around $V_{BG} = 0.2$ V significantly reduces the leakage current. In the Read mode, V_{BG} is switched to 0.8–1 V to improve the Read stability and performance. In the standby mode, results from mixed-mode simulations indicate that the leakage current is reduced by $\sim 4\times$ for a sub-array consisting of two "16 $\times$ 144" banks. Note that, in Fig. 26, SRAM cells can be used with either independent-gate controlled device or tied DG device [12].

To optimize Read/Write margins for stable SRAM operation, one would like to have a higher supply voltage during the Read operation to maintain an adequate noise margin, and a lower supply voltage during the Write operation to facilitate writing [22]. The scheme can be effectively extended to a dynamic Read/Write supply voltage (V_{DD}) for SRAM cells with a header scheme, as illustrated in Fig. 27a.

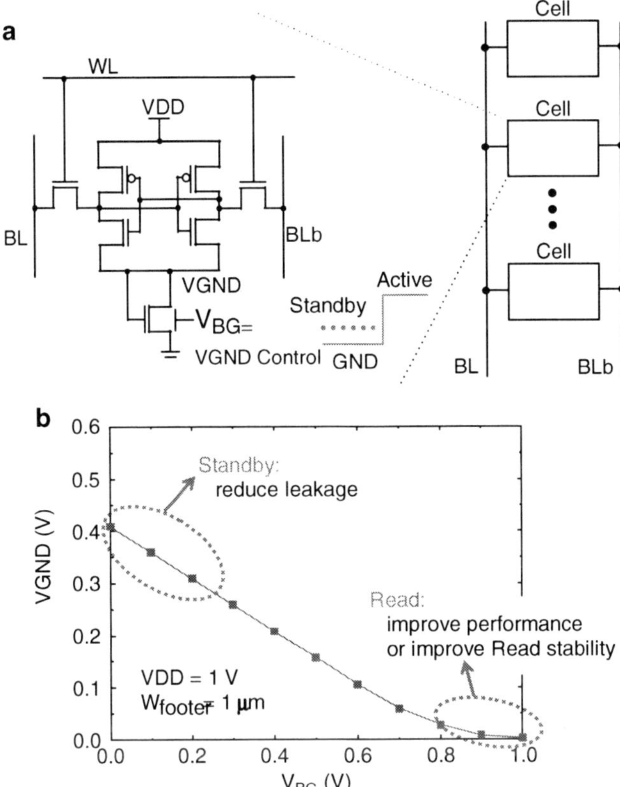

Fig. 26 (**a**) Footer-based scheme for low-leakage SRAM cells, and (**b**) MEDICI-simulated V_{GND} versus V_{BG} in a simplified SRAM cell array at $V_{DD} = 1$ V

Figure 27b shows the MEDICI-predicted V_{VDD} versus V_{BG} for the proposed dynamic Read/Write supply scheme. Around 0.35 V range of V_{VDD} can be achieved for a "2 × 16 × 144" sub-array. In the Read mode, the lower back-gate bias (0.2 V) significantly improves Read stability (this reduces PFET V_t). During the Write operation, V_{BG} is switched to a higher voltage around 0.8–1.0 V to facilitate the Write operation. In the standby mode, lower V_{VDD} reduces leakage. Note that the column or sub-array-based back-gate control is timing-feasible for high-speed array access given the fact that only a small amount of supply voltage level adjustment is required during mode switching, resulting in a measurable Read/Write margin improvement in SRAMs.

The proposed scheme requires only one V_{DD} with the back-gate bias to control/ change the voltage across the SRAM cells. Note that the conventional scheme requires either two external power supplies or on-chip voltage generator/regulator to provide the extra supply level [22]; it is also necessary to route the two supply lines. In the proposed scheme, it is only necessary to route the virtual V_{DD} control line.

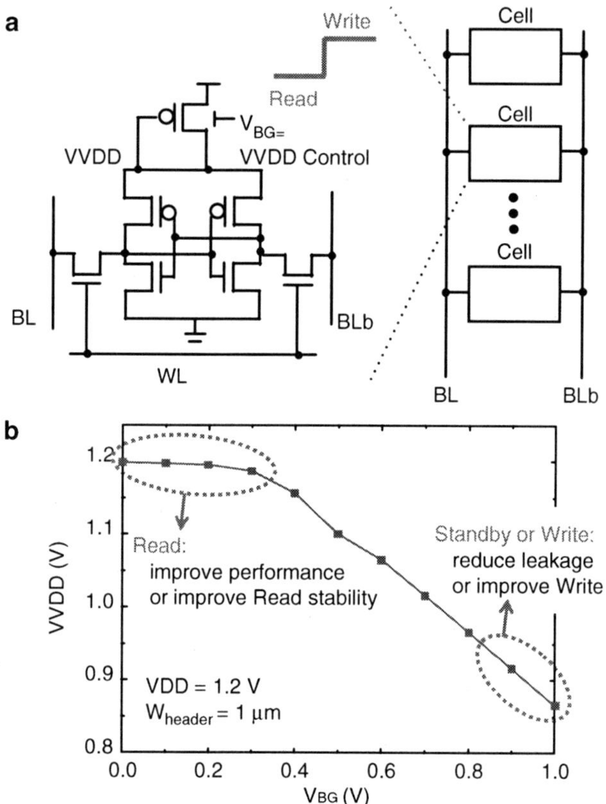

Fig. 27 (a) Proposed dynamic V_{DD} scheme for SRAM cells to improve Read/Write margin, and (b) MEDICI-predicted V_{VDD} versus V_{BG} at $V_{DD} = 1.2$ V for the proposed dynamic Read/Write scheme

Furthermore, the proposed scheme requires only one header diode (while the conventional scheme uses two pass transistors to perform the MUX function [22]).

Figure 28 compares the settling behavior of V_{VDD} for the proposed and conventional schemes. In the conventional scheme, the drain of the pass transistors connects directly to V_{VDD}. Hence, a large amount of charge has to be moved to charge/discharge the voltage rail capacitance. In the proposed scheme, the virtual supply control line only needs to charge/discharge the back-gate capacitance, and the voltage rail is charged/discharged by the front-gate current. As a result, the virtual supply settling time is shorter in the proposed scheme.

8.2 Leakage Current of Footer Device

The leakage current for the DG-SRAM schemes can be derived. When the subthreshold diffusion current is the predominant off-state leakage current, the total

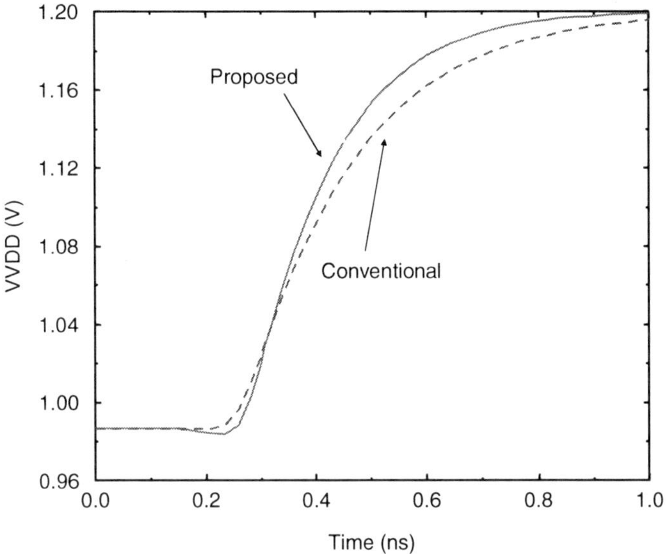

Fig. 28 MEDICI-predicted V_{VDD} settling waveforms for the proposed and conventional [22] schemes

leakage current of the proposed scheme for the latch in the standby mode is determined by the power switch and diode clamping device regardless of core transistors. Therefore, it can be expressed as the sum of the power switch-off current and diode clamping current:

$$I_{\mathrm{tot}} = W_{\mathrm{pg(off)}} I_{\mathrm{o(off)}} 10^{\lambda_{\mathrm{d}} V_{\mathrm{GND}}/S} + W_{\mathrm{pg(diode)}} I_{\mathrm{o(diode)}} 10^{(V_{\mathrm{GND}} - V_{\mathrm{t0}} + r V_{\mathrm{BG}})/S}, \tag{24}$$

where $W_{\mathrm{pg(off)}}$ and $W_{\mathrm{pg(diode)}}$ are the device width of the power switch and diode clamping devices, respectively. $I_{\mathrm{o(off)}}$ and $I_{\mathrm{o(diode)}}$ are the sub-threshold leakage current per unit micrometer device width for the power switch and diode clamping devices, respectively, when V_{GND} is very low ($\sim$0.05 V, close to the ground voltage). λ_{d} is the DIBL factor [8]. S is the sub-threshold swing. Finally, V_{t0} is the front-gate threshold voltage when the back-gate voltage (V_{BG}) is at ground, and r is the gate–gate coupling factor, which is a function of the device structure $[r = 3t_{\mathrm{oxf}}/(3t_{\mathrm{oxb}} + t_{\mathrm{Si}})]$ [2, 9].

Based on the physical equation (24), the power switch leakage current is not a direct function of V_{BG}, but it weakly depends on V_{BG} since V_{GND} is slightly increased for lower V_{BG}. However, the diode clamping current, which is the larger leakage component, is mainly controlled by V_{BG}. For lower V_{BG}, the diode clamping current is reduced even if V_{GND} slightly increases. Note that the proposed (split single-gate mode) DG device has higher S relative to tied-gate-mode DG device (95 mV vs. 65 mV in this case). The higher S will reduce the statistical leakage distribution since $\Delta I_{\mathrm{tot(leakage)}}$ is lower for higher S under the same V_{t} fluctuation, based on (24). This is one of the important advantages of the split-gate device in low-leakage applications. Note that the gate-induced drain leakage (GIDL) current from the back-gate would be significant in asymmetrical DG devices due to the higher electric field between the p^+-poly back-gate and n^+ drain [23]. However, in the proposed scheme, the positively biased V_{BG} reduces the voltage drop between the drain and back-gate, which significantly reduces the GIDL current [24].

9 Other Mixed Split/DG Designs

In this section, we provide a quick overview of existing SRAM designs. In the following sections, we will present more detailed case studies.

It is possible to represent the cell with different mixed/split-gate combinations. Often the goal is to improve stability and/or performance of the design. Figure 29 illustrates a design where the pass-gate (PG) strength may be adjusted by grounding its back-gate (see Fig. 29a), thereby increasing the pull-down (PD) to PG beta ratio. This focus is more on stability as the PG back-gate grounding can impact cell performance. Writability performance may be further enhanced by connecting the pull-up (PU) device back-gate to V_{DD} (Fig. 29b); this improves the PU to PG ratio and was discussed in [25].

The authors in [12] proposed a design that improves on that of Fig. 29b. They connect the PG back-gate to the nearby storage node, as illustrated in Fig. 30, thereby improving stability. The inherent feedback helps reduce the impact on writability compared to grounding the PG back-gate node. The PU devices are also connected to the write wordline, thereby enabling similar writability gains as that in Fig. 29b.

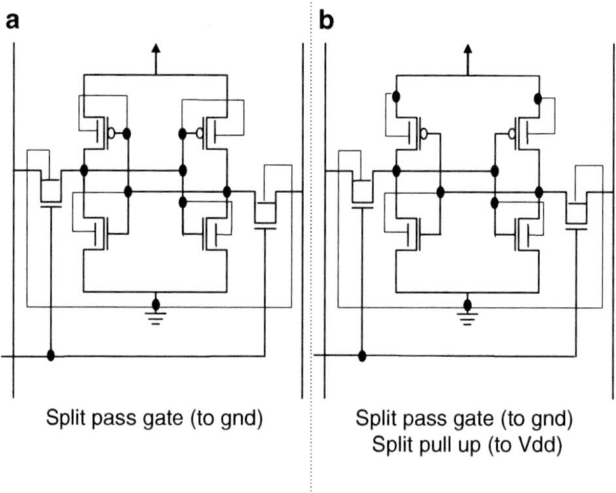

Fig. 29 Mixed/split gate SRAM cell. (**a**) PG with grounded back-gate, (**b**) both PG and PU in split mode [25]

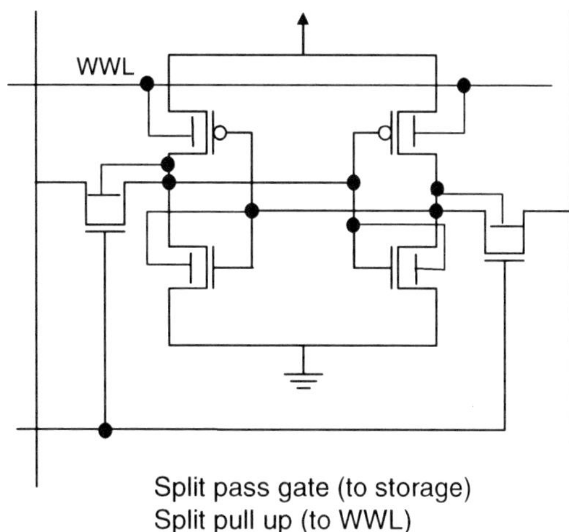

Fig. 30 Mixed/split gate SRAM cell [12]

The above methods, however, maintain a weak PG device during Read/Write operations; in Fig. 30, the PG next to the "0" storage node is weak and the design suffers writability degradation compared to a traditional DG design. Furthermore, split-gate-mode pull-up devices add to the complexity of the design. A favorable solution would thus enable strong performance and improved stability. The key to this lies in extrinsic dynamic control of the PG, as will be discussed in the following sections.

9.1 Column-Decoupled Design

9.1.1 8T-Decoupled Cell

Figure 31 presents the FinFET implementation of the 8T-decoupled design [26]. By relying on an AND function, the WLS signal is ON only when both the bit-select (column enable) and wordline signals are ON. This completely eliminates half-select stability problems that arise when the wordline of an unselected cell is ON. Read stability can be minimized by relying on a differential sense amplifier topology that assists in discharging the bitlines. Overall, cell stability is improved. This is achieved without impacting the performance of the design; the PG operates in full capacity during Read/Write.

9.1.2 6T-Decoupled Cell

The authors in [27] proposed a single high-V_t device that can offer the functionality of two stacked devices and, hence, can be used to model the AND function. Given such device designs, a novel low-voltage 6T-column-decoupled design shown in Fig. 32 was proposed in [28].

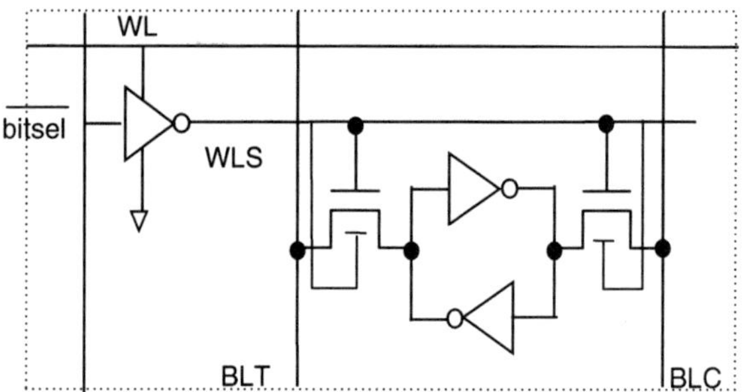

Mode	BitSelect	WL	WLS	Disturb	Performance
Read	ON	ON	ON	minimized by the Sense Amp	Good
Write	ON	ON	ON	---------	Good
Half-select	OFF	ON	OFF	ELIMINATED	-------

Fig. 31 FinFET implementation of half-select free 8T-column-decoupled SRAM cell [26] that uses only tied-gate devices

6T_column_decoupled SRAM cell
Using 'Stacked-And' devices

mode	(WL, Bitsel)	Threshold Voltage
Single gate	(off, on) I (on, off)	V_{t_SG}~ 0.7V
Double Gate	(on, on)	V_{t_DG}~0.4V

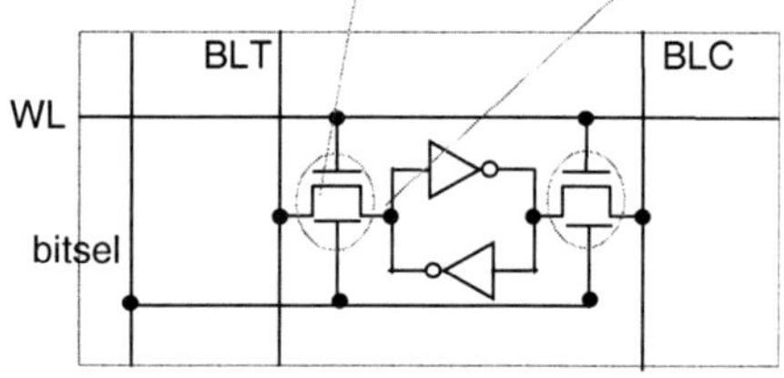

Fig. 32 6T-decoupled design. The first bit-select-gated 6T cell design

Both the bitselect and the wordline signals control the PG. For low-voltage operation, the PG is OFF unless it is operating in the DG mode, i.e., when both inputs are ON. The design yield and functionality gains are similar to the 8T-decoupled cell shown in Fig. 31. In what follows, we evaluate the 6T-decoupled design requirements and sensitivities to the device design in the extreme condition when $V_D = V_{t\text{-SG}} = 0.7$ V. $V_{t\text{-SG}}$ is the single gate mode threshold voltage; i.e., it is the threshold voltage when one of the gates of the DG device is turned off.

For the 6T-decoupled design, we studied writability and stability as a function of $V_{t\text{-DG}}$, the threshold voltage of the DG device. Statistical simulations show that the design works well for $V_{t\text{-DG}} \sim 0.4$ V. When studying the half-select stability as a function of the cell $\sigma_{V_{t\text{-SG}}}$, the decoupling proves to be efficient even in the presence of high $\sigma_{V_{t\text{-SG}}}$ with very good half-select stability. False Read/Write operations are also studied in [28].

Exercises

1. Figure 18 shows the layout that uses individual back-gate biases with an asymmetrical FinFET technology based on 2-D rule checks.

   (a) Analyze and draw the layout of the FinFET SRAM in tied-gate (i.e., DG-mode) configuration.
   (b) Analyze and draw the layout of a "row-based" FinFET SRAM in "back-gated asymmetrical" FinFET technology.
   (c) Compare area for cases (a) and (b).

2. Discuss problems that arise with the individual back-gate and row-based FinFET SRAM schemes shown in Sect. 7 if they are implemented with a symmetrical FinFET technology.

3. For the discussion in 2, evaluate the I_{on}/I_{off} ratio for the back-gate scheme when implemented using symmetrical FinFET technology, and discuss which FinFET technologies are more suitable for the back-gate schemes.

4. Suppose there are sixteen cells on a bitline. Assume that during read "0" we get negligible noise on the "0" storage node. Also, assume that the cell only sees the device capacitance.
   Suppose that $I_{PG} = C_{PG}(\text{input}) \times dV/dt_{read}$. For the same dV, compare t_{read} (cell read time) for

   (a) bulk-Si planar device
   (b) tied-gate symmetric FinFET device
   (c) split-gate symmetric FinFET device
   (d) split-gate asymmetric FinFET device

   All devices have the same width.

5. For each case in 4 (b)–(d), find the equivalent bulk device width that would match the read current of the other devices.

6. Repeat 4 for 32 cells/bitlines. Is the gain of cases (b)–(d) compared to the bulk device in (a) still the same as 4.

7. From (2)–(5) for DG devices:

   1. Find the percent reduction or increase in the FinFET device current as the equivalent back-gate voltage changes from -0.2 to 1 V.

2. Derive V_t for the back-gate FinFET device.
3. What would be the equivalent beta ratio (PD to PG) compared to tied-gate FinFET device for

   (a) $V_{BG} = 0$ V
   (b) $V_{BG} = -0.2$ V
   (c) $V_{BG} = 1$ V

References

1. E.J. Nowak, I. Aller, T. Ludwig, K. Kim, R.V. Joshi, C.T. Chuang, K. Bernstein, and R. Puri, "Turning silicon on its edge," *IEEE Circ Dev Mag* **20**(1): 20–31, Jan./Feb. 2004.
2. K. Kim and J.G. Fossum, "Double-gate CMOS: Symmetrical versus asymmetrical-gate devices," *IEEE Trans Electron Dev* **48**: 294–299, Feb. 2001.
3. Y. Taur and T.H. Ning, Fundamentals of Modern VLSI Devices, Cambridge, UK: Cambridge University Press, 1998.
4. D. Vasileska and Z. Ren, "SCHRED-2.0 Manual," Arizona State University, Tempe, AZ, and Purdue University, West Lafayette, IN, Feb. 2000.
5. K. Kim, K. Das, R.V. Joshi, and C.-T. Chuang, "Leakage power analysis of 25-nm double-gate CMOS devices and circuits," *IEEE Trans Electron Dev* **52**(5): 980–986, May 2005.
6. D.J. Frank, S.E. Laux, and M.V. Fischetti, "Monte Carlo simulation of a 30-nm dual-gate MOSFET: How short can Si go?," in *Proc. Int. Electron Devices Meeting*, pp. 553–556, 1992.
7. H.-S.P. Wong et al., "Nanoscale CMOS," in *Proc. IEEE*, pp. 537–570, Apr. 1999.
8. MEDICI: Two-dimensional device simulation, Mountain View, CA: Synopsys, Inc., 2003. http://www.synopsys.com/products/mixedsignal/taurus/device_sim_ds.html.
9. H.-K. Lim and J.G. Fossum, "Threshold voltage of thin-film silicon-on-insulator (SOI) MOSFET's," *IEEE Trans Electron Dev* **18**: 1244–1251, Oct. 1983.
10. M.-H. Chiang, J.-N. Lin, K. Kim, and C.-T. Chuang, "Optimal design of triple-gate devices for high-performance and low-power applications," *IEEE Trans. Electron Dev* **55**(9): 2423–2428 Sep. 2008.
11. R.V. Joshi, S. Mukhopadhyay, D.W. Plass, Y.H. Chan, C.-T. Chuang, and Y. Tan, "Design of sub-90 nm low-power and variation tolerant PD/SOI SRAM cell based on dynamic stability metrics," *IEEE J Solid-State Circuits* **44**: 965–976, Mar. 2009.
12. Z. Guo, S. Balasubramanian, R. Zlatanovici, T.-J. King, and B. Nikolic, "FinFET-based SRAM design," in *Proc. Int. Symp. Lower Power Elec. Des.*, pp. 2–7, Aug. 2005.
13. R.V. Joshi, R.Q. Williams, E. Nowak, K. Kim, J. Beintner, T. Ludwig, I. Aller, and C. Chuang, "FinFET SRAM for high-performance and low-power applications," in *Proc. European Solid-State Device Research Conference*, pp. 69–72, Sep. 2004.
14. C.-T. Chuang, K. Bernstein, R.V. Joshi, R. Puri, K. Kim, E.J. Nowak, T. Ludwig, and I. Aller, "Scaling planar silicon devices," *IEEE Circuits Dev Mag* **20**(1): 6–19, Jan./Feb. 2004 (invited).
15. Y. Liu, M. Masahara, K. Ishii, T. Sekigawa, H. Takashima, H. Yamauchi, and E. Suzuki, "A high-threshold voltage-controllable 4T FinFET with an 8.5-nm-thick Si-fin channel," *IEEE Electron Dev Lett* **25**(7): 510–512, Jul. 2004.
16. R.A. Thakker, C. Sathe, A.B. Sachid, M.S. Baghini, V.R. Rao, and M.B. Patil, "A table-based approach to study the impact of process variations on FinFET circuit performance," *IEEE Trans CAD* **29**(4): 627–631, July 2010.
17. R.V. Joshi, J.A. Pascual-Gutiérrez, and C.T. Chuang, "3-D thermal modeling of FinFET," in *Proc. European Solid-State Device Research Conference*, pp. 77–80, 2005.

18. J.A. Pascual-Gutiérrez, J.Y. Murthy, R. Viskanta, R.V. Joshi, and C. Chuang, "Simulation of nano-scale multi-fingered PD/SOI MOSFETs using Boltzmann transport equation," in *ASME Heat Transfer/Fluids Engineering Summer Conference*, July 2004.
19. BSIMPD MOSFET Model User's Manual: http://www.device.eecs.berkeley.edu/~bsimsoi/get.html.
20. R.V. Joshi, K. Kim, R.Q. Williams, E.J. Nowak, and C.-T. Chuang, "A high-performance, low-leakage, and stable SRAM row-based back-gate biasing scheme in FinFET technology," in *Proc. Int. Conf. on VLSI Design*, pp. 665–672, Jan. 2007.
21. K. Kim, C.-T. Chuang, J.B. Kuang, H.C. Ngo, and K. Nowka, "Low-power high-performance asymmetrical double-gate circuits using back-gate-controlled wide-tunable-range diode voltage," *IEEE Trans Electron Dev* **54**(9): 2263–2368, Sep. 2007.
22. K. Zhang, U. Bhattacharya, Z. Chen, F. Hamzaoglu, D. Murray, N. Vallepalli, Y. Wang, B. Zheng, and M. Bohr, "A 3-GHz 70 Mb SRAM in 65-nm CMOS technology with integrated column-based dynamic power supply," in *Proc. IEEE Int. Solid-State Circuits Conference*, pp. 474–475, Feb. 2005.
23. J.G. Fossum, K. Kim, and Y. Chong, "Extremely scaled double-gate CMOS performance projections, including GIDL-controlled off-state current," *IEEE Trans Electron Dev* **46**: 2195–2200, Nov. 1999.
24. M. Yamaoka, K. Osada, R. Tsuchiya, M. Horiuchi, S. Kimura, and T. Kawahara, "Low-power SRAM menu for SOC application using Yin-Yang-feedback memory cell technology," in *Proc. Symp. VLSI Circuits*, pp. 288–291, Jun. 2004.
25. V. Ramadurai, R.V. Joshi, and R. Kanj, "A disturb decoupled column select 8T SRAM cell," in *Proc. Custom Integrated Circuits Conference*, pp. 25–28, Sep. 2007.
26. M.H. Chiang, K. Kim, C.-T. Chuang, and C. Tretz, "High-density reduced-stack logic circuit techniques using independent-gate controlled double-gate devices," *IEEE Trans Electron Dev* **53**: 2370–2377, Sep. 2006.
27. R. Kanj, R.V. Joshi, K. Kim, R. Williams, and S. Nassif, "Statistical evaluation of split gate opportunities for improved 8T/6T column-decoupled SRAM cell yield," in *Proc. Int. Symp. Quality Electronic Design*, pp. 702–707, Mar. 2008.
28. International Technology Roadmap for Semiconductors, 2004 http://public.itrs.net/
29. J.A. López-Villanueva et al., "Effects of inversion-layer centroid on the performance of double-gate MOSFET's," *IEEE Trans. Electron Devices*, **47**:141–146, Jan. 2000.
30. H. Ananthan, C.H. Kim, and K. Roy, "Larger-than-Vdd Forward Body Bias in Sub-0.5V Nanoscale CMOS" International Symposium on Low Power Electronics and Design, 8–13, 2004.
31. H. Ananthan and K. Roy, "Technology-circuit co-design in width-quantized quasi-planar double-gate SRAM", International Conference on Integrated Circuit Design and Technology, pp. 155–160
32. L. Chang, D.M. Fried, J. Hergenrother, J.W. Sleight, R.H. Dennard, R.K. Montoye, L. Sekaric, S.J. McNab, A.W. Topol, C.D. Adams, K.W. Guarini, and W. Haensch, "Stable SRAM cell design for the 32 nm node and beyond", 2005 Symposium on VLSI Technology, pp. 128–129

A Hybrid Nano/CMOS Dynamically Reconfigurable System

Wei Zhang, Niraj K. Jha, and Li Shang

Abstract Rapid progress is being made in the area of nanoelectronic circuit design. However, nanofabrication techniques are not mature yet. Thus, large-scale fabrication of such circuits is not feasible. To ease fabrication and overcome the expected high defect levels in nanotechnology, hybrid nono/CMOS reconfigurable architecture are attractive, especially if they can be fabricated using photolithography. This chapter describes one such architecture called NATURE. Unlike traditional reconfigurable architectures that can only support partial or coarse-grain dynamic reconfiguration, NATURE can support cycle-level dynamic reconfiguration. This allows the amount of functionality mapped in the same chip area to increase by more than an order of magnitude. The chapter also discusses how arbitrary logic circuits can be efficiently mapped to NATURE.

Keywords Dynamic reconfiguration · FPGA · Logic folding · Nano RAM · NATURE

1 Introduction

Since CMOS is expected to approach its physical limits in the coming decade [23], to enable future technology scaling, intensive research has been directed towards the development of nanoscale molecular devices, such as carbon nanotubes [24, 52], nanowires [11], resonant tunneling diodes [7], etc., which demonstrate superior characteristics in terms of integration density, performance, power consumption, etc. However, a lack of a mature fabrication process is a roadblock in implementing chips using these nanodevices. In addition, if nanoelectronic circuits/architectures are implemented using bottom-up chemical self-assembly techniques, then the chip defect levels are expected to be high (between 1 and 10%) [6]. To be able to deal

N.K. Jha, (✉)
Department of Electrical Engineering, Princeton University, NJ, USA
e-mail: jha@princeton.edu

N.K. Jha and D. Chen (eds.), *Nanoelectronic Circuit Design*,
DOI 10.1007/978-1-4419-7609-3_4, © Springer Science+Business Media, LLC 2011

with such high defect levels, regular architectures, such as a field-programmable gate array (FPGA), are favored. In addition to being regular, reconfigurable architectures allow reconfiguration around fabrication defects.

Recent research on reliable nanoscale circuits and architectures has resulted in a variety of nanoelectronic and hybrid nano/CMOS reconfigurable designs. DeHon and Wilson [15] and DeHon [13] proposed a nanowire-based programmable logic structure, using multilayer crossed nanowires connected by reconfigurable diode switches. Snider et al. [42] proposed a defect-tolerant nanoscale fabric using nanowire-based FETs and reconfigurable switches. Goldstein and Budiu [18] proposed a nanofabric using chemically assembled electron nanotechnology. Strukov and Likharev [44] and Tu et al. [50] proposed CMOL, a hybrid nanowire/CMOS reconfigurable architecture. In this fine-grain nanofabric, parallel nanowires are connected through molecular-scale nanoswitches to perform logic circuit operation, and four-transistor CMOS cells serve as data and control I/Os. These advanced hybrid designs achieve significant performance and logic density improvement. However, to fabricate molecular-scale nanodevices, they require self-assembly to produce a highly regular, well-aligned architecture. Even though rapid advancements are being made in self-assembly so that such molecular-scale fabrics may be feasible in this decade, they are not ready for large-scale fabrication today.

In this chapter, we present a hybrid CMOS/NAnoTechnology reconfigURablE architecture, called NATURE, which can be fabricated with CMOS-compatible manufacturing processes and leverage the beneficial aspects of both technologies immediately. NATURE is based on CMOS logic and nano RAMs.

The nano RAMs can be carbon nanotube-based RAM (NRAMs) [37], phase change memory (PCMs) [27], magnetoresistive RAM (MRAMs) [47], CMOS embedded DRAMs or other such RAMs, which are high density, high speed, and low power. Such RAMs have already been prototyped or are already commercially available. Besides the outstanding properties of nano RAMs in terms of density, speed or power, they have another important advantage in that they are compatible with the CMOS fabrication process. Thus, NATURE can also be fabricated using photolithography, alleviating the concerns related to the ability to fabricate large chips and presence of high defect levels.

In NATURE, nano RAMs are exploited by distributing them in the reconfigurable fabric to act as the main storage for storing large sets of reconfiguration bits. Thus, the functionality implemented in the logic elements (LEs) of the reconfigurable architecture and interconnects can be reconfigured every few cycles, or even every cycle, making both coarse-grain and fine-grain runtime reconfiguration possible. Note that traditional reconfigurable architectures do allow partial dynamic reconfiguration [21], in which only a part of the architecture is reconfigured at runtime. However, the reconfiguration latency is still high, allowing it to be used only sparingly. They may also allow only coarse-grain [32] or multicontext reconfiguration (with very few contexts) [49], due to the area overhead associated with the configuration SRAMs. In NATURE, nano RAMs store a large set of configuration bits at a small area overhead. At the same time, on-chip nano RAMs make reconfiguration very fast. Hence, NATURE makes it possible to perform fine-grain cycle-level reconfiguration.

The ability to reconfigure NATURE every few cycles leads to the concept of temporal logic folding, that is, the possibility of folding the logic circuit in time and mapping each fold to the same LEs in the architecture. This provides significant gains (an order of magnitude or more for large circuits) in the area-delay product compared to traditional reconfigurable architectures, while allowing the flexibility of trading area for performance. The high logic density provided by logic folding makes it possible to use much less expensive chips with smaller capacities to execute the same applications, hence making them attractive for use in cost-conscious embedded systems. Note that other reconfigurable architectures, such as PipeRench [19], allow later stages of a pipeline to be folded back and executed in a set of LEs that executed an earlier stage. This can be thought of as coarse-grain temporal folding. However, such architectures are largely limited to stream media applications. NATURE does not have these limitations and is very general in the applications that can be mapped to it. The concept of temporal folding is also similar to the concept of temporal pipelining used in the dynamically programmable gate array (DPGA) design [12]. However, the use of nano RAMs enables this concept to be fully exploited through various architectural innovations. The preliminary design of NATURE was presented in Zhang et al. [53, 54]. In this chapter, the design is extended to include other mature nano RAMs, the architecture is parameterized and optimized, and its power consumption is analyzed for different logic folding configurations.

A design automation tool plays a key role in efficiently exploring the potential of reconfigurable architectures and implementing applications on them. In order to fully utilize the potential of NATURE, an integrated design optimization platform, called NanoMap, has been developed. NanoMap conducts design optimization from register-transfer level (RTL) down to the physical level. Preliminary work on NanoMap was presented in Zhang et al. [55, 56]. In the past, numerous design tools have been proposed for reconfigurable architectures with various optimization targets, including area/delay minimization and routability maximization (see Cong [9] for an excellent survey). They mainly perform technology mapping from Boolean network to netlist of look-up tables. NanoMap differs fundamentally from previous work by supporting the temporal logic folding model and targets design flexibility enabled by logic folding. Given an input design specified in RTL and/or gate-level netlist (netlist of look-up tables from the output of technology mapping), NanoMap further optimizes and implements the design on NATURE through logic mapping, temporal clustering, temporal placement, and routing.

Given user-specified area/delay constraints, the mapper can automatically explore and identify the best logic folding configuration based on the chosen instance of the NATURE architecture family and make appropriate trade-offs between delay and area. It uses a force-directed scheduling (FDS) technique [35] to balance resource usage across different logic folding cycles. Traditional place-and-route methods are modified to integrate the temporal logic folding model. Combining NanoMap with existing commercial synthesis tools, such as Synopsys Design Compiler [46], provides a complete design automation flow for NATURE. We introduce the NATURE architecture and NanoMap in detail next.

2   Background

In this section, we first provide background material on the properties of some recently developed nano RAMs that are promising candidates for on-chip configuration memory in NATURE, including NRAMs, PCMs, and MRAMs. We omit embedded DRAMs from this discussion since their design is well-known.

2.1   NRAMs

The theory behind NRAMs was first proposed in Rueckes et al. [37]. It is being developed by a startup company, called Nantero. Figure 1 [34] shows the basic structure of an NRAM. NRAMs are considerably faster than DRAM and denser than SRAM, have much lower power consumption than DRAM or flash, have similar speed to SRAM, and are highly resistant to environmental forces (temperature, magnetism). Memory cells are fabricated in a two-dimensional array using photolithography. Each memory cell comprises multiple suspended nanotubes, interconnect, and electrode. A small dot of gold is deposited on top of the nanotubes on one of the supports, providing electrical connection, or terminal. The cell size is less than $6F^2$, where F is the technology node.

The memory state of an NRAM cell is determined by the state of the suspended nanotubes; whether they are bent or not leads to well-defined electrostatically switchable ON/OFF states. When opposite voltages are applied to the interconnect and electrode of a memory cell, the suspended nanotubes are bent due to coulombic force, reducing the resistance between the nanotubes and electrode to as low as several hundred ohms, corresponding to the "1" state. On the other hand, when the same high voltage is applied to the interconnect and electrode, the nanotube remains straight or "springs" back from the "1" state, resulting in a resistance of several gigaohms, which is defined as the "0" state. Such ON/OFF states have been shown to be both electrically and mechanically very stable. After writing, the voltages can be

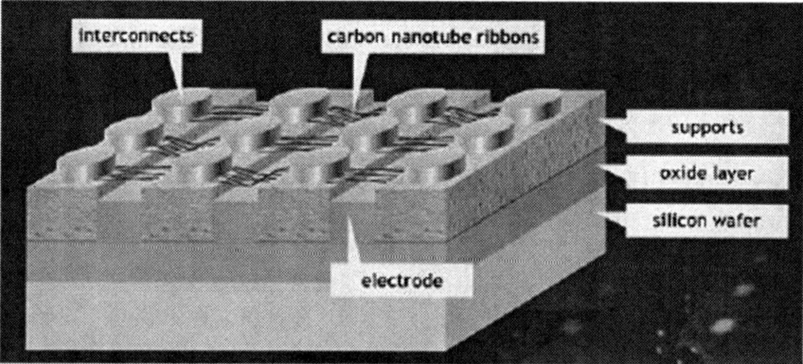

Fig. 1 Structure of an NRAM [34]

NRAM

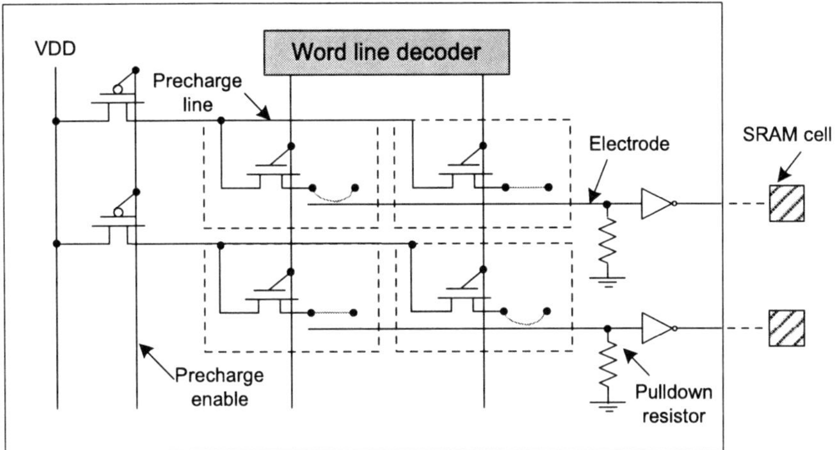

Fig. 2 Circuit model for reading from NRAM

removed without changing the cell state, which is preserved by van der Waals forces. Hence, NRAMs are nonvolatile. When the reading voltage is applied to the support and electrode, the different resistances corresponding to the two states result in different output voltages. The simple circuit model for reading bits out of an NRAM can be built, based on information provided by Nantero [34], as shown in Fig. 2.

The nanotube is modeled according to the circuit model given in Burke [5]. Before reading, the precharge line is first charged to the source voltage. Then when reading begins, the control transistors are closed to separate the precharge line and source voltage and an NRAM word is selected. If an NRAM cell stores a "1," the small resistance between the nanotube and electrode quickly discharges the precharged voltage through a small pulldown resistor. The SRAM cell is then written with a "1." Otherwise, the large resistance between the nanotube and electrode resulting from a "0" of the NRAM cell cannot discharge the precharged voltage. Thus, the electrode is kept at a high voltage and a "0" is then written into the SRAM cell. An NRAM cell can be scaled following the same trajectory as a CMOS process. A 22-nm device has been tested [41]. The access time for a small NRAM is as small as around 160 ps.

2.2 MRAMs

MRAM [47] is a nonvolatile memory technology under development since the 1990s. In an MRAM, data are stored in magnetic storage elements. The elements are formed from two ferromagnetic plates, separated by a thin insulating layer, as shown in Fig. 3. There is a magnetic field in each of plates. One of the plates is a permanent magnet set to a particular polarity. The other's field changes

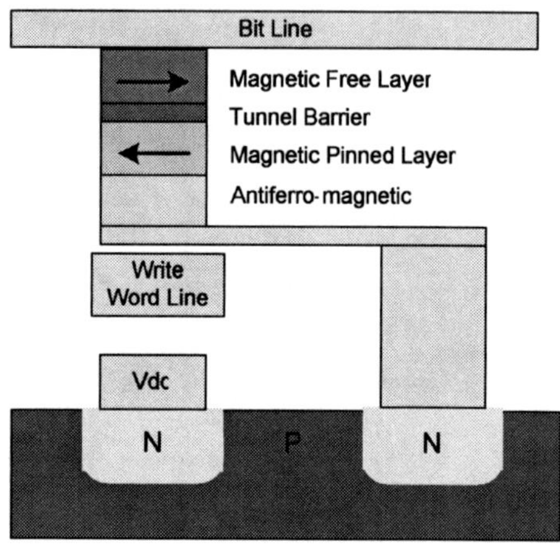

Fig. 3 Structure of an MRAM cell

corresponding to the external field. If the two plates have the same polarity, it stores a "0," else a "1." Different polarities result in different output voltages while reading. Data are written to the cells in various ways, such as toggle mode and spin-transfer-switching [16, 51]. MRAMs have the potential for deployment from the 65-nm technology node onwards. An MRAM cell size is ideally $8F^2$ [45].

Overall, an MRAM may have a similar speed to an SRAM and similar density to a DRAM, but much lower power consumption than a DRAM. Recently, NEC announced the fastest 1 Mb SRAM-compatible MRAM with access time as small as 3.7 ns, almost equivalent to that of recent embedded SRAMs. When reading from MRAMs, the current needs to be low to maintain the polarity of the MRAM cell. Based on the macromodel in Kim et al. [26], simulation of a small 512 Mb MRAM block suitable for on-chip distributed reconfiguration showed its read time to be around 400 ps. The current long access time of an MRAM is mainly due to its long write time, which, fortunately, will only increase the initialization time if used in NATURE. However, its fast read time is enough to support on-chip runtime reconfiguration.

2.3 PCMs

PCMs [27] were first proposed in the 1970s. They employ chalcogenide phase change materials, such as Ge2Sb2Te5 (GST), and various dopants as storage material. The crystalline and amorphous states of chalcogenide glass have dramatically different electrical resistivity. This forms the basis for data storage. The PCM cell functions by applying appropriate electrical pulses through Joule heating to quench or crystallize a dome-shaped "active volume" of GST for resetting (amorphous) or setting (crystalline) states [39]. The difference in resistance between the two states is read with a low voltage pulse without disturbing the GST.

The PCM cell structure is a 1T/1R (one-transistor/one-resistor) structure, where the select transistor can be a MOSFET or bipolar junction transistor (BJT), and the

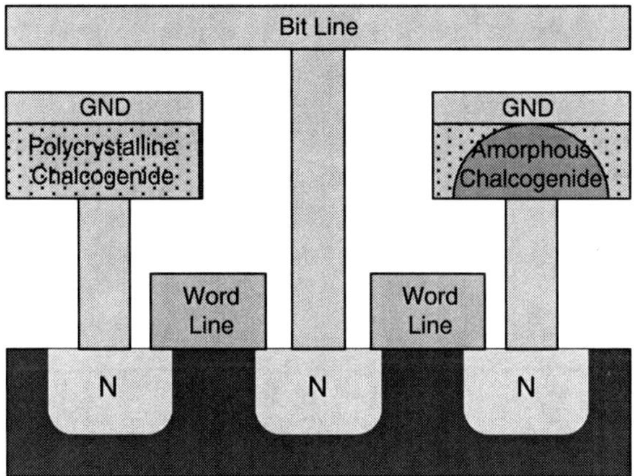

Fig. 4 Structure of PCM cells

resistor is the storage element. Figure 4 shows the structure of two PCM cells. Many new PCM designs are under development. Its cell can be as small as 0.0968 μm^2 ($12F^2$) using the 90-nm process [28]. Its access time for a 256–512 Mb prototype is around 8–20 ns [20]. The reading current required that ensures that the status of chalcogenide does not change is low. Based on the low current (≤ 60 μA) derived from the simulation model in Salamon and Cockburn [38], the read time for a small 512 Mb PCM suitable for on-chip reconfiguration is around 370 ps. Hence, a PCM is also a promising candidate for on-chip RAM utilization.

3 Temporal Logic Folding

The concept of temporal logic folding is crucial to the NATURE architecture. We introduce it next.

The basic idea behind logic folding is that one can use on-chip RAM-enabled runtime reconfiguration, which is described in Sect. 4, to realize different Boolean functions in the same LE every few cycles. Let us consider an example. Suppose a subcircuit can be realized as a series of n serially connected look-up tables (LUTs). Traditional reconfigurable architectures will need n LUTs to implement this subcircuit. However, using runtime reconfiguration, at one extreme all these LUTs can be mapped to a single LE, which is configured to implement LUT1 in the first cycle, LUT2 in the second cycle, and so on, requiring n cycles for execution.

In this case, it would need to store n configuration sets in the nano RAM associated with this LE. Moreover, since all communications between the LUTs mapped to the same LE are local, global communication is reduced and the routing delay is significantly reduced, too. Note that traditional reconfigurable architectures only support partial dynamic reconfiguration or coarse-grain multicontext reconfiguration, and do not allow such fine-grain temporal logic folding. The price paid for logic folding is reconfiguration time. However, through SPICE simulation [5], it has been found that the time required to read the reconfiguration bits from an on-chip nano RAM and store them in the SRAM, that is, the reconfiguration time to configure the LUT to implement another function, is only around hundreds of picoseconds. This is small compared to the routing delays.

Logic folding can be performed at different levels of granularity, providing flexibility to perform area-delay trade-offs. As an example, consider the LUT graph (in which each node denotes a LUT and lines between nodes are input/output connections of the LUTs) shown in Fig. 5a, which denotes the so called level-1

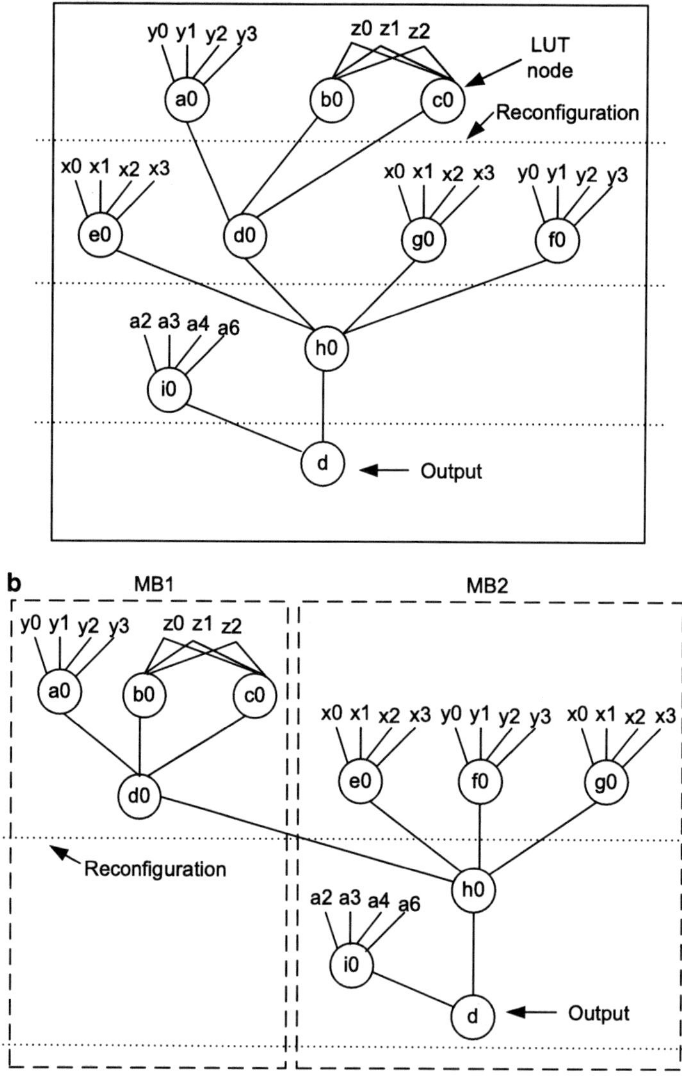

Fig. 5 (a) Level-1 folding and (b) level-2 folding

folding. Such a folding implies that the LEs are reconfigured after the execution of each LUT mapped to it. Thus, in this example, four LEs are enough to execute the whole LUT graph. Here, four LEs are assumed to belong to a MacroBlock (MB) that is described in Sect. 4. On the other hand, Fig. 5b shows level-2 folding which implies reconfiguration of the LE after the execution of two LUT computations. In this case, two MBs are required. A level-p folding can be similarly defined as reconfiguration of the LE after p executions of LUT computation. If p is allowed to be arbitrarily large, it leads to no-folding, which corresponds to mapping of circuits to a traditional reconfigurable architecture in which spatial partitioning of the LUT graph is feasible, but not temporal folding. In this case, 10 LEs spread over 3 MBs would be required.

There are various trade-offs involved in the choice of the folding level, since different folding levels result in different circuit delays and area. Given a logic circuit, increasing the folding level leads to a higher clock period, but smaller cycle count, since a larger number of logic operations need to be performed within a single clock cycle. The overall circuit delay typically decreases as the folding level increases. This would generally be true when communications between LEs are local in the folded circuit. If the area required for implementing the circuit is such that long global communication is required in some cycle, then a small folding level may give better performance. Since circuit execution is synchronous, the total circuit delay equals the clock period times the total number of clock cycles. Since the clock period is determined by the longest delay needed in each cycle, which is mainly due to interconnect delay, the worst interconnect delay can significantly increase total delay.

On the other hand, as far as the implementation area is concerned, increasing the folding level results in much higher resource usage in terms of the number of LEs. For example, in Fig. 5, level-1 folding requires four cycles to execute, and each clock cycle is composed of reconfiguration delay, LUT computation delay, and interconnect delay. For level-2 folding, the clock cycle similarly comprises reconfiguration delay, two LUT computation delays, and interconnect delay. Since reconfiguration delay is very short and level-2 folding contains fewer clock cycles, it leads to a slight performance improvement but occupies more area than required by level-1 folding. Hence, it is crucial to choose an appropriate folding level to achieve the desired area-delay trade-off and balance the logic well. This problem is discussed in detail later in the chapter.

4 Architecture of Nature

In this section, the architecture of NATURE is discussed in detail. The architecture is parameterized first and then the best value for each parameter is discussed. A high-level view of NATURE's architecture is shown in Fig. 6. It contains island-style logic blocks (LBs) [3], connected by various levels of interconnect, including length-1 wire, length-4 wire, long wire, and direct link, supporting local and global communications among LBs. Connection and switch blocks are as shown in the figure. $S1$ and $S2$ refer to switch blocks that connect wire segments. An LB contains a Super-MacroBlock (SMB) and a local switch matrix. The inputs

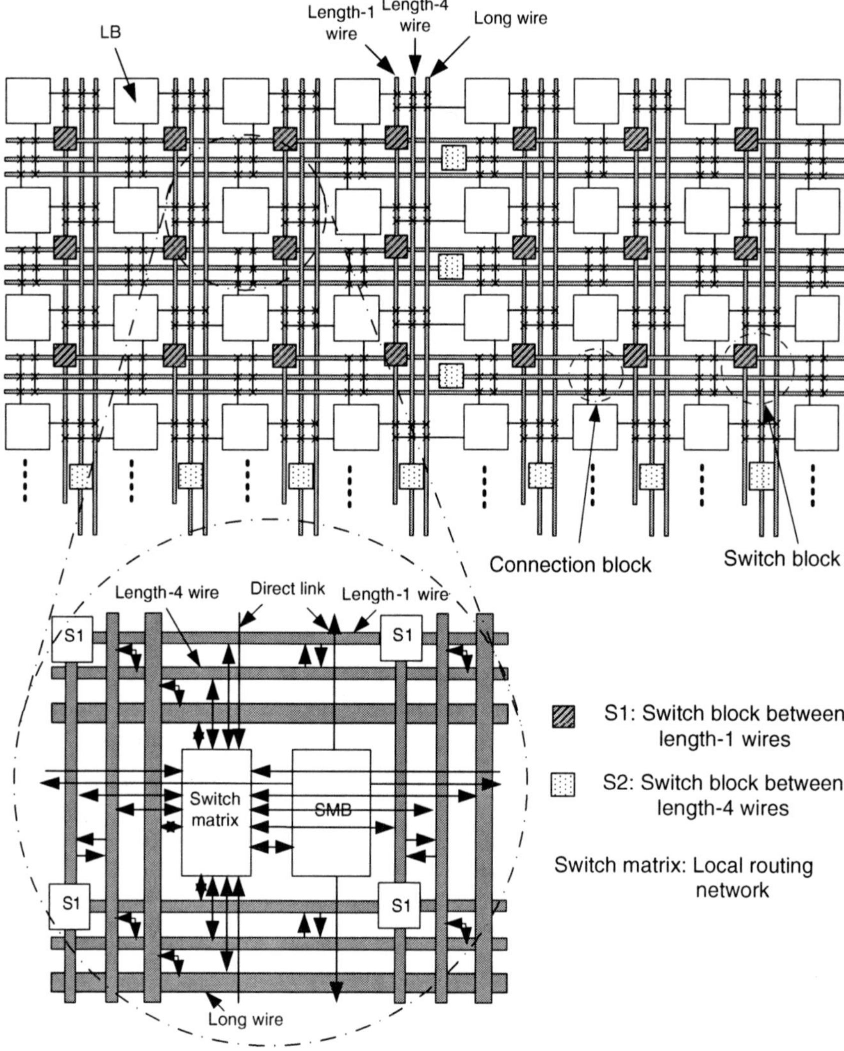

Fig. 6 High-level view of the architecture of NATURE

and outputs of an SMB are connected to the interconnection network through a switch matrix. Neighboring SMBs are also connected through direct links. We discuss the design issues involving the SMB and interconnect in detail next.

4.1 The SMB Architecture

Two levels of logic clusters in an LB are used to facilitate temporal logic folding of circuits and enable most interblock communications to be local, as will be clearer

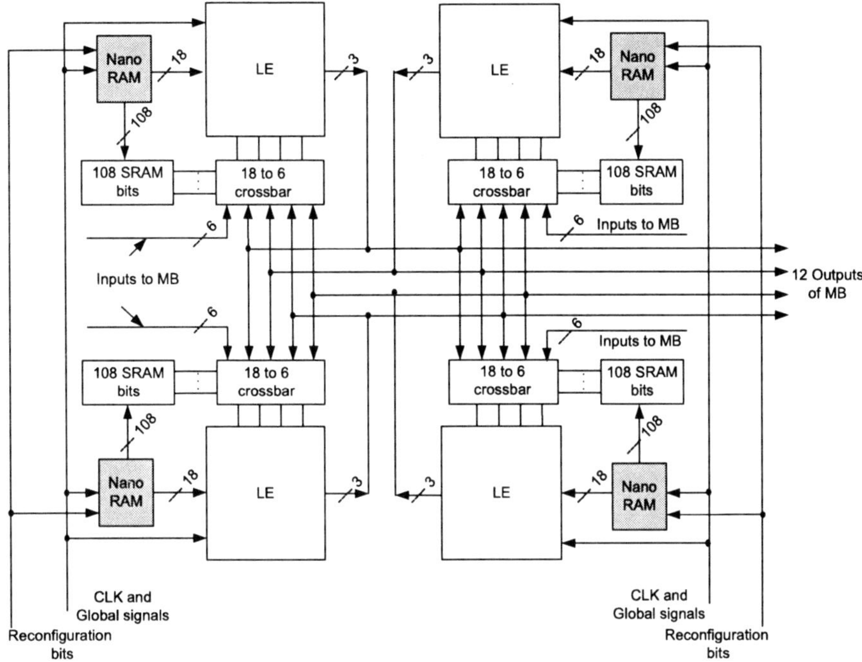

Fig. 7 Architecture of an MB with four LEs

later. The first (i.e., lower) level, called the MB level, is shown in Fig. 7. An MB contains n_1 m-input reconfigurable LUTs (in this figure, $n_1 = 4$, $m = 4$). In the second level, n_2 MBs comprise an SMB, as shown in Fig. 8 (in this figure, $n_2 = 4$). Within an MB or SMB, communications among various components take place through a local crossbar/multiplexer (MUX) to speed up local communications. A MUX is used at the input of an MB, instead of a crossbar as in the original design [53], in order to reduce the number of reconfiguration bits needed.

Full connectivity is provided among various components inside an MB and SMB. As shown in Fig. 7, the m inputs of the LUT within an LE can arrive from the outputs of other LEs, or from its own feedback, or from the primary inputs to the MB. Similarly, the inputs of an MB can arrive from the outputs of other MBs or the inputs to the SMB through the switch matrix. The outputs (three in this instance) from an LE can be used within the MB or go to the upper-level SMB or go to other SMBs through the routing network. This logic/interconnect hierarchy maximizes local communications and provides a high level of design flexibility for mapping circuits to the architecture.

An LE implements a basic computation. The structure of an LE is shown in Fig. 9. It contains an m-input LUT and l (in this figure, $l = 2$) flip-flops. The m-input LUT can implement any m-variable Boolean function. The flip-flop can be used to store the computation result of the LUT for future use (the result of a previous stage is often needed by a subsequent stage) or store some internal result from other LUT outputs. Pass transistors controlled by SRAM cells decide if the internal results are stored or not for the folding cycle.

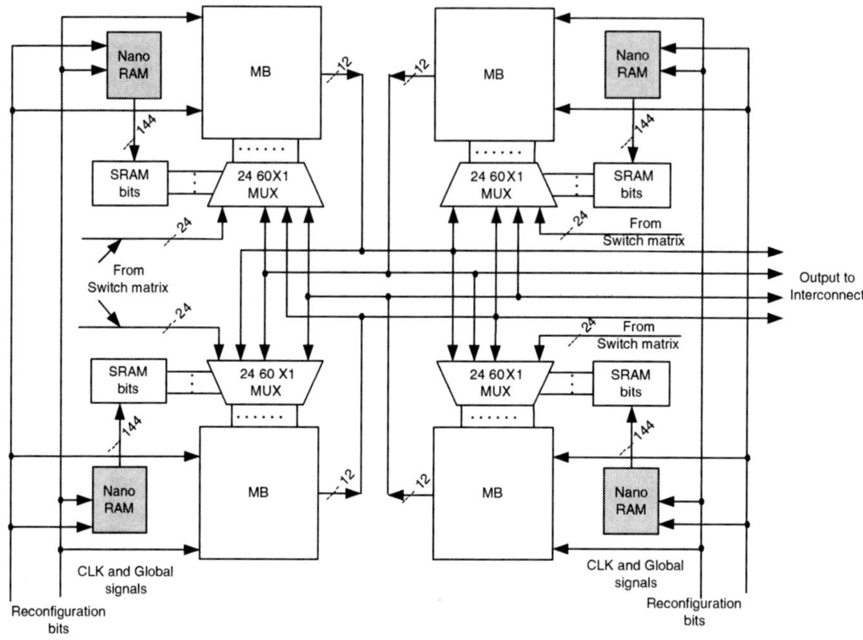

Fig. 8 Architecture of an SMB with four MBs

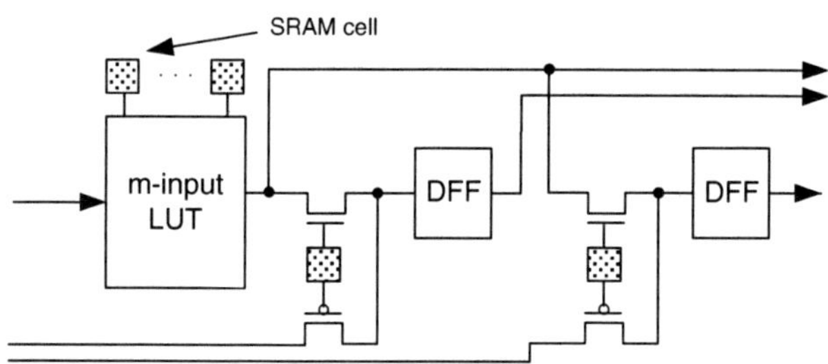

Fig. 9 Architecture of an LE with two flip-flops

4.2　Support for Reconfiguration

In this section, we dwell on how nano RAMs provide support for reconfiguration. As shown in the previous subsection, a nano RAM block is associated with each reconfigurable block and stores its runtime reconfiguration bits. If k configuration sets are stored in a nano RAM, then the associated components can be reconfigured k times during execution. As an example, for the MB architecture shown in Fig. 7, 126 reconfiguration bits are required for one configuration set (when $m = 4, l = 2$). In this set, 16 bits are required for each 4-input LUT, and two bits for determining whether to store the internal results in the flip-flops or not, and the remaining 108 bits to reconfigure the crossbar. Hence, when $n_1 = 4$, and k configuration sets are used, the total number of nano RAM bits required for one MB is 126 kn_1. Note that the configuration sets for different LE-crossbar pairs are different, providing immense flexibility to NATURE for reconfiguring such pairs independently of each other.

Let us see next how reconfiguration is performed. Reconfiguration bits are written into nano RAMs only at the time of initial configuration from off-chip storage. For runtime reconfiguration, reconfiguration bits are read out of nano RAMs sequentially and placed into SRAM cells to configure the LEs and crossbar/switches to implement different logic functionalities or interconnections. At the rising edge of the clock signal CLK, reconfiguration commences, followed by computation. At the same time, the prior computation result is stored into flip-flops to support computation in the next cycle. The timing graph for the clock cycle is shown in Fig. 10.

A cycle is composed of reconfiguration delay, LUT computation delay, interconnect delay, and setup delay for the flip-flop. The figure shows only one instance of time distribution. The LUT computation delay and interconnect delay may vary for different folding levels and different designs. However, usually, the reconfiguration delay only occupies a small fraction of the cycle, and interconnect delay plays the dominant role. If different sections of the chip use different folding levels, their clock frequencies will also differ. In this case, each such section can be controlled by a different clock signal. A limited number of global clock signals can be generated by a clock manager module, as in current commercial reconfigurable chips. Note that in this design, since reconfiguration bits for different logic configurations are read out in order, counters can be used to control the nano RAMs instead of complex address-generating logic.

Inclusion of nano RAMs in the LB incurs an area overhead. The success of NATURE is not tied to any specific nano RAM. Any high-density and high-performance nano RAM, such as NRAM, PCM, MRAM, or embedded DRAM, can be used as on-chip memory, depending on which one turns out to be the best in terms of fabricatability, yield, and economics of manufacturing. If NRAMs are used, then from the parameters given in Nantero [34] and Stix [43] for an NRAM cell, assuming 70-nm technology for implementing CMOS logic, 70-nm nanotube length, and $k = 16$, the NRAMs occupy roughly 10.6% of the LB area. However, through NRAM-enabled logic folding, the number of LBs required to implement a circuit is reduced nearly k-fold. Hence, the small overhead due to the NRAMs is not only compensated, but also at the same time more area saving is obtained.

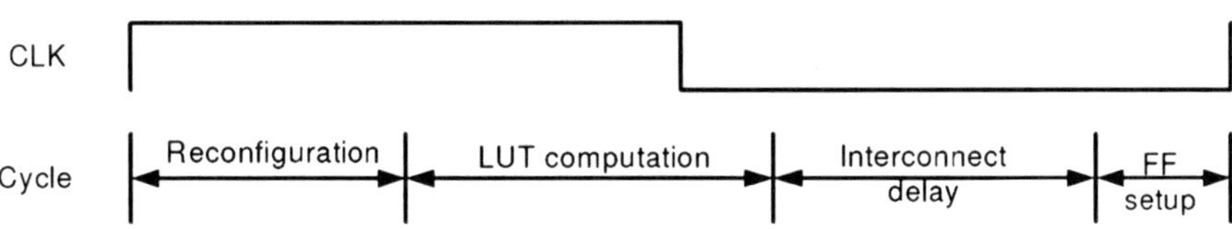

Fig. 10 Timing graph for the clock cycle

To account for these facts, it is helpful to define the concept of relative logic density. This is the ratio of the amount of logic NATURE can implement in a given amount of area compared to the amount of logic a traditional reconfigurable architecture can. When $k = 16$ and assuming the circuit being implemented can use 16 configurations (as most large circuits would), the relative logic density can be calculated as $16(1 - 0.106) = 14.3$. This means that in the same area, NATURE can implement roughly 14 times more logic than a traditional architecture, or equivalently needs 14 times less area to implement the same functionality.

4.3 Parameter Optimization for the SMB Architecture

Next, we will focus on the search for the best values of n_1, n_2, m, l in the architecture family. The $n_1 = 4$, $n_2 = 4$, $m = 4$, $l = 2$ instance is used as the baseline design. In this exploration, one parameter is varied and the others are kept unchanged. The interconnect structure is assumed to be unchanged and the number of tracks needed for each variation is evaluated.

4.3.1 Number of LEs (n_1) per MB

Changing the value of n_1 leads to area-delay trade-offs. A larger n_1 results in larger crossbars within the MB whose size is $[(l + 1) * n_1 + m + l] * (m + l)$. The size of input MUXes to the MB also increases, which selects one input from a total of $(l + 1) * (n_2 - 1) * n_1 + n_1 * (m + l)$ inputs. Hence, the MB and SMB size increases greatly with an increase in n_1. Moreover, since the number of inputs and outputs of an SMB increases linearly with n_1, the size of the switch matrix also grows fast. In terms of timing, a change in n_1 affects different folding levels differently. Generally speaking, for circuit mapping based on small folding levels, for example, level-1 or level-2 folding, the circuit delay tends to increase with a growth in SMB size. This is because, as we discussed in Sect. 3, the interconnect delay plays a key role in total delay. When the folding level is small, the cycle time is dominated by the interconnect delay between SMBs, which increases with an increase in the switch matrix size when n_1 increases. For mapping with middle to large folding levels or no-folding, the cycle time includes both delays between SMBs and inside SMBs.

Since increasing the SMB size makes more communications local, overall the circuit delay is improved. The previous arguments are buttressed by experimental results shown in Table 1. The results are normalized to the $n_1 = 4$ case. The delay is averaged for level-1, level-2, level-4, and no-folding cases. It can be seen that $n_1 = 2$ minimizes the SMB area; however, it requires a relatively large number of interconnect tracks to accomplish routing. In addition, from the circuit delay point of view, $n_1 = 4$ is the best. Hence, based on the area-delay product, n_1 is chosen to be from 2 to 4.

Table 1 Evaluation of different numbers of LEs in an MB

#LEs per MB	2	3	4	5	6
SMB area	0.35	0.65	1	1.41	1.86
#SMBs	2.00	1.33	1	0.80	0.67
Total area	0.70	0.86	1	1.13	1.24
#Tracks needed	1.20	1.01	1	1.18	1.35
Circuit delay	1.06	1.03	1	1.05	1.04
Area-delay prod.	0.84	0.89	1	1.19	1.29

Table 2 Evaluation of different numbers of MBs in an SMB

#MBs per SMB	2	3	4	5	6
SMB area	0.40	0.67	1	1.34	1.69
#SMBs	2.00	1.33	1	0.80	0.67
Total area	0.80	0.89	1	1.07	1.13
#Tracks needed	1.20	1.03	1	1.22	1.39
Circuit delay	1.02	1.04	1	1.02	1.02
Area-delay prod.	0.82	0.93	1	1.09	1.15

4.3.2 Number of MBs (n_2) per SMB

Similarly to n_1, varying n_2 also results in area/delay trade-offs. The variation in area and delay with n_2 is similar to that seen for n_1, since the number of LEs in an SMB is the same for the two cases, and so is the input/output switch matrix size. The difference between the two cases is that when varying n_2, the local crossbar inside the MB is unaffected, and the size of the input MUX to an MB does not increase as fast as with an increase in n_1. The mapping results are shown in Table 2. We can see that n_2 can also be chosen to be from 2 to 4.

4.3.3 Number of Inputs (m) per LUT

This is a very important consideration for any FPGA architecture. With an increase in m, the size of the local crossbar, input MUX to the MB, and switch matrix for the SMB all grow in size. If m is too large, and the application cannot always make use of all the inputs of each LUT, area is wasted. On the other hand, if m is too small, it leads to a larger number of LUTs, MBs, and SMBs, as well as more communications. For conventional FPGAs with no-folding, prior work [2] suggests that four-input LUTs provide the best area-delay trade-off. As can be seen from Table 3 (where results are normalized relative to the $m = 4$ case), $m = 4$ is also best in the case of logic folding.

4.3.4 Number of Flip-flops (l) per LE

In traditional FPGAs, usually only one flip-flop is associated with a LUT. In NATURE, due to the use of temporal logic folding, the area for implementing

Table 3 Evaluation of different numbers of inputs for a LUT

#Inputs per LUT	3	4	5
SMB area	0.83	1	1.20
#SMBs	1.50	1	0.87
Total area	1.25	1	1.04
Circuit delay	1.20	1	1.03
Area-delay prod.	1.50	1	1.07

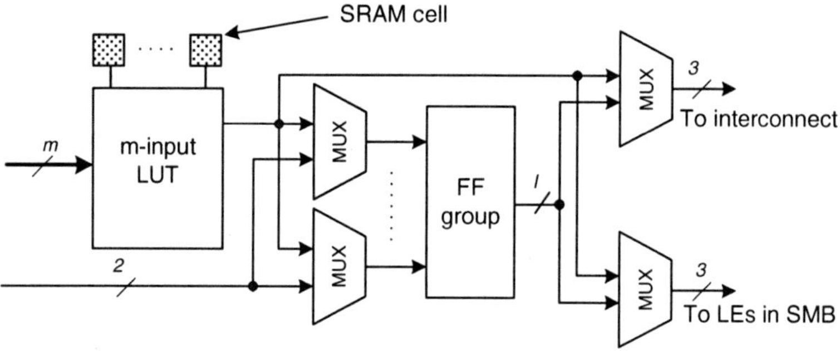

Fig. 11 Architecture of LE with limited inputs/outputs

combinational logic is reduced so much that the number of registers in the design becomes the bottleneck for further area reduction. Hence, to reduce the area even more, as opposed to traditional FPGAs, more flip-flops are needed in an LE. We can see from Fig. 9 that to give enough flexibility and improve the utilization of flip-flops, a separate input is provided to each flip-flop such that the flip-flops can store different values at the same time. However, since the number of inputs to an LE is $m + l$ and the number of outputs from each LE is $l + 1$, and the communication network within and outside the SMB grows polynomially with an increase in the number of inputs/outputs per LE, per MB, and per SMB, as discussed earlier, this significantly increases the SMB size. Hence, another approach is to limit the total number of inputs/outputs per LE while increasing the number of flip-flops in the LE, as shown in Fig. 11. Outputs are separated into two parts; three outputs to the interconnect and three outputs to the inside of the SMB. Hence, the size of the local interconnect inside the SMB and the number of outputs per SMB both do not change when l increases. From design space exploration, $l = 2$–4 is found to be best for area-delay product optimization.

4.3.5 Number of Reconfiguration Sets (k)

It can be seen that both the nano RAM size and relative logic density vary with the value of k. How many copies of reconfiguration bits are stored in the nano RAM determines how large the nano RAM needs to be. If k is too small, the benefit of logic folding is significantly curtailed. If it is too large, it may not be possible to make use of the extra configurations, thus leading to wasted nano RAM area that could have been put to other use. When k varies, nano RAM size and logic density also vary. In Table 4, the relationship between logic density and area overhead for different k is given, assuming NRAMs. Since the best k value varies with the specific design, it can be obtained through design space exploration. It is found that $k = 16$ or 32 works well in practice.

To further improve the performance of the architecture at the expense of increased area, one can use a shadow reconfiguration SRAM to hide the reconfiguration latency for transferring bits from the nano RAMs to the SRAMs. This allows one group of SRAM bits to load reconfiguration bits from nano RAMs, while another SRAM group supports current computation. The performance improvement due to this feature depends on the level of logic folding.

4.3.6 Different Types of Nano RAMs

Finally, in order to illustrate the impact of using other types of nano RAMs, such as MRAMs and PCMs, performance comparisons are presented in Table 5. We can see that since the interconnect delay dominates, the small increase in the reconfiguration delay does not affect the total delay much, especially for a large folding level. Hence, the improvement in area-delay product is still significant when deploying MRAMs or PCMs, and the advantages of NATURE remain.

4.4 Interconnect Design

Interconnect design is very important because of its large routing delay and area. Consequently, the routing architecture must be designed to be both fast and area-efficient. In this case, it should also aid logic folding and local communication as

Table 4 Logic density and area overhead for different k

k	Area overhead (%)	Logic density
8	5.6	7.6X
16	10.6	14.3X
32	19.2	25.9X

Table 5 Performance comparison using different nano RAMs

Nano RAM	Reconf. delay (ps)	Area (F^2)	Delay for different folding levels		
			Level-1	Level-2	Level-4
NRAM	160	6	1	1	1
MRAM	420	8	1.25	1.16	1.09
PCM	370	12	1.20	1.13	1.07

much as possible. In NATURE, a hybrid interconnect is used. Within the SMB, the interconnect is hierarchical to aid logic clusters and local communication. To connect SMBs, wire segments of various lengths are used.

To give a better picture of the interconnect structure, we next address the following routing architecture features [3]:

- the length of each routing wire segment (i.e., how many LBs a routing wire spans before terminating);
- the number of wires (tracks) in each routing channel;
- location of routing switches and which routing wires they connect;
- the type of each routing switch: pass transistor, tristate buffer, or MUX;
- size of transistors in the switches, and the metal width and spacing of the routing wires.

For the first feature, since too many short wires decrease circuit performance, while too many long wires provide little routing flexibility and may waste area, a mixed wire segment scheme is implemented including length-1, length-4, and long wires. Length-1 (length-4) wire segments span one (four) LB(s) before connecting to a switch block, while long wires traverse the chip horizontally and vertically, connecting to each LB along the way. There are also direct links from the outputs of an LB to its four neighboring LBs, further facilitating local communications.

To address the second feature, based on the observations in Betz and Rose [3], DeHon and Rubin [14], and Bozorgzadeh et al. [4], first, the total number of routing tracks on one side of the LB, $W_{\min}$, should satisfy the equation

$$W_{\min} \geq \frac{2 * I + O}{4}. \tag{1}$$

Here, I/O refers to the number of inputs/outputs of the SMB. For the architectural instance in which $m = n_1 = n_2 = 4, l = 2, I = 96$, and $O = 48$, $W_{\min}$ equals 60. In addition, based on the discussion in Bozorgzadeh et al. [4] and experiments, usually 20% additional tracks can ease routing further. Through experiments, it is found that a total of 160 horizontal and vertical tracks can satisfy the routing demand quite well. A distribution of 30–40% for length-1, 40–60% for length-4 and 10–20% for long wires was found to be the best in Zhang et al. [57]. These 160 tracks are shared by the LBs on both sides, hence 80 tracks are provided per LB per side. In addition, 48 tracks for direct links between adjacent SMBs are provided (since $O = 48$).

Figure 12 illustrates how the inputs/outputs of an SMB are connected to the routing network. Note that this architecture caters to all levels of folding, including the no-folding case. When employing logic folding, global communication can be reduced significantly. Consequently, the number of required global routing tracks decreases and the routing area can be reduced considerably.

Next, let us consider the design of the connection block, characterized by Fc, and switch block, characterized by Fs (Fc refers to the number of adjacent tracks a pin of an LB can connect to and Fs the number of tracks to which each track entering

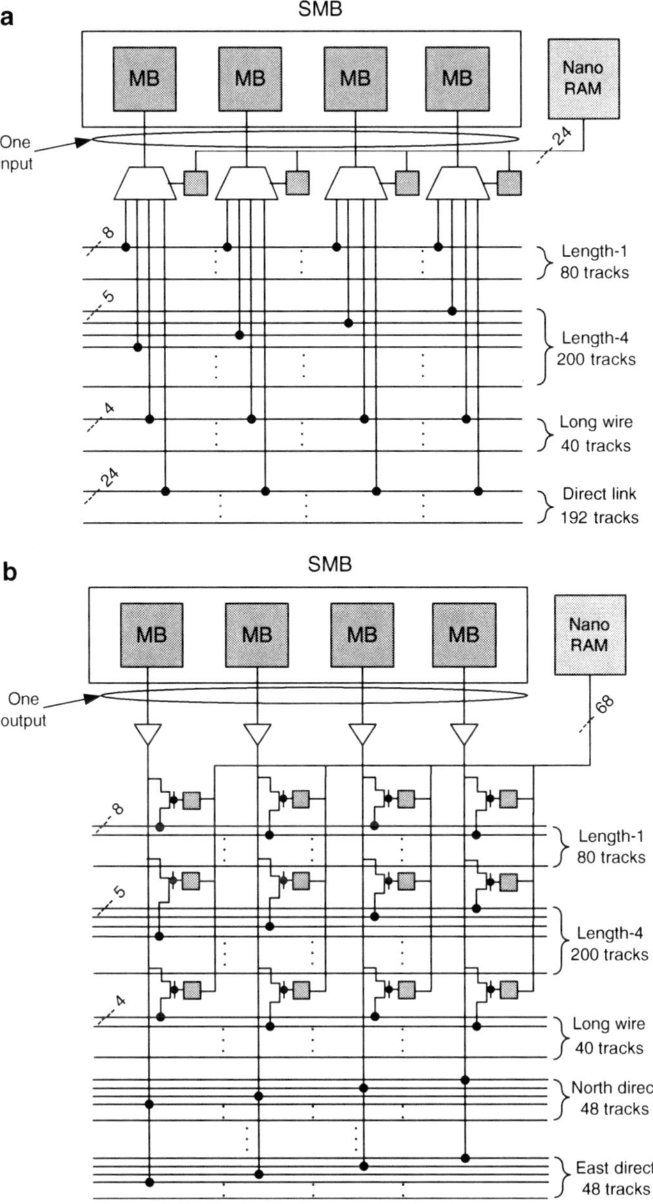

Fig. 12 Connection block for (**a**) one input of an MB, and (**b**) one output from an MB

the switch block can connect [36]). Higher values of Fc and Fs result in higher routing flexibility, however, at the expense of a higher number of switches and hence more routing area. For a cluster of N LUTs, Fc can be chosen as $1/N$ of the total number of tracks and Fs should be no less than three in order to be able to achieve routing completion while maintaining area efficiency [2]. $Fc = 1/N \approx 0.1$ and $Fs = 3$ are used in NATURE.

Another related important issue is whether or not the internal connection blocks or switch blocks should be populated [3, 8] (such a block is said to be populated if it is possible to make connections from the middle of the block to LBs or other such blocks). When they are both fully populated, the number of routing tracks required to achieve routing completion can be reduced, at the expense of a larger number of switches being attached to a wire (resulting in more capacitance and, hence, decrease in speed). Since the investigation in Chow et al. [8] demonstrates that when the connection blocks are depopulated, only a minor increase in the number of tracks is experienced, but when the switch block is unpopulated, it results in a significant number of excess tracks, the connection blocks are depopulated and the switch blocks are populated to provide the best performance-area advantage. Fig. 13 shows an example connection of disjoint switch blocks.

Next, we discuss the types of switches. There are typically three types of switches: pass transistor, MUX, and tristate buffer. Since using pass transistors results in the fastest implementations of small crossbars, pass transistors are chosen for the local crossbars within the MB. A MUX has longer delay, but needs fewer reconfiguration bits. Therefore, it is used in the SMB and the switch matrix to connect to the inputs of an SMB, as shown in Fig. 12a. The outputs of an SMB are

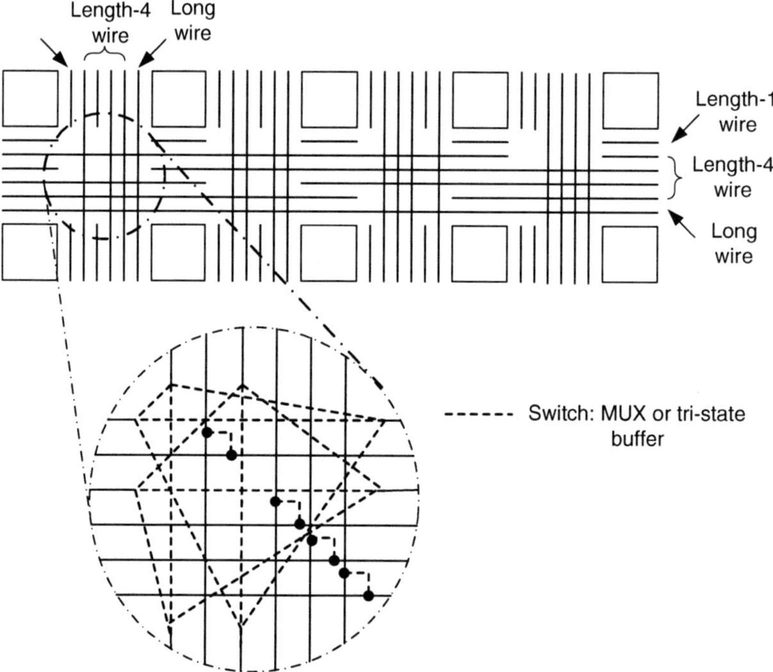

Fig. 13 Illustration of a switch block

connected to long interconnects through tristate buffers in the switch matrix, as shown in Fig. 12b. For the switch blocks, both the MUX and tristate buffer are applicable.

For the last feature, the size of pass transistors is 10 (5) times the size of a minimum-sized transistor for tristate buffers (MUXes) [3]. A minimum width and spacing are used for the metal wires.

5 Power Estimation

In this section, we analyze the power consumption of NATURE to compare different logic folding scenarios. There are two dominant types of power consumption in FPGAs: switching and leakage. Switching power is the combined power from charging all capacitive nodes. In the CMOS part of the FPGA, the leakage power consists mostly of subthreshold leakage.

Power is consumed in various components of NATURE. As discussed in previous sections, inside the SMB, there are a LUT and flip-flops in the LEs, input MUXes to MBs (MBmux), and local crossbar (Lbar) between LEs. The segmented routing architecture between SMBs includes direct links, length-1 wires, length-4 wires, and long wires, in both vertical and horizontal dimensions. There are also pass transistors and buffers associated with each set of wires, which need to be considered during power estimation. Besides, there are two sets of switches that connect wire segments to the inputs and outputs of each SMB. We refer to them as the input matrix (IMatrix) and output crossbar (OXbar), respectively. In addition to the logic and interconnect, global resources and switches that accommodate clocking, which are necessary for synchronization of the folded circuit, are also considered.

5.1 *Switching Power Estimation*

Switching power dissipation P is caused by signal transitions in the circuit. In CMOS circuits, the power is consumed in charging a capacitance. It can be derived using

$$P = \frac{1}{2}V^2 f \sum_i C_i S_i, \tag{2}$$

where V is the supply voltage, f the operating frequency, C_i the capacitance, and S_i the switching activity (i.e., the number of $0{\rightarrow}1$ and $1{\rightarrow}0$ transitions) of resource i, respectively. C_i refers to *effective capacitance*, which models all the lumped capacitances for this resource, and is defined as the sum of parasitic effects due to interconnect wires and transistors, and the emulated capacitance due to the short-circuit current [40].

5.1.1 Effective Capacitance

The *effective capacitance* for each type of resource is obtained through HSPICE simulation. Each resource is separately modeled and simulated. Table 6 summarizes the effective capacitance for the major resources in NATURE.

Based on the layout, the LUT and MUX inputs can be classified into two groups; the one driving a smaller gate capacitance switches faster, while the one driving a larger gate capacitance switches slower. Hence, the individual input capacitance varies. Table 6 presents their average value. However, during power estimation, the exact capacitance value is used based on the layout. "internode" in the table denotes the intermediate node inside the LUT or MUX, whose capacitance is usually the drain/source capacitance of the pass transistor. When the input values of the LUT/MUX change, not only may the output value of the LUT/MUX switch, but some of the intermediate nodes may also switch. Thus, there is a need to account for switching activity at intermediate nodes as well.

5.1.2 Switching Activity

Figure 14 shows the flow for switching power estimation. First, through the automatic design flow, NanoMap [55, 56], the circuit is mapped to NATURE and its netlist and routing result are obtained. Then logic simulation, using Synopsys Design Compiler [46], is performed on the netlist with the available input traces. After the simulation profile is back-annotated to the routed circuit, the switching activity for each resource can be obtained.

Switching activity calculation requires consideration of two types of elements in the design: nets and logic. Nets often have one source and multiple destinations, and signals maintain their switching activity when traversing the nets. However, due to

Table 6 Effective capacitance summary

Type	Resource	Capacitance (*fF*)
Interconnect across SMBs	IMatrix input	17.6
	IMatrix internode	1.1
	OXbar	46.4
	Direct link	32.5
	Length-1 wire	29.5
	Length-4 wire	118
	Long wire	590
Logic/interconnect inside SMBs	LUT input	17.6
	LUT internode	1.1
	FF input	4.5
	Lbar	38.4
	MBmux input	65
	MBmux internode	1.1
Clocking	Global wiring	1939
	Local	7.7

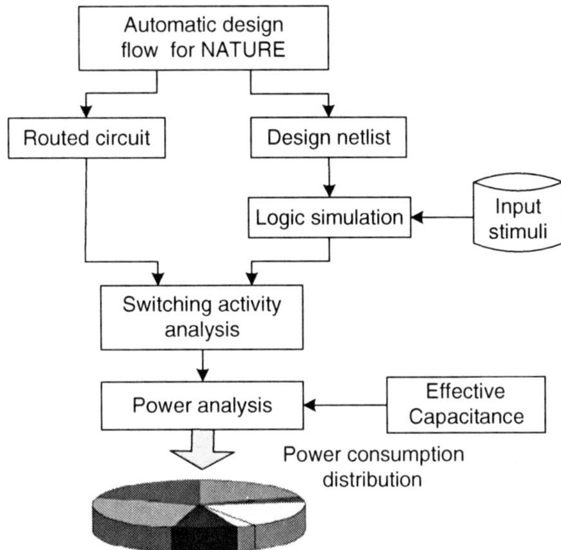

Fig. 14 Power estimation flow

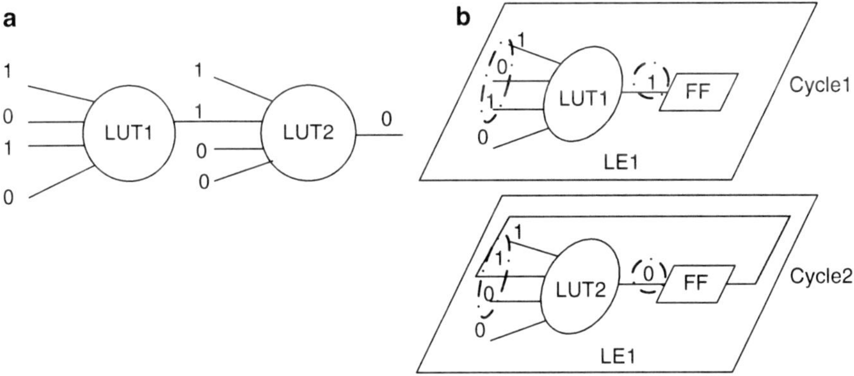

Fig. 15 Illustration of (**a**) logic implementation, and (**b**) switching of signal between folding cycles

the logic manipulation that occurs in LUTs and MUXes, the switching activity is altered at their intermediate nodes. Logic simulation accounts for this.

When logic folding is used, switching activity calculation is slightly more complex than when no-folding (i.e., traditional mapping) is used. Figure 15 illustrates the logic folding scenario. The two-LUT circuit is implemented in one LE executing over two cycles. There is switching activity at the LUT inputs and output. Hence, the switching activity between folding cycles also needs to be accounted for.

5.2 Leakage Power

With technology scaling, leakage power is assuming increasing importance [25]. The leakage power is dependent on the input state of the circuit component. The average leakage power is obtained using HSPICE over all the vectors in the input trace.

6 Motivational Example for NanoMap

In order to fully utilize the potential of NATURE, NanoMap, an integrated design optimization platform for optimizing implementation of applications on NATURE, has also been developed. In this section, we first use an RTL example to demonstrate the design optimization flow of NanoMap, that is, how NanoMap maps the input design to NATURE. First, we introduce some concepts for ease of exposition. Given a circuit, such as the one shown in Fig. 16a, the registers contained in it are first levelized. The logic between two levels of registers is referred to as a *plane*. For instance, the example circuit contains two planes. The registers associated with

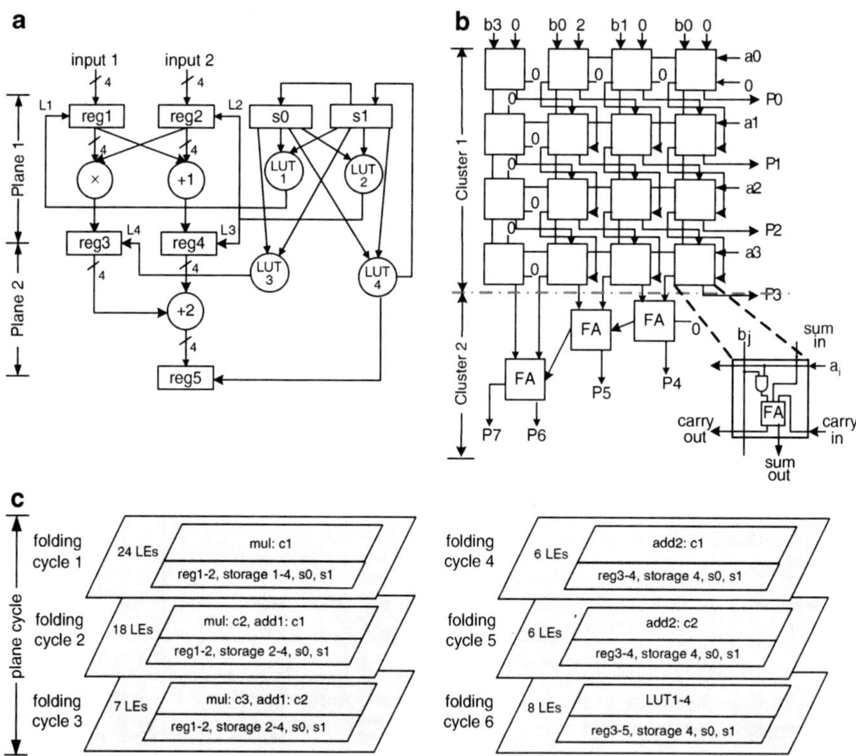

Fig. 16 Illustration of (**a**) an example circuit, (**b**) module partition, and (**c**) mapping result

the plane are called *plane registers*. The propagation cycle of a plane is called the *plane cycle*. Using temporal logic folding, each plane is further partitioned into *folding stages*. Resources can be shared among different folding stages within a plane or across planes. The propagation cycle of a single folding stage is defined as a *folding cycle*. Note that different planes should consist of the same number of folding stages to guarantee global synchronization. Thus, the key issue is to determine how many planes are folded together and the appropriate folding level, that is, the number of folding stages in one plane, to achieve the best area-delay trade-off under specified design constraints.

The example shown in Fig. 16a comprises a four-bit controller-datapath. An adder and a multiplier are included in the first plane and a single adder in the second plane. The controller consists of flip-flops s0 and s1 and LUTs LUT1–LUT4 that control loading of registers reg1–reg5 in the datapath. Assuming ripple-carry adders and parallel multipliers, the circuit requires 58 LUTs in all (50 LUTs for plane 1 and 8 LUTs for plane 2) and 22 flip-flops. The adder consists of eight LUTs with a logic depth, that is, the number of LUTs along the critical path, of four. The multiplier consists of 38 LUTs with a logic depth of seven. Hence, the logic depth is seven for plane 1 and four for plane 2. The mapping procedure can target many different optimization objectives. For the example, suppose the objective is to minimize delay under a total area constraint of 30 logic elements (LEs). Here, delay indicates the register output to register input delay. It is assumed that each LE contains one LUT and two flip-flops. Hence, 30 LEs contain 30 LUTs along with 60 flip-flops. Since the number of available flip-flops is more than required, we concentrate on the LUT constraint.

NanoMap uses an iterative optimization flow. It first chooses an initial folding level based on the constraints and input design structure and then gradually adjusts the folding level to achieve the optimization goal. As discussed in Sect. 3, fewer folding stages generally lead to better delay. Thus, NanoMap can start with a folding level that results in a minimal number of folding stages. In the example, since the design allows cross-plane folding (since it is not pipelined), one plane can be folded on top of the other. The minimal number of folding stages in each plane is equal to the maximum number of LUTs among all the planes divided by the LUT constraint: $\lceil 50/30 \rceil = 2$, that is, at least two folding stages are required to meet the LUT constraint. Since there are two planes, two folding stages are needed for each plane. The initial folding level is then obtained by the maximum logic depth among all the planes divided by the number of folding stages, which equals $\lceil 7/2 \rceil = 4$.

Next, based on the chosen folding level, the adder and multiplier are partitioned into a series of connected LUT clusters in such a way that if the folding level is p, then all the LUTs at a depth less than or equal to p in the module are grouped into the first cluster, all the LUTs at a depth larger than p but less than or equal to $2p$ are grouped into the second cluster, and so on. The LUT cluster can be considered in its entirety and contained in one folding stage. By dealing with LUT clusters instead of a group of single LUTs, the logic mapping procedure can be greatly sped up.

Figure 16b shows the partition for the multiplier using level-4 folding. However, the first LUT cluster of the multiplier needs 32 LUTs, already exceeding the area constraint. This is because the logic is not balanced well between the two planes. Thus, the folding level has to be further decreased to level-3 in order to guarantee that each LUT cluster can be accommodated within the available LEs. Correspondingly, the number of folding stages increases to $\lceil 7/3 \rceil = 3$ for each plane.

Next, FDS is used to determine the folding cycle assignment of each LUT and LUT cluster to balance the resource usage across the three folding stages in each plane. If the number of LUTs and flip-flops required by every folding stage is below the area constraint, that is, 30 LEs, the solution is valid and offers the best possible delay. Otherwise, the folding level is reduced by one, followed by another round of optimization until the area constraint is met, assuming it can be satisfied.

Figure 16c shows the mapping result for level-3 folding for the six folding stages in two planes. Note that plane registers, which provide inputs to the plane, need to exist through all the folding stages in the plane. The first folding cycle requires 24 LEs, which is the most number of LEs among the six folding stages. They are required for mapping the first cluster of the multiplier (*mul:c*1). Four-bit registers reg1-reg2, LUT computation results for LUT1-LUT4, and one-bit state registers s0 and s1 are mapped to the available flip-flops inside the LEs assigned to the multiplier. The results from the LUT computations are also stored in the available flip-flops inside the LEs. Note that the computation result of LUT1 is not needed after folding cycle 1, and hence not kept. reg3–reg5 are not created until they are used in order to save area. From the figure, we can see that 18, 7, 6, 6, and 8 LEs are needed for the following folding cycles, respectively. Hence, the number of LEs for mapping this RTL circuit is the maximum required across all the folding cycles, that is, 24, which is less than the area constraint.

Next, in each folding stage, clustering, which groups LEs into SMBs, placement, and routing are performed to produce the final layout of the implementation.

7 NanoMap: Overview of the Optimization Flow

In this section, we discuss the different steps of NanoMap, an integrated design optimization flow developed for NATURE, in detail and discuss its complexity. Fig. 17 illustrates the complete flow. Given an RTL or gate-level design, NanoMap performs logic mapping, temporal clustering, temporal placement and routing, and produces the configuration bitmap for NATURE. If the power consumption of the design needs to be evaluated, a power simulation (Step 16) can also be performed according to the method described in Sect. 5.

- *Logic Mapping.* (Steps 2–7). These steps use an iterative approach to identify the best folding level based on user-specified design constraints, optimization objectives, and input circuit structure. The input network can be obtained with the help of tools like Synopsys Design Compiler and FlowMap [10]. First, an appropriate initial folding level is computed, based on which RTL modules are partitioned into connected LUT clusters. An area check is performed here in order to verify if the module partition meets the area constraint. Then NanoMap

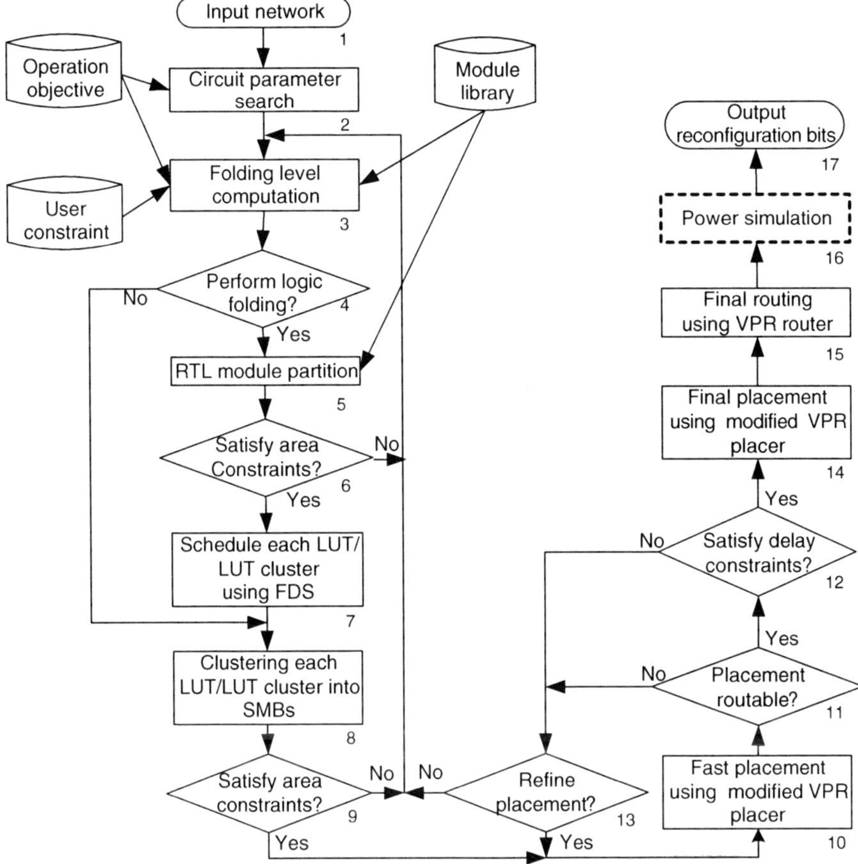

Fig. 17 Automatic design flow

uses FDS [35] to assign LUTs and LUT clusters to folding stages and balance interfolding stage resource usage, and produces the LUT network of each temporal folding stage.

- *Temporal Clustering* (Steps 8–9). These steps take the flattened LUT network as input and cluster the LUTs into MBs and SMBs to minimize the need for global interconnect and simplify placement and routing. As opposed to the traditional clustering problem, each hardware resource, that is, LE, MB, or SMB, is temporally shared by logic from different temporal folding stages. Temporal folding necessitates that both intrastage and interstage data dependencies be jointly considered during LUT clustering. Note that the folding stages need not be limited to one plane, that is, temporal clustering can span planes. Verifying if the area constraint is met is done after clustering. If it is met, placement is invoked. Otherwise, NanoMap returns to the logic mapping step.

- *Temporal Placement* (Steps 10–14). These steps perform physical placement and minimize the average length of inter-SMB interconnects. They are implemented on top of an FPGA place-and-route tool, VPR [1], to provide inter-folding stage resource sharing. Placement is performed in two steps. First, a fast placement is used to derive an initial starting point. A low-precision routability and delay analysis is then performed. If the analysis indicates success, a detailed placement is invoked to derive the final placement. Otherwise, several attempts are made to refine the placement and if the analysis still does not indicate success, NanoMap returns to the logic mapping step.

- *Routing* (Step 15). This step uses the VPR router to generate intra-SMB and inter-SMB routing. After routing, the layout for each folding stage is obtained and the configuration bitmap generated for each folding cycle.
- *Complexity of the Algorithm.* In each iteration of the loop, the most computationally intensive steps are FDS and temporal placement. If n is the number of nodes (LUT/LUT cluster) to be scheduled in the design, then the complexity of FDS is $O(n^3)$ [35]. If r is the number of SMBs after clustering, the complexity of placement is $O(r^{4/3})$ [31]. Since $n = Cr$, where C is a constant, the total complexity of FDS and placement is still $O(n^3)$. We will see later that the maximum number of iterations required is related to the maximum logic depth of the circuit. If the logic depth is s, the complexity of NanoMap is $O(sn^3)$. If in the FDS step, sorting is used first, the run time of FDS can be reduced to $O(n^2)$. Hence, the total complexity of NanoMap in this case is reduced to $O(sn^2)$. Note that plane separation and circuit parameter search are performed only once before the iterations start and their complexity is $O(Bn)$, where B denotes the number of inputs of the circuit.

In the following sections, we describe these steps in detail.

8 Logic Mapping

In this section, we discuss the steps involved in logic mapping.

8.1 Plane Separation

Plane separation is the key part of circuit parameter search, which collects the necessary circuit parameters for choosing an appropriate folding level. In this step, it first needs to analyze the input design structure, such as how many planes are contained in the design, the logic depth in each plane, plane width, which is the maximum number of LUTs at any logic depth in the plane, etc. First, the circuit structure analysis involves identifying each plane and labeling each module/LUT belonging to it. Then each plane is searched and the above-mentioned parameters obtained.

To identify each plane, a depth-first search algorithm is used. The algorithm begins the search by enumerating every input to the circuit as the root node, and assigning it plane index 1. Then from each root node, the algorithm explores the

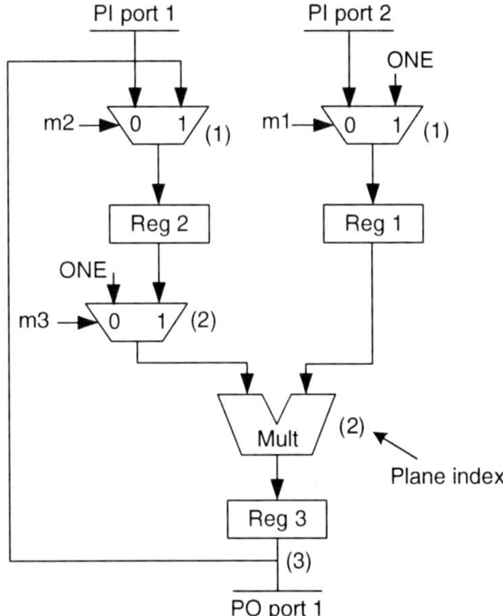

Fig. 18 Plane separation for multiple paths using the least plane index

structure, as far as possible, along each branch and assigns the node (module or LUT) along its path the current plane index. The plane index is increased by one when the algorithm reaches plane registers. If a node is on multiple paths with different plane indices, the least index is used since it guarantees the correct execution sequence and produces a minimal number of planes, that is, results in a smaller delay. Fig. 18 gives an example illustrating plane separation. The number inside the parentheses indicates the plane index of the node. We can see that for the MUX selected by *m2* on the path from *PI port1*, the plane index is 1. However, on the path from *reg3*, its plane index is 3.

Based on the policy that the least plane index is assigned, the above MUX is labeled with plane index 1. The search algorithm is summarized in Algorithm 1. Note that path index is used in the algorithm to differentiate among multiple paths in case there is a loop along one path. We can see that this plane separation algorithm favors delay optimization. To adjust it for area optimization, modules with multiple plane indices can be assigned to the appropriate plane to balance the number of modules or total number of LUTs among the planes.

8.2 Folding Level Computation

After obtaining the circuit parameters, they are used to compute the initial folding level and adjust the folding level gradually later. Choosing an appropriate folding level is critical to achieving the best area-delay trade-off for the specific optimization objective. As discussed earlier, folding level computation depends on the input circuit structure, which is obtained by identifying each plane and obtaining the circuit parameters within each plane. We summarize the necessary circuit parameters below:

- Number of planes in input circuit: *num_plane*
- Number of LUTs in plane i: *num_LUT$_i$*
- Maximum number of LUTs among all the planes: $LUT\_max = max\{num\_LUT_i\}$, $i=1,\dots,num\_plane$
- Plane width in plane i: *width_plane$_i$*
- Maximum plane width among all the planes: $Width\_max = max\{width\_plane_i\}$, $i=1,\dots,num\_plane$
- Logic depth of plane i: *depth$_i$*
- Maximum logic depth among all the planes: $depth\_max = max\{depth_i\}$, $i=1,\dots,num\_plane$
- Area constraint, for example, the available number of LEs: *Available_LE*
- Delay constraint, or sample period, for example, how often new data arrive at the circuit inputs: *sample_period*
- Number of reconfiguration copies in each nano RAM: *num_reconf*

Given the specified optimization objective and constraint, the best folding level is computed using the given parameters. There can be various optimization objectives and constraints. We show how to target the following three design objectives. Similar procedures can target other objectives.

- minimize delay with/without area constraint;
- minimize area with/without timing constraint;
- minimize area-time (AT) product;

Algorithm 1. Algorithm for plane separation.

```
1. Plane_search (circuit_graph G)
2. /*If there are multiple plane indices for one nodechoose the least*/
3. listL=empty;
4. boolean next=false;
5. foreach input ei of the circuit do
6.     set current plane inde px=1;
7.     add edge with current plane index (ei, p)to L;
8.     whileL is not empty do
9.         remove pair(ei, p)from the end of L;
10.        foreach leaf node n of edge e do
11.            if(the plane index pn of node n is not set) or((the current path index ‡
12.            the label of the node)and(p<pn))then
13.                set pn = p;
14.                set next = true;
15.                set the label of node n to be the current path index i;
16.            if next = true then
17.                for each output edge en of node n do
18.                    if node n is a plane register then add pair(en,p+1)to L;
19.                    else add pair(en,p)to L;
```

8.2.1 Delay Minimization

Suppose the optimization objective is to minimize delay. If there is no area constraint, it can first use no-folding and then search around the neighboring folding levels to obtain the shortest delay, as discussed in Sect. 3. If an area constraint is given, it needs to be satisfied first, then the best possible delay obtained. There are two scenarios that need to be considered.

Multiple planes are allowed to share resources: Such a scenario is possible when there is no feedback across planes. Cross-plane folding is first performed in which all the planes are stacked together, that is, resources are shared across all planes, since this does not increase circuit delay but reduces area. Then, if *LUT_max* is larger than *available_LE*, each plane is folded internally to further reduce area. Since circuit delay is equal to the plane cycle times the number of planes in the circuit, the plane cycle has to be minimized under the area constraint.

Initial Folding Level Selection

Initially, it is assumed that the logic is balanced well in the plane and we compute the minimum number of folding stages required within each plane as follows:

$$\#folding\_stage\_min = \left\lceil \frac{LUT\_max}{available\_LE} \right\rceil. \tag{3}$$

Since the number of folding cycles should be kept the same in each plane, the maximum logic depth is used to compute the maximum initial folding level:

$$initial\_level\_max = \left\lceil \frac{depth\_max}{\#folding\_stage\_min} \right\rceil. \tag{4}$$

Then, since for most circuits, the logic is not well-balanced along the critical path, it tries to push to another extreme and obtain a minimum initial folding level using the maximum plane width:

$$initial\_level\_min = \left\lceil \frac{available\_LE}{width\_max} \right\rceil. \tag{5}$$

It also needs to find the minimum folding level, *min_level*, allowed by *num_reconf*:

$$min\_level = \left\lceil \frac{depth\_max * num\_plane}{num\_reconf} \right\rceil. \tag{6}$$

Hence, the final minimum initial folding level is:

$$initial\_level\_min = max\{min\_level, initial\_level\_min\}. \tag{7}$$

The initial folding level can be chosen between *initial_level_min* and *initial_level_max*. It was found through experiments that for RTL mapping, the best folding level is close to *initial_level_min*, while for gate-level mapping, the best folding level is close to *initial_level_max*. Hence, the initial folding level can be selected based on the input circuit or can just be computed as the average of *initial_level_min* and *initial_level_max* for simplicity.

Folding Level Adjustment

Using the chosen folding level, NanoMap performs FDS and temporal clustering to obtain the area required. If the area constraint is not satisfied, the folding level is decreased by one. Otherwise, the folding level can be increased by one and it can be verified if the area constraint is still satisfied and delay reduced. NanoMap iterates until the area constraint is met or the folding level reduces to the minimum allowed, that is, *min_level*. Figure 19 illustrates one of the optimization approaches.

Multiple planes are not allowed to share resources: Such a scenario is possible if the RTL circuit is pipelined and, hence, the different pipeline stages need to be resident in the FPGA simultaneously. In this scenario, temporal logic folding can only be performed within each plane. Then *LUT_max* or *Width_max* are replaced with $\sum_i num\_LUT_i$ or $\sum_i width\_plane_i$ in the folding level computation.

8.2.2 Area Minimization

If the objective is to minimize the mapped area without any timing constraint, then as much logic folding as possible should be performed. Hence, inside the plane, the minimal folding level, for example, folding level-1 or one bounded by *num_reconf*, should be used. If a timing constraint is specified, for example, a sample period

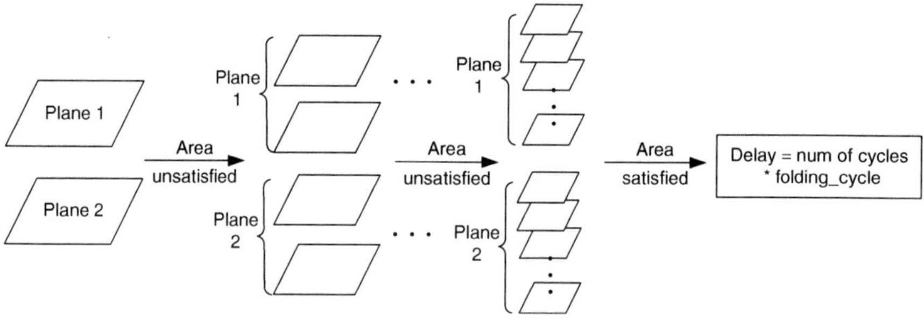

Fig. 19 An example for delay optimization under area constraint, assuming cross-plane resource sharing

constraint, denoting the allowed time between successive input samples, NanoMap ensures that the pipeline stage delay of the mapped circuit is less than *sample_period* to guarantee the required throughput. It then obtains the best area. It again considers two scenarios.

Multiple planes are allowed to share resources: It first stacks the planes and tries to use the minimum folding level in each plane and then gradually increase it to meet the timing constraint, since the sample period cannot be used to compute the initial folding level directly.

Initial Folding Level Selection

This includes two parts: number of planes stacked together for one pipeline stage and the folding level used inside the plane. As discussed above, it first tries the minimum folding level inside the plane. We can see that if *num_reconf* is larger than *depth_max*, level-1 folding is allowed in each plane. Then the number of planes contained in each pipeline stage, *num_folds*, can be obtained as:

$$num\_folds = \left\lfloor \frac{num\_reconf}{depth\_max} \right\rfloor. \tag{8}$$

Otherwise, if *num_reconf* is less than or equal to *depth_max*, one plane is contained in one pipeline stage (*num_folds* = 1). The minimal folding level inside the plane is set by *depth_max* divided by *num_reconf*.

Folding Level Adjustment

The delay for one pipeline stage is equal to *num_folds* × *plane_cycle*. If the stage delay is larger than *sample_period*, *num_folds* has to be decreased first, then the folding level has to be increased to reduce the pipeline stage delay until the timing constraint is satisfied. Then the smallest possible area under the timing constraint can be obtained. Figure 20 illustrates the discussed procedure.

Multiple planes are not allowed to share resources: Logic folding is only performed within planes, as discussed before. The NanoMap iteration starts from the minimum folding level and the folding level is increased by one each time, until the timing constraint is met.

8.3 Module Library

If the input design is at the RTL, the RTL modules need to be expanded to the LUT level. This expansion can be obtained from the module library, which may

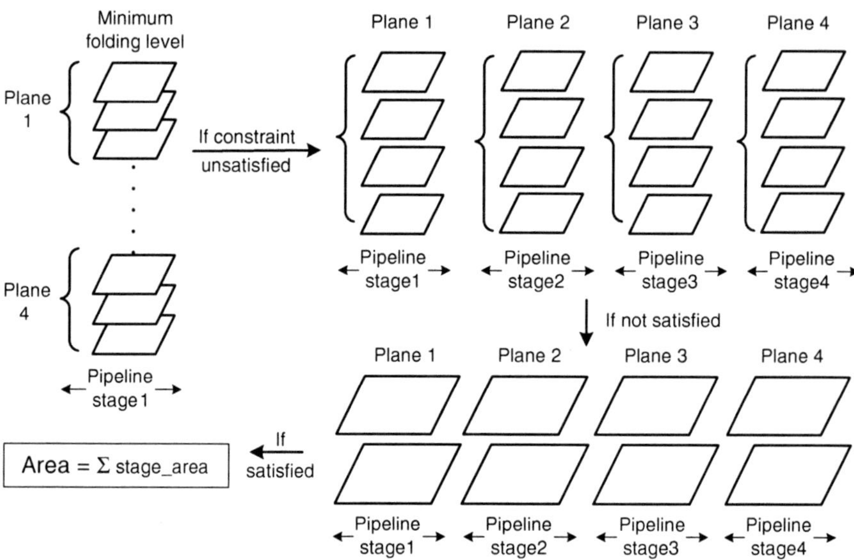

Fig. 20 An illustration for area optimization under timing constraint assuming cross-plane resource sharing

Fig. 21 Part of the module library

```
name add2
operation add
type RC
total_LUT 2
width 2
depth 1
input a[0:1] b[0:1]
output c[0:1]
cluster 1
cluster_input a[0] b[0] a[1] b[1]
cluster_output c[0] c[1]
LUT1 (a[1] a[0] b[1] b[0];c[0]) : 0001001101101100
LUT2 (a[1] a[0] b[1] b[0];c[1]) : 1110110010000000
reg 2
```

be manually designed, targeting area or delay optimization, or automatically generated for regular architectures. The module implementations for required folding levels are provided in the library. Figure 21 shows a part of the library for level-1 folding. It can be seen that for each module, the library provides information on the module name, module function and type (for example, adder can be ripple-carry or carry-select or another type), its bit-width, logic depth, input/output, number of LUT clusters, realization of each LUT cluster, and the number of LUTs used. For each LUT contained in the LUT cluster, the labels inside the parentheses specify the m inputs of the LUT ($m = 4$ in the example), followed by its output, separated by a semicolon. The 16-bit binary number indicates the SRAM configuration for the LUT, which is precomputed to realize the module function. The value *reg num* gives the number of flip-flops needed to store the temporary computation results from the LUT cluster. This information is used in the LUT scheduling step later.

8.4   RTL Module Partitioning

As discussed in Sect. 6, to speed up the mapping process, NanoMap can take advantage of the RTL module hierarchy and simply slice it into several connected LUT clusters, with each cluster's logic depth being less than or equal to the chosen folding level. During the partitioning process, if the folding level is p, the LUTs with logic depth more than $(n - 1)p$ and no more than np are grouped into the nth cluster. Then each cluster is processed in its entirety in the following scheduling step. At the same time, the inputs/outputs of the module are separated into bit lines that connect various LUT clusters. Based on each selected folding level, the mapper can simply fetch the corresponding cluster based module implementation from the library (this is introduced later). Fig. 22 shows a simple module partitioning example. The partition results in a network of LUTs/(LUT clusters) which is input to the scheduling step. Note that if there is an area constraint, NanoMap first checks to see whether the size of the partitioned LUT cluster exceeds the area constraint. If so, the folding level has to be decreased and partitioning repeated.

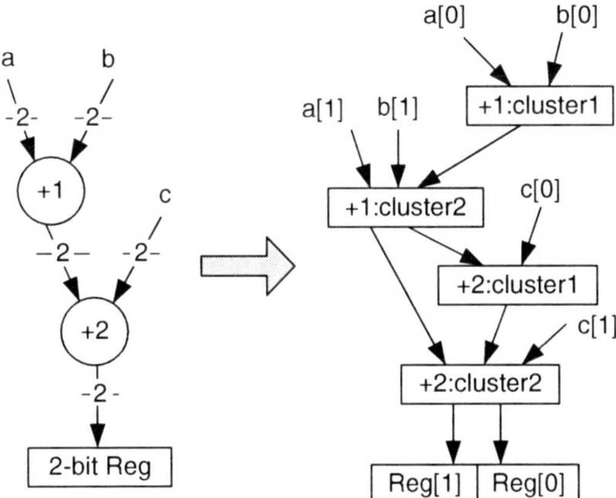

Fig. 22 Illustrating module partitioning

8.5 Force-Directed Scheduling

FDS [35] is a popular scheduling technique in high-level synthesis. It uses an iterative approach to obtain the schedule of operations in order to minimize overall resource usage. Resource usage is modeled as a force. Scheduling of an operation to some time slot, which results in minimum force, indicates a minimum increase in resource usage. Force is calculated based on distribution graphs (DGs), which describe the probability of resource usage for a type of operation in each time slot.

In this work, FDS is used in another scenario. It is modified to assign the LUT or LUT cluster to folding stages and balance the resource usage of the folding stages. Based on the underlying NATURE architecture instance, the LE usage in each folding cycle is computed differently.

- Since the number of flip-flops inside each LE is limited, the LE usage in each folding cycle is dependent on both the LUT computations and register storage operations performed in parallel. Thus, two DGs, one describing the resource usage of LUT computations and another for register storage usage, have to be built.
- When the number of flip-flops inside each LE is adequate for storing all temporal results for all folding stages, then LE usage is only determined by LUT computations in each folding cycle. For this case, only the LUT computation DG is needed.

Next, we first discuss how to create the two DGs, and then how the forces are calculated based on the two DGs.

8.5.1 Creation of LUT Computation DG

A LUT computation DG models the aggregated probability distribution of the potential concurrency of LUT/(LUT cluster) computations within each folding cycle. To compute the probability distribution of each LUT or LUT cluster, their time frames, $time\_frame_i$, or feasible time interval, are first obtained. Time frames are defined as the span from the folding cycle it is assigned to in the as-soon-as-possible (ASAP) schedule to the folding cycle it is assigned to in the as-late-as-possible (ALAP) schedule. To illustrate this, we use the example shown in Fig. 23. As shown in the figure, since the folding level can be larger than one, the chaining of LUTs is allowed in order to provide more flexibility to the mapper. We can see that $time\_frame_{LUT2}$ spans folding cycles 1 to 3, denoted as [1, 3]. Here, $clus_i$ denotes LUT cluster i. Figure 24 illustrates the $time\_frame$ of each LUT and LUT cluster, as shown in the example of Fig. 23, at their initial status before scheduling.

If a uniform probability distribution is assumed, the probability that a computation i is assigned to a feasible folding cycle j within its time frame equals $1/|time\_frame_i|$ for $j \in time\_frame_i$ as the number on top of the bars shown in Fig. 24 indicates. Then the LUT computation DG value in folding cycle j,

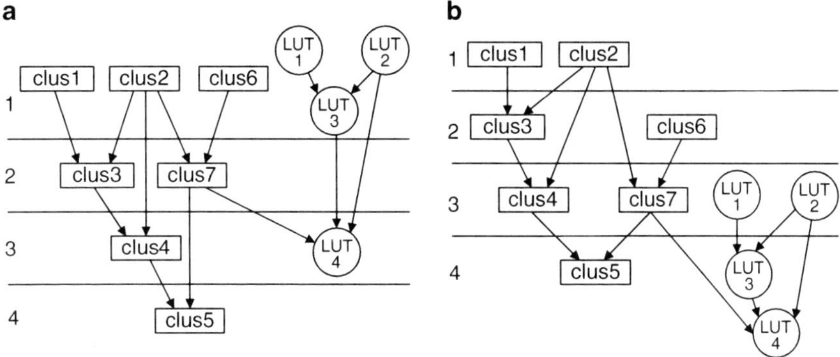

Fig. 23 Schedules of LUTs and LUT clusters in a plane: (**a**) ASAP schedule, (**b**) ALAP schedule

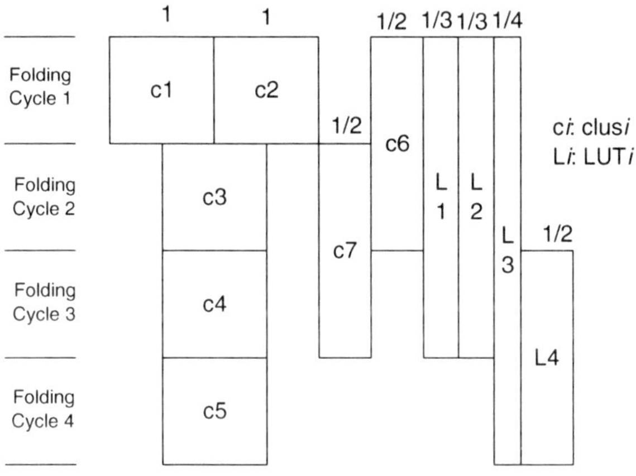

Fig. 24 Time frames of the LUT/(LUT cluster) computations (initial status)

LUT_DG(j), is the sum of the probabilities of all the computations assigned to this folding cycle:

$$LUT\_DG(j) = \sum_{i=1}^{N} \frac{1}{|time\_frame_i|} * weight_i, \tag{9}$$

where $weight_i$ is 1 for a LUT or equal to the number of LUTs in a LUT cluster.

8.5.2 Creation of Register Storage DG

Register storage DG models the distribution of register storage usage. A similar procedure as above is adopted to model the flip-flop usage inside the LE. A storage operation is created at the output of every source computation that transfers a value to one or more destination computations in a later folding cycle. If both the source and destinations of a storage operation are scheduled, the distribution of the storage operation equals its *lifetime*, which begins from the folding cycle of the source and ends at the folding cycle of the last destination. It is assumed that the results are stored at the beginning of each folding cycle. If one or more of the source or destinations are not scheduled, a probabilistic distribution has to be obtained. If the storage must exist in some cycle, its distribution probability will be 1 in that cycle.

The storage operations in one part of the example in Fig. 23 are shown in Fig. 25. The following heuristic is used to quickly estimate the resulting storage distribution. First, *ASAP_life* and *ALAP_life* of a storage operation are defined as its lifetime in the ASAP and ALAP schedules, respectively. For example, in Fig. 25, the output of source computation LUT2, that is, storage **S**, transfers the value to destination computations LUT3 and LUT4. In the ASAP schedule, **S** begins at folding cycle 2 and ends at folding cycle 3. Hence, $ASAP\_life_{\mathbf{S}} = [2, 3]$ and the length of $ASAP\_life_{\mathbf{S}}$: $|ASAP\_life_{\mathbf{S}}| = 2$. Similarly, in the ALAP schedule, **S** begins at folding cycle 4 and ends at folding cycle 4, which results in $|ALAP\_life_{\mathbf{S}}| = 1$.

The longest possible lifetime *max_life* for the storage operation is the union of its *ASAP_life* and *ALAP_life*, whose length is obtained as:

$$|max\_life| = (ALAP\_life\_end - ASAP\_life\_begin + 1). \tag{10}$$

For the example, **S** begins in folding cycle 2 in the ASAP schedule, that is, $ASAP\_life\_begin_{\mathbf{S}} = 2$. Its lifetime ends in cycle 4 in the ALAP schedule, that is, $ALAP\_life\_end_{\mathbf{S}} = 4$. Thus, the length of the maximum lifetime for **S** is $|max\_life_{\mathbf{S}}| = 3$.

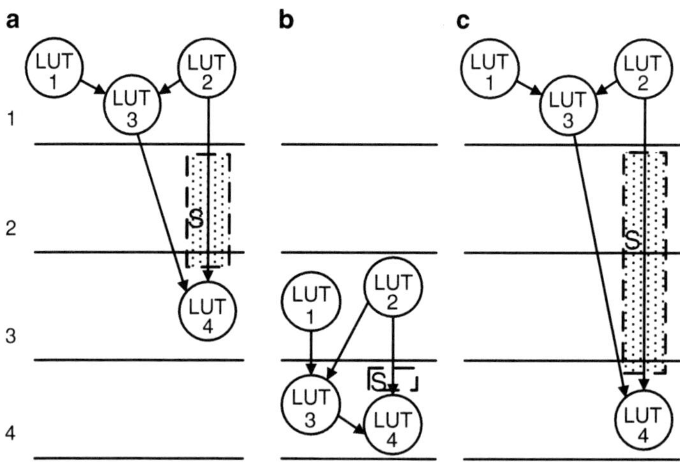

Fig. 25 Storage lifetimes for all schedules: (**a**) ASAP lifetime, (**b**) ALAP lifetime, and (**c**) maximum lifetime

If *ASAP_life* overlaps with *ALAP_life*, the overlap time, *overlap*, is the intersection of *ASAP_life* and *ALAP_life*, whose length is similarly obtained as:

$$|overlap| = (ASAP\_life\_end - ALAP\_life\_begin + 1). \tag{11}$$

Within the overlap time, a storage operation must exist with probability 1. For the example, there is no overlap time for **S**. Then an estimate of the average length of all possible lifetimes can be obtained by:

$$avg\_life = \frac{|ASAP\_life| + |ALAP\_life| + |max\_life|}{3}. \tag{12}$$

Next, the probability of a storage operation performed for a LUT or LUT cluster computation i in folding cycle j can be calculated as follows.

When j is outside of $overlap_i$ and $j \in max\_life_i$,

$$storage_i(j) = \frac{avg\_life_i - |overlap_i|}{|max\_life_i| - |overlap_i|} * weight_i. \tag{13}$$

When j is within $overlap_i$, which means a storage operation must be performed, then

$$storage_i(j) = weight_i. \tag{14}$$

The process is carried out for all storage operations, and separate probabilities due to N LUTs and LUT clusters in folding cycle j are added to obtain a single storage DG as follows.

$$storage\_DG(j) = \sum_{i=1}^{N} storage_i(j), \quad j \in max\_life_i. \tag{15}$$

8.5.3 Calculation of Forces

The FDS algorithm uses force to model the impact of scheduling of operations on resource usage. It is much like that exerted by a spring that obeys Hooke's law: $F = Kx$, where K is given by the value of $DG(j)$ which presents the probability of resource usage concurrency. A higher force implies higher concurrency of runtime operations, which requires more resources in parallel. The force is contributed by both self-force and the forces of predecessors and successors.

Self-Force

The force in cycle j is calculated based on DGs as:

$$force\ (j) = DG(j) * x(j), \tag{16}$$

where $DG(j)$ is either $LUT\ DG(j)$ or $storage\ DG(j)$, and $x(j)$ is the increase (or decrease) in the probability of computation in cycle j due to the scheduling of the computation. For example, before scheduling, the computation has a uniform probability of being scheduled in each folding cycle in its time frame. If in a scheduling attempt, the computation is scheduled in folding cycle a, the probability of the computation being scheduled in folding cycle a will increase to 1 and the probability of it being scheduled in other folding cycles will decrease to 0. The self-force associated with the assignment of a computation i, whose time frame spans folding cycles a to b, to folding cycle j is defined as the sum of all the resulting forces in its time frame:

$$self\_force_i(j) = DG(j) * x(j) + \sum_{k=a,k\neq j}^{b} [DG(k) * x(k)],$$
$$x(j) = (|time\_frame_i| - 1)/|time\_frame_i|, \tag{17}$$
$$x(k) = -1/|time\_frame_i|.$$

Resource usage can be dictated by either LUT computations or storage operations. Assuming there are h LUTs and l flip-flops in one LE, the self-force for scheduling a LUT or LUT cluster i in folding cycle j is determined by both the self-force from the LUT computation and self-force from storage operations as

$$max\{LUT\_self\_force_i(j)/h, storage\_self\_force_i(j)/l\}. \tag{18}$$

when $LUT\_self\_force_i(j)$ and $storage\_self\_force_i(j)$ are positive. $LUT\_self\_force_i(j)$ and $storage\_self\_force_i(j)$ are computed using (17), based on the LUT computation and storage DGs. If the forces are negative, the equation is adjusted to:

$$min\{LUT\_self\_force_i(j)/h, storage\_self\_force_i(j)/l\}. \tag{19}$$

Note that it can never be the case that the LUT self-force is negative and the corresponding storage self-force is positive or vice versa. To illustrate the procedure, let us schedule clus6 (see Fig. 23) as an example. If it is scheduled in cycle 1, the probability of the cluster will change from 1/2 to 1 in cycle 1 and from 1/2 to 0 in cycle 2. Assuming there are four LUT computations and one output in clus6, the resulting $self\_force$ in cycle 1 is $self\_force(1)$ due to $x(1)$:

$$LUT\_DG(1) * weight * x(1) = 5.92 * 4 * 0.5 = 11.84,$$

$$Storage\_DG(1) * weight * x(1) = 0 * 1 * 0.5 = 0.$$

According to (18), $self\_force(1)$ is 11.84. $self\_force(1)$ due to $x(2)$:

$$LUT\_DG(2) * weight * x(2) = 4.92 * 4 * (-0.5) = -9.84,$$

$$Storage\_DG(2) * weight * x(2) = 4.88 * 1 * (-0.5) = -2.44.$$

Therefore, the self-force due to scheduling of clus6 in cycle 1 is as $11.84 - 9.84 = 2$. Similarly, the self-force from assigning clus6 to cycle 2 can be computed. *self_force*(2) is then $9.84 - 11.84 = -2$.

Comparing the self-forces under the two assignments, we can see that scheduling clus6 in cycle 1 results in more force than in cycle 2. Since less the force is, less the concurrency the assignment will result in, the computation should be scheduled into cycle 2.

Predecessor and Successor Forces

The second part of the force is the predecessor and successor forces. Assigning a LUT computation to a specific folding cycle often affects the time frame of its predecessors and successors, which in turn creates additional forces affecting the original move. Equation (17) is used to compute the force exerted by each predecessor or successor. The overall force is then the sum of the self-force and forces of predecessors and successors. Then the total forces under each schedule for a computation are compared and the computation is scheduled into the folding cycle with the lowest force. For example, if clus6 was tentatively assigned to cycle 2, the succeeding clus7 would implicitly be assigned to cycle 3 and LUT4 to cycle 4. The time frames of the other nodes would not be affected. Assuming four LUTs and one output in clus7 too, the successor force from implicitly assigning clus7 to cycle 3 can be obtained as:

$$LUT\_DG(3) * weight * x(3) = 4.92 * 4 * 0.5 = 9.84,$$

$$Storage\_DG(3) * weight * x(3) = 4.55 * 1 * 0.5 = 2.28,$$

$$LUT\_DG(2) * weight * x(2) = 4.92 * 4 * (-0.5) = -9.84,$$

$$Storage\_DG(2) * weight * x(2) = 4.88 * 1 * (-0.5) = -2.44.$$

Hence, the successor force due to clus7 is $9.84 - 9.84 = 0$. Similarly, the successor force due to LUT4 can be obtained as:

$$LUT\_DG(4) * weight * x(4) = 2.75 * 1 * 0.5 = 1.38,$$

$$Storage\_DG(4) * weight * x(4) = 3.67 * 1 * 0.5 = 1.84,$$

$$LUT\_DG(3) * weight * x(3) = 4.92 * 1 * (-0.5) = -2.46,$$

$$Storage\_DG(3) * weight * x(3) = 4.55 * 1 * (-0.5) = -2.28.$$

Thus, the successor force from assigning LUT4 is $1.84 - 2.46 = -0.62$. Hence, the total successor force is -0.62. Then considering all the related forces, clus6 should be scheduled to cycle 2 to give the least concurrency.

Algorithm 2. Force-directed scheduling.

```
1. while there are LUT/(LUT cluster)computations to be scheduled do
2.     evaluate its time frame using ASAP and ALAP scheduling;
3.     create the LUT computation and register storage DGs;
4.     foreach unscheduled LUT/(LUT cluster)computation i do
5.         foreach feasible clock cycle j it can be assigned to do
6.             calculate the self-force of assigning node i to cycle j;
7.              add predecessor and successor forces to self-force to get the total force
8.              for node i in cycle j;
9.         select the cycle with the lowest total force for node i;
10.    pick the node with the lowest total force and schedule it in the selected cycle;
```

8.5.4 Summary of the FDS Algorithm

FDS uses an iterative approach to schedule one computation in each iteration. In each iteration, the LUT computation and register storage DGs are obtained. The LUT or LUT cluster with the minimum force is chosen, and assigned to the folding cycle with the minimum force, which potentially leads to a minimal increase in resource usage. This process continues until all the LUT or LUT cluster computations are scheduled. The pseudo-code of the proposed FDS technique is shown in Algorithm 2. This algorithm is called Heuristic 1. To improve the run time, another approach is to sort the LUT or LUT cluster according to its weight (number of LUTs) first. The node with the largest weight is scheduled first. Since smaller the weight of the node, smaller its impact, the experiments show that the scheduling result is degraded only slightly, but the run-time is improved significantly. This approach is referred to as Heuristic 2.

9 Temporal Clustering

During scheduling, nodes of LUTs or LUT clusters are assigned to each folding stage. In the temporal clustering step, for each folding stage, a constructive algorithm is used to assign the network of LUTs to LEs and pack LEs into MBs and SMBs. To construct each SMB, it first chooses a LUT cluster with a maximal number of inputs and chooses a LUT, which uses a maximal number of its inputs, within that cluster as an initial seed. Then, new LUTs with high attractions to the seed LUT are chosen and assigned to the SMB, until the SMB is fully packed. Then a new LUT seed is selected. The attraction between a LUT i and the seed LUT, $Attraction_{i,seed}$, depends on timing criticality and input pin sharing [30], as follows:

$$Attraction_{i,seed} = \alpha * Criticality_i + (1 - \alpha) * \frac{Nets_{i,seed}}{G}. \tag{20}$$

The first term on the right-hand side models the timing criticality and the second term models the net congestion. In the first term, the timing criticality of LUT i, $Criticality_i$, can be further expanded as:

$$Criticality_i = max_j \{Connection\_Criticality(j)\} + \varepsilon * total\_paths\_affected_i$$
$$j \in \{\text{inputs of } LUT \ i \text{ connected to } seed\}, \quad \text{and} \quad Connection\_Criticality(j)$$
$$= 1 - \frac{slack(j)}{MaxSlack}. \tag{21}$$

Here, *slack* is defined as the amount of delay between the actual arrival time and the required arrival time, and *MaxSlack* is the maximum *slack*. The second part of (21) is used for tie-breaking purposes when multiple nodes have the same criticality. The tie is broken in a manner to most effectively reduce the number of nodes remaining on the critical paths. *Total_paths_affected_i* tracks how many critical paths the node is on and ε is a parameter with a small value that ensures the *total_paths_affected* value acts only as a tie-breaker ($\varepsilon = 0.01$ worked well in the experiments).

In the second term, $Nets_{i,seed}$ is the number of shared inputs/outputs between the two LUTs, and G is the number of inputs/outputs per LE. α is a parameter between 0 and 1 that allows a trade-off between timing criticality and interconnect demand and can be tuned to the specific optimization. For example, for routing area optimization, it can mainly lay emphasis on the interconnect, and for delay optimization, timing criticality plays the key role.

As opposed to traditional clustering methods, to support temporal logic folding, inter-folding stage resource sharing needs to be considered during clustering. Since due to logic folding, several folding stages may be mapped to a set of LEs, some of the LEs may be used to store the internal results and transfer them to another folding cycle. Hence, there are LEs spanning several cycles and connecting with other LEs in different cycles. Fig. 26a illustrates an LE with one LUT and two flip-flops. In folding cycle 1, the LUT implements function A and stores the computation result in the first flip-flop. The result is maintained until folding cycle n. In the second cycle, the LUT implements function B and if the result also needs to be stored, it has to be stored in flip-flop 2 since flip-flop 1 is already occupied.

Suppose in some cycle between the second and nth, all the flip-flops inside the LE are occupied and only the LUT is available. If it is a LUT computation requiring result storage, a new LE is needed to accommodate the LUT computation, since separating the computation and result storage will worsen the delay. Otherwise, if the objective is to optimize area, the LUT computation and its result can be separately mapped to a LUT and flip-flop in different currently available LEs in order to reduce total LE usage. Hence, in temporal clustering, both the LUT mapping and the lifetime of temporal result storage need to be carefully tracked.

Moreover, since two LEs (e.g., C and D in Fig. 26b) may have multiple attractions between them across several folding cycles, when performing temporal clustering, the attractions between them should be set to the maximum attraction over all the cycles. The pseudo-code for temporal clustering algorithm is shown in Algorithm 3.

Algorithm 3. Temporal clustering.

```
1. foreach folding stage i do
2.        while there are unclustered LUTs in the folding stage i do
3.               found = true;
4.               if current SMB curr_SMB not full then
5.                     if there are LUTs which can be accommodated in curr_SMB then
6.                            foundLUTa with the highest attraction with curr_SMB;
7.                            select_LUT = LUTa;
8.                     else
9.                            found = false;
10.              else
11.                     found = false;
12.              if found == false then
13.                     select the seed LUT  LUTb
14.                     select_LUT = LUTb
15.                     create a new SMB and assign it to  curr_SMB;
16.              assign select_LUT to the available LE in the  curr_SMB;
17.              make select_LUT as clustered;
18.              record LUT mapping and flip -flop usage ;
19.              reduce the number of available LEs in  curr_SMB by one;
```

10   Temporal Placement and Routing

VPR is modified to perform placement and routing. Placement uses a two-step simulated annealing approach. It starts with a fast low-precision placement, which has $10\times$ speed-up and only 5% performance degradation. VPR's routability delay analysis is then used to evaluate the quality of this initial placement, which determines whether a high-precision placement or another round of optimization using adjusted logic folding should be invoked.

VPR uses simulated annealing for placement. The functional form of its cost function targeted at net congestion is

$$\text{Cost} = \sum_{n=1}^{N_{\text{nets}}} \frac{q(n)}{C_{\text{av}}} [bb_x(n) + bb_y(n)], \tag{22}$$

where the summation is over all the nets in the circuit. For each net n, $bb_x(n)$ and $bb_y(n)$ denote the horizontal and vertical spans of its bounding box, respectively. The value $q(n)$ is the pin-count net-weight, which is 1 for nets with three or fewer terminals, and slowly increases to 2.79 for nets with 50 terminals. C_{av} is the average

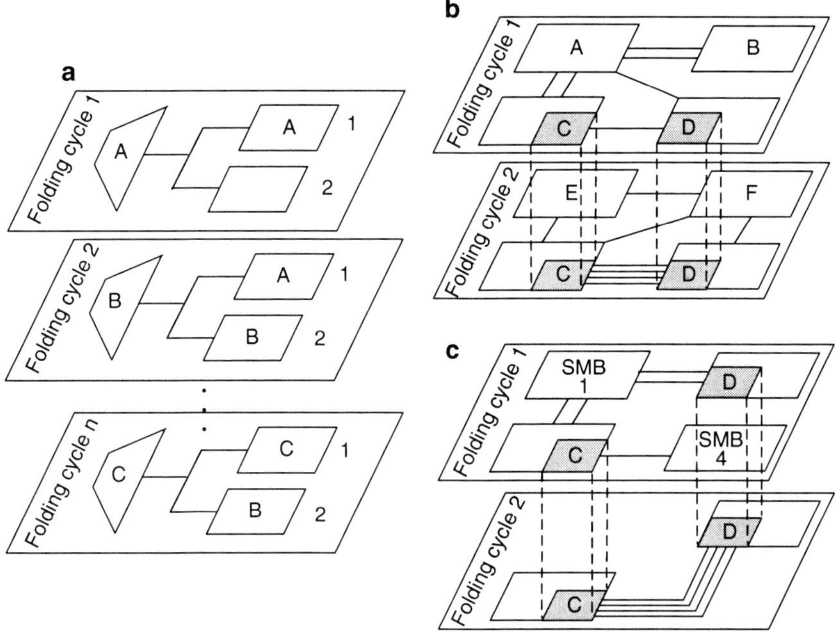

Fig. 26 Illustration of (**a**) flip-flop usage in temporal clustering, (**b**) attraction computation in temporal clustering, and (**c**) placement with logic folding

channel capacity (in terms of the number of tracks). Since in NATURE, all channels have the same capacity, C_{av} is a constant. Hence, the congestion cost function is equivalent to a bounding box cost function. The VPR placer is modified to support temporal logic folding, which introduces interfolding stage dependencies.

Consider the example in Fig. 26c. In folding cycle 1, since there are few connections between C and D, they may be placed far apart. However, such a placement would not be good for folding cycle 2 in which C and D communicate a lot. Hence, the pin-count for such a pair of SMBs should consider the connections in all the folding stages. When swapping such pairs, they should be swapped in all the folding stages where they appear. Recently, timing-driven placement has also been added to VPR, and it has been similarly modified to support logic folding.

After the placement is done, detailed routing can be performed using VPR to connect all the SMBs in each folding stage and finish the mapping process. Direct links are also targeted in the router in order to support the interconnect structure in NATURE. Routing in VPR can be conducted in a timing-driven fashion. Hence, the router tends to first use direct links, then length-1 and length-4 wire segments and finally global interconnects (these are the four types of interconnects available in NATURE). Note that a length-i interconnect spans i SMBs. Finally, based on the placement and routing result in each folding cycle, the reconfiguration bits for each LE/switch can be generated.

11 Simulation-Based Analysis

In this section, to show the advantages of NATURE and efficiency of NanoMap, we analyze simulation results for the mapping of RTL/gate-level benchmarks to typical instances of NATURE using NanoMap. Among the twelve benchmarks targeted, GCD is Euclid's greatest common divisor implemented in Muttreja et al. [33]. ex1 is the circuit shown in Fig. 16, but with a bit-width of 16. ex2 is an RTL circuit from [48]. Paulin and diffeq are two differential-equation solvers implemented in [48] and [58], and FIR and Biquad are two types of digital filters. ASPP4 is an application-specific programmable processor synthesized in Ghosh et al. [17]. des is a combinational benchmark from Lisanke [29]. Wavelet and DCT are two mathematical functions implementing wavelet transform and discrete cosine transform computations, respectively. Finally, b20 is a gate-level benchmark from the ITC'99 benchmark suite [22].

NATURE is a family of architectures, which may vary in the number of inputs (m) and flip-flops (l) in an LE, number of LEs (n_1) in an MB, number of MBs (n_2) in an SMB, and even the type of on-chip nano RAM. To set up the simulation, based on the discussion in Sect. 4, the chosen architectural instance has one four-input LUT, two flip-flops in an LE, four LEs in an MB and four MBs in an SMB to obtain good area-delay trade-offs. NRAM is assumed as on-chip nano RAM. All simulations are based on a 70-nm technology (i.e., both the CMOS logic and NRAM are implemented in this technology).

First, parameter k is varied in order to compare implementations corresponding to different folding levels, such as level-1, level-2, level-4, and no logic folding (note that the number of NRAM bits increases going from no-folding to level-4 folding and towards level-1 folding since the number of LE configurations increases). It is assumed that k is as large as needed for each case. We can observe the area/delay trade-offs that become possible because of the use of logic folding. From level-1 folding to level-4 folding, the area typically increases. However, sometimes, level-1 folding requires too many flip-flops to store the internal computation results such that the area used by level-1 folding exceeds the area required by level-2 folding.

Consider the example of a 64-bit ripple-carry adder. Its LUT graph has 64 LUTs on the critical path. Using level-1 logic folding, the complete adder can be mapped to only two LEs. This, of course, requires reconfiguration of the LEs from the local NRAMs at each cycle. Traditional reconfigurable architectures will require 128 LEs for such an adder (some architectures incorporate a carry generation circuit with each LE; in such a case, they will require 64 LEs, although each LE will be larger due to the carry generation circuit overhead) because they cannot perform any temporal logic folding.

As the number of required LEs increases, the need for using higher-level (i.e., more global) interconnects to connect them also increases, which is one of the reasons traditional reconfigurable architectures are not competitive with ASICs in terms of performance. Hence, significant area reduction is achieved. If more LEs are allowed (as in level-2, level-4, and no-folding cases), the circuit delay typically goes down because fewer reconfigurations are required. Since the reconfiguration delay is short, even for the worst case of level-1 folding, generally, the reconfiguration delay only contributes to less than 10% to the total delay.

Next, consider the area-delay product. For larger, more serially connected circuits of larger depth, the area-delay product advantage of level-1 folding relative to no-folding is typically larger. For example, for the 64-bit ripple-carry adder, we found this advantage to be about $52\times$. This results from a large saving in area while maintaining competitive performance. On an average, the area-delay product improvement achieved using level-1 folding compared to no-folding is over $8\times$ for the 12 benchmarks.

NanoMap also provides significant flexibility in performing area-delay trade-offs when mapping applications to NATURE. It can target many different optimization objectives. Typical objectives are: (1) minimization of circuit delay with or without an area constraint, (2) minimization of area with or without a delay constraint, (3) minimization of the area-delay product, and (4) finding a feasible implementation under both area and delay constraints. To save space, different optimization objectives are chosen for different benchmarks and the results presented in Table 7. In Column 2, the optimization objectives are listed. In columns 3 and 4, the constraints (area or delay or neither) are listed. These results show the versatility of NATURE and NanoMap. A significant side-benefit of the dramatic area reduction made possible by logic folding is the associated reduction in the need for a deep interconnection hierarchy in NATURE. Since cycle-by-cycle reconfiguration makes LE utilization very high, approaching 100%, the need for global communication greatly reduces. It was found that global interconnect usage went down by more than 50% when using level-1 folding as opposed to no-folding. This points to trading interconnect area for increased nano RAM area as an attractive alternative for NATURE.

Since NanoMap can start with an input design specified at the RTL or gate-level, there are also two ways to implement a design, thus providing flexibility to designers. For RTL descriptions, since the modules are partitioned into LUT clusters and mapped as one entity, there is less scheduling freedom for each LUT. However, the manually designed library enables optimization inside LUT clusters. Hence, this may result in some variation in circuit delay when comparing the mappings from a gate-level description and from an RTL description. The average

Table 7 Circuit mapping results for typical optimizations

Circuit	Opt. obj.	Area const. (#LEs)	Delay const. (ns)	Num_folds	Folding level	#LEs	Delay (ns)
GCD	Area	–	–	2	2	34	44.08
Ex1	Delay	–	–	1	No	667	39.86
Diffeq	Area	–	70	2	4	128	68.80
Biquad	Delay	110	–	1	1	102	55.16
Paulin	–	250	80	2	4	240	70.84

reduction in the number of LEs required using gate-level mapping increases with a folding level increase due to its scheduling freedom advantage.

However, the use of LUT clusters in RTL mapping speeds up the mapping process. The average reduction in the number of LEs required using gate-level mapping is 11.9%, while RTL mapping gives 3.3% delay improvement compared to gate-level mapping. NanoMap was run on a 2 GHz PC with 1 GB DRAM under RedHat Linux 9. RTL mapping is on an average 1.5× faster than gate-level mapping using Heuristic 2 and 30× faster using Heuristic 1. Comparing the two heuristics, Heuristic 1 gives only slightly better results in mapping area and delay than Heuristic 2. However, its run-time is more than an order of magnitude longer than that of Heuristic 2 for large circuits.

Finally, power estimation was performed at folding level-1, level-2, level-4, and no-folding. Since no-folding is akin to traditional mapping, its switching power consumption includes the power for logic, interconnect, and clock. While using logic folding, since there is reconfiguration every few clock cycles, reconfiguration power has the most contribution to overall power. It increases with an increase in frequency. For clock power, it is assumed that parts of the chip not being used are shut down using power management. Hence, clock power is small since the area used is small. Interconnect power includes all the power consumed in the interconnects inside and outside the SMBs. Logic power includes the power for LUT computation and flip-flop storage. We observe that the interconnect power is much larger than logic power. Under leakage power, it only considers subthreshold leakage in the logic, interconnect, and SRAMs, since it is the dominant leakage component. Here, NRAM leakage is not considered in the power consumption for the no-folding case, since traditional FPGAs do not contain NRAMs.

Figure 27. presents the power consumption for different folding levels. We can see that for the logic folding case, with an increase in folding level, the power may go up or down, since a level increase typically results in an area increase and frequency decrease at the same time, which affect the total power consumption in opposite directions (an area increase results in increased leakage power whereas a frequency decrease results in decreased switching power). Comparing the logic folding and no-folding scenarios, no-folding typically results in more logic,

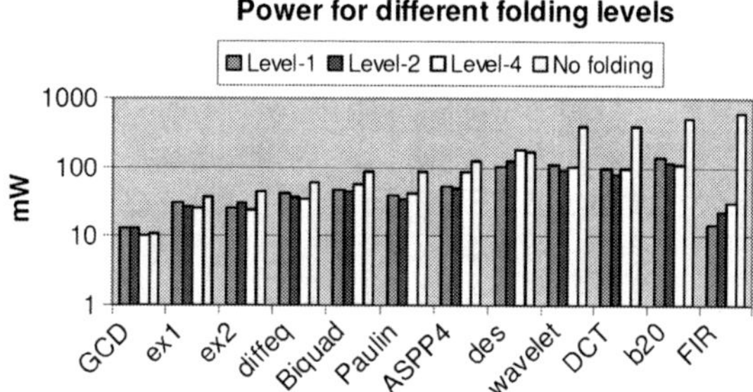

Fig. 27 Power consumption of benchmarks for different folding levels

interconnect, and leakage power than logic folding, since no-folding needs more resources to implement a design. For most benchmarks, since logic folding dramatically reduces the area required, thus significantly reducing leakage power, logic folding significantly reduces total power consumption compared to no-folding case. Level-1, level-2 and level-4 folding, on an average, require 49, 48, and 52% of the power consumption, respectively, compared to the no-folding case.

12 Conclusions

In this chapter, we presented a novel high-performance run-time reconfigurable architecture, called NATURE. It exploits recently proposed high-density, high-speed nano RAMs in the architecture to enable cycle-by-cycle reconfiguration and logic folding. The choice of different folding levels allows the designer flexibility in performing area-delay trade-offs. The significant increase in relative logic density (more than an order of magnitude for large circuits) made possible by NATURE can allow the use of less expensive reconfigurable architectures with smaller logic capacities to implement the same functionality, thus giving a boost to their use in cost-conscious embedded systems. Power estimation results show that logic folding can also significantly reduce the power consumption.

To fully exploit the features of NATURE, a tool to automatically map different functionalities to NATURE, called NanoMap, has been implemented. It incorporates temporal logic folding during the logic mapping, temporal clustering and placement steps. It can automatically select the best folding level and use FDS to balance resources across different folding stages. The mapping can be targeted at various optimization objectives and design constraints, providing significant design flexibility.

Exercises

1. How is fine-grain run-time reconfiguration enabled in NATURE, and why is such a reconfiguration difficult in current reconfigurable architectures?
2. Why are reconfigurable architectures favored for nanoelectronic designs?
3. What salient features are required in a nano RAM for it to be useful as on-chip reconfiguration memory in NATURE?
4. What is the concept of logic folding and what are its benefits?
5. Discuss the area-delay trade-offs involved in the use of different folding levels.
6. Given the circuit shown in the figure below, determine the minimum LUT usage and the number of folding cycles required for (a) level-1 folding, and (b) level-2 folding.

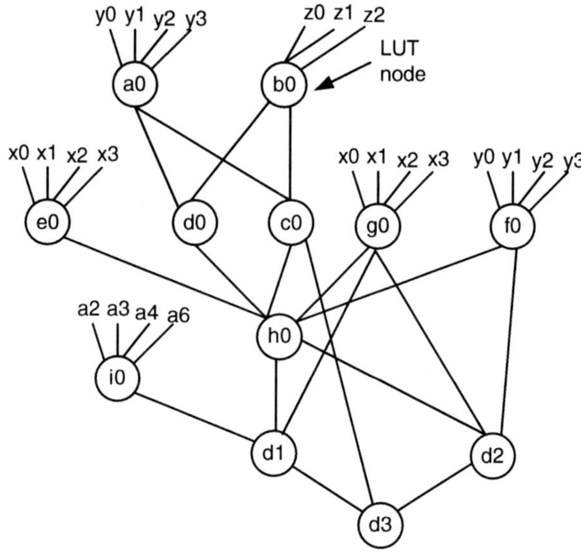

7. What is a folding cycle composed of? For the circuit mapping shown in the figure below, how can the clock cycle be determined?

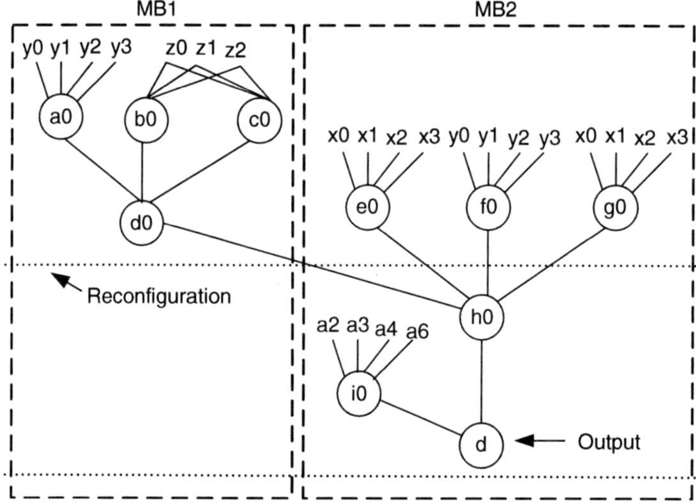

8. Suppose there are $k = 32$ copies of reconfiguration bits stored for each programmable element. The SMB architecture instance has $n_1 = 4$, $n_2 = 4$, $m = 4$, and $l = 2$, as shown in Figs. 7 and 8. Calculate the total number of

reconfiguration bits required for an SMB (not including the switch matrix that connects it with outside interconnect), and the logic density improvement compared with the no-folding case.

9. Suppose there are 40 tracks in each horizontal and vertical interconnect channel and an LB can connect to all four channels around it. The interconnect hierarchy includes length-1, length-4 and long wires with a distribution of 30, 60, and 10% for each type, respectively. The connection box is unpopulated. F_c is set to 0.2. How many tracks of each type of wire does an input/output of the LB connect to?

10. What are the main steps in the NanoMap flow? Specify which step the following operations are performed in. (1) Computing the folding level according to the optimization objectives and constraints. (2) Assigning LUTs to appropriate LEs. (3) Assigning each LUT/LUT cluster to appropriate folding stages. (4) Placing the SMB in a suitable location in NATURE. (5) Connecting inputs/outputs between SMB blocks in NATURE. (6) Packing LEs into MB and SMBs. (7) Partitioning RTL modules into LUT clusters based on the selected folding level.

11. Suppose the objective is to optimize the circuit delay with an area constraint of 30 LEs. An LE contains one LUT and four flip-flops. The circuit contains three planes. The maximum number of LUTs in one plane is 200 and the maximum plane depth is 14. There are a maximum of four 16-bit registers per plane that need to be mapped. The number of reconfigurable copies on-chip is 32. Compute the initial folding level for the optimization objective.

12. Discuss the power consumption characteristics for the logic folding and no-folding cases.

13. Suppose the circuit mapping results are as shown in Fig. 5, and the SMB architecture is as shown in Figs. 7 and 8. Suppose the timing conditions are as follows. (a) The reconfiguration time is 200 ps. (b) The LUT computation time is 150 ps. (c) The flip-flop setup time is 30 ps. (d) The local crossbar delay in an MB is 180 ps. (e) The MUX delay in an SMB is 240 ps. Calculate the clock period for folding level-1 and level-2 and their total circuit delay.

14. According to the discussions in the chapter, if the number of inputs/outputs per LE increases, the SMB size increases quickly. For the SMB and LE architectures shown in Figs. 7, 8 and 9, if four flip-flops are included in an LE, calculate the number of inputs/outputs per SMB, and the size of the crossbar at the input of an LE and the MUX at the input of an MB. Also, how many MUXes are needed in an SMB?

15. From Exercise 14, we can see that the SMB size grows quickly with an increase in the number of flip-flops in an LE. In order to limit the SMB size increase when the number of flip-flops in an LE increases, the LE architecture in Fig. 11 can be used. Suppose the SMB architecture shown in Figs. 7 and 8 and LE architecture shown in Fig. 11 are used. Each LE contains four flip-flops. The number of inputs and outputs per LE are still limited to 6 and 3, respectively. Calculate how many reconfiguration bits per copy are needed for an SMB?

16. Consider the circuit shown below. Compute its DG for LUT computation for both level-1 and level-2 folding cases. Suppose the depth of the clusters is the same as the specified folding level and their weights are 4.

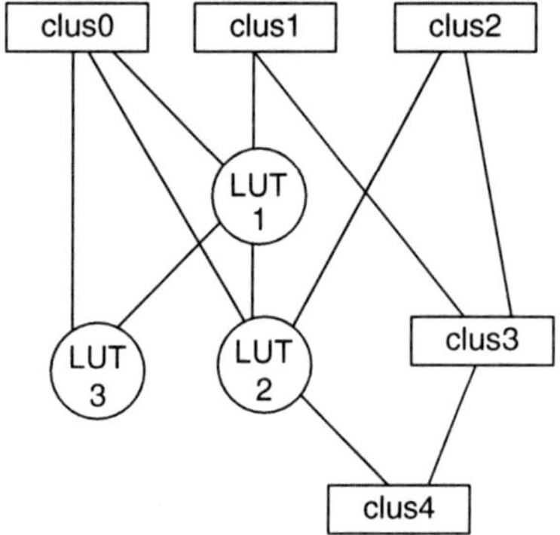

17. Suppose the circuit shown below needs to be mapped to NATURE. FDS is performed to schedule the LUTs into appropriate folding levels for level-1 folding. Suppose after scheduling clus1-clus5, the DG graph is as follows: DG (1) = 5, DG(2) = 6, DG(3) = 2, and DG(4) = 6. Suppose the next node to be scheduled is LUT2. Then if only considering the force for LUT computation, which clock cycle should LUT2 be scheduled in?

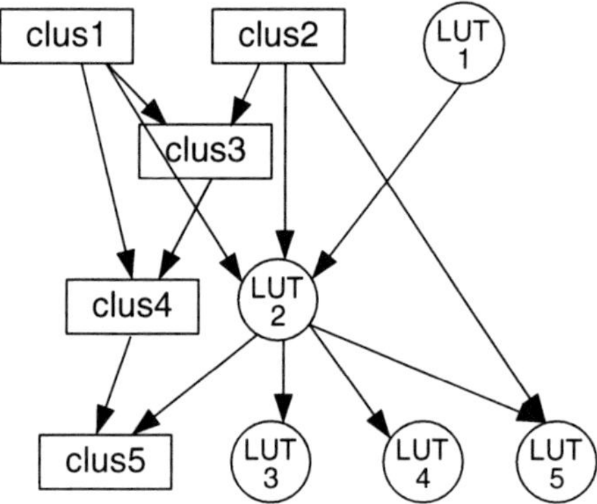

18. As discussed in the chapter, in the temporal clustering step, the maximum slack of each LUT needs to be computed to obtain the criticality of the LUT. Consider the circuit shown below. The required arrival time is 7 at the output of LUT10. Assuming the actual arrival time is 1 at the outputs of LUT1–LUT3, compute the minimum slack of LUT6 and LUT7.

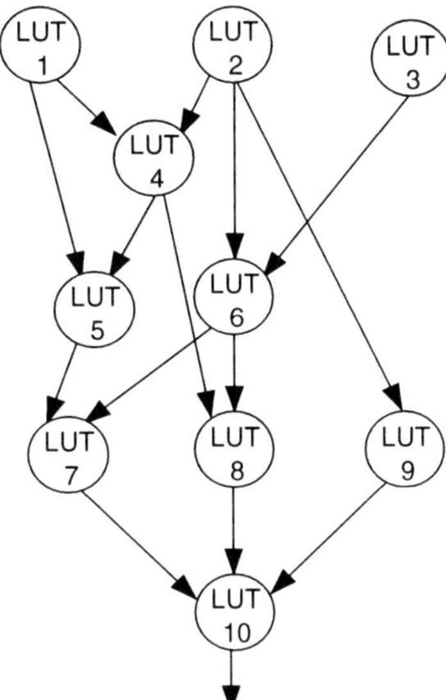

Required arrival time: 7

References

1. Betz, V. and Rose, J. 1997. VPR: A new packing, placement and routing tool for FPGA research. In *Proceedings of the International Workshop on Field-Programmable Gate Arrays.* 213–222.
2. Betz, V. and Rose, J. 1998. How much logic should go in an FPGA logic block. *IEEE Des. Test Comput. 15*, 10–15.
3. Betz, V. and Rose, J. 1999. FPGA routing architecture: Segmentation and buffering to optimize speed and density. In *Proceedings of the International Symposium on Field Programmable Gate Arrays.* 59–68.
4. Bozorgzadeh, E., Memik, S. O., Yang, X., and Sarrafzadeh, M. 2004. Routability-driven packing: Metrics and algorithms for cluster-based FPGAs. *J. Circ. Syst. Comput. 13*, 77–100.
5. Burke, P. J. 2003. An RF circuit model for carbon nanotubes. *IEEE Trans. Nanotechnol. 2*, 55–58.
6. Butts, M., Dehon, A., and Goldstein, S. C. 2002. Molecular electronics: Devices, systems and tools for gigagate, gigabit chips. In *Proceedings of the International Conference on Computer-Aided Design.* 433–440.
7. Capasso, F. and Kiehl, R. A. 1985. Resonant tunneling transistor with quantum well base and high-energy injection: A new negative differential resistance device. *J. Appl. Phys. 58*, 1396.
8. Chow, P., Seo, S. O., Rose, J., Chung, K., Paez–Monzon, G., and Rahardja, I. 1999. The design of an SRAM-based field-programmable gate array – Part I: Architecture. *IEEE Trans. VLSI Syst. 7*, 191–197.
9. Cong, J. 1996. Combinational logic synthesis for LUT based field-programmable gate arrays. *ACM Trans. Des. Automat. Electron. Syst. 1*, 145–204.
10. Cong, J. and Ding, Y. 1994. FlowMap: An optimal technology mapping algorithm for delay optimization in lookup-table-based FPGA designs. *IEEE Trans. Comput.-Aid. Des. 13*, 1–12.
11. Cui, Y., Zhong, Z., Wang, D., Wang, W. U., and Lieber, C. M. 2003. High performance silicon nanowire field effect transistors. *Nano Lett. 3*, 149–152.

12. DeHon, A. 1996. Dynamically programmable gate arrays: A step toward increased computational density. In *Proceedings of the 4th Canadian Workshop of Field-Programmable Devices*. 47–54.

13. DeHon, A. 2006. 3D nanowire-based programmable logic. In *Proceedings of the International Conference on Nano-Networks*. 1–5.

14. DeHon, A. and Rubin, R. 2004. Design of FPGA interconnect for multilevel metallization. *IEEE Trans. VLSI Syst. 12*, 1038–1050.

15. DeHon, A. and Wilson, M. J. 2004. Nanowire-based sublithographic programmable logic arrays. In *Proceedings of the International Symposium on Field Programmable Gate Arrays*. 123–132.

16. Fukµmoto, Y., Nebashi, R., Mukai, T., Tsuji, K., and Suzuki, T. 2008. Toggle magnetic random access memory cells scalable to a capacity of over 100 megabits. *Am. Inst. Phys. 103*, 40–48.

17. Ghosh, I., Raghunathan, A., and Jha, N. K. 1999. Hierarchical test generation and design for testability methods for ASPPs and ASIPs. *IEEE Trans. Comput.-Aid. Des. 18*, 357–370.

18. Goldstein, S. C. and Budiu, M. 2001. Nanofabrics: Spatial computing using molecular nanoelectronics. In *Proceedings of the International Symposium on Computer Architecture*. 178–189.

19. Goldstein, S. C., Schmit, H., Budiu, M., Cadambi, S., Moe, M., and Taylor, R. R. 2000. PipeRench: A reconfigurable architecture and compiler. *IEEE Comput. 33*, 70–77.

20. Ha, D. and Kim, K. 2007. Recent advances in high-density phase change memory (PRAM). In *Proceedings of the Conference on VLSI Technology, System and Applications*. 1–4.

21. Hauck, S., Fry, T. W., Hosler, M. M., and Kao, J. P. 2004. The Chimaera reconfigurable functional unit. *IEEE Trans. VLSI Syst. 12*, 206–217.

22. ITC. 1999. ITC'90 benchmarks. http://www.eerc.utexas.edu/itc99–benchnmarks/bench.html.

23. ITRS. 2007. International Technology Roadmap for Semiconductors. http://public.itrs.net.

24. Javey, A., Guo, J., Farmer, F. B., Wang, Q., and Wang, D. 2004. Carbon nanotube field-effect transistors with integrated ohmic contacts and high-k gate dielectrics. *Nano Lett. 4*, 447–450.

25. Kao, J., Naren, S., and Chandrakasan, A. 2002. Subthreshold leakage modeling and reduction techniques. In *Proceedings of the International Conference on Computer-Aided Design*. 141–148.

26. Kim, J.-H., Lee, J.-W., Lee, S.-J., and Shin, H. 2002. Macro model and sense amplifier for an MRAM. *J. Korean Phys. Soc. 41*, 896–901.

27. Lai, S. 2003. Current status of the phase change memory and its future. In *Proceedings of the International Electronic Devices Meeting*. 10.1.1–10.1.4.

28. Lee, K.-J., Cho, B.-H., Cho, W.-Y., Kang, S., and Choi, B.-G. 2007. A 90 nm 1.8 V 512 Mb diode-switch PRAM with 266MB/s read throughput. In *Proceedings of the IEEE International Solid-State Circuits Conference*. 472–473.

29. Lisanke, R. 1988. Logic synthesis and optimization benchmarks. Tech. rep., Microelectronics Center of North Carolina.

30. Marquardt, A. S., Betz, V., and Rose, J. 1999. Using cluster-based logic blocks and timing-driven packing to improve FPGA speed and density. In *Proceedings of the International Symposium on FPGAs*. 37–46.

31. Marquardt, A. S., Betz, V., and Rose, J. 2000. Timing-driven placement for FPGAs. In *Proceedings of the International Symposium on FPGA*. 203–213.

32. Mei, B., Vernalde, S., Verkest, D., Man, H. D., and Lauwereins, R. 2003. ADRES: An architecture with tightly coupled VLIW processor and coarse-grained reconfigurable matrix. In *Proceedings of the International Conference on Field-Programmable Logic and Applications*. 61–70.

33. Muttreja, A., Ravi, S., and Jha, N. K. 2008. Variability-tolerant register-transfer level synthesis. In *Proceedings of the International Conference on VLSI Design*. 621–628.

34. NANTERO. 2008. Nantero. http://www.nantero.com.

35. Paulin, P. G. and Knight, J. P. 1989. Force-directed scheduling for the behavioral synthesis of ASIC's. *IEEE Trans. Comput.-Aid. Des. 8*, 661–679.
36. Rose, J., Gamal, A. E., and Sangiovanni-Vincentelli, A. 1993. Architecture of field-programmable gate arrays. *Proc. IEEE 81*, 1013–1029.
37. Rueckes, T., Kim, K., Joselevich, E., Tseng, G., Cheung, C., and Lieber, C. M. 2000. Carbon nanotube-based nonvolatile random access memory for molecular computing. *Science 289*, 94–97.
38. Salamon, D. and Cockburn, B. F. 2003. An electrical simulation model for the chalcogenide phase change memory cell. In *Proceedings of the International Workshop on Memory Technology, Design and Testing*. 86–91.
39. Sarkar, J. 2007. Evolution of phase change memory characteristics with operating cycles: Electrical characterization and physical modeling. *Appl. Phys. Lett. 91*, 89–93.
40. Shang, L., Kaviani, A. S., and Bathala, K. 2002. Dynamic power consumption in Virtex-II FPGA family. In *Proceedings of the FPGA Conference*. 157–164.
41. Smith, R. F., Rueckes, T., Konsek, S., Ward, J. W., and Brock, D. K. 2007. Carbon nanotube based memory development and testing. In *Proceedings of the Aerospace Conference 1–5*.
42. Snider, G., Kuekes, P., and Williams, R. S. 2004. CMOS-like logic in defective, nanoscale crossbars. *Nanotechnology 15*, 881–891.
43. Stix, G. 2005. Nanotubes in the clean room. *Sci. Am.* 82–85.
44. Strukov, D. B. and Likharev, K. K. 2005. CMOL FPGA: A reconfigurable architecture for hybrid digital circuits with two-terminal nanodevices. *Nanotechnology 16*, 888–900.
45. Sugibayashi, T., Honda, T., Sakimura, N., Tahara, S., and Kasai, N. 2007. MRAM applications using unlimited write endurance. *IEEE Trans. Electron 10*, 1936–1940.
46. SYNOPSYS. 2009. Synopsys. http://www.synopsys.com.
47. Tehrani, S., Slaughter, J. M., Deherrera, M., Engel, B. N., and Rizzo, N. D. 2003. Magnetoresistive random access memory using magnetic tunnel junctions. *Proc. IEEE 91*, 703–714.
48. Lingappan, L., Ravi, S., and Jha, N. K. 2006. Satisfiability-based test generation for non-separable RTL controller-datapath circuits. *IEEE Trans. Comput.-Aid. Des.*, 25 544–557.
49. Trimberger, S., Carberry, D., Johnson, A., and Wong, J. 1997. A time-multiplexed FPGA. In *Proceedings of the Symposium on FPGAs for Custom Computing Machines*. 22–28.
50. Tu, D., Liu, M., and Haruehanroengra, S. 2007. Three-Dimensional CMOL: Three-dimensional integration of CMOS/nanomaterial hybrid digital circuits. *Micro Nano Lett. 2*, 40–45.
51. Wang, J. P. and Meng, H. 2007. Spin torque transfer structure with new spin switching configurations. *Eur. Phys. J. B 59*, 471–474.
52. Zhang, W. and Jha, N. K. 2005. ALLCN: An automatic logic-to-layout tool for carbon nanotube based nanotechnology. In *Proceedings of the International Conference on Computer Design*. 281–288.
53. Zhang, W., Jha, N. K., and Shang, L. 2006. NATURE: A hybrid nanotube/CMOS dynamically reconfigurable architecture. In *Proceedings of the Design Automation Conference*. 711–716.
54. Zhang, W., Jha, N. K., and Shang, L. 2009. A hybrid nano/CMOS dynamically reconfigurable system – Part I: Architecture. *ACM J. Emerg. Technol. Comput. Syst. 5*, 16.1–16.30.
55. Zhang, W., Shang, L., and Jha, N. K. 2007. NanoMap: An integrated design optimization flow for a hybrid nanotube/CMOS dynamically reconfigurable architecture. In *Proceedings of the Design Automation Conference*. 300–305.
56. Zhang, W., Jha, N. K., and Shang, L. 2009a. A hybrid nano/CMOS dynamically reconfigurable system – Part II: Design optimization flow. *ACM J. Emerg. Technol. Comput. Syst. 4*, 13.1–13.31.
57. Zhang, W., Jha, N. K., and Shang, L. 2009b. Design space exploration and data memory architecture design for a hybrid nano/CMOS dynamically reconfigurable architecture. *ACM J. Emerg. Technol. Comput. Syst. 5*, 17.1–17.27.
58. Zhong, L. and Jha, N. K. 2005. Interconnect-aware low-power high-level synthesis. *IEEE Trans. Comput.-Aid. Des. Integ. Circ. Syst. 24*, 336–351.

Reliable Circuits Design with Nanowire Arrays

M. Haykel Ben Jamaa and Giovanni De Micheli

1 Introduction

The emergence of different fabrication techniques of *silicon nanowires (SiNWs)* raises the question of finding a suitable architectural organization of circuits based on them. Despite the possibility of building conventional CMOS circuits with SiNWs, the ability to arrange them into *regular arrays*, called *crossbars*, offers the opportunity to achieve higher integration densities. In such arrays, molecular switches or phase-change materials are grafted at the *crosspoints, i.e.,* the crossing nanowires, in order to perform computation or storage. Given the fact that the technology is not mature, a hybridization of CMOS circuits with nanowire arrays seems to be the most promising approach.

This chapter addresses the impact of variability on the nanowires in circuit designs based on the hybrid CMOS-SiNW crossbar approach. A large part of this chapter has been published in [1].[1] The variability stemming from the shrinking nanowire dimensions is modeled and its impact on the interface between the CMOS circuit and the nanowire arrays, the *decoder*, is investigated. The approach presented is based on the abstract representation of nanowires as a sequence of codes. Based on the impact of variability on codes, optimized design methodologies for encoding the nanowires and for testing the array decoder are derived.

2 Fabrication Technologies

Nanowire crossbars have attracted increasing interest over the last few years because the fabrication techniques have become more mature and versatile. Parallel research works have been carried out at different levels of the IC design hierarchy, ranging from device to circuit and system level, in order to identify and address the

---

[1] The authors would like to acknowledge those who contributed to [1].

G.D. Micheli (✉)
Institute of Electrical Engineering, EPFL, Lausanne, Switzerland
e-mail: giovanni.demicheli@epfl.ch

N.K. Jha and D. Chen (eds.), *Nanoelectronic Circuit Design*,
DOI 10.1007/978-1-4419-7609-3_5, © Springer Science+Business Media, LLC 2011

challenges facing the utilization of this emerging paradigm in the future. Circuit design depends on properties of the fabrication techniques. Thus, understanding the fabrication techniques and device properties enables a better assessment of the global problem. In the following discussion, we survey the different fabrication techniques for bare nanowires and nanowire crossbars.

2.1 Nanowire Fabrication Techniques

The existing nanowire fabrication techniques follow two main paradigms: the so-called bottom-up and top-down approaches. Bottom-up approaches are based on the growth of nanowires from nanoscale metallic catalysts. In contrast, top-down approaches use various types of patterning techniques.

2.1.1 Bottom-Up Techniques

One of the widely used bottom-up techniques is the *vapor–liquid–solid (VLS)* process, in which the generally very slow adsorption of a silicon-containing gas phase onto a solid surface is accelerated by introducing a catalytic liquid alloy phase. The latter can rapidly adsorb vapor to a supersaturated level; then crystal growth occurs from the nucleated catalytic seed at the metal–solid interface. Crystal growth with this technique was established in the 1960s [2] and silicon nanowire growth is today mastered with the same technique. A related technique to VLS is the laser-assisted catalytic growth. The silicon-containing gas is generated by irradiating a Si substrate with high-powered, short laser pulses [3]. On the other hand, the *chemical vapor deposition (CVD)* method uses materials that can be evaporated at moderate temperatures [4, 5].

2.1.2 Top-Down Techniques

The top-down fabrication approaches have in common the utilization of CMOS steps or hybrid steps that can be integrated into a CMOS process, while keeping the process complexity low and the yield high enough. They also have in common the ability to define the functional structures (nanowires) directly onto the functional substrate.

Standard photolithography techniques: These techniques use standard photolithography to define the position of the nanowire. Then, by using smart processing techniques, including the accurate control of the etching, oxidation and deposition of materials, it is possible to scale the dimensions down far below the photolithographic limit [6–11].

Miscellaneous mask-based techniques: The electron-beam lithography [12, 13] offers a higher resolution below 20 nm than standard photolithography.

It, however, has a lower throughput. The highest resolution can be achieved by using *extreme ultraviolet interference lithography (EUV-IL)* [14]. However, this approach needs a highly sophisticated setup in order to provide the required EUV wavelength.

The stencil technique [15] is a different approach that requires no photoresist patterning. It is based on the definition of a mask that is fully open in the patterned locations. The mask is subsequently clamped onto the substrate, and the material to be deposited is evaporated or sputtered through the mask openings onto the substrate.

Spacer techniques: The spacer technique is based on the idea of transforming thin lateral dimensions, in the range of 10–100 nm, into a vertical dimension by means of an anisotropic etch of the deposited materials. In [16], spacers with a thickness of 40 nm were demonstrated with a line-width roughness of 4 nm and a low variation across the wafer. The nanowire count can be duplicated by using the spacers themselves as sacrificial layers for a following set of spacers [17].

Nanomold-based techniques: Alternative techniques use *nanoimprint lithography (NIL)*, which is based on a mold with nanoscale features [18] that is pressed onto a resist-covered substrate in order to pattern it. The substrate surface is scanned by the nanomold in a stepper fashion. The as-patterned polymer resist is processed in a similar way to photolithographically patterned photoresist films. The *super-lattice nanowire pattern transfer technique (SNAP)* and the *planar edge defined alternate layer (PEDAL)* [19] are examples of NIL.

2.2   Crossbar Technologies

The previously surveyed techniques yield parallel or mashed nanowires. In order to arrange them into arrays that are generally called *crossbars*, additional techniques are required and will be explained in the following.

2.2.1   Crossbars with Bottom-Up Nanowires

Nanowires fabricated with bottom-up processes have the property of generally being grown on a different substrate from the functional one. Consequently, they need to be dispersed into a solution and transferred onto the substrate to be functionalized. The iteration of the transfer operations with different directions may lead to a crossbar structure [20]. It has been demonstrated that the application of an electrical field to the substrate improves the directionality of the assembled nanowires [21, 22]. The approach is, however, limited by the electrostatic interference between nearby electrodes, and the requirement for an extensive lithography to fabricate the electrodes [20].

2.2.2   Nanomold-Based Nanowire Crossbars

NIL was used in [23, 18] in order to define two orthogonal layers of metallic nanowires. First, the nanomold was fabricated by electron-beam lithography and *reactive ion etching (RIE)* of a SiO_2-covered silicon substrate. The mold was then pressed onto a spin-coated polymer to define a lift-off mask for Ti/Pt nanowires. A layer of molecular switches, [2]rotaxane, was deposited over the entire substrate using the *Langmuir–Blodgett (LB)* method [24]. Then, the fabrication of the top Ti/Pt nanowire layer was performed in a similar way as explained for the lower nanowire layer. High-density crossbars were also demonstrated with the SNAP technique that was explained in Section. 2.1 [25], yielding 160-kb molecular memories with a density up to 10^{11} bit/cm^2 [26].

2.2.3   Crossbar Switches

Many attempts have been carried out in the last few decades to design molecules comprising a donor-(σ bridge)-acceptor, which have an asymmetric behavior, allowing the current to flow in a preferential direction [27–29]. Another class of switching molecules is represented by bistable molecules, such as [2]rotaxanes, pseudorotaxanes and [2]catenanes. They consist of two mechanically interlocked, or threaded, components having two stable states and can be switched between them when the appropriate bias voltage is applied [30, 31]. Other research groups have focused on phase change materials as a switching material at the nanowire crosspoints [32] operating as diodes.

3   Architecture of Nanowire Crossbars

Nanowire crossbars are defined on a scale that can be far below the lithographic limit. The ability to hybridize CMOS technology with the previously surveyed nanowire techniques (Section. 2) promotes the organization of the overall system in a regular way, where globally CMOS parts are operational, while locally nanowire crossbars are used. This raises the questions of the way in which crossbars should be connected to the outer CMOS circuit on the one hand, and the type of functions that crossbars can execute on the other hand. This section introduces the crossbar organization, surveying some emerging crossbar architectures and focuses on the design of the decoder.

3.1   Organization of Nanowire Crossbars

The baseline organization of a nanowire crossbar circuit is depicted in Fig. 1a. An arrangement of two orthogonal layers of parallel nanowires defines a regular grid of intersections called crosspoints. The separation between the two layers can be filled

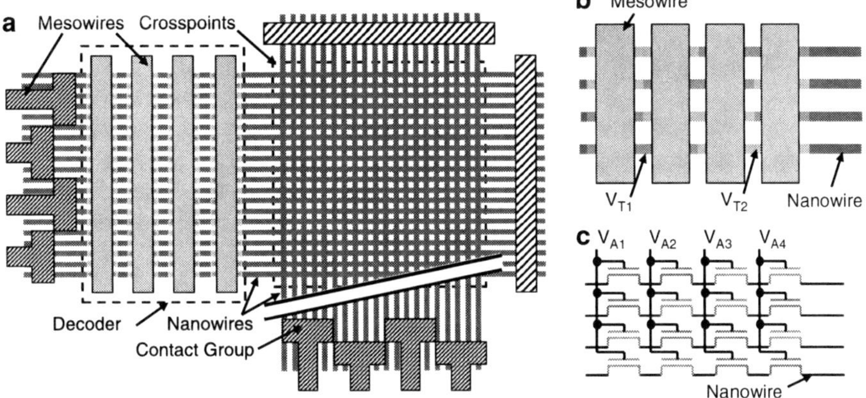

Fig. 1 Baseline organization of a crossbar circuit and its decoder. (**a**) Architecture of a crossbar circuit. (**b**) Decoder layout. (**c**) Decoder circuit design

with phase change material or molecular switches at the crosspoints. Information storage, interconnection or computation can be performed with these crosspoints [30, 33]. A set of contact groups is defined on top of the nanowires. Every contact group makes an ohmic contact to a corresponding distinct set of nanowires, which represents the smallest set of nanowires that allows contacts with lithographically defined lines, called *mesowires (MWs)*. The *mesoscale* corresponds in the context of this chapter to the lithography scale; while the *nanoscale* corresponds to the sublithographic scale.

This configuration bridges every set of nanowires within a contact group to the outer CMOS circuit. In order to *fully bridge the scales* and make every nanowire within this set uniquely addressable by the outer circuit, a decoder is needed. It is formed by a series of transistors along the nanowire body, controlled by the mesowires and having different threshold voltages V_{th} (Fig. 1b). The distributions of V_{th}'s is called the *nanowire pattern*. Depending on this pattern and the pattern of applied voltages in the decoder (V_A's), one single nanowire in the array can be made conductive (Fig. 1c). In this case, this nanowire is said to be addressed by the applied voltage pattern.

It is possible to think of replacing each transistor at the diagonal crosspoints by an ohmic contact and to eliminate all other transistors; thus, mapping each horizontal wire onto a vertical one. However, this method is technologically difficult, because the nanowire pitch is defined below the photolithographic limit, which justifies the proposed decoder design.

3.2 Architectures Based on Nanowire Crossbars

Before the emergence of the crossbar architecture, many experiments were performed with a massively parallel computer built at Hewlett-Packard laboratories, the Teramac [34], in state-of-the-art CMOS technology. Despite the high defect rate affecting single components in the Teramac, the approach seemed to be efficient, resulting in $100\times$ faster robust operation than a high-end single-processor computer in some configurations. The required architectural elements are a large number of computing instances, parallelism of their operation and high bandwidth. Since these elements naturally exist in the crossbar architecture, this architectural paradigm emerged as a possible approach for reliable massively parallel computing with highly defective basic components [35], where molecular devices can perform a logic operation or information storage at the crosspoints.

Some crossbar prototypes were fabricated with different sizes [18, 26, 30], and the basic function that these prototypes implemented is information storage. Crossbars implementing computational units, such as the *nanoBlock* [36, 37], are also conceptually possible. However, they need restoration stages and latches that can be implemented using *resonant tunneling diodes (RTDs)* or by hybridizing crossbars with CMOS. The CMOS part can also provide the necessary gain and input/output interface. It is not excluded that the CMOS part performs more functions than the crossbars in a hybrid architecture, however, the parallelism, reconfigurability and high connectivity will be the main advantages provided by crossbars owing to their matrix form, in addition to their ability to scale down below the limit imposed by photolithography.

The *nanoPLA* architecture is a concept based on semiconducting SiNWs organized in a crossbar fashion with molecular switches at their crosspoints. The switches can be programmed in order to perform either signal routing or wired-OR logic function. The input of the crossbar represents a decoder, which is used in order to uniquely address every nanowire independently of the others. The decoder design assumes that the nanowires are differentiated by a certain doping profile [38]. This will be explained in more detail in Sect. 3.3. The output of the crossbar is routed to a second crossbar, in which the signals can be inverted by gating the nanowires carrying the signals. A cascade of these two planes is equivalent to a NOR plane [39]. Two back-to-back NOR planes can implement any logic function in two-level form.

3.3 Decoding Nanowires

The decoder is the element of the crossbar circuit that bridges the meso- to nanoscale. Even though the structure of the decoder circuit is simple, its reliable fabrication and design are challenging. The need to use different transistors necessitates different doping levels in specific regions on the nanowires whose location cannot be controlled precisely because the nanowire scale is below the lithographic limit. Thus, nanowires that are already doped during the fabrication process may simplify the task. We distinguish, therefore, between fabrication techniques that yield differentiated and those that yield undifferentiated nanowires. Differentiated nanowires have a certain doping profile; they are generally fabricated using a bottom-up approach and the doping profile is defined during nanowire growth. Undifferentiated nanowires have no specific doping profile; they are generally fabricated using a top-down approach.

3.3.1 Decoders for Differentiated Nanowires

Differentiated NWs have an axial or radial doping profile which is defined during the NW growth process. An axial decoder was presented in [40], in which the distribution of the V_{th}'s is fully random. The NWs are dispersed parallel to each other and they are addressable when they have different V_{th} patterns. The probability that their addresses are different may be increased by increasing the number of addressing wires. On the other hand, the radial decoder [41] relies on NWs with several radial doping shells. The remaining shells after a sequence of etchings depends on the etching order in every region. The suite of shells along the NW after all etching steps defines the NW patterns. While both axial and radial decoders require the same estimate of the number of MWs needed to address the available NWs; the radial decoder has the advantage of being less sensitive to misalignment of NWs. To address N NWs, M MWs are needed, $M = [2.2 \times \log_2(N)] + 11$. With these dimensions, the decoders address every nanowire with a probability greater than 99%.

3.3.2 Decoders for Undifferentiated Nanowires

On the other hand, for undifferentiated nanowires, namely those fabricated in a top-down process, a mask-based decoder was presented in [42] and its ability to control undifferentiated NWs was proven. The MWs are separated from the NWs by a nonuniform oxide layer: in some locations, a high-κ dielectric is used, in others, a low-κ dielectric. The high-κ dielectric amplifies the electric field generated by the MWs relatively to the low-κ dielectric. Consequently, the field-effect control by the MWs happens only at the NW regions lying under the high-κ dielectric. The oxide mask is lithographically defined; making the decoder dependent on lithography limits. In order to address N nanowires, the mask-based decoder necessitates the use of $M = 2 \times \log_2(N) + \varepsilon$ mesowires, with ε a small constant ≥ 1, which depends on the fabrication technique and the degree of redundancy to be achieved.

For undifferentiated NWs, a random contact decoder has been presented in [43, 44]. Unlike the other decoders for which the NW codes are among a known set of codes, the connections established between MWs and NWs for this decoder are fully random. It results from a deposition of gold particle onto the NWs, where the only controlled parameter is the density of particles. In order to control each of the N NWs uniquely with a high probability, $M = 4.8 \times \log_2(N) + C$ mesowires are needed, with C being a large constant that depends on the design parameters.

4 Decoder Logic Design

The sequence of V_{th}'s along every NW defining the NW pattern associates a unique code word with the NW that can activate it. Previously explored nanowire encoding schemes, *i.e.*, codes, are binary. The code length impacts the decoder size and the overall crossbar area. It is, therefore, interesting to investigate the benefits of reducing the code length by using multivalued logic (MVL) codes. The generalization of the usual codes to MVL produces novel code families that have not been explored before. In this section, the construction rules for new code families are presented. Defects that can affect them are modeled. Then, the fault tolerance of the considered codes and their impact on the crossbar circuit in terms of reliability and area are investigated.

4.1 Semantic of Multi-valued Logic Addressing

In the following discussion, we generalize the notion of encoding to multiple-valued bits by first defining some basic relations needed to identify possible codes. Some basic concepts used in coding theory are generalized from the binary definitions stated in [45] to multiple-valued logic. The matching of a code word and its pattern corresponds here to conduction. Before introducing the impact of defects, we consider the code (Ω) and pattern (A) spaces to be identical, realizing a one-to-one mapping between each other. Algebraic operations are performed as defined in the ring of integers.

Definition 1. *A multiple-valued pattern* **a**, or simply a pattern **a**, is a suite of M digits a_i, in the n-valued base $\mathbb{B}$; *i.e.*, $\mathbf{a} = (a_0,\ldots, a_{M-1}) \in \mathbb{B}^M$, $\mathbb{B} = \{0,\ldots, n-1\}$. A multiple-valued *code word* **c**, or simply a code word **c**, is defined the same way as a pattern.

A pattern represents a serial connection of M transistors in the silicon nanowire core; each digit a_i of the code word represents a threshold voltage $V_{th,i}$, with the convention $a_i < a_j \Leftrightarrow V_{th,i} < V_{th,j}$, $\forall i, j = 0,\ldots, M-1$. An analogous equivalence holds for $a_i = a_j$ and, Consequently, for $a_i > a_j$. This convention is equivalent to discretizing the M values of V_{th} and ordering them in an increasing order. In Fig. 2a, b, we illustrate the pattern 002120 representing the V_{th} sequence (0.2 V, 0.2 V, 0.6 V, 0.4 V, 0.6 V, 0.2 V).

A code word represents the suite of applied voltages V_A at the M mesowires. These are defined such that every $V_{A,i}$ is slightly higher than $V_{th,i}$, and lower than $V_{th,i+1}$ (Fig. 2c, d).

Definition 2. *A complement* of digit x_i in a code word or pattern **x** is defined as: $\text{NOT}(x_i) = \overline{x_i} = (n-1) - x_i$. The operator NOT can be generalized to vector **x**, acting on each component as defined above. Note that $\text{NOT}(\text{NOT}(\mathbf{x})) = \mathbf{x}$.

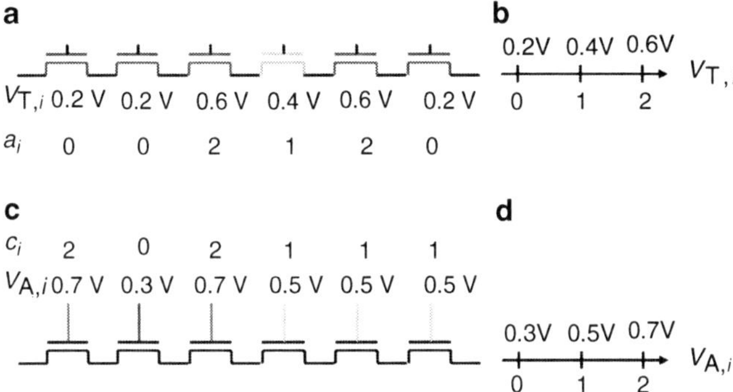

Fig. 2 Mapping of threshold and applied voltages onto discretized values. (**a**) Pattern 002120 and its V_{th} sequence. (**b**) Discretization of V_{th} values. (**c**) Code word 202111 and its V_A sequence. (**d**) Discretization of V_A values

Definition 3. *A pattern* **a** *is covered* by a code word **c** if and only if the following relation holds: $\forall i = 0,\ldots, M - 1, c_i \geq a_i$. By using the sigmoid function:

$$\sigma(x) = \begin{cases} 0 & x \leq 0 \\ 1 & x > 0 \end{cases}$$

generalized to vectors: $\sigma(\mathbf{x}) = (\sigma(x_0),\ldots, \sigma(x_{M-1}))$, the definition above becomes: **a** is covered by $\mathbf{c} \Leftrightarrow \|\sigma(\mathbf{a} - \mathbf{c})\| = 0$. Alternatively, we can define the order relations on vectors **c** and **a**:

$$\mathbf{c} < \mathbf{a} \Leftrightarrow \forall i, \quad c_i < a_i$$
$$\mathbf{c} > \mathbf{a} \Leftrightarrow \forall i, \quad c_i > a_i.$$

The relation becomes relaxed (*i.e.*, $\leq$ or $\geq$) if there exists i such that $c_i = a_i$. Then, a pattern **a** is covered by a code word **c** if and only if $\mathbf{a} \leq \mathbf{c}$. The same definition for covering can be generalized to two patterns or two code words.

Covering a given pattern with a certain code word is equivalent to applying a suite of gate voltages making every transistor conductive. Then, the nanowire is conducting and we say that it is controlled by the given sequence of gate voltages. Figure 3a illustrates the case in which the code word covers the pattern and the nanowire is conducting, while Fig. 3b illustrates the opposite case.

Definition 4. *A pattern* **a** *implies* a pattern **b** if and only if $\|\sigma(\mathbf{b} - \mathbf{a})\| = 0$; *i.e.*, **b** is covered by **a**. We note this as follows: $\mathbf{a} \Rightarrow \mathbf{b}$. Since a one-to-one mapping between the patterns and codes was assumed, we generalize this definition to code words: $(\mathbf{c}^a \Rightarrow \mathbf{c}^b) \Leftrightarrow \|\sigma(\mathbf{c}^b - \mathbf{c}^a)\| = 0$; *i.e.*, $\mathbf{c}^b$ is covered by $\mathbf{c}^a$.

This means that if a nanowire with pattern **a** corresponding to code word $\mathbf{c}^a$ is covered by a code word $\mathbf{c}^*$, then the nanowire with pattern **b** corresponding to code

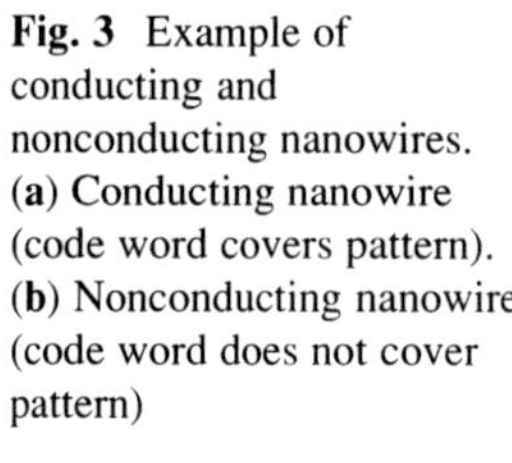

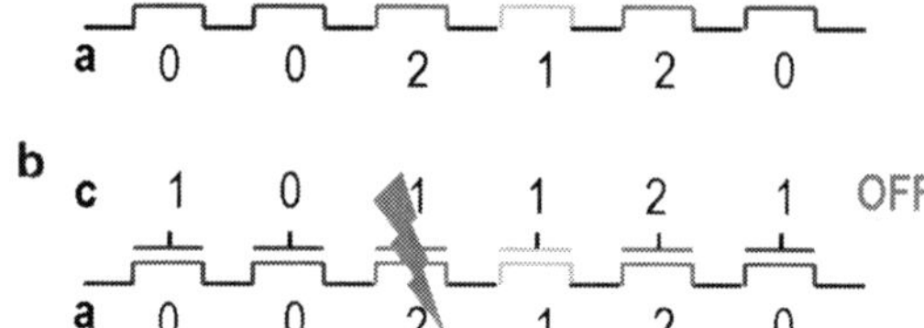

Fig. 3 Example of conducting and nonconducting nanowires. (**a**) Conducting nanowire (code word covers pattern). (**b**) Nonconducting nanowire (code word does not cover pattern)

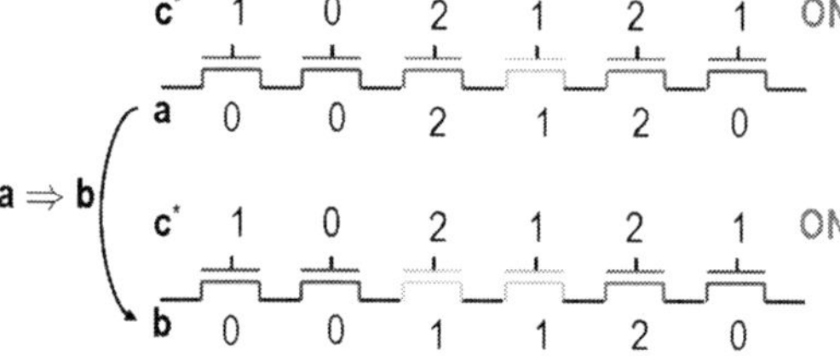

Fig. 4 Example of implication between patterns: c^* covers **a**; since $a \Rightarrow b$, then c^* also covers **b**

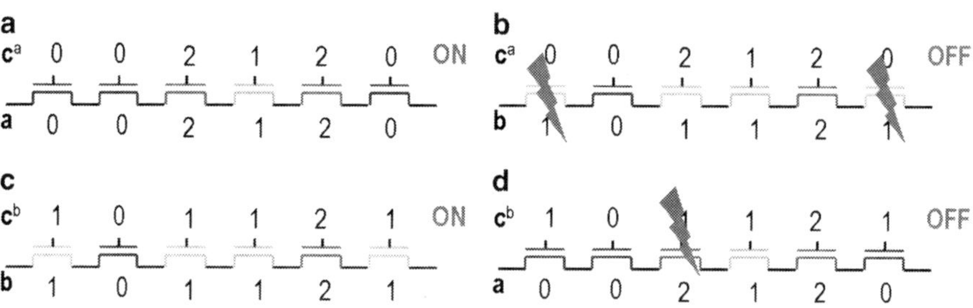

Fig. 5 Example of independent covering: code words c^a and c^b are independently covered

word c^b is also covered by the same code word c^*. Applying the voltage suite c^* will result in turning on the nanowires with either pattern (see Fig. 4).

Definition 5. *Code words c^a and c^b are independently covered* if and only if c^a does not imply c^b and c^b does not imply c^a.

This definition means that there exists a voltage suite that turns on the nanowire with pattern **a** corresponding to c^a, but not with pattern **b** corresponding to c^b (see Figs. 5a, b). Reciprocally, there exists a second voltage pattern that turns on the nanowire with pattern **b** corresponding to c^b, but not with pattern **a** corresponding to c^a (see Figs. 5c, d).

Definition 6. *Code word c^a belonging to set Ω is addressable* if and only if it does not imply any other code word in $\Omega \backslash \{c^a\}$. We define set Ω to be addressable if and only if every code word in Ω is addressable.

Assuming that there is a one-to-one mapping between code space Ω and pattern space A, then saying that a code word $\mathbf{c}^a$ implies no other code word in $\Omega \backslash \{\mathbf{c}^a\}$ is equivalent to saying that it covers only pattern $\mathbf{a}$ and no other pattern in $A \backslash \{\mathbf{a}\}$. Thus, there exists a voltage sequence that activates only the nanowire with pattern $\mathbf{a}$ and no other nanowire having its pattern in $A \backslash \{\mathbf{a}\}$.

Proposition 1. *A set Ω of code words is addressable if and only if every code word in Ω is independently covered with respect to any other code word in Ω.*

Proof. This follows directly from Defs. 5 and 6.

Consequently, an admissible set of applied voltages that uniquely addresses each nanowire corresponds to the set of code words Ω that independently covers every pattern in A. This set of patterns can be simply taken as Ω itself, if Ω is addressable.

4.2 Code Construction

4.2.1 Hot Encoding

In binary logic, the (k, M) hot code space is defined as the set of code words with length M having k occurrences of bit '1' and $(M - k)$ occurrences of bit '0' in every code word ($k \leq M$). It is also known as the k-out-of-M code; which was first used as a defect-tolerant encoding scheme [46]. This definition can be generalized to the n-valued logic. We first define $\mathbf{k}$ as an n-dimensional vector $(k_0, \ldots, k_{n-1})$, such that $\sum_i k_i = M$. Then, the multivalued $(\mathbf{k}, M)$-hot encoding is defined as the set of all code words having length M such that each k_i represents the occurrence of digit i, $i = 0, \ldots, n - 1$. We consider, for instance, the ternary logic ($n = 3$), and we set $\mathbf{k} = (4, 3, 1)$ and $M = 8$. Then, every code word in the considered $(\mathbf{k}, M)$-hot space contains $4\times$ the digit '0', $3\times$ the digit '1' and $1\times$ the digit '2'. The considered code space includes, for instance, code words 00001112 and 00210110. The code space defined by a multi-valued $(\mathbf{k}, M)$-hot encoding is addressable and its size is maximal for $k_i = M/n$, $\forall$ $i = 0, \ldots, n - 1$. The size of the maximal-sized space is asymptotically $\propto n^M / M^{(n-1)/2}$ for a given n. In this chapter, it is implicitly understood that the $(\mathbf{k}, M)$-hot code with the maximal-sized space is used, even when just $(\mathbf{k}, M)$-hot code is mentioned.

4.2.2 N-ary Reflexive Code

The binary tree code with length M is a 2-to-2^M encoder representing the 2^M binary numbers $0 \cdots 0$ to $1 \cdots 1$. Similarly, an n-ary tree code with length M is defined as the set of n^M numbers ranging from $0 \cdots 0$ to $(n - 1) \cdots (n - 1)$. For instance, the ternary ($n = 3$) tree code with length $M = 4$ includes all ternary logic numbers ranging from 0000 to 2222. As one can easily see, some code words imply many others from the same space: for instance, 2222 implies all other code words. It is possible to prevent the inclusive character of the n-ary tree code by attaching the complement of the code word (*i.e.*, 2222 becomes 22220000). The as-constructed code is the *N-ary Reflexive Code (NRC)*. The code space defined by the NRC is addressable and its size is n^M. In a similar fashion, the reflection principle works for any other code (*e.g.*, Hamming code), making the whole code space addressable. However, in return it doubles the code length.

4.3 Defect Models

4.3.1 Basic Error Model

Figure 6 illustrates the main assumptions behind basic error models. We assume that the threshold voltages $V_{\mathrm{th},i}$ are equidistant; *i.e.*, $V_{\mathrm{th},i+1} - V_{\mathrm{th},i} = 2\alpha V_0$, V_0 being a given scaling voltage and α is given by the technology. The applied voltages $V_{\mathrm{A},i}$ are set between every two successive threshold voltages $V_{\mathrm{th},i}$ and $V_{\mathrm{th},i+1}$, not necessarily in the middle, rather shifted by υV_0 towards $V_{\mathrm{th},i}$; where υ is a design parameter.

If the variability of $V_{\mathrm{th},i}$ is high or the spacing between two successive $V_{\mathrm{th},i}$'s is low due to the large number of doping levels, then $V_{\mathrm{th},i}$ may exceed a voltage $V_{\mathrm{X},i}$ given by $V_{\mathrm{A},i} - \delta \cdot V_0$; where δ will be derived later. When V_{th} increases, the sensed current, while a_i is applied to digit c_i, decreases ($a_i = c_i$) and the sensed current,

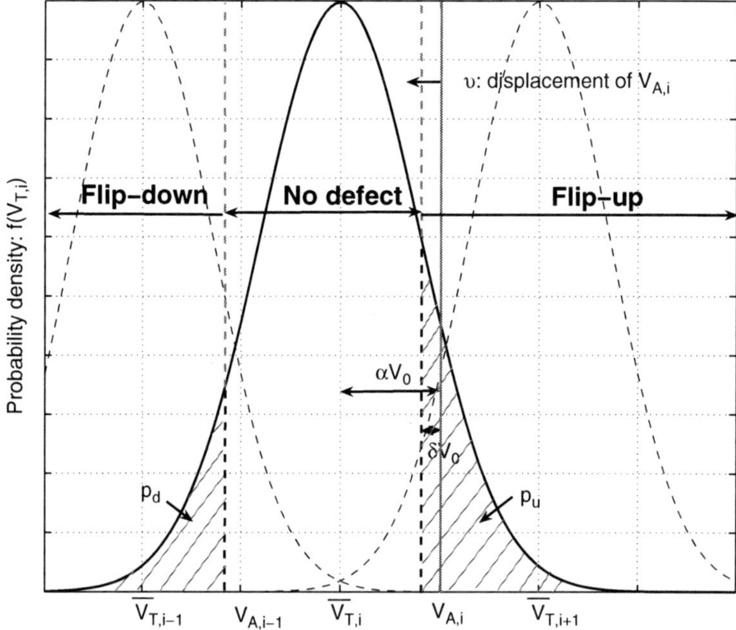

Fig. 6 Coding defects induced by V_{th} variability

while $a_i + 1$ is applied to the same digit, increases. Voltage $V_{X,i}$ is defined as the gate voltage which results in the decrease of the sensed current for a_i by a factor q from its value at $\overline{V}_{th,i}$. Higher the q, more accurate the sensing. Thus, q is also considered a design parameter. Assuming that the transistors are saturated, then the current in the saturation region is proportional to $(V_{A,i} - V_{th})^2$, where V_{th} is the actual threshold voltage. Consequently, the following condition on V_X must hold: $(V_{A,i} - \overline{V}_{th,i})^2/(V_{A,i} - V_{X,i})^2 = q$; which gives: $\delta = (\alpha + \upsilon)/\sqrt{q}$ for long-channel transistors.[2] This fixes the values of $V_{X,i}$; when $V_{th,i}$ exceeds $V_{X,i}$, digit a_i acts as $a_i + 1$; its address becomes $c_i + 1$ and we call this case the *flip-up defect*.

Now, consider the case when $V_{th,i}$ falls below $V_{A,i-1} - \delta \cdot V_0 = V_{X,i-1}$, then the current flowing while $a_i - 1$ is applied is not ~ 0 anymore, and always greater than q times the current flowing while $c_i - 2$ is applied. Then, a_i is implied by c_i and $c_i - 1$ but not by $c_i - 2$; its address is $c_i - 1$ which means that a_i acts as $a_i - 1$; this case is called the *flip-down defect*. The probabilities of flip-ups and flip-downs are given by the following expressions, which are independent of i. Here, f_i is the probability density function of $V_{th,i}$:

$$p_u = \int_{V_{X,i}}^{\infty} f_i(x)\mathrm{d}x \quad p_d = \int_{-\infty}^{V_{X,i-1}} f_i(x)\mathrm{d}x$$

When $V_{th,i}$ falls within the range between the threshold values for flip-up and flip-down defects, the digit is correctly interpreted. We notice that the flip-down error never occurs at digits having the smallest value, 0, since the corresponding $\overline{V}_{th,i}$ is by definition smaller than the smallest $V_{A,i}$ available. For the same reason, the flip-up error never occurs at digits having the largest value, $n - 1$. In order to study the size of the addressable code space, we consider flip-up and flip-down errors in the code space instead of flip-up and flip-down defects at the nanowires, since the two considerations are equivalent.

4.3.2 Overall Impact of Variability

If V_{th} varies within a small range close to its mean value, then the pattern does not change, since the nanowire still conducts under the same conditions. Then, a one-to-one mapping between the code and the pattern space holds, which is shown in Fig. 7 for a ternary hot code with $M = 3$. On the contrary, if the V_{th} variation is large, then some digits may be shifted up or down, as explained above. When a pattern has a sequence of errors, it can be either covered by one or more code words or it can be uncovered. When we consider the code words, some of them cover one or more patterns and some cover no pattern under the error assumptions. The following example explains this conjecture:

---

[2] If we consider short-channel transistors, then the saturation current is proportional to $(V_{A,i} - V_{th})$ and $\delta = (\alpha + \upsilon)/q$.

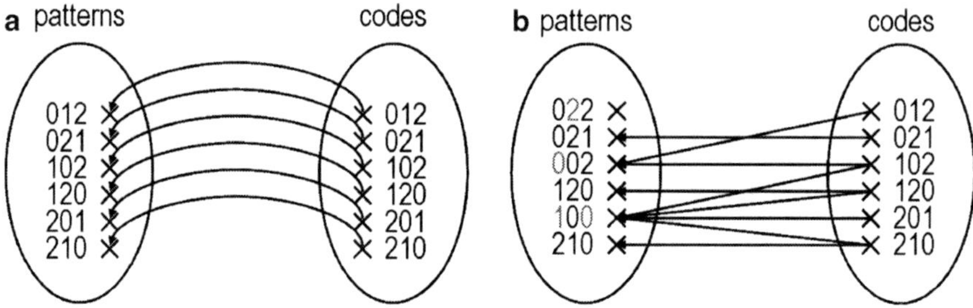

Fig. 7 Mapping of the code space onto the pattern space. (**a**) Mapping in the defect-free case. (**b**) Mapping and in the case of defects

Example 1. *Figure 7b illustrates the digit shift at some patterns. We notice that the first pattern 022 (which underwent a defect) is not covered by any code word anymore. Thus, its nanowire cannot be addressed. All the other patterns are covered at least by one code word. Two categories among these covered patterns can be distinguished. On the one hand, the fourth pattern 120 is covered by the fourth code word, which in turn covers another pattern (the fifth). Thus, by activating the fourth nanowire, the required control voltages activate either the fourth or fifth nanowire. Consequently, the fourth nanowire cannot be addressed uniquely. This case represents the patterns covered only by code words covering more than a single pattern. On the other hand, the complementary case is illustrated by the fifth pattern 100, which is covered by many code words. However, one of these code words (201) covers no other pattern except the considered one. Thus, it is possible to uniquely activate the fifth nanowire by applying the voltage sequence corresponding to code word 201.*

The examples shown in Fig. 7 demonstrate that a pattern undergoing defects can be either (1) not covered by any valid code word, in which case the nanowire cannot be identified as addressable and the pattern is useless; or (2) covered by at least one valid code word. In the second case, if two patterns or more are covered by the same code word, then this code word cannot be used because more than one nanowire would have the same address. Thus, in the second case, the pattern is only useful if at least one code word covering it covers no other pattern, insuring that the covered pattern can be addressed.

Assuming that, on an average, every code word covers v patterns when errors occur, let p_I be the probability that a pattern becomes uncovered, and p_U the probability that a code word covers a unique pattern ($\bar{p}_U = 1 - p_U$). Let $|\Omega|$ be the original size of the code space and $|\Omega'|$ the size after errors occur. Set Ω' contains useful addresses under defect conditions, *i.e.*, those that address unique nanowires even though the nanowires are defective. The size of Ω' indicates the number of nanowires that remain useful under high variability conditions. Then:

$$|\Omega'| = |\Omega| \cdot (1 - p_I)(1 - \bar{p}_U^v) \tag{1}$$

A model for multidigit errors in multivalued logic codes was presented in [47] and gives an estimate of p_I and p_U for both types of code space. Parameter v is estimated as a fit parameter from Monte Carlo simulations.

4.4 Impact of the Encoding Scheme

In order to assess the variation of the addressable code space under variable V_th, we plotted separately the uncovered part $|\Omega|_\mathrm{un} = p_\mathrm{I} \cdot |\Omega|$, the addressable part $|\Omega'|$, and the immune part $|\Omega|_\mathrm{im}$ in which no defects occur. The fit parameter v was estimated with Monte-Carlo simulations. Figure 8 shows the sizes of these subspaces for a ternary (3, 14)-reflexive code depending on the 3σ-value of V_th. The Monte-Carlo simulation confirms in the same figure the analytical results and gives the value 2.8 for the fit parameter v. The size of the addressable space $|\Omega'|$ drops quickly when 3σ reaches 0.4 V. At the same time, more patterns become uncovered. Interestingly, there are more addressable than immune patterns, because some defective patterns can be randomly addressed. This tendency increases for unreliable technologies, and around 10% of the original code space size can be randomly addressed under extreme conditions. The simulation of hot codes was not shown, because the result

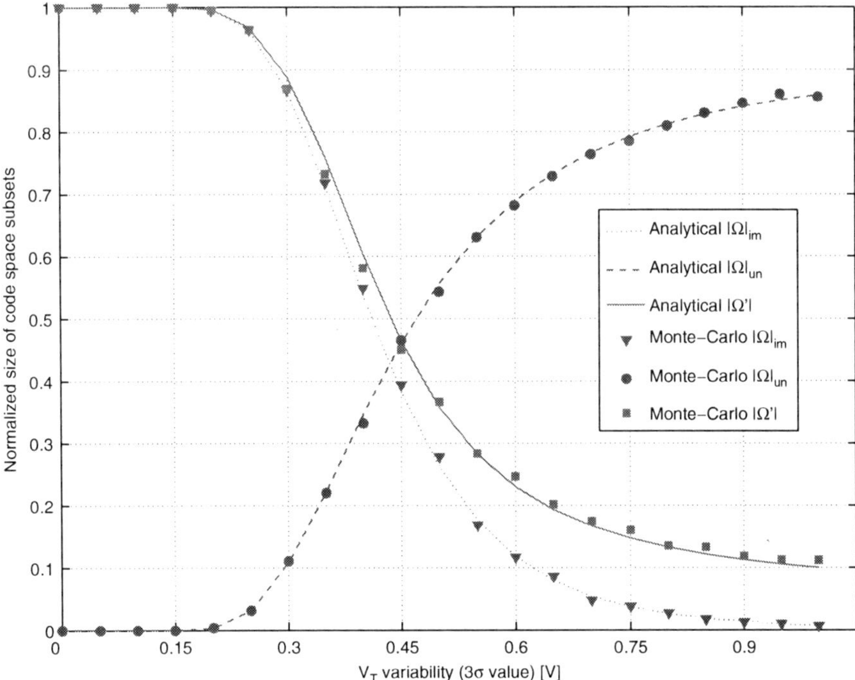

Fig. 8 Dependency of different code space subsets on V_th variability

is similar, except for large defect probabilities: under these conditions, the size of the addressable space goes faster towards 0 because the construction of hot codes imposes more constraints than the NRC.

The sizing of memory blocks (*i.e.,* the size of contact groups in Fig. 1) and the number of V_{th}'s are interdependent. As a matter of fact, Fig. 9 shows that increasing the number of V_{th}·s has two opposite effects: on one hand, it enables the addressing of more wires with the same code length; on the other hand, it makes the transistors more vulnerable to defects and increases the number of lost code words. A typical trade-off situation is illustrated in Fig. 9 with the ternary (3,9) and binary (2,12) hot codes (with $(n, k) = (3,3)$ and $(2,6)$, respectively) yielding almost the same number of addressable nanowires for 3σ around 0.4 V. The first one saves area because it has shorter code words, whereas the second one is technologically easier to realize (only two different V_{th}'s). The use of the ternary decoder is recommended for reliable technologies (ensuring less area and more code words), but when the technology becomes more unreliable, there is a trade-off between area savings and easier fabrication process.

The benefits of using multivalued logic to design bottom-up decoders is summarized in Table 1. Among the decoders presented in Sect. 3.3, the radial decoder would need several oxide shell thicknesses and the random contact decoder would

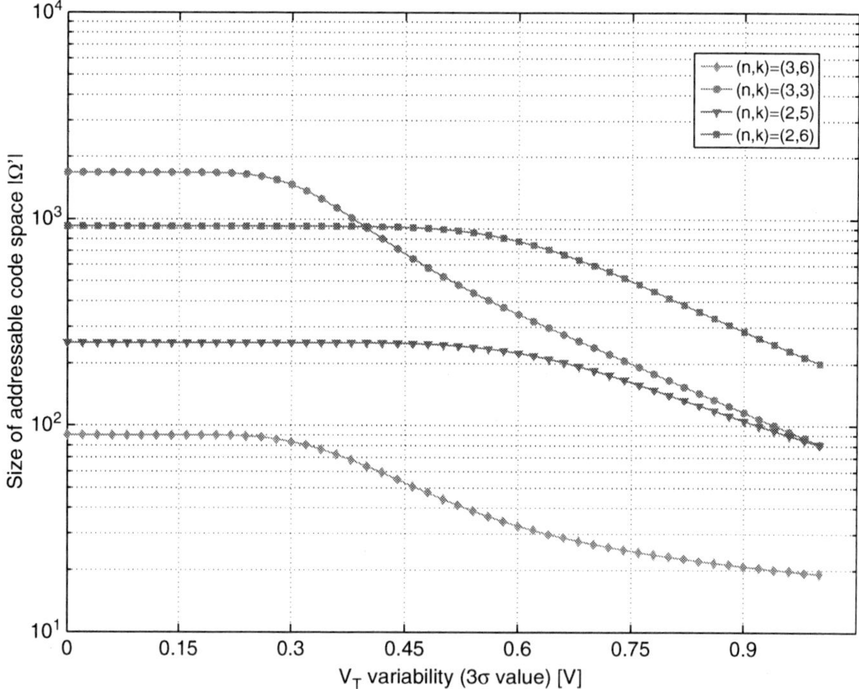

Fig. 9 Number of addressable nanowires for different hot codes

Table 1 Yield of different decoders in terms of area per working bit (nm^2) at the technology node 45 nm

Raw size (kB)	Base	Axial decoder	Mask-based
8	2	1,576	622
8	3	1,196	550
8	Δ	24.1%	11.5%
32	2	846	423
32	3	676	373
32	Δ	20.2%	11.8%

need more than one level of conduction in order to be extended to n-ary logic. These features are not inherent to the decoders, as shown in [41] and [44]; thus they cannot be extended to multilevel logic. On the contrary, it is possible to assume more than two levels of doping for the axial decoder and more than one oxide thickness for the mask-based decoder in order to perform MVL addressing without altering the underlying decoding paradigm. Consequently, only these two decoders were extended to MVL addressing. The bottom-up approaches promise a high effective density under technological assumptions that are still to be validated. The use of ternary logic in 32 kB raw area memories saves area up to 20.2% for memories with axial decoder, and up to 11.8% for memories with mask-based decoder.

5 Testing Crossbars

The physical defects affecting the nanowires have been modeled at a high abstraction level as changes in the nanowire addresses. A defect can cause a change of the nanowire address such that the nanowire becomes unaddressable in the considered code space, or it shares the same address with another nanowire. In these cases, it is required that defective nanowire addresses are detected and discarded from the used set of addresses. This task can be performed by testing the decoder circuit.

Testing the decoder, in order to keep only defect-free parts of the code space, highly simplifies the test procedure of the whole crossbar circuit. This section proposes a test method that identifies the defective codes. The method quantifies the test quality, measured as the probability of test error, and it investigates the dependency of the test quality on the decoder design parameters. Without loss of generality, crossbar circuits considered in the following implement a memory function.

5.1 Testing Procedure

This section presents an overview of a test method that can be applied to nanowire arrays. This is an exhaustive method used to illustrate the testing principle. More efficient pseudo-random techniques also exist. However, the focus here is only on the thresholder design and the test quality.

The nanowire testing is performed for every layer separately. Thus, we depicted a single nanowire layer with its additional test circuitry in Fig. 14. Besides the nanowire layer, the system comprises the interfacing circuit (decoder) and a CMOS part formed by a thresholder, a control unit and a *lookup table (LUT)*. The thresholder measures the output current and indicates whether a single nanowire is detected. The control unit regulates the execution of the testing phase and other functions, such as the reading and writing operations. The LUT stores the valid addresses, *i.e.*, those that activate a single nanowire each.

The test can be performed by applying the following exhaustive procedure. First, the two nanowire layers are disconnected by setting the power (V_P) and sense (GND) electrodes of every layer to the same voltage, such that a large voltage drop is created between the two layers. Then, we consider every layer separately. By going through all possible addresses, a voltage V_P is applied; then the address is stored in the LUT if the sensed current indicates the activation of a single nanowire. The same procedure is repeated for the second layer and it is linear with N.

The output of the nanowire layer (I_s) is sensed by the thresholder. We assume that the variability mainly affects the sublithographic part of the memory representing the nanowire array. This part is fabricated using an unreliable technology, unlike the rest of the circuit, defined on the lithography scale and assumed to be more robust. Thus, we consider that the thresholder, the control circuit or the LUT are defect-free. The thresholder senses I_s, it possibly amplifies it, then it compares I_s to two reference values (I_0 and I_1 with $I_0 < I_1$). If the sensed current is smaller than I_0, then no nanowire is addressed. If the sensed current is larger than I_1, then at least two nanowires are activated with the same address. If the sensed current is between the reference current levels, then only one nanowire is activated and the address is considered to be valid. Given the statistical variation of the threshold voltages, the ability to correctly detect addresses can be expressed with the following probabilities (Fig. 10):

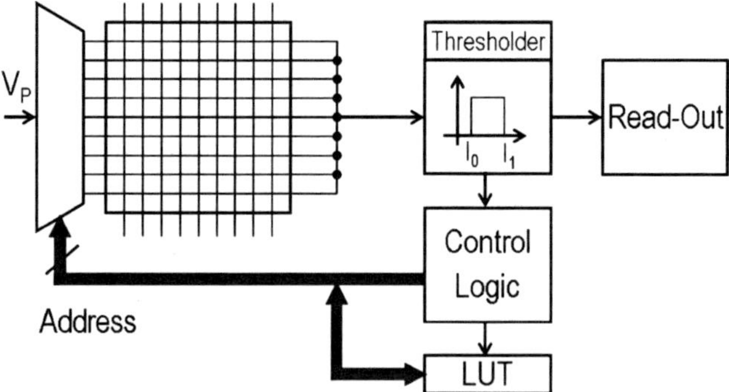

Fig. 10 Crossbar memory and testing unit: besides the memory array and the decoder, the system comprises a CMOS part formed by a thresholder that detects the bit state, a control unit that synchronizes the test operation, and a LUT that saves correct addresses

$$\begin{cases} P_0 = \Pr\{(I_s \leq I_0) \text{ given that no nanowire is addressed}\} \\ P_1 = \Pr\{(I_0 < I_s < I_1) \text{ given that 1 nanowire is addressed}\} \\ P_2 = \Pr\{(I_1 \leq I_s) \text{ given that } \geq 2 \text{ nanowires are addressed}\} \end{cases} \quad . \qquad (2)$$

Then, the probability that all three events happen simultaneously is given by: $P_0 \times P_1 \times P_2$, assuming that the considered events are independent. We can define the error probability of this test procedure as follows:

$$\varepsilon = 1 - P_0 \times P_1 \times P_2 \qquad (3)$$

The purpose of the following is to design the thresholder in order to obtain the best test result with the smallest ε. In the next sections, we derive the analytical expressions of P_0, P_1 and P_2, then we optimize I_0 and I_1 in order to minimize ε.

5.2 Perturbative Current Model

During the code testing phase, every nanowire is disconnected from the crossing nanowires. It can be modeled as a wire connecting the power electrode to the sensing electrode and formed by two parts (see Fig. 15): the decoder part that is a series of M pass transistors, and the memory part. Since the memory part is disconnected from the second layer of nanowires, it can be modeled as a resistive load R_M. We model the devices (SiNWFETs) in this section in a general way as a voltage-controlled current sources, *i.e.,*: $I = f(V_{DS}, V_{GS}, V_T)$ where I is the drain-source current, V_{DS}, V_{GS} and V_T are respectively the drain-to-source, gate-to-source and threshold voltages. The decoder design is based on two different V_T's ($V_{T,Ref0}$ and $V_{T,Ref1}$ such that $V_{T,Ref0} < V_{T,Ref1}$, and we define $\Delta V_T = V_{T,Ref1} - V_{T,Ref0}$). When a nanowire is addressed, every variation of V_T results in a variation of the current through the nanowire, which can be noted the following way (Fig. 11):

$$I = I^{OP} + \delta I \qquad (4)$$

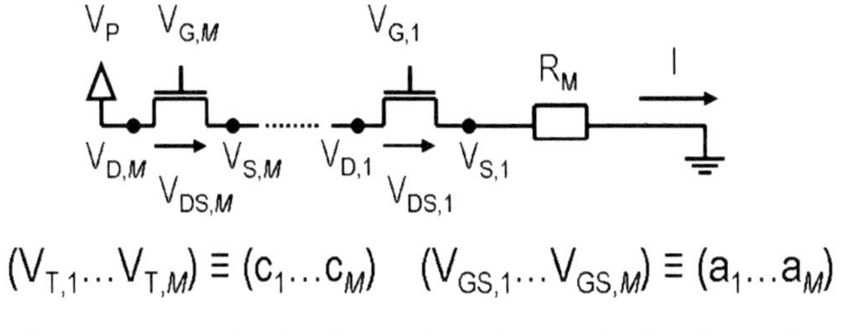

Fig. 11 Electrical parameters of a biased nanowire under test: the decoder part is represented by M transistors in series, and the memory part is represented by a resistance R_M. Notice that the perpendicular nanowire layer is disconnected from the nanowire under test

The signal I is linearized around the *operating point (OP)* and divided into a *large I^{OP}* and a *small signal δI*. This approach is widely used in circuit and network theory and in sensitivity analysis [48]. The large signal can be estimated with a SPICE simulator. The small signal can be calculated by linearizing all the equations describing the circuit around the OP:

$$\delta I = -\frac{1}{R_M} \cdot v^T \cdot A^{-1} \cdot B \cdot \delta V_T \tag{5}$$

with the variational vector $\delta \mathbf{V}_T = [\delta V_{T,1},\ldots, \delta V_{T,M}]^T$ for the threshold voltages, and the small signal matrices $\mathbf{A}$ and $\mathbf{B}$ given by:

$$A = \begin{bmatrix} 1 + r_1 \cdot g_{DS,1} & 1 & \cdots & 1 \\ 1 - r_2 \cdot g_{m,2} & 1 + r_2 \cdot g_{DS,2} & \cdots & 1 \\ \vdots & & & \vdots \\ 1 - r_M \cdot g_{m,M} & 1 - r_M \cdot g_{m,M} & \cdots & 1 + r_M \cdot g_{DS,M} \end{bmatrix},$$

$$B = \begin{bmatrix} -r_1 \cdot g_{T,1} & 0 & \cdots & 0 \\ 0 & -r_2 \cdot g_{T,2} & \cdots & 0 \\ \vdots & & & \vdots \\ 0 & 0 & \cdots & -r_M \cdot g_{T,M} \end{bmatrix}.$$

We used the following notations: $g_{DS,i} = \partial f_i / \partial V_{DS,i}$, $g_{m,i} = \partial f_i / \partial V_{GS,i}$, $g_{T,i} = \partial f_i / \partial V_{T,i}$ and $r_i = R_M \| g_{m,i}^{-1}$ (parallel resistance connection). All the components of the matrices $\mathbf{A}$ and $\mathbf{B}$ are considered at the operating point.

5.3 Stochastic Current Model

We divide the sensed current into a useful and a noisy part. The useful signal (I_u) is the current that flows through a nanowire when the code corresponding to its pattern is applied. On the other hand, the noise can be generated by two different processes: intrinsically (I_i), or defect-induced (I_d). The intrinsic noise is generated by nanowires that are switched off, which generate subthreshold current. The defect-induced noise is generated by unintentionally addressed nanowires. Their number is denoted by N_{def}, while the number of nanowires generating intrinsic noise is N_{off}. Since the total number of nanowires is N, the following equation must hold $N_{use} + N_{off} + N_{def} = N$, where $N_{use} = 0$ if no nanowire is activated by the applied code, and $N_{use} = 1$ otherwise.

5.3.1 Distribution of the Useful Signal

Every V_T is considered as an independent and normally distributed stochastic variable with mean value $\overline{V}_T$ and standard deviation σ_T: $V_T \sim N(\overline{V}_T, \sigma_T^2)$. If the nanowire pattern is correct, then the operating point of V_T coincides with its mean value. If a defect happens so that the bit representing V_T flips, then the operating point of V_T is shifted from the mean value of V_T by $-\Delta V_T$.

We consider a nanowire with a defect-free pattern **a**, which is controlled by its corresponding code $\mathbf{c}^a$, and which generates the useful signal I_u. Then, $V_T^{OP} = \overline{V}_T$ and $\delta \mathbf{V}_T \sim N(0, \sigma_T^2 \cdot v)$ hold. Then, a useful signal follows the distribution resulting from (Eq. 10) to (Eq. 11). The operating point is the on-current of the transistors I_{on}, which is calculated with SPICE simulator; whereas the variable part is given by (Eq. 11). We obtain the following mean value and standard deviation of I_u:

$$\begin{cases} \bar{I}_u = I_{on} \\ \sigma_u = \dfrac{\sigma_T}{R_M} \cdot \| v^T A^{-1} B \| \end{cases} \tag{6}$$

5.3.2 Distribution of the Defect-Induced Noise

Now we consider a nanowire NW^b with the pattern **b** that undergoes some defects and turns into $\mathbf{b}^*$. This defective nanowire can be activated by the code $\mathbf{c}^a$ of another nanowire NW^a having the pattern **a**. In this case, NW^b generates a defect-induced noise I_d. This defect can be described by a series of shifts at the digits of **b** represented by the vector $\mathbf{s} \in \{0, 1\}^M$, where $\Delta V_T \cdot s_i \in \{0, \Delta V_T\}$ indicates whether a threshold voltage shift happened at the transistor i ($i = 1,\ldots, M$). Assuming that N_{def} nanowires generate a defect-induced noise, then every one of them is characterized by a given threshold voltage shift vector $\mathbf{s}_i$, $i \in \{1,\ldots, N_{def}\}$. By applying the summation rule of independent Gaussian distributions, we obtain the following mean value and standard deviation of I_d:

$$\begin{cases} \bar{I}_d = N_{def} \cdot I_{on} - \dfrac{\Delta V_T}{R_M} \cdot v^T \mathbf{A}^{-1} \mathbf{B} \cdot \displaystyle\sum_{i=1\ldots N_{def}} \mathbf{s}_i \\ \sigma_d = \dfrac{\sqrt{N_{def}} \cdot \sigma_T}{R_M} \cdot \| \mathbf{v}^T \mathbf{A}^{-1} \mathbf{B} \| \end{cases} \tag{7}$$

5.3.3 Distribution of the Intrinsic Noise

The intrinsic noise is generated in the subthreshold regime of the transistors forming the decoder part of the nanowire. If N_{off} nanowires are not conducting, then $I_i = N_{\text{off}} \times I_{\text{off}}$ is the maximum expected intrinsic noise, assumed to be an additive constant to the total sensed current.

5.4 Test-Aware Design Optimization

The model was implemented using the bulk MOSFET model for the considered SiNWFET, as described in [49]. The linearization around the operating point was performed in the linear region, in order to keep $V_{\text{DS},i}$, and consequently V_P, as low as possible. It is desirable to obtain a symmetrical device operation, *i.e.*, the same value of the operating point at all transistors, in order to simplify the matrices **A** and **B**. We assumed also the simple case of a binary reflexive code with the length M, and we set $V_\text{P} = 0.9\ V$.

The thresholder parameters that we are investigating in this work are I_0 and I_1. The minimal value of I_0 has to be greater than $N \cdot I_{\text{off}}$ in order to insure that $P_0 = 1$. While keeping I_0 larger than this critical value, we plotted I_1 that gives the best test quality (*i.e.*, the minimal error ε). The results are shown in Fig. 16 for different technology and design parameters. Among the considered technology parameters, β has the strongest influence on I_1. However this influence is globally weak: for $R_\text{M} = 10\ k\Omega$, increasing β by a factor of 10, adds just 4% to I_1. It is unlikely to have both β and R_M large; because β increases with the nanowire width W; while the opposite happens to R_M. Increasing the design parameter M from 12 to 18 has less impact than increasing β by a factor of $10\times$, because $\sigma \sim \beta/\sqrt{M}$, showing that the dependency on M is weaker. Consequently, I_1 has a robust value $\sim 1.2 \times I_{\text{on}}$ with respect to design and technology variation. On the other hand, I_0 should be large enough compared to the intrinsic noise I_i, but not too large, in order to separate the useful signal from the intrinsic noise. For a wide range of reasonable technological assumptions and array size, $I_0 \sim 0.66 \times I_{\text{on}}$ holds.

By using these optimized thresholder parameters, we investigated the test quality under different conditions. The test quality is improved by reducing the minimum test error, as plotted in Fig. 17. As expected, the best test quality is obtained for $I_1 \sim 1.2 \times I_{\text{on}}$. For a small-granularity array with $M = 12$, the test error is $\varepsilon \sim 10^{-4}$. Reducing the power level from 0.9 V down to 0.6 V reduces the current level at the operating point without reducing its variable part. Thus, it increases the noise level in the sensed current, and the test quality degrades by a factor of $22\times$. The variability level is the most critical parameter: increasing σ_T to 100 mV degrades the test quality by a factor larger than $50\times$. Improving the transistor gain factor β by $10\times$ enhances the test quality by a factor of $3\times$. Our analytical model and results show that a better strategy is to increase the number of addressing wires M by using redundant decoders (Figs. 12 and 13).

The physical defects affecting the nanowires have been modeled at a high abstraction level as changes in nanowire addresses. A defect can cause a change of the nanowire address such that the nanowire becomes unaddressable in the considered code space, or it shares the same address with another nanowire.

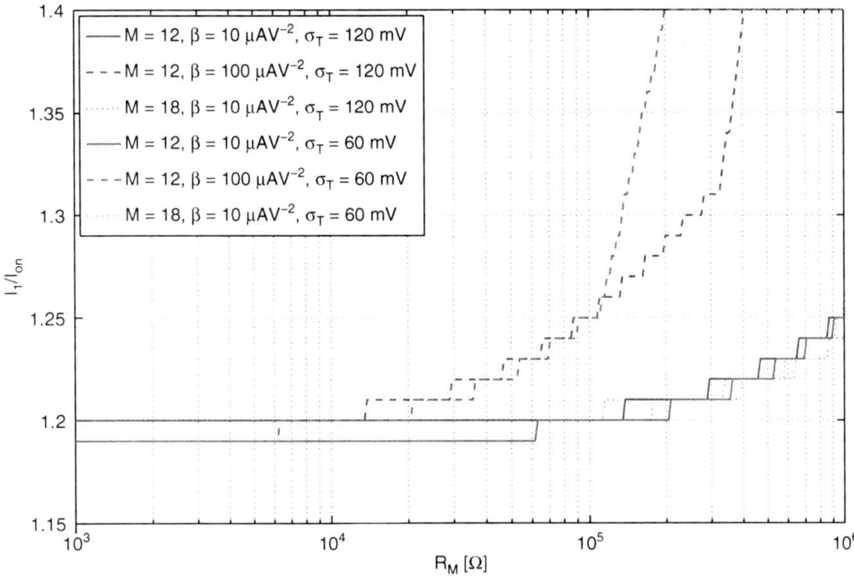

Fig. 12 Optimal value of I_1 vs. design and technology

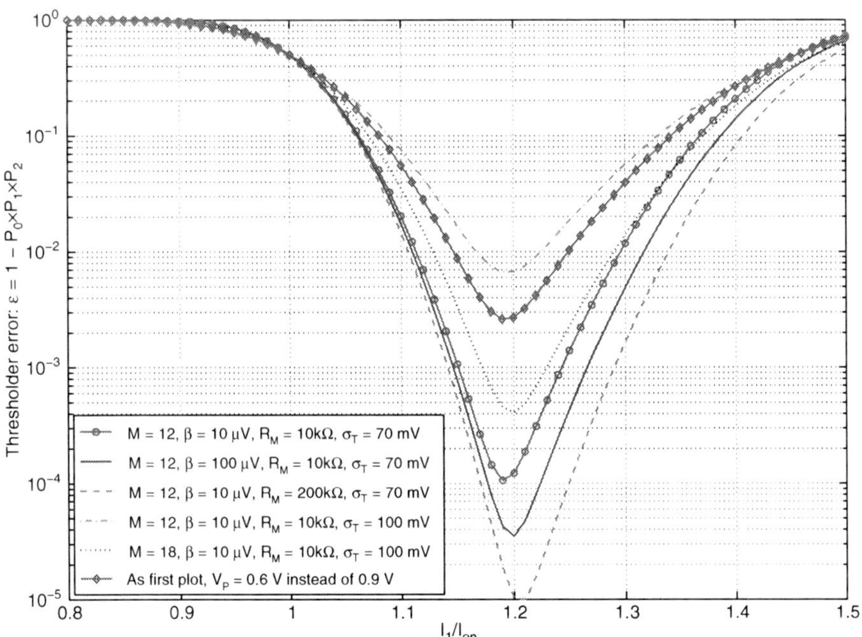

Fig. 13 Test quality vs. thresholder parameter I_1

In these cases, it is required that defective nanowire addresses be detected and discarded from the used set of addresses. This task can be performed by testing the decoder circuit.

Testing the decoder, in order to keep only defect-free parts of the code space, highly simplifies the test procedure of the whole crossbar circuit. This section proposes a test method that identifies the defective code words. The method quantifies the test quality, measured as the probability of test error, and investigates the dependency of the test quality on the decoder design parameters. Without loss of generality, crossbar circuits considered in the following discussion implement a memory function.

5.5 Testing Procedure

This section presents an overview of a test method that can be applied to nanowire arrays. This is an exhaustive method used to illustrate the testing principle. More efficient pseudo-random techniques also exist. However, the focus here is only on the thresholder design and test quality.

Nanowire testing is performed for every layer separately. We depict a single nanowire layer with its additional test circuitry in Fig. 14. Besides the nanowire layer, the system comprises the interfacing circuit (decoder) and a CMOS part formed by a thresholder, control unit and *look-up table (LUT)*. The thresholder measures the output current and indicates whether a single nanowire is detected. The control unit regulates the execution of the testing phase and other functions, such as the reading and writing operations. The LUT stores the valid addresses, *i.e.,* those that activate a single nanowire each.

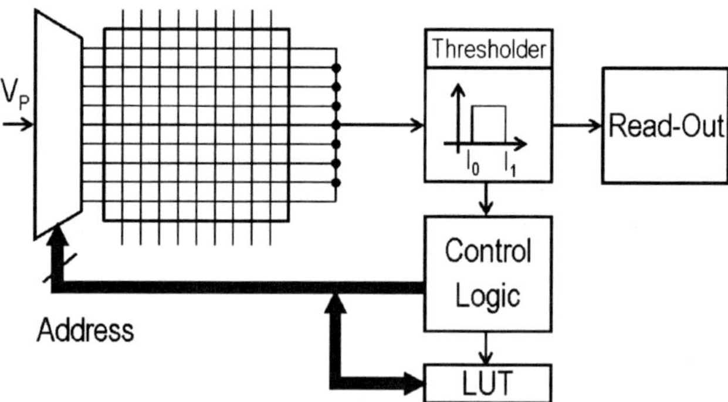

Fig. 14 Crossbar memory and testing unit: besides the memory array and the decoder, the system comprises a CMOS part formed by a thresholder that detects the bit state, a control unit that synchronizes the test operation, and a LUT that saves correct addresses

The test can be performed by applying the following exhaustive procedure. First, the two nanowire layers are disconnected by setting the power (V_P) and sense (GND) electrodes of every layer to the same voltage, such that a large voltage drop is created between the two layers. Then, each layer is considered separately. By going through all possible addresses, a voltage V_P is applied; then the address is stored in the LUT if the sensed current indicates the activation of a single nanowire. The same procedure is repeated for the second layer. The procedure is linear in N.

The output of the nanowire layer (I_s) is sensed by the thresholder. We assume that the variability mainly affects the sub-lithographic part of the memory representing the nanowire array. This part is fabricated using an unreliable technology, unlike the rest of the circuit, defined on the lithography scale and assumed to be more robust. Thus, we consider that the thresholder, the control circuit or the LUT are defect-free. The thresholder senses I_s, possibly amplifies it, then compares I_s to two reference values (I_0 and I_1 with $I_0 < I_1$). If the sensed current is smaller than I_0, then no nanowire is addressed. If the sensed current is larger than I_1, then at least two nanowires are activated with the same address. If the sensed current is between the reference current levels, then only one nanowire is activated and the address is considered to be valid. Given the statistical variation of the threshold voltages, the ability to correctly detect addresses can be expressed using the following probabilities:

$$\begin{cases} P_0 = \Pr\{(I_s \leq I_0) \text{ given that no nanowire is addressed}\} \\ P_1 = \Pr\{(I_0 < I_s < I_1) \text{ given that 1 nanowire is addressed}\} \\ P_2 = \Pr\{(I_1 \leq I_s) \text{ given that } \geq 2 \text{ nanowires are addressed}\} \end{cases} \quad (8)$$

Then, the probability that all three events occur simultaneously is given by $P_0 \times P_1 \times P_2$, assuming that the considered events are independent. We can define the error probability of this test procedure as follows:

$$\varepsilon = 1 - P_0 \times P_1 \times P_2 \quad (9)$$

The purpose of the following discussion is to design the thresholder in order to obtain the best test result with the smallest ε. Next, we derive the analytical expressions of P_0, P_1 and P_2, then we optimize I_0 and I_1 in order to minimize ε.

5.6 Perturbative Current Model

During the code testing phase, every nanowire is disconnected from the crossing nanowires. It can be modeled as a wire connecting the power electrode to the sensing electrode and formed by two parts (see Fig. 15): the decoder part that is a series of M pass transistors, and the memory part. Since the memory part is

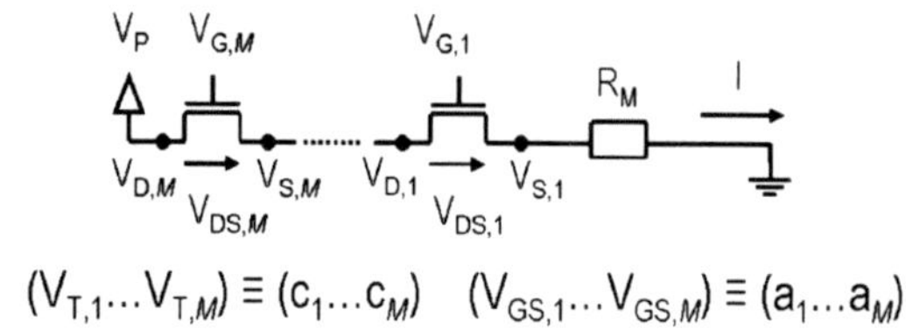

Fig. 15 Electrical parameters of a biased nanowire under test: the decoder part is represented by M transistors in series, and the memory part is represented by a resistance R_M. Note that the perpendicular nanowire layer is disconnected from the nanowire under test

disconnected from the second layer of nanowires, it can be modeled as a resistive load R_M. We model the devices (SiNWFETs) in this section in a general manner as voltage-controlled current sources, i.e., $I = f(V_{DS}, V_{GS}, V_{th})$ where I is the drain-source current, V_{DS}, V_{GS}, and V_{th} are, respectively, the drain-to-source, gate-to-source and threshold voltages. The decoder design is based on two different V_{th}s ($V_{th,Ref0}$ and $V_{th,Ref1}$ such that $V_{th,Ref0} < V_{th,Ref1}$; we define $\Delta V_{th} = V_{th,Ref1} - V_{th,Ref0}$). When a nanowire is addressed, every variation of V_{th} results in a variation of the current through the nanowire, which can characterized as:

$$I = I^{OP} + \delta I \tag{10}$$

Signal I is linearized around the *operating point (OP)* and divided into a *large* I^{OP} and a *small signal δI*. This approach is widely used in circuit and network theory and sensitivity analysis [48]. The large signal can be estimated with a SPICE simulator. The small signal can be calculated by linearizing all the equations describing the circuit around OP:

$$\delta I = -\frac{1}{R_M} \cdot v^T \cdot A^{-1} \cdot B \cdot \delta V_{th} \tag{11}$$

where variational vector $\delta \mathbf{V}_{th} = [\delta V_{th,1}, \cdots, \delta V_{th,M}]^T$ for the threshold voltages, and the small signal matrices **A** and **B** are given by:

$$A = \begin{bmatrix} 1 + r_1 \cdot g_{DS,1} & 1 & \cdots & 1 \\ 1 - r_2 \cdot g_{m,2} & 1 + r_2 \cdot g_{DS,2} & \cdots & 1 \\ \vdots & & & \vdots \\ 1 - r_M \cdot g_{m,M} & 1 - r_M \cdot g_{m,M} & \cdots & 1 + r_M \cdot g_{DS,M} \end{bmatrix},$$

$$B = \begin{bmatrix} -r_1 \cdot g_{th,1} & 0 & \cdots & 0 \\ 0 & -r_2 \cdot g_{th,2} & \cdots & 0 \\ \vdots & & & \vdots \\ 0 & 0 & \cdots & -r_M \cdot g_{th,M} \end{bmatrix}.$$

We have used the following notations: $g_{DS,i} = \partial f_i/\partial V_{DS,i}$, $g_{m,i} = \partial f_i/\partial V_{GS,i}$, $g_{th,i} = \partial f_i/\partial V_{th,i}$ and $r_i = R_M || g_{m,i}^{-1}$ (parallel resistance connection). All components of matrices $\mathbf{A}$ and $\mathbf{B}$ are computed at the operating point.

5.7 Stochastic Current Model

We divide the sensed current into a useful and noisy part. The useful signal (I_u) is the current that flows through a nanowire when the code word corresponding to its pattern is applied. On the other hand, the noise can be generated by two different processes: intrinsically (I_i) or defect-induced (I_d). The intrinsic noise is generated by nanowires that are switched off, which generate subthreshold current. The defect-induced noise is generated by unintentionally addressed nanowires. Their number is denoted by N_{def}, while the number of nanowires generating intrinsic noise is N_{off}. Since the total number of nanowires is N, the following equation must hold: $N_{use} + N_{off} + N_{def} = N$, where $N_{use} = 0$ if no nanowire is activated by the applied code word, and 1 otherwise.

5.7.1 Distribution of the Useful Signal

Every V_{th} is considered as an independent and normally distributed stochastic variable with mean value $\overline{V}_{th}$ and standard deviation σ_{th}: $V_{th} \sim N(\overline{V}_{th}, \sigma_{th}^2)$. If the nanowire pattern is correct, then the operating point of V_{th} coincides with its mean value. If a defect occurs such that the bit representing V_{th} flips, then the operating point of V_{th} is shifted from the mean value of V_{th} by $-\Delta V_{th}$.

We consider a nanowire with a defect-free pattern $\mathbf{a}$, which is controlled by its corresponding code word $\mathbf{c}^a$, and which generates the useful signal I_u. Then, $V_{th}^{OP} = \overline{V}_{th}$ and $\delta V_{th} \sim N(\mathbf{0}, \sigma_{th}^2 \cdot v)$ hold. Thus, a useful signal follows the distribution resulting from (10) to (11). The operating point is the on-current, I_{on}, of the transistors, which is calculated using the SPICE simulator; whereas the variable part is given by (11). We obtain the following mean value and standard deviation of I_u:

$$\begin{cases} \overline{I}_u = I_{on} \\ \sigma_u = \dfrac{\sigma_{th}}{R_M} \cdot \| v^T A^{-1} B \| \end{cases} \qquad (12)$$

5.7.2 Distribution of Defect-Induced Noise

Next, we consider a nanowire NW^b with pattern $\mathbf{b}$ that undergoes some defects and the pattern turns into $\mathbf{b}^*$. This defective nanowire can be activated by the code word $\mathbf{c}^a$ of another nanowire NW^a with pattern $\mathbf{a}$. In this case, NW^b generates a defect-induced noise I_d. This defect can be described by a series of shifts at the digits of $\mathbf{b}$ represented by vector $\mathbf{s} \in \{0, 1\}^M$, where $\Delta V_{th} \cdot s_i \in \{0, \Delta V_{th}\}$ indicates whether a threshold voltage shift occurred at transistor $i(i = 1,\ldots, M)$. Assuming that N_{def} nanowires generate a defect-induced noise, then every one of them is characterized by a given threshold voltage shift vector $\mathbf{s}_i$, $i \in \{1,\ldots, N_{\text{def}}\}$. By applying the summation rule of independent Gaussian distributions, we obtain the following mean value and standard deviation of I_{d}:

$$
\begin{cases}
\bar{I}_{\text{d}} = N_{\text{def}} \cdot I_{\text{on}} - \dfrac{\Delta V_{\text{th}}}{R_{\text{M}}} \cdot v^T \mathbf{A}^{-1} \mathbf{B} \cdot \displaystyle\sum_{i=1\ldots N_{\text{def}}} \mathbf{s}_i \\[4mm]
\sigma_{\text{d}} = \dfrac{\sqrt{N_{\text{def}}} \cdot \sigma_{\text{th}}}{R_{\text{M}}} \cdot \| \, \mathbf{v}^T \mathbf{A}^{-1} \mathbf{B} \, \|
\end{cases}
\tag{13}
$$

5.7.3 Distribution of the Intrinsic Noise

The intrinsic noise is generated in the subthreshold regime of the transistors forming the decoder part of the nanowire. If N_{off} nanowires are not conducting, then $I_i = N_{\text{off}} \times I_{\text{off}}$ is the maximum expected intrinsic noise, and assumed to be a an additive constant to the total sensed current.

5.8 Test-Aware Design Optimization

The model was implemented using the bulk MOSFET model for the considered SiNWFET, as described in [49]. The linearization around the operating point was performed in the linear region, in order to keep $V_{\text{DS},i}$, and consequently V_{P}, as low as possible. It is desirable to obtain a symmetrical device operation, *i.e.*, the same value of the operating point at all transistors, in order to simplify matrices $\mathbf{A}$ and $\mathbf{B}$. We assume the simple case of a binary reflexive code with length M, and we set $V_{\text{P}} = 0.9$ V.

The thresholder parameters that we have investigated in this work are I_0 and I_1. The minimal value of I_0 has to be greater than $N \cdot I_{\text{off}}$ in order to ensure that $P_0 = 1$. While keeping I_0 larger than this critical value, we plotted I_1 that gives the best test quality (*i.e.*, minimal error ε). The results are shown in Fig. 16 for different technology and design parameters. Among the considered technology parameters, β (the transistor gain factor) has the strongest influence on I_1. However this influence is globally weak: for $R_{\text{M}} = 10$ $k\Omega$, increasing β by a factor of 10 adds just 4% to I_1. It is unlikely to have both β and R_{M} large; because β increases with nanowire width W, whereas the opposite happens to R_{M}. Increasing the design parameter M (the number of addressing wires) from 12 to 18 has less impact than increasing β by a factor of $10\times$, because $\sigma \sim \beta/\sqrt{M}$, showing that the dependency on M is weaker. Consequently, I_1 has a robust value $\sim 1.2 \times I_{\text{on}}$ with respect to design and technology variation. On the other hand, I_0 should be large enough compared to intrinsic noise I_i, but not too large, in order to separate the useful signal

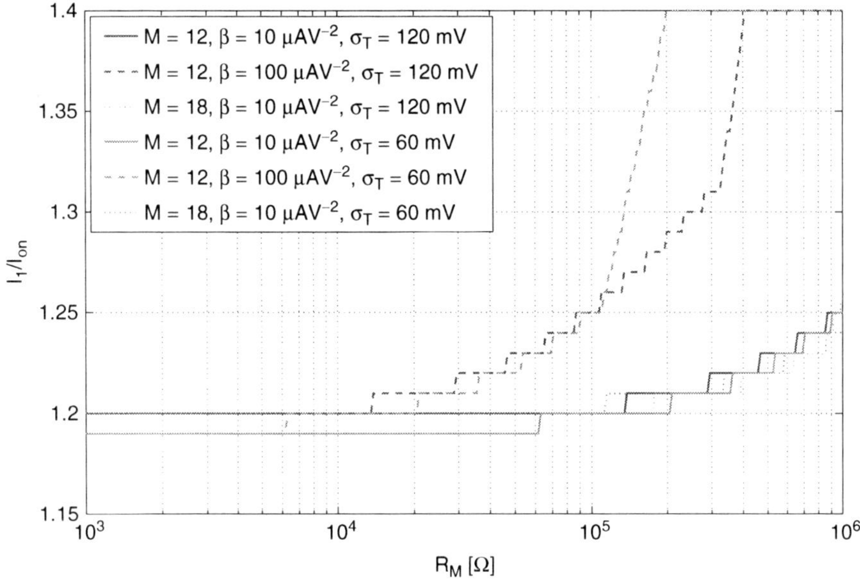

Fig. 16 Optimal value of I_1 vs. design and technology

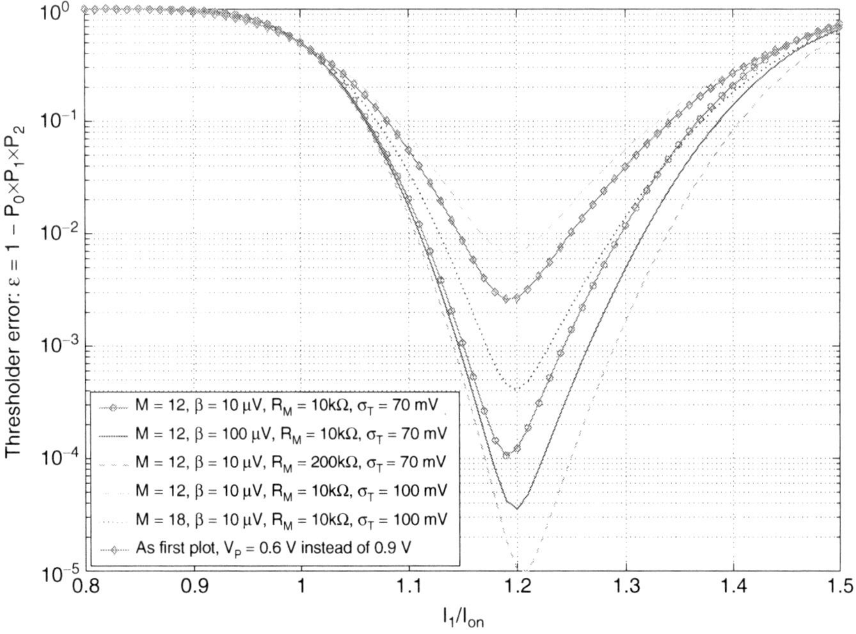

Fig. 17 Test quality vs. thresholder parameter I_1

from the intrinsic noise. For a wide range of reasonable technological assumptions and array size, $I_0 \sim 0.66 \times I_{on}$ holds.

By using these optimized thresholder parameters, we investigated the test quality under different conditions. The test quality is improved by reducing the minimum test error, as plotted in Fig. 17. As expected, the best test quality is obtained for $I_1 \sim 1.2 \times I_{on}$. For a small-granularity array with $M = 12$, the test error is $\varepsilon \sim 10^{-4}$. Reducing the power level from 0.9 V down to 0.6 V reduces the current level at the operating point without reducing its variable part. Thus, it increases the noise level in the sensed current, and the test quality degrades by a factor of $22\times$. The variability level is the most critical parameter: increasing σ_{th} to 100 mV degrades the test quality by a factor larger than $50\times$. Improving the transistor gain factor β by $10\times$ enhances the test quality by a factor of $3\times$. Our analytical model and results show that a better strategy is to increase the number of addressing wires M by using redundant decoders.

6   Conclusions

The crossbar architecture is a possible architectural paradigm for high-density integration of SiNWs into CMOS chips, and can be applied to a wide range of NW fabrication technologies. The variability of the nanowires due to their shrinking dimensions has an impact on the operation of the arrays. In this chapter, the focus was on the impact of variability on decoder design, which is the part of the circuit that bridges the array to the CMOS part of the chip. In order to address this problem, an abstract model of nanowires as a set of code words in a code space was introduced and the impact of variability was modeled as a sequence of errors that affect the code words. Based on this model, the decoder design was shown to be made more reliable by optimizing the choice of encoding scheme. Detection of errors can be performed by carrying out a test procedure. It was demonstrated that the decoder design can be optimized with respect to the test procedure in order to minimize the test error probability.

Exercise 1   Delay in a Crossbar

Consider the following NxN crossbar circuit with N nanowires in every plane and $M = \log_2(N)$ access transistors in the decoder of every plane. Determine the delay through the crossbar when the address corrsponding to the crosspoint (X,Y) is activated (X and Y are between 1 and M). Assume the following parameters:

- Decoder parameters:
  - On-resistance of an access transistor: $R_{on} = 10$ kΩ
  - Off-resistance of an access transistor: $R_{off} = 100$ MΩ
  - Drain/source capacitances of an access transistor: $C_{D/S} = 1$ fF

– Parameters of the functional part of the crossbar
  • Resistance of a nanowire length unit equal to the nanowire pitch: RNW = 100 Ω
  • Capacitance of a molecular switch: C_S = 2 fF
  • Resistance through a molecular switch: R_S = 1 kΩ
  • Parasistic capacitance between crossing nanowires, parallel nanowires and between the nanowires and the substrate: not included

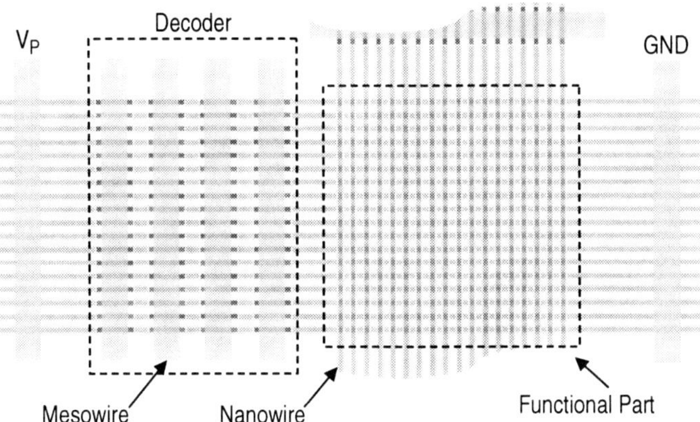

Ex. Figure 1 Baseline crossbar architecture

Exercise 2   Process Optimization

In goal of this exercise is to optimize the geometry of mask used in the MSPT process. The MSPT can used iteratively, starting with a given sacrificial layer, in order to define the spacers that may be used as sacrifical layers in the following steps.

This techniques envolves the deposition of a first sacrificial layer (1st step) with the width W and pitch P. Then, a sacrificial layer with the height W_1 = qW is deposited (2nd step) and etched (3rd step) in order the form the sacrifical layer with the width W1 and a smaller pitch than P (4th step). These steps can be repeated with a following deposition of a layer with a height W_2 = q W_1 (step 5 to 7) in order to decrease the pitch further.

Questions taken from [50]:

1. Calculate the extension of the spacer underneath and beyond the first sacrificial layer after n iterations ($l_{out}(n)$ and $l_{in}(n)$ respectively).
2. For a large number of iterations, calculate the optimal values for q and W/P.

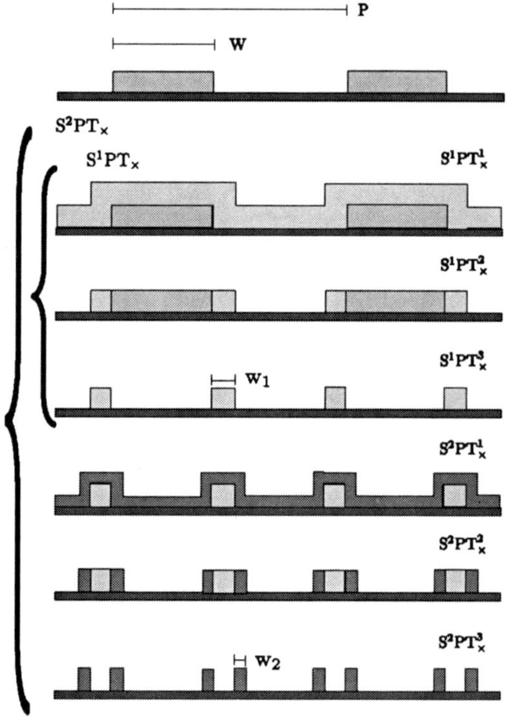

Ex. Figure 2 Multiplicative road of the multi-spacer technique [50]

References

1. M. H. Ben Jamaa, "Fabrication and Design of Nanoscale Regular Circuits," *PhD Thesis at Ecole Polytechnique Fédérale de Lausanne (EPFL)*, Sept. 2010.
2. R. S. Wagner and W. C. Ellis, "Vapor-liquid-solid mechanism for single crystal growth," *Applied Physics Letters*, vol. 4, no. 5, pp. 89–90, 1964.
3. Y. Cui, X. Duan, J. Hu, and C. M. Lieber, "Doping and electrical transport in silicon nanowires," *The Journal of Physical Chemistry B*, vol. 4, no. 22, pp. 5213–5216, 2000.
4. R. He and P. Yang, "Giant piezoresistance effect in silicon nanowires," *Nature Nanotechnology*, vol. 1, no. 1, pp. 42–46, 2006.
5. L. J. Lauhon, M. S. Gudiksen, D. Wang, and C. M. Lieber, "Epitaxial core-shell and core-multishell nanowire heterostructures," *Nature*, vol. 420, pp. 57–61, 2002.
6. K. E. Moselund, D. Bouvet, L. Tschuor, V. Pot, P. Dainesi, C. Eggimann, N. L. Thomas, R. Houdré, and A. M. Ionescu, "Cointegration of gate-all-around MOSFETs and local silicon-on-insulator optical waveguides on bulk silicon," *IEEE Transactions on Nanotechnology*, vol. 6, no. 1, pp. 118–125, 2007.
7. K.-N. Lee, S.-W. Jung, W.-H. Kim, M.-H. Lee, K.-S. Shin, and W.-K. Seong, "Well controlled assembly of silicon nanowires by nanowire transfer method," *Nanotechnology*, vol. 18, no. 44, p. 445302 (7pp), 2007.
8. S. D. Suk, S.-Y. Lee, S.-M. Kim, E.-J. Yoon, M.-S. Kim, M. Li, C. W. Oh, K. H. Yeo, S. H. Kim, D.-S. Shin, K.-H. Lee, H. S. Park, J. N. Han, C. Park, J.-B. Park, D.-W. Kim, D.

Park, and B.-I. Ryu, "High performance 5nm radius twin silicon nanowire MOSFET (TSNWFET): fabrication on bulk Si wafer, characteristics, and reliability," in *Proc. IEEE Int. Electron Devices Meeting*, pp. 717–720, Dec. 2005.

9. D. Sacchetto, M. H. Ben-Jamaa, G. De Micheli, and Y. Leblebici, "Fabrication and characterization of vertically stacked gate-all-around Si nanowire FET arrays," in *Proc. European Solid-State Device Research Conference*, 2009.

10. L. Doherty, H. Liu, and V. Milanovic, "Application of MEMS technologies to nanodevices," *Circuits and Systems, 2003. ISCAS '03.* Proceedings of the 2003 International Symposium on, vol. 3, pp. III–934–III–937, May 2003.

11. R. M. Y. Ng, T. Wang, and M. Chan, "A new approach to fabricate vertically stacked single-crystalline silicon nanowires," in *Proc. IEEE Conf. on Electron Devices and Solid-State Circuits*, pp. 133–136, Dec. 2007.

12. S.-M. Koo, A. Fujiwara, J.-P. Han, E. M. Vogel, C. A. Richter, and J. E. Bonevich, "High inversion current in silicon nanowire field effect transistors," *Nano Letters*, vol. 4, no. 11, pp. 2197–2201, 2004.

13. J. Kedzierski and J. Bokor, "Fabrication of planar silicon nanowires on silicon-on-insulator using stress limited oxidation," *Journal of Vacuum Science and Technology B*, vol. 15, no. 6, pp. 2825–2828, 1997.

14. V. Auzelyte, H. H. Solak, Y. Ekinci, R. MacKenzie, J. Vrs, S. Olliges, and R. Spolenak, "Large area arrays of metal nanowires," *Microelectronic Engineering*, vol. 85, no. 5–6, pp. 1131–1134, 2008.

15. O. Vazquez-Mena, G. Villanueva, V. Savu, K. Sidler, M. A. F. van den Boogaart, and J. Brugger, "Metallic nanowires by full wafer stencil lithography," *Nano Letters*, vol. 8, no. 11, pp. 3675–3682, 2008.

16. J. Hållstedt, P.-E. Hellström, Z. Zhang, B. Malm, J. Edholm, J. Lu, S.-L. Zhang, H. Radamson, and M. Östling, "A robust spacer gate process for deca-nanometer high-frequency MOS-FETs," *Microelectronic Engineering*, vol. 83, no. 3, pp. 434–439, 2006.

17. Y.-K. Choi, J. S. Lee, J. Zhu, G. A. Somorjai, L. P. Lee, and J. Bokor, "Sublithographic nanofabrication technology for nanocatalysts and DNA chips," *Journal of Vacuum Science Technology B: Microelectronics and Nanometer Structures*, vol. 21, pp. 2951–2955, 2003.

18. W. Wu, G.-Y. Jung, D. L. Olynick, J. Straznicky, Z. Li, X. Li, D. A. A. Ohlberg, Y. Chen, S.-Y. Wang, J. A. Liddle, W. M. Tong, and R. S. Williams, "One-kilobit cross-bar molecular memory circuits at 30-nm half-pitch fabricated by nanoimprint lithography," *Applied Physics A: Materials Science and Processing*, vol. 80, no. 6, pp. 1173–1178, 2005.

19. G.-Y. Jung, E. Johnston-Halperin, W. Wu, Z. Yu, S.-Y. Wang, W. M. Tong, Z. Li, J. E. Green, B. A. Sheriff, A. Boukai, Y. Bunimovich, J. R. Heath, and R. S. Williams, "Circuit fabrication at 17 nm half-pitch by nanoimprint lithography," *Nano Letters*, vol. 6, no. 3, pp. 351–354, 2006.

20. Y. Huang, X. Duan, Q. Wei, and C. M. Lieber, "Directed assembly of one-dimensional nanostructures into functional networks," *Science*, vol. 291, no. 5504, pp. 630–633, 2001.

21. P. A. Smith, C. D. Nordquist, T. N. Jackson, T. S. Mayer, B. R. Martin, J. Mbindyo, and T. E. Mallouk, "Electric-field assisted assembly and alignment of metallic nanowires," *Applied Physics Letters*, vol. 77, pp. 1399–1401, 2000.

22. X. Duan, Y. Huang, Y. Cui, J. Wang, and C. M. Lieber, "Indium phosphide nanowires as building blocks for nanoscale electronic and optoelectronic devices," *Nature*, vol. 409, pp. 66–69, 2001.

23. Y. Chen, D. A. A. Ohlberg, X. Li, D. R. Stewart, R. Stanley Williams, J. O. Jeppesen, K. A. Nielsen, J. F. Stoddart, D. L. Olynick, and E. Anderson, "Nanoscale molecular-switch devices fabricated by imprint lithography," *Applied Physics Letters*, vol. 82, pp. 1610–1612, 2003.

24. J. A. Zasadzinski, R. Viswanathan, L. Madsen, J. Garnaes, and D. K. Schwartz, "Langmuir-Blodgett films," *Science*, vol. 263, no. 5154, pp. 1726–1733, 1994.

25. N. A. Melosh, A. Boukai, F. Diana, B. Gerardot, A. Badolato, P. M. Petroff, and J. R. Heath, "Ultrahigh-density nanowire lattices and circuits," *Science*, vol. 300, no. 5616, pp. 112–115, 2003.

26. J. E. Green, J. W. Choi, A. Boukai, Y. Bunimovich, E. Johnston-Halperin, E. Deionno, Y. Luo, B. A. Sheriff, K. Xu, Y. S. Shin, H.-R. Tseng, J. F. Stoddart, and J. R. Heath, "A 160-kilobit molecular electronic memory patterned at 10^{11} bits per square centimetre," *Nature*, vol. 445, pp. 414–417, 2007.

27. G. Ho, J. R. Heath, M. Kondratenko, D. F. Perepichka, K. Arseneault, M. Pézolet, and M. R. Bryce, "The first studies of a tetrathiafulvalene-sigma-acceptor molecular rectifier," *Chemistry – A European Journal*, vol. 11, no. 10, pp. 2914–2922, 2005.

28. R. L. McCreery, "Molecular electronic junctions," *Chemistry of Materials*, vol. 16, no. 23, pp. 4477–4496, 2004.

29. G. J. Ashwell, B. Urasinska, and W. D. Tyrrell, "Molecules that mimic Schottky diodes," *Physical Chemistry Chemical Physics (Incorporating Faraday Transactions)*, vol. 8, pp. 3314–3319, 2006.

30. Y. Luo, C. P. Collier, J. O. Jeppesen, K. A. Nielsen, E. DeIonno, G. Ho, J. Perkins, H.-R. Tseng, T. Yamamoto, J. F. Stoddart, and J. R. Heath, "Two-dimensional molecular electronics circuits," *Journal of Chemical Physics and Physical Chemistry*, vol. 3, pp. 519–525, 2002.

31. C. P. Collier, G. Mattersteig, E. W. Wong, Y. Luo, K. Beverly, J. Sampaio, F. M. Raymo, J. F. Stoddart, and J. R. Heath, "A [2]catenane-based solid state electronically reconfigurable switch," *Science*, vol. 289, pp. 1172–1175, 2000.

32. Y. Zhang, S. Kim, J. McVittie, H. Jagannathan, J. Ratchford, C. Chidsey, Y. Nishi, and H.-S. Wong, "An integrated phase change memory cell with Ge nanowire diode for cross-point memory," in *Proc. IEEE Symp. on VLSI Technology*, pp. 98–99, June 2007.

33. A. DeHon, "Design of programmable interconnect for sublithographic programmable logic arrays," in *Proc. Int. Symp. on Field-Programmable Gate Arrays*, 2005, pp. 127–137.

34. W. Culbertson, R. Amerson, R. Carter, P. Kuekes, and G. Snider, "Defect tolerance on the Teramac custom computer," in *Proc. IEEE Symp. on FPGAs for Custom Computing Machines*, pp. 116–123, Apr. 1997.

35. J. R. Heath, P. J. Kuekes, G. S. Snider, and R. S. Williams, "A defect-tolerant computer architecture: Opportunities for nanotechnology," *Science*, vol. 280, no. 5370, pp. 1716–1721.

36. S. C. Goldstein and M. Budiu, "NanoFabrics: Spatial computing using molecular electronics," in *Proc. Int. Symp. on Computer Architecture*, 2001, pp. 178–189.

37. S. Goldstein and D. Rosewater, "Digital logic using molecular electronics," in *Proc. IEEE Int. Solid-State Circuits Conference*, vol. 1, pp. 204–459, 2002.

38. M. S. Gudiksen, L. J. Lauhon, J. Wang, D. C. Smith, and C. M. Lieber, "Growth of nanowire superlattice structures for nanoscale photonics and electronics," *Nature*, vol. 415, pp. 617–620, 2002.

39. A. DeHon and K. K. Likharev, "Hybrid CMOS/nanoelectronic digital circuits: devices, architectures, and design automation," in *Proc. IEEE/ACM Int. Conf. on Computer-aided Design*, 2005, pp. 375–382.

40. A. DeHon, P. Lincoln, and J. Savage, "Stochastic assembly of sublithographic nanoscale interfaces," *IEEE Trans. on Nanotechnology*, vol. 2, no. 3, pp. 165–174, 2003.

41. J. E. Savage, E. Rachlin, A. DeHon, C. M. Lieber, and Y. Wu, "Radial addressing of nanowires," *ACM Journal on Emerging Technologies in Computing Systems*, vol. 2, no. 2, pp. 129–154, 2006.

42. R. Beckman, E. Johnston-Halperin, Y. Luo, J. E. Green, and J. R. Heath, "Bridging dimensions: demultiplexing ultrahigh density nanowire circuits," *Science*, vol. 310, no. 5747, pp. 465–468, 2005.

43. P. J. Kuekes and R. S. Williams, "Demultiplexer for a molecular wire crossbar network (MWCN DEMUX)," US Patent 6,256,767, 2001.

44. T. Hogg, Y. Chen, and P. Kuekes, "Assembling nanoscale circuits with randomized connections," *IEEE Trans. on Nanotechnology*, vol. 5, no. 2, pp. 110–122, 2006.
45. E. Rachlin, "Robust nanowire decoding," 2006. [Online]. Available: http://www.cs.brown.edu/publications/theses/masters/2006/eerac.pdf
46. D. A. Anderson and G. Metze, "Design of totally self-checking check circuits for m-out of-n codes," in *Proc. Int. Symp. on Fault-Tolerant Computing, "Highlights from Twenty-Five Years,"* pp. 244–248, June 1995.
47. M. H. Ben Jamaa, D. Atienza, K. E. Moselund, D. Bouvet, A. M. Ionescu, Y. Leblebici, and G. De Micheli, "Variability-aware design of multilevel logic decoders for nanoscale crossbar memories," *IEEE Trans. on Computer-Aided Design*, vol. 27, no. 11, pp. 2053–2067, Nov. 2008.
48. R. K. Brayton, *Sensitivity and Optimization*. Elsevier, 1980.
49. J. M. Rabaey, *Digital Integrated Circuits: A Design Perspective*. Prentice-Hall International Editions, 1996.
50. C. F. Cerofolini, "The multi-spacer patterning technique: a non-lithographic technique for terascale integration," *Semiconductor Science and Technology*, 23(2008) 075020.

Leveraging Emerging Technology Through Architectural Exploration for the Routing Fabric of Future FPGAs

Soumya Eachempati, Aman Gayasen, N. Vijaykrishnan,
and Mary Jane Irwin

Abstract Field-programmable gate arrays (FPGAs) have become very popular in recent times. With their regular structures, they are particularly amenable to scaling to smaller technologies. They form an excellent platform for studying emerging technologies. Recently, there have been significant advances in nanoelectronics fabrication that make them a viable alternative to CMOS. One such promising alternative is bundles of single-walled carbon nanotube (SWCNT).

In this chapter, we explore several different architectural options for implementing the routing fabric of future FPGA devices. First, we present the benefits obtained from directly replacing the copper interconnect wires by SWCNT bundle interconnect. The architectural parameters are tuned to leverage this new interconnect. Second, the routing fabric is completely redesigned to use nanowires arranged in a crossbar with molecular switches placed at their junctions to provide programmability. A thorough evaluation of this architecture considering various wire parameters is conducted and compared to the traditional SRAM-based FPGA. The chapter presents attractive area-delay product advantages, thus setting the motivation for using these emerging technologies in the FPGA domain. Most of these benefits are due to the area reduction that nanoscale technologies can provide (nearly $100\times$ smaller area). On average for MCNC benchmarks, a critical path delay benefit of 21% was observed when SWCNT nanowire crossbar and molecular switches were used (SWCNT-Arch1) over the traditional buffered-switch SRAM FPGA architecture.

Keywords FPGA · SWCNT · Routing Fabric

N. Vijaykrishnan, (✉)
Department of Computer Science and Engineering, The Pennsylvania
State University, University Park, PA, USA
e-mail: vijay@cse.psu.edu

N.K. Jha and D. Chen (eds.), *Nanoelectronic Circuit Design*,
DOI 10.1007/978-1-4419-7609-3_6, © Springer Science+Business Media, LLC 2011

1 Introduction

For the past decade, it has been accepted that one of the big roadblocks in continuing to apply Moore's law is the global interconnect and its delay, which is increasing with technology scaling [1]. Technological advancements have been the major propellant for architectural innovations. Several nanoscale technologies have emerged as alternatives for interconnects as well as logic.

One of the most promising new options for interconnects is bundled SWCNT [2–4]. Copper interconnect is reaching its physical limits, and has shown to be plagued by electromigration at very small dimensions, which results in permanent faults in the copper interconnect. The resistivity of copper is also increasing rapidly. Large-scale research efforts have been undertaken by both academia and industry to analyze feasible options for replacing or supplementing copper in CMOS [40, 41, 44].

Fabrication is one of the key concerns for the nanoscale proposals. There are two approaches to fabrication, i.e., top-down and bottom-up. Both are being studied. Nanoimprint and conventional lithography are examples of top-down approaches, while self-assembly techniques, where molecules assemble to form nanostructures, are instances of bottom-up methodologies [5]. At present, the latter techniques are limited to very regular structures, and unlikely to produce the complex structures that current lithography can produce. Some studies show that nanoimprint lithography is (comparatively) low-cost and reliable [6, 7]. This raises hopes for cost-effective nanoscale fabrication.

The effectiveness of the emerging nanoscale technologies is demonstrated by applying them to FPGAs [39, 41, 42]. Due to their reconfigurability property, FPGAs are very amenable to experimentation and can be deployed for demonstrating prototypes. A contemporary island-style FPGA consists of RAM modules, hard-coded blocks such as multipliers, and some processors apart from the basic logic blocks. Additionally, the basic logic blocks that consist of lookup tables (LUTs) have been enhanced with fast carry-chain circuits. These additional features are filling the gap between FPGAs and ASIC designs for various applications [8].

Figure 1 shows a Virtex-4 FPGA architecture. It stores the configuration information in SRAM cells, each of which typically consists of six transistors. The basic logic element in a Virtex-4 is called a slice. A slice consists of two LUTs, two flip-flops, fast carry logic, and some wide multiplexers (MUXes) [8]. A configurable logic block (CLB) in turn consists of four slices and an interconnect switch matrix. The interconnect switch matrix consists of large MUXes controlled by configuration SRAM cells. Note that Fig. 1 is not drawn to scale, and in reality, the interconnect switches account for nearly 70% of the CLB area. The FPGA contains an array of such CLBs along with block RAMs (BRAMs), multipliers, and I/O blocks as shown in Fig. 1.

Another kind of FPGA is the accelerator FPGA from Actel, which consists of antifuse-based interconnects. The main drawback of these FPGAs is that they are not reprogrammable. Another category of FPGAs from Actel is the ProASIC family, which uses flash technology to implement reprogrammable interconnects.

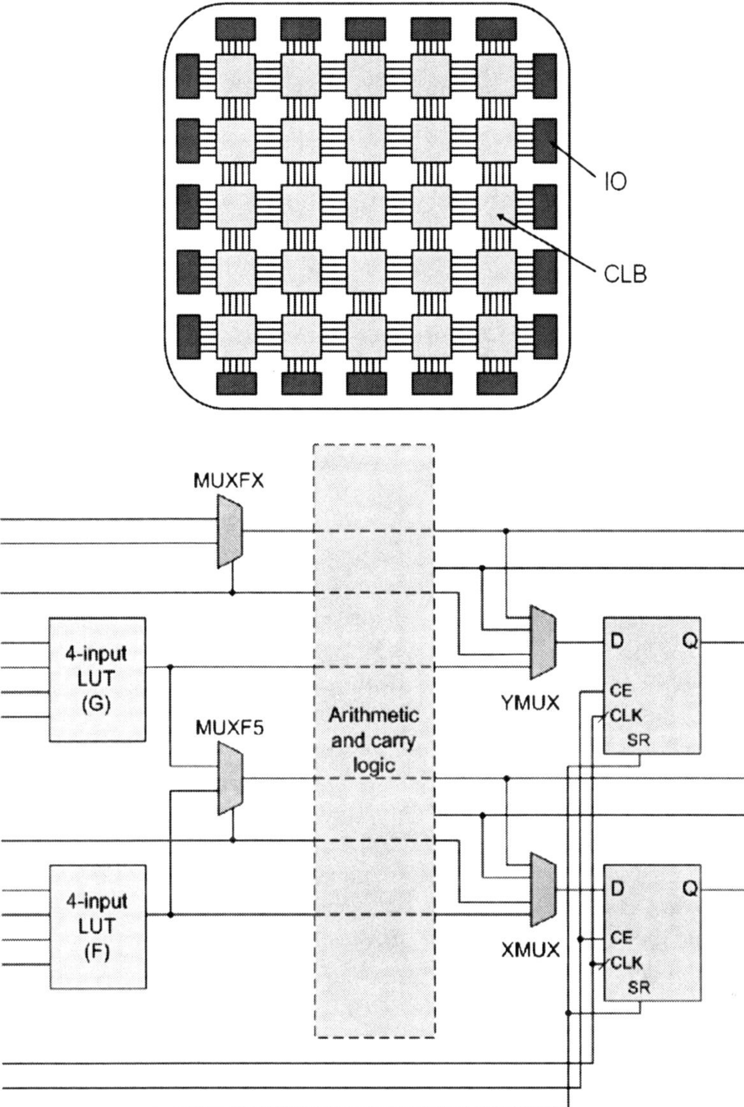

Fig. 1 *Top*: Virtex-4 architecture. CLB is made of 4 slices. *Bottom*: a simplified slice (*Source*: 1-core.com)

Although it mitigates the reprogrammability problem, there are issues with the scaling of the flash-based process. As reported in [9], it is difficult to sustain the programming voltages when the oxide used for the floating gate is made very thin.

It is widely accepted that interconnects are a major bottleneck in FPGAs. Even after careful timing-driven packing and placement, interconnects are the dominant source of delay for most designs. Interconnects also contribute to nearly 85% of the power consumption in FPGAs [10]. The interconnect MUXes in Xilinx's Virtex-4 FPGAs occupy around 70% of the CLB area. Consequently, FPGAs form an excellent platform for working on the interconnect problem. With the rising delay of copper, the interconnect problem is exacerbated. Newer materials, such as bundles of SWCNT that have lower resistivities, might aid in alleviating the interconnect problem. Contemporary FPGAs use a segmented architecture where the interconnect fabric consists of different-length segments and different connection and switch box configurations. For such architectures, these parameters relevant to the routing fabric need to be reevaluated when new material is used.

We evaluated the simplest case of directly replacing the inter-CLB copper wires by bundles of SWCNT. The routing fabric is then tuned to fully harness the new technology. Then, a new architecture is built guided by the lessons learnt from earlier architectural explorations. Nanowires are arranged as a crossbar and molecular switches are used at the cross-point junctions in the proposed architecture. These molecular switches replace the pass-transistors in the FPGA fabric and they are reprogrammable [5, 11]. This alleviates the need for SRAM cells to control the state of the switch, since these molecules store the state within themselves. This is similar to anti-fuse FPGAs, but in contrast to anti-fuse technology, these molecules are reprogrammable.

We explore nanowires of different widths and materials as interconnect. Specifically, SWCNT and metal nanowires were analyzed as candidate materials for the crossbar. Furthermore, we expect the structure of the CLB to be more difficult to realize efficiently in a technology that is more suited to the implementation of regular structures. Therefore, the logic blocks in our architecture are fabricated using lithographic techniques.

The chapter is organized into eight sections. Section 6.2 provides the background of the technologies used in this chapter. Section 6.3 provides the model for SWCNT bundles. The architectural exploration for SWCNT interconnect is detailed in Sect. 6.4. In Sect. 6.5, the implementation and details of the nanowire crossbar and molecular switch-based architectures are presented. Section 6.6 provides experimental results for the nanowire architectures. Section 6.7 presents the prior work. Section 6.8 concludes the chapter.

2   Primitives

2.1   Single-Walled Carbon Nanotube Bundled Interconnect

SWCNTs have been proposed as a possible replacement for on-chip copper interconnects due to their large conductivity and current-carrying capabilities [12]. SWCNTs are rolled graphitic sheets that can either be metallic or semiconducting, depending on their chirality (Fig. 2) [13]. Due to their covalently bonded structure, CNTs are extremely resistant to electromigration and other sources of physical

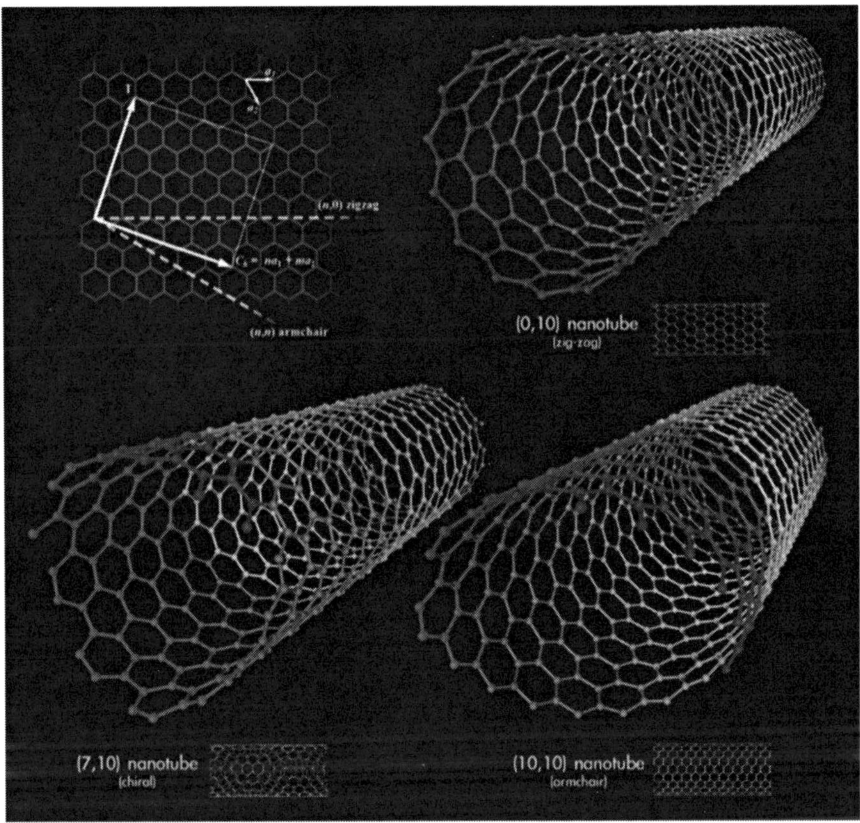

Fig. 2 SWCNT is either metallic or semiconducting based on chirality. (*Source*: Wikimedia)

breakdown [14]. SWCNT fabrication technology will play a crucial role in determining the viability and performance of nanotube bundles in future integrated circuits [15]. SWCNT bundles consist of metallic nanotubes that are randomly distributed within the bundle.

With no special separation techniques, the metallic nanotubes are distributed with probability $P_m = 1/3$ since approximately one-third of possible SWCNT chiralities are metallic [13]. However, techniques, such as alternating current (AC) dielectrophoresis [16] and ion-exchange chromatography [17] have the potential to increase the proportion of metallic nanotubes. The interface between a copper interconnect and SWCNT bundle interconnect is modeled as a metal contact. Consequently, the increased lumped resistance due to imperfect metal contacts to the nanotubes is significant for local and intermediate interconnect applications [2].

A recent work has demonstrated the growth of aligned CNT arrays on commonly used metal substrates, such as Al, Si [18]. Previously, the growth of CNT was restricted to nonconductive substrates owing to significant catalyst–metal interactions.

This growth process yields significantly lower contact resistance for Au, Ag, and Al metals. The ability to grow CNT directly on copper eliminates the metal contact interface, resulting in the possibility of seamless integration of the SWCNT-Cu interconnect in the future. For global interconnects, the ohmic resistance dominates over the contact resistance, making the impact of variation in bonding techniques negligible. In addition, fanout in SWCNT interconnect has been conservatively modeled through metal contacts due to a lack of known fabrication techniques currently.

2.2 *Nanowire-Based Crossbar Interconnect*

Several nanostructure fabrication techniques have been proposed over the past few years. Among them, nanoimprint [6, 7] and "dip pen" nanolithography (DPN) [19] as shown in Fig. 3a, b, respectively, are promising techniques. In the case of nanoimprint technology [6, 7], e-beam lithography (or any other technique) is used to create a mold, which is subsequently used to emboss the circuit on other chips for mass production. The mold can be made very fine, and the technique is expected to scale up to a few nanometers of feature size.

DPN [19], on the other hand, uses an atomic force microscope (AFM) to write the circuit on the die. Although inherently slower than nanoimprint, using multiple AFM tips improves the writing speed significantly. This has been demonstrated to produce very small features and is expected to fabricate features smaller than 10 nm. Directed self-assembly [5] is another approach for making nanostructures. Although this may be the cheapest way for fabricating circuits, it suffers from very high defect rates. Note that all these (nanoimprint, DPN, and self-assembly) technologies are expected to be limited to very simple geometries. It has been shown that it is possible to get sets of parallel wires using any of the above techniques. Therefore, we propose to use them (preferably nanoimprint) to make only wires in the FPGA. These wires could be made using a single crystal of metal-silicide (e.g., NiSi nanowires [20]), made out of metal or SWCNT.

In addition to the wires, we also need some sort of programmable switches to provide programmable connections among the wires and between wires and logic

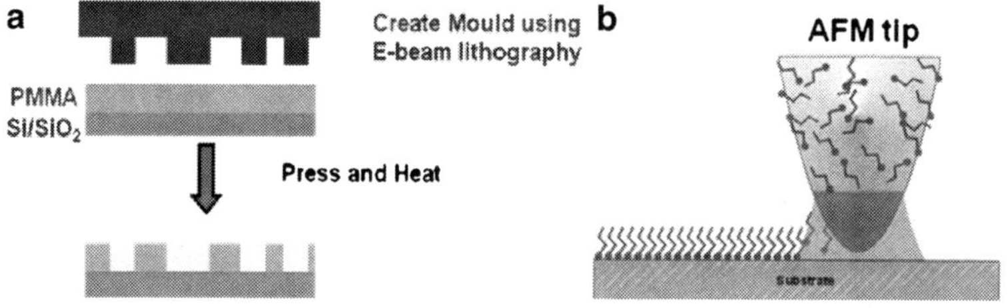

Fig. 3 Fabrication techniques: (**a**) Nanoimprint and (**b**) dip pen nanolithography

pins. In the FPGAs of Xilinx and Altera, these connections are made using pass transistors and SRAM cells, while accelerator FPGAs use one-time programmable anti-fuse material. At the nanoscale, we can use single-molecule switches that exhibit reversible switching behavior [21]. These molecules self-assemble at the cross-points of nanowires, and can be switched between ON and OFF states by the application of a voltage bias. It is desirable that these switches have very low ON resistance and a very large OFF resistance. ON resistances of hundreds of ohms and OFF/ON ratios of 1,000 have been observed recently [11]. Note that very fast switching characteristics are not essential for FPGAs, because these switches will not be configured very frequently and the FPGA configuration time is normally not critical. In this chapter, a fixed vertical separation between nanowires of 30 nm is assumed.

3 Modeling of SWCNT Interconnect Bundles

Even though SWCNTs have desirable electrical properties, individual carbon nanotubes have very high contact resistance that is independent of the length [22]. To overcome this limitation, bundles of SWCNT in parallel have been physically demonstrated as a possible interconnect medium [12]. We utilize the circuit model in [2] that can characterize their behavior in a scalable manner. Each SWCNT has lumped ballistic ($R \sim 6.5$ kΩ) and contact (R_c) resistances. The contact resistance is due to imperfect SWCNT–metal contacts, which are typically constructed using gold, palladium, or rhodium [23]. The nanotubes also have a distributed ohmic resistance (R_o), which is determined by the length L_b and mean free path of acoustic phonon scattering (λ_{ap}) in the individual nanotubes. Since the individual SWCNTs have a maximum saturation current (I_o), the overall resistance also depends on the applied bias voltage ($R_{hb} = V_{bias}/I_o$) [22]. The total resistance of a nanotube bundle (R_b) is R_t/n_b where R_t is the resistance of an individual nanotube and n_b is the total number of nanotubes in the bundle. The capacitance of a nanotube bundle consists of both a quantum capacitance and an electrostatic capacitance, which can be modeled using the techniques presented in [2]. The inductance of an SWCNT bundle can be modeled using the scalable inductance model presented in [24]. The inductance can have a relatively large impact for global interconnect, the magnetic inductance being the dominant source [25, 26]. Table 1 shows the values of various parameters.

Table 1 SWCNT technology assumptions

Technology assumption	P_m	$R_i + R_c$ (kΩ)	C_λ	I_o (μA)	Min d_t (nm)
Conservative	0.33	20	889	20	1
Moderate	0.6	10	889	25	0.8
Aggressive	0.9	6.5	1333	30	0.4

4 Replacing Copper by SWCNT Bundle-Based Interconnect

The base architecture is the simple island-style FPGA with copper interconnect (CMOS). The architecture under exploration is made up of SWCNT interconnect (CMOS+CNT). The logic blocks and routing switches in the base case as well as in the SWCNT case are in CMOS scaled to the 22 nm technology. The goal of these experiments is to search for the routing configuration that will fully leverage the new technology while working around its inherent limitations. The routing architecture in an FPGA is tailored to support both long and short connections and provide sufficient flexibility to connect different logic blocks. Consequently, FPGA interconnect in a channel is composed of segments that are of different lengths. For example, a channel with ten wires has three wires spanning a single logic block (single-length segment), four double-length segments spanning two logic blocks, and three hex-length segment spanning six logic blocks as shown in Fig. 4a. Wire segments can also span the entire row of logic blocks (long lines).

In Table 2, the second column provides a typical distribution of wire lengths in a Virtex-4 FPGA. Wire length segment distributions chosen for architecture are influenced by the underlying wire performance. Consequently, when adopting SWCNT-based interconnect in FPGAs, we need to explore how the segmentation distribution would change.

The level of routing flexibility in the interconnect fabric is determined by the switch box (SB) and connection box (CB) configurations. The SB population fraction, displayed in Table 2 (column 4) for the segmentation experiments performed in Sect. 6.4.2, is the ratio of the actual number of SBs to the maximum number of SBs.

Figure 4b illustrates an example of how the population fraction of CB and SB in wiring segments is evaluated. In the SWCNT-based architecture, the additional connectivity imposes challenges since the contact resistance of the connections to the SBs and CBs is significant. Consequently, the SB and CB configurations for a SWCNT-based interconnect fabric have different constraints than traditional Cu-based architectures. In this section, we present and explain the explorations performed in the context of segmentation and routing flexibility.

4.1 Assumptions and Experimental Setup

For the following set of experiments, a logic block consists of a cluster of four LUTs, with ten inputs and four outputs. Both SWCNT and Cu are assumed to have same wire dimensions, such as width, height, and interwire spacing, as predicted at the 22-nm node. The diffusion capacitance was obtained through HSPICE simulations, and the process parameters were obtained from ITRS [1]. The SWCNT bundle interconnect model presented in Sect. 6.3 was used to characterize SWCNT-based wires. The best bundle configurations were selected based on the

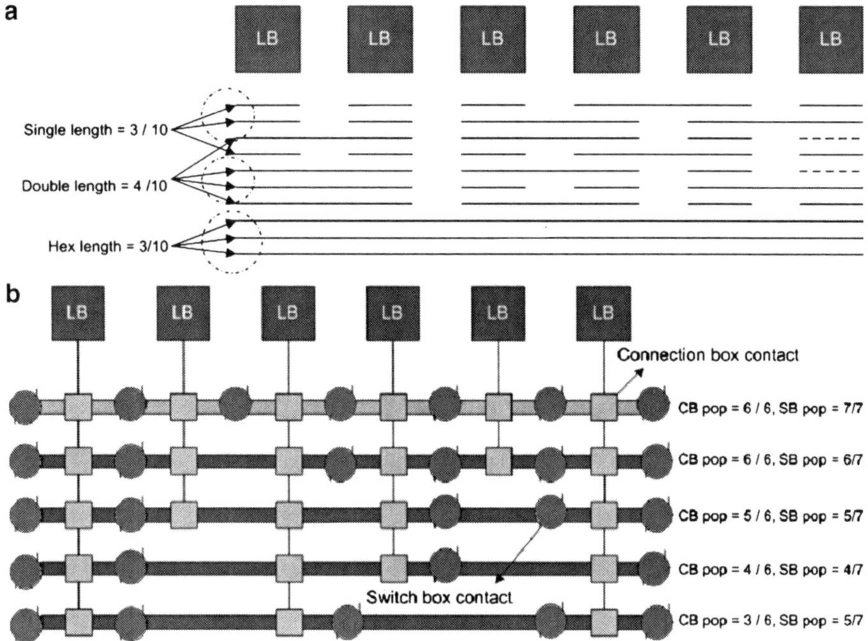

Fig. 4 Example of (**a**) segmentation and (**b**) connection and switch box population

Table 2 Typical routing fabric configuration

Segment length	% of tracks in a channel	CB population	SB population
(Single)	0.08	1	1
2 (Double)	0.20	1	0.66
6 (Hex)	0.60	0.5	0.5
Long	0.12	0.5	0.5

length of the wire segment and the number of contacts due to the SB/CB configuration. Our simulations were performed for the three different assumption sets for SWCNT technology listed in Table 1. These assumptions vary from being conservative to aggressive in nature in order to span the evolution of this technology. Using twenty MCNC benchmarks and VPR [27], we evaluated the area and delay performance of the various architectural options in the design space.

4.2   Segmentation Experiments

In these experiments, four different segment lengths – namely single, double, hex, and long – were used. We simulated 65 different segmentation options for both

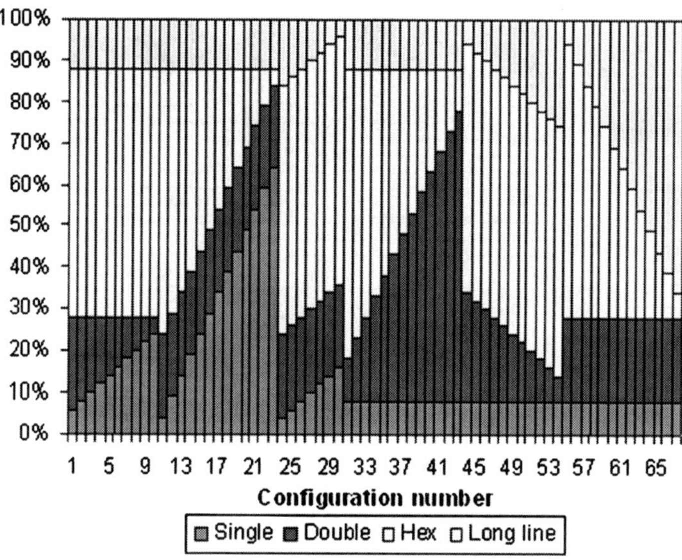

Fig. 5 (*Left*) Pseudocode for segmentation fraction generator. (*Right*) The outcome

Cu- and SWCNT-based architectures. These were generated using the algorithm as shown on the left side of Fig. 5. Let vector w be the initial segmentation ratio $(w_1, \ldots, w_4)$ given in Table 2. Let v be the final array of segmentation vectors. The outcome of the generator is shown on the right-hand side of Fig. 5. For these cases, the CB/SB configuration is shown in Table 2 (third and fourth columns) and these CB/SB configurations were kept constant for this set of experiments.

Figure 6 shows the area-delay product (ADP) for the different configurations over 20 benchmarks for the moderate technology assumption at the 22 nm technology node. The results indicate that a larger benefit can be obtained from longer-length wires (double and hex) for an SWCNT-based architecture. There is an average improvement of 12.8 and 13.8% over Cu-based architectures in ADP and critical path delay (CPD), respectively.

Similar trends were obtained for the other technology assumptions and a maximum improvement of 17.5% in CPD and 19.0% in ADP was obtained for the aggressive technology. The benchmarks that experienced the least performance improvement, e.g., "diffeq," "frisc," and "tseng" (Fig. 6), utilized a significant number of short interconnect segments. The shorter-length segments have higher per unit length resistance due to the higher SWCNT–metal contact resistance.

In contrast, an improvement in ADP of as high as 54% was seen for the "bigkey" benchmark, which maps relatively well to the long wires. Based on the distribution of wire segment lengths for the best-performing SWCNT-based FPGA architecture (minimum average CPD/ADP), displayed in Table 3, we can infer that the SWCNT-based architectures prefer longer wire segments than Cu-based architectures. The reason for this trend is the significant nanotube contact resistance. Also,

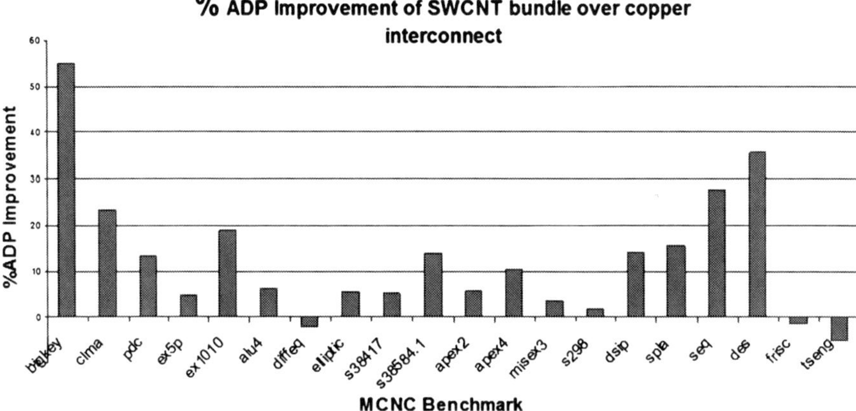

Fig. 6 CMOS+CNT architecture ADP improvement over CMOS

Table 3 Architectures with lowest average CPD and ADP

Length	Best CPD CMOS (%)	Best CPD CMOS+CNT (%)	Best ADP CMOS (%)	Best ADP CMOS+CNT (%)
1	39	16	8	16
2	20	20	65	20
6	29	60	15	60
Long	12	4	12	4

notice a slight reduction in the fraction of long lines, which do not benefit as much from SWCNT interconnects since they are less flexible due to the limited connectivity they can offer. Table 4 summarizes the segmentation experiment results.

4.3 SB/CB Configurations

From the earlier discussion, we understand that one of the constraints of SWCNT bundled interconnect that we need to work around is the high contact resistance. This limits the routing flexibility that SWCNT can offer. In this section, the results of the switch population experiments are presented. In order to keep the design space tractable, we use a nonsegmented architecture, i.e., every wire in the architecture is of the same length. We experiment with wire lengths from 2 to 6. A total of 33 different architectures derived by varying the fraction of CBs and SBs were simulated. As expected, a smaller number of CBs and SBs are preferred for SWCNT, but it significantly degrades routability. In fact, for some benchmarks, few of the architectures were unable to route them because of a very low number of

Table 4 Segmentation experiment results

	% Improvement in ADP	% Improvement in CPD
Max	54 (bigkey)	23 (bigkey)
Min	−5 (tseng)	−8.7 (dsip)
Average	12.7	13.8

Table 5 Best-performing architecture for the CB/SB experiments

Arch for min average ADP	Moderate SWCNT	Aggressive SWCNT	Copper
Length	3	5	2
Fraction of CBs	0.75	0.66	0.66
Fraction of SBs	1	0.8	1

Table 6 Results for the CB/SB population experiments

CNT tech. assumption	% Improvement in ADP	% Improvement in CDP
Conservative	1.03	0.39
Moderate	9.66	6.24
Aggressive	14.0	8.81

contacts. Consequently, longer-length wire segments based on SWCNT wires with moderate connectivity provide a good solution that satisfies both flexibility and performance.

The longer SWCNT wire segments help balance the increased contact resistance due to increased CB/SB connections with decreased ohmic resistance. For SWCNT wires, the ohmic resistance does not grow linearly with length. The best combination of wire segments, CB and SB configurations for CNT achieved an 8% improvement in CPD and a 14% improvement in ADP over the best configuration of these parameters for the Cu wire. The best performing architectures are given in Table 5.

Table 6 shows the improvements with the best combination of segment lengths and SB/CB configurations of CNT over that of Cu averaged over all benchmarks. For more aggressive technology assumptions, the preferred wire lengths further increase (see Table 5). For aggressive assumptions, both ohmic and contact resistances decrease. Yet the decreased ohmic resistance has a bigger impact resulting in the increased optimal length for balancing the contact resistance.

5 Routing Fabric Design Using Crossbar and Molecular Switches

Having understood what would be required to adopt nanotechnology in FPGA interconnect, we take the next steps of conducting an exploration for designing an architecture built out of a crossbar of nanowires with molecular switches at the crosspoint. Such an architecture is shown in Fig. 7.

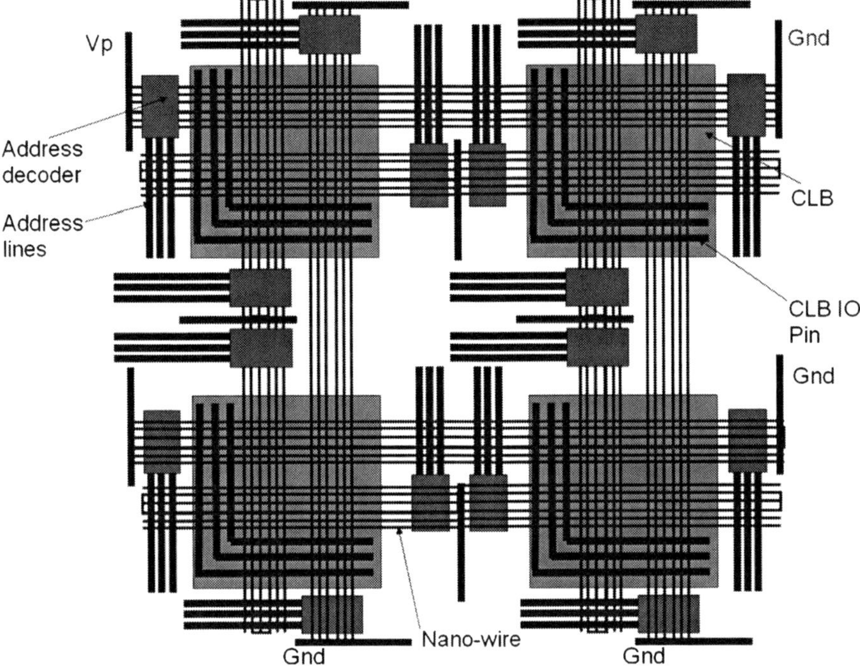

Fig. 7 FPGA using nanowires and molecular switches

These molecular switches remember the state and, thus we can save area and power by eliminating the SRAM cells. As in the previous approach, the logic blocks are in silicon at the 22 nm technology node. Two different architectures were explored. In the first architecture (Arch1), we use nanowires that are fabricated using nanofabrication technology and, thus, have high density and pitch as small as 10 nm. The interconnect switches at the junctions are self-assembled molecular switches. These switches provide connections between the nanowires as well as connections to the logic block (i.e., CB and SB). For this architecture, decoders are required to uniquely address a crosspoint for programming. In the second architecture (Arch2), the inter-CLB wires are made from lithography that will be used at the 22 nm technology node and programming connections are provided by the molecular switches.

The main difference between the first and second architecture is the wire-pitch that can be achieved between nanowires, which is 10 and 54 nm, respectively. In addition, the molecular switches do not have signal restoration (unbuffered). These are compared to the traditional Xilinx-style FPGA where pass transistors and SRAM cells are used for programming the routing fabric. For comparison, we simulated both the unbuffered switches (Arch3) as well as buffered switches (buf-Arch3).

5.1 Architecture 1

Figure 8 shows the 3D organization of this architecture. The intra-logic block interconnect is confined to M1 and M2 layers. The I/O pins of the logic block are in the M2 layer and the nanowires are on top of this. Each layer in Fig. 8 is isolated from its adjacent layers by a dielectric. The nanowires can be made of any material. In particular, copper, metal silicide, and carbon nanotubes bundles have been analyzed here. Typically, an interconnect should possess low resistivity and high current-carrying capability, and need a small pitch to allow higher integration densities. Cu is expected to have resistivity of the order of 2.2 $\mu\Omega$/cm at 22 nm. Yet, its current-carrying capabilities are limited by electromigration at these small pitches. SWCNT bundles not only have lower resistivity when compared to Cu but also have high current-carrying capabilities. However, SWCNT interconnect is limited by the contact resistance. Consequently, we evaluate copper vs. SWCNT architectures.

Next, the routing architecture needs to be chosen. A segmented routing architecture with three segments of length single, double, and triple are used. The logic block (8 LUTs + FFs) in the 22 nm technology is expected to be around 12.5 μm $\times$ 12.5 μm. In addition to this, the decoders take some space. Therefore, a single-length wire in our architecture needs to run 25 μm, a double-length wire 38 μm, triple-length wire 50 μm. Routing with 20% single-length, 30% double-length, and 50% triple-length wires is chosen. This choice is favorable to SWCNT, as evident from the earlier exploration. The decoders are made using advanced nanoimprint technology. Use of $\sqrt{N}$ control signals for a decoder to address N wires is proposed in [28]. We use a similar technique and, therefore, account for 15 decoder control signals for 200 wires in the FPGA channel. Note that these decoders are needed only to configure the switches and are switched off at operation time.

The switches can be configured by applying correct voltages at the wires (similar to anti-fuse FPGAs). The logic functionality of this FPGA can be easily programmed using SRAM cells. Programming the routing is similar to antifuse FPGAs, except that we need decoders to address the nanowires and that the switches are reprogrammable. The main concept is that the wires should be activated in some particular order to avoid impacting the wrong switches. A way

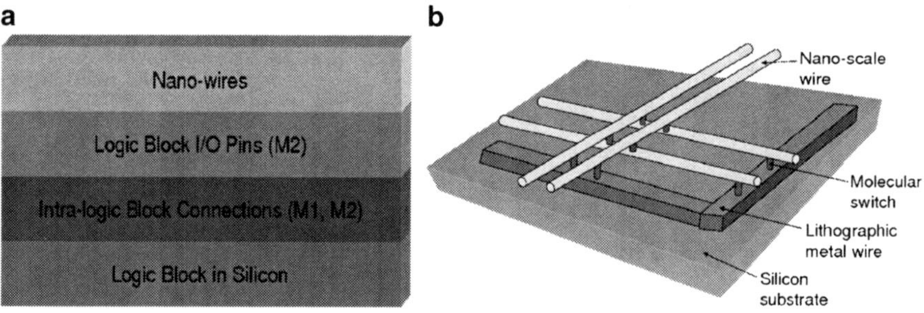

Fig. 8 3D organization of nanowires

to program the anti-fuses in an anti-fuse FPGA is presented in [29]. This is directly applicable to our architecture too. Initially, all the molecular switches are off and all the wires are precharged to a voltage $V_p / 2$. This is required to ensure that the voltage difference of V_p is applied only to the desired switch. Then the two wires that need to be connected through a switch are addressed using a decoder and pulled to V_p and ground, respectively, thus applying a voltage difference of V_p to the molecular switch that needs to be turned on. Note that V_p needs to be larger than the operating voltage. Experiments with molecular switches have shown a value of 1.75 V [21], which is more than double that of the operating voltage at the 22-nm node.

5.2   Architecture 2

In this architecture, the nanowires are replaced by lithographic wires to form a crossbar and keep the molecular switches at the junction. As explained earlier, the main difference between Architectures 1 and 2 is the wire pitch that can be achieved by the fabrication technologies. In addition, this architecture does not require decoders for addressing purposes. The configuration bits present in the island-style architecture can be used for programming the molecular switches. In Architecture 1, the nanowires are packed together and, thus, fewer configuration bits are used for performing the addressing of all the molecular switches using decoders.

Note that assuming a channel width of 200, the area of the CLB will be determined by the wires instead of the logic. For the 22 nm technology, ITRS predicts a wire pitch of 54 nm. For a channel width of 200, we will need 400 wires within the CLB pitch. This comes out to be $40 \times 54 = 21.6\,\mu\text{m}$ long. In addition, we will need space for the logic pins, which leads to $40 \times 54 = 2.16\,\mu\text{m}$. Therefore, the CLB dimensions in this case are projected to be $23.76\,\mu\text{m} \times 23.76\,\mu\text{m}$, which is only slightly smaller than the current Xilinx CLB scaled to the 22 nm technology ($25\,\mu\text{m} \times 25\,\mu\text{m}$). We explore both carbon nanotube bundled interconnect and Cu interconnect for the inter-CLB wires.

5.3   Evaluation

In order to model the proposed architectures in VPR, we modeled a new type of SB that allows a wire to connect only to the wires at right angles to it. This was done because in Architecture 1, molecules assemble only at wire crossover points and not between two wires running in the same direction. In order to account for the large defect rates expected at this scale, we started with the assumption that only half of the switches are operational, but due to the very large number of programmable switches in our architecture (even when only half of the switches are visible), VPR takes extremely long (>2 days on a SunBlade-2000, for a 191 CLB design) to finish the placement and routing of the designs. In order to facilitate experimentation with multiple designs, we limited the number of switches in VPR to only about 1% of the total physically present switches. Consequently, in VPR, the CLB outputs have switches to only half of the wires in the channel, and a wire can connect to only four other wires in the switch box, two in each of the perpendicular directions.

The performance obtained by limiting the number of switches was not very different from that obtained by keeping all the switches for the few designs we initially experimented with. Since the flexibility provided by our SB is still greater than the switches built in VPR, we expect that our SB is still not very limiting, and similar results will be obtained when all switches are considered too. Note that the junction capacitances between all crossing wires are still counted. MCNC benchmark circuits were used for all experimentation. These designs varied in size from 131 to 806 CLBs. The area used in the results that follow are per CLB area, as the design size is the same across all the architectures.

6 Results

Figure 9 shows that Architecture 1 (where nanowires and molecular switches are used), performs better for all benchmarks when compared to Architecture 2 (lithographic wires + molecular switches) and 3 (original – lithographic wires + SRAM based). The wire width of nanowires is 15 nm and spacing between adjacent wires is 10 nm (a total pitch of 25 nm), and for the lithographic wire, the wire pitch is 54 nm (pitch is divided equally between the width and the interwire spacing), as predicted by ITRS. At smaller widths, the increased resistivity of Cu was taken into account. The model described earlier in Sect. 6.4 for SWCNT was used. The molecular switch ON resistance was assumed to be 1 KΩ.

The CPD of SWCNT-based Architecture 1 is the best among all. SWCNT-Architecture1 performs on an average 13% better than SWCNT-buf-Arch3, 21% better than SWCNT-Architecture2, and 42% better than SWCNT-Architecture 3. In comparison to Cu architectures, the improvements are 22% (Cu-buf-Arch3), 46% (Cu-Arch1), 54% (Cu-Arch2), and 72% (Cu-Arch3), respectively. The area of Architecture 1 is only 30% of that of Architecture 2. We see that a buffered Xilinx-style architecture (buf-Arch3) is superior to the unbuffered architecture (Arch3).

The performance gains obtained by using the proposed architecture depend on the ON resistance of the molecular switch. The ON resistance is varied from 100 KΩ down to 100 Ω. The average CPD normalized to Cu-Architecture 3 is shown in Fig. 10. It shows that Architectures 1 and 2 are only beneficial over the regular Cu and SRAM-based buffered FPGA architecture when the ON resistance is less than 1 KΩ. In fact, for the 100 KΩ switch resistance, the CPD is 10× that of the base Cu-Architecture3.

In addition, we can see that these nanowire architectures are only beneficial when SWCNT wires are used. Even if we can build Cu nanowires that can

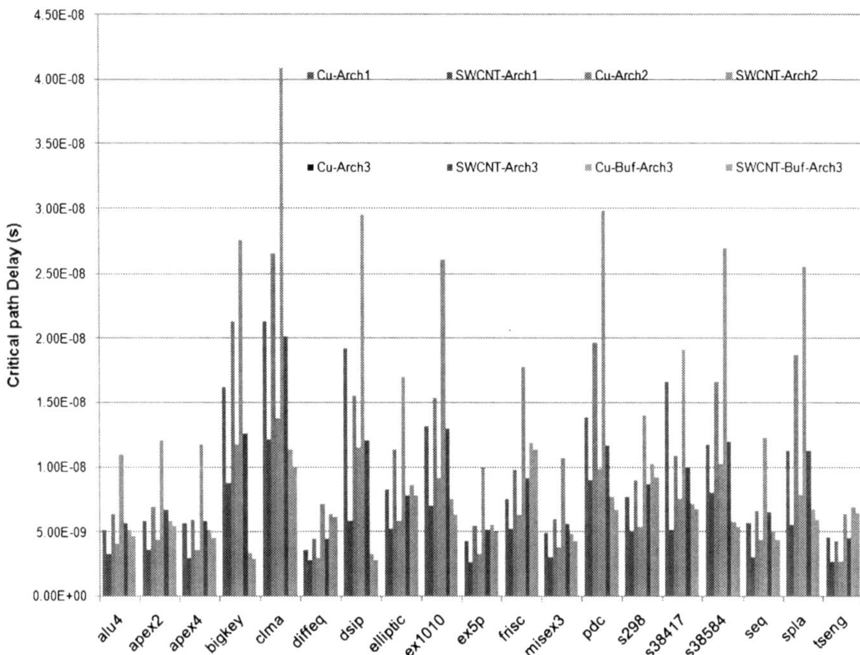

Fig. 9 CPD: Arch1 (nanowires + molecular switches), Arch2 (lithographic wires + molecular switches) and Arch3 (lithographic wires + pass transistor switches). (Switch resistance 1 KΩ)

withstand electromigration, they will not provide any benefit in terms of the CPD. However, if the area is considered, then even the Cu-nanowire-based architectures (Cu-Arch1) will be 100× smaller than any of the Arch3 configurations. Next, we will delve into the configurations with molecular switch architectures and analyze the results in terms of area delay product. Compared to Cu-Arch2, Cu-Arch1 yields ADP improvements ranging from 85% when 100 KΩ switch is used to 77% when 100 Ω is used (see Fig. 11). These benefits increase from 85% (100 KΩ) to 89% (100 Ω) for corresponding SWCNT Architecture 1. Thus, when the molecular switch is at its early phase and has a high ON resistance, the use of SWCNT could lead to higher benefit than when copper is used with 1,000× smaller ON resistance switch (advanced). For SWCNT-Arch2, the ADP improvement increase from 16 to 50% over Cu-Arch2. This indicates that the contact resistance constrained SWCNT interconnect benefits from smaller molecular switch resistance.

Simply using a buffered switch in the Cu + SRAM-based architecture gives 53% better ADP, and 64% CPD improvement compared to the unbuffered case. This can be seen from the Figs. 10 and 12 in tandem. For the buffered configuration, direct replacement of Cu by SWCNT yields 31% improvement in ADP. Additionally, replacing the pass transistor switches by molecular switches provides 100× area reduction. Thus, for reaping early benefits before the technology to build efficient molecular switches is ready, Architecture 3 with SWCNT wires and buffers is a

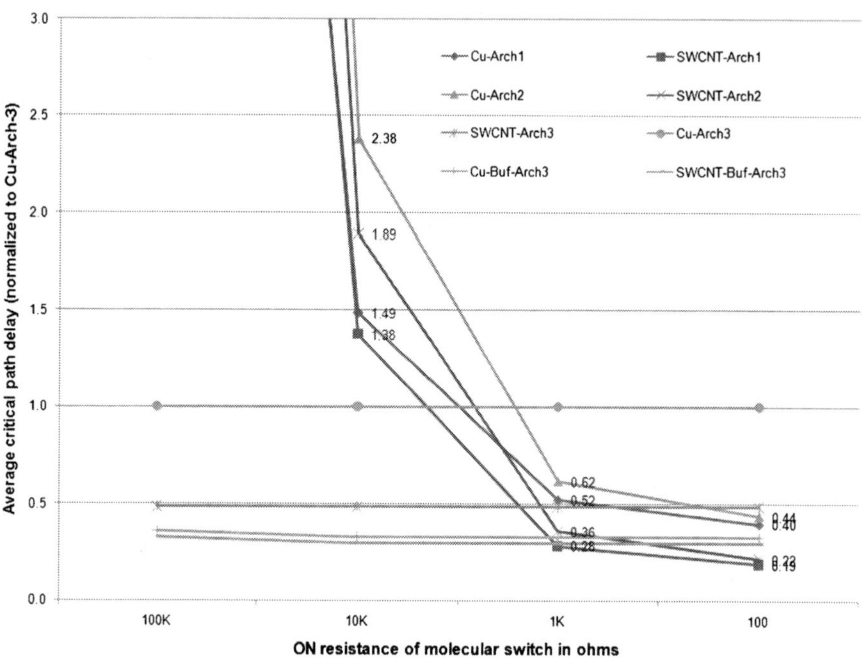

Fig. 10 CPD dependence on ON resistance of molecular switch

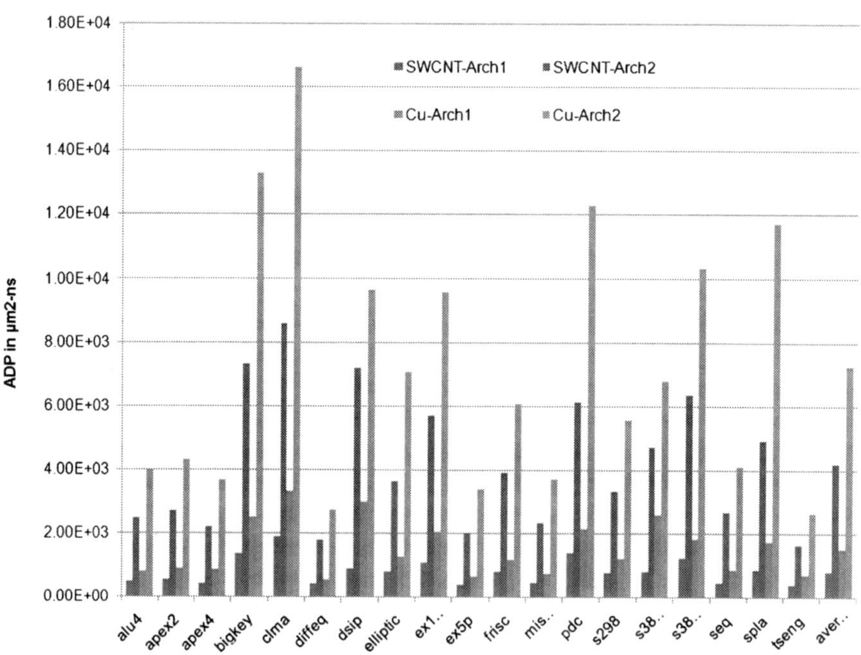

Fig. 11 ADP for Architectures 1 and 2 for 1 KΩ ON resistance

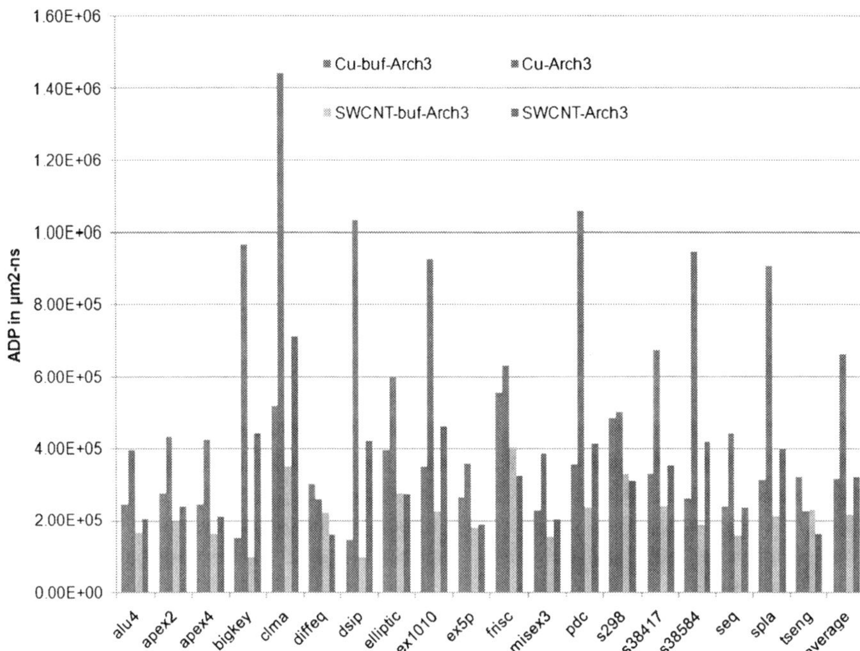

Fig. 12 Architecture 3: ADP

good option. Such a direct replacement of Cu with SWCNT gives an average of 10% improvement in CPD. When SWCNT is used, the number of contacts needs to be considered and longer length segments are preferable in order to take full advantage of the technology.

7 Related Work

The last decade has seen a number of research projects towards realizing logic and/ or interconnects with emerging technologies. A less disruptive architecture, when deploying the emerging technology, is more likely to be chosen in the early stages. Goldstein [30] proposed to build crossbar-based devices by aligning the nanowires at right angles and placing molecules at the crosspoint. The molecules at the crosspoints provided for both programmable logic as well as interconnections.

However, these devices suffered from signal degradation and there was no way to restore the signal using only two terminals. DeHon [28] solved this problem by using SiNW-based FET for restoration and designed a PLA structure. DeHon's methodology requires selective doping over the nanowires, which is challenging. The logic functionality is limited to OR and inversion. A nanoPLA block was designed, which is similar to our logic block, and a 3D logic chip was proposed [31].

Semiconducting nanowires, made up of a variety of materials such as Si, Ge, GaAs, were evaluated in this work. The nanoPLA methodology can also be applied to the architecture evaluated in this chapter. Our emphasis here is on the interconnection between the logic blocks as that is a more near-term problem.

Likharev [32] proposed a CMOL circuit that is programmable logic built from a hybrid of CMOS and slightly rotated nanowire crossbar with latched switches at crosspoints. The two-terminal latched switches were made from single-electron tunneling nanodevices. This methodology was evaluated for small circuits. The fault tolerance in the presence of high defect rates was also analyzed. It used the nanowires for interconnection by selectively configuring crosspoints. Note that here the nanowires are used for the local interconnect (inside the logic block). This methodology also focuses on the programmability inside the logic block.

The logic architecture proposed by Likharev was modified to be fabrication friendly. A revised architecture on the lines of CMOL called FPNI (field-programmable nanowire interconnect) was proposed by Snider [33]. It contained a hyper cell completely implemented in a CMOS layer that was intended to mimic the logic block. The nanowire-based devices were used only for interconnection. It lifted the configuration bit and the associated components out of the semiconducting plane and placed them in the interconnect by using nonvolatile switches. As a result, a level of intelligence is introduced in the interconnect. Results showed that their design had very high defect tolerance and very small area (almost one-eighth that of the CMOS circuit), while running 22% slower. Our design has a smaller area as well as yields better performance of 73% on an average. This suggests that it is important to optimize the routing fabric.

Tour [21] also used nanowire-based crossbar for building logic and placed the interconnect in CMOS. Another work also uses nanowire-based logic and retains the interconnect in CMOS [34]. These proposals do not target scaling of the main bottleneck, which is the global interconnect. Most of the above proposals use nanowire-based crossbar with switches for building logic in nanoscale FPGAs. Researchers have also proposed using CNT-based memories for storing the FPGA configuration bits [35]. There have been several nanodevice proposals, such as NW-FET using a single nanowire and redox active molecules, CNT-FET, etc., that hold promise for replacing CMOS technology [36]. A field programmable carbon nanotube array was proposed in [37], where the LUT was built using CNTFET devices and NRAM. The intra-logic block and inter-logic block interconnect are all SWCNT bundles. The proposed designs in this chapter, on the other hand, take advantage of the mature CMOS technology and only scale the interconnect to nanoscale.

Dong et al. [38] used a suite of nanomaterials for the interconnect and memory in a 3D-FPGA and showed promising results. The logic block in this proposal is retained in CMOS, but the intra-logic block connections are implemented with a nanowire crossbar. This work also targets the global interconnect problem. It uses SWCNT for both vias and intra-layer interconnect. In addition, it uses the nanowire crossbar and molecular switch structure for providing memory, inter-logic block connections as well as intra-logic block connections. It also uses NRAM, which is made up of CNT, for additional memory. The proposed architecture shows attractive results of 4× reduction in footprint with 2.6× performance benefits over the baseline 2D architecture. Our chapter can provide an additional point of comparison as all the 3D designs proposed were compared with 2D CMOS alone. We provide an enhanced 2D architecture using the emerging technologies.

8 Conclusions

Moore's law is facing a significant roadblock because of the increasing global interconnect delay. Cu interconnect resistivity is increasing with decreasing wire width. In addition, it is predicted that Cu interconnect will be plagued by hard faults caused by electro-migration at very low widths. These problems motivate the exploration of new materials for interconnects in the nanoscale regime. SWCNT is a promising candidate for implementing interconnects. It has a very low resistance and high current-carrying capability. However, a single nanotube has very high contact resistance. Consequently, bundles of SWCNT that decreases the effect of contact resistance have been proposed for achieving superior delay characteristics.

In this chapter, we first explored the architectural options that can fully leverage the advantages offered by the SWCNT bundle interconnect in an island-style FPGA. Note that only the inter-CLB wires are replaced by SWCNT. All the experiments in this chapter assume that the intra-CLB wires (local wires) are made of Cu and the logic block is implemented in the 22 nm CMOS technology. The exploration encompasses varied segmentation ratios and CB/SB population ratios. These experiments showed that SWCNT interconnects favor long distance wires and fewer contacts, implying lower CB/SB population.

Using the lessons learnt from the previous exploration, we experimented with inter-CLB routing fabric made from an array of nanowires laid at right angles (Architecture 1). Molecular switches were placed at their cross-points and provided the programmable connections between the nanowires as well as connections to the logic blocks. We assumed nanowires of small wire-width of 15 nm. Both Cu nanowires and SWCNT wires were evaluated. Cu nanowires at this width might be limited by electro-migration. Yet, they were used optimistically as a possible candidate for the nanowires for the sake of comparison. Another design option, where the width of the wire is kept at the lithographic scale, was also considered for exploration for near-term benefits (Architecture 2). Molecular switches were used at the cross-points of the lithographic wires.

Finally, these architectures were compared to an architecture consisting of pass transistors and SRAM cells for configurability and lithographic wires for inter-CLB connections. Both buffered and unbuffered pass transistor switches were simulated. Note that the molecular switches used in the nanoscale architectures (Architectures 1 and 2) are nonrestoring. For the molecular switch architectures, SWCNT-based nanoscale architectures (SWCNT-Arch1/Arch2) yield significant ADP benefit (upto 89%) over the Cu-Arch1/Arch2. The CPD improvement obtained from

SWCNT-Arch1 over Cu-Arch2 is 57%. Architecture 1 was found to be only 30% in area when compared to Architecture 2. Architectures 1 and 2 provide benefits only if the molecular switch ON resistance is much less than 1 KΩ. Yet, efforts directed towards making a good molecular switch is very attractive with respect to ADP. A simple direct replacement of Cu with a SWCNT bundle at the lithographic scale using buffered pass transistor switches can provide 31% benefit in ADP and 10% in CPD on an average for all the MCNC benchmarks [39–44].

Exercises

This set of exercise problems requires the VPR tool that can be downloaded from http://www.eecg.toronto.edu/~vaughn/vpr/vpr.html

VPR is a placement and routing tool for FPGAs that is easy to learn and is very fast. It is quite portable across platforms.

One of the input files for the tool is an architecture file (.arch). This file takes the technology parameters for the both the logic and the routing fabric. These can be changed in order to experiment with newer technologies or for architectural changes in the FPGA design. The exercise problems below focus on how this tool can be used for these purposes.

1. There are three types of programmable switches in SRAM-based FPGAs: pass transistor-based switch, buffered switch, and mux-based switches. What is the equivalent RC circuit model used in VPR for a buffered switch?
2. Define segmentation distribution.
3. Define the terms *fraction of CB population* and *SB*.
4. What is CB and SB flexibility?
5. Assume that the CLB area is determined by the interconnect area. Let the size of the CLB be 21.6×21.6 μm^2 when the channel width is 200 and the wire pitch is 54 nm. What is the CLB size when the channel width is halved and the wire-pitch is also halved? Ignore the area occupied by the logic pins.
6. Let the logic block size be 25×25 μm^2. Let the wire pitch be 10 nm and interwire spacing is 15 nm. How many tracks can fit in the channel?
7. Read the VPR manual and understand the architecture file parameters. Change the default segmentation distribution to any value of your choice (change the segment frequency). Run VPR for at least five MCNC benchmarks and report the critical path delay, segment utilization, area (optimistically assuming buffer sharing), and channel width for both the base case and your new architecture.
8. In the base segmented architecture where we have four segments, change the CB and SB populations for one segment. Generate at least four pairs of values for CB and SB population. Run VPR and report the critical path delay, total net delay, and total logic delay for at least five MCNC benchmarks. Let VPR choose the routing channel width. *Note*: Sometimes the benchmarks cannot be routed if the CB and SB populations are too low.

9. Change the default VPR file to use nonbuffered switches. You can define a new switch or make an existing switch nonbuffered. There are several parameters that need to be changed.

   (*Hint*: All delays that include a switch need to be changed.) Assume that gate capacitance of NMOS $= 5.25e-16F/\mu m$ and that the diffusion capacitance is the same as the gate capacitance. Let resistance per unit length of NMOS be 75 Ω μm and gate length $= 9$ nm.

   (*Hint*: You have to change the R, Cin and Cout, and Tdel of the switches. In addition, all the delay values such as T_ipin_cblock, Tclb_ipin_to_sblk_ipin, etc., need to be changed.)

10. Switches are used in the wires and at the output pad. Run VPR for at least five MCNC benchmarks and report the channel width, area, critical path delay, and segment utilization for the following scenarios: (a) only nonbuffered switches are used and (b) only buffered switches are used.

11. Perform the experiments similar to Exercise 10 for (a) a mix of buffered and unbuffered switches: when buffered switches are used in the wire and opin switches are unbuffered, and (b) mix of buffered and unbuffered switches: when unbuffered switches are used in the wire and opin switches are buffered.

12. Repeat the experiments in Exercise 11 for (a) a mix of buffered and unbuffered switches: where all the short segments have unbuffered switches and the segments with length >6 use buffered switches, and (b) all the short segments have buffered switches and the segments with length >6 use unbuffered switches.

Acknowledgments This work was supported in part by NSF CCF award 0829926 and CNS 0916887.

References

1. International technology roadmap for semiconductors. http://public.itrs.net/Files/2007ITRS/Home2007.htm 2007

2. Y. Massoud and A. Nieuwoudt, "Modeling and evaluating carbon nanotube bundles for future VLSI interconnect" in *Proceedings of IEEE International Conference on Nano-Networks*, 2006.

3. N. Srivastava and K. Banerjee, "Performance analysis of carbon nanotube interconnects for VLSI applications," in *Proceedings of International Conference on Computer-Aided Design*, 2005.

4. A. Naeemi, R. Sarvari, and J. D. Meindl, "Performance comparison between carbon nanotube and copper interconnects for gigascale integration (GSI)," *IEEE Electron Device Letters*, vol. 26, Issue 2, 84–86, 2005.

5. B.A. Mantooth and P.S. Weiss, "Fabrication, assembly, and characterization of molecular electronic components," *Proceedings of the IEEE*, vol. 91, no. 11, 1785–1802, 2003.

6. Y. Chen et al., "Nanoscale molecular-switch devices fabricated by imprint lithography," *Applied Physics Letters*, vol. 82, no. 10, 1610–1612, 2003.

7. M.C. McAlpine, R.S. Friedman, and C.M. Lieber, "Nanoimprint lithography for hybrid plastic electronics," *Nano Letters*, vol. 3, 443–445, 2003.

8. Xilinx Product Datasheets, http://www.xilinx.com/literature

9. "Floating gate flash demise sees Actel update ProASIC," http://www.electronicsweekly.com, March 19, 2003.
10. L. Shang, A.S. Kaviani, and K. Bathala, "Dynamic power consumption in Virtex–II FPGA family," in *Proceedings of ACM/SIGDA International Symposium on FPGA*, 2002.
11. D.R. Stewart, D.A.A. Ohlberg, P.A. Beck, Y. Chen, R.S. Williams, J. O. Jeppesen, K. A. Nielsen, and J. F. Stoddart, "Molecule-independent electrical switching in P_t/organic monolayer/T_i devices," *Nano Letters*, vol. 4, no. 1, 133–136, 2004.
12. P.L. McEuen and J.Y. Park, "Electron transport in single-walled carbon nanotubes," *MRS Bulletin*, vol. 29, no. 4, 272–275, 2004.
13. J.W.G. Wildoeer et al., "Electronic structure of atomically resolved carbon nanotubes," *Nature*, vol. 391, 59–62, 1998.
14. R.H. Baughman, A.A. Zakhidov, and W.A. de Heer, "Carbon nanotubes: The route toward applications," Science, vol. 297, no. 5582, 787–792, 2002.
15. M. Liebau et al., "Nanoelectronics based on carbon nanotubes: Technological challenges and recent developments," *Fullerenes, Nanotubes, and Carbon Nanostructures*, vol 13, no. S1, 255–258, 2005.
16. N. Peng et al., "Influences of AC electric field on the spatial distribution of carbon nanotubes formed between electrodes," *Journal of Applied Physics*, vol. 100, Issue 2 on pages 024309–024309-5, 2006.
17. S. R. Lustig et al., "Theory of structure-based carbon nanotube separations by ion-exchange chromatography of DNA/CNT hybrids," *Journal of Physical Chemistry B*, vol. 109(7), 2559–2566, 2005.
18. P.M. Parthangal et al., "A generic process of growing aligned carbon nanotube arrays on metals and metal alloys," *Nanotechnology*, vol. 18, 185605.1–185605.5, 2007.
19. J.-H. Lim, C. A. Mirkin, "Electrostatically driven Dip-Pen nanolithography of conducting polymers," *Adv. Mat.*, vol. 14, no. 20, 1474–1477, 2002.
20. Y. Wu, J. Xiang, C. Yang, W. Lu, and C.M. Lieber, "Single-crystal metallic nanowires and metal/semiconductor nanowire heterostructures," *Nature*, vol. 430, 61–65, 2004.
21. J.M. Tour, W.L. Van Zandt, C.P. Husband, S.M. Husband, L.S. Wilson, P.D. Franzon, and D. P. Nackashi, "Nanocell logic gates for molecular computing," *IEEE Transactions on Nanotechnology*, vol. 1, 100–109, 2002.
22. Z. Yao et al., "High-field electrical transport in single-wall carbon nanotubes," *Physical Review Letters*, vol. 84, Issue 13, 2941–2944, 2000.
23. W. Kim and A. Javey, "Electrical contacts to carbon nanotubes down to 1 nm in diameter," *Applied Physics Letters*, vol. 87, Issue 17, 173101-1–173101-3, 2005.
24. A. Nieuwoudt and Y. Massoud, "Scalable modeling of magnetic inductance in carbon nanotube bundles for VLSI interconnect," in *Proceedings of Conference on Nanotechnology*, 2006.
25. Y. Massoud et al., "Managing on-chip inductive effects," *IEEE Transactions on VLSI Systems*, vol. 10, no. 6, 789–798, 2002
26. A. Nieuwoudt and Y. Massoud, "Performance implications of inductive effects for carbon nanotube bundle interconnect," *IEEE Electron Device Letters*, vol. 28, pp. 305–307, 2007.
27. V. Betz and J. Rose, "VPR: A new packing, placement and routing tool for FPGA research," in *Proceedings of the International Workshop on Field-Programmable Logic and Applications*, 1997.
28. A. DeHon and M.J. Wilson, "Nanowire-based sublithographic programmable logic arrays," in *Proceedings of the International Symposium on Field Programmable Gate Arrays*, 2004.
29. J. Greene, E. Hamdy, and S. Beal, "Antifuse field programmable gate arrays," *Proceedings of the IEEE*, vol. 81, no. 7, 1042–1056, 1993.
30. S.C. Goldstein and M. Budiu, "Nanofabrics: Spatial computing using molecular electronics," in *Proceedings of the International Symposium on Computer Architecture*, July 2001.
31. B. Gojman, R. Rubin, C. Pilotto, A. DeHon, and T. Tanamoto, "3-D nanowire-based programmable logic," in *Proceedings of the Nano–Networks Workshop*, pp. 1–5, Sept. 2006.

32. D.B. Strukov and K.K. Likharev, "CMOL FPGA: A reconfigurable architecture for hybrid digital circuits with two-terminal nanodevices," *Nanotechnology*, vol. 16, 888–900, 2005.
33. G.S. Snider and R.S. Williams, "Nano/CMOS architectures using a field-programmable nanowire interconnect," *Nanotechnology*, vol. 18, 035204, 2007.
34. R.M.P. Rad and M. Tehranipoor, "Evaluating area and performance of hybrid FPGAs with nanoscale clusters and CMOS routing," *ACM Journal of Emerging Technologies in Computing Systems*, vol. 3, no. 3, Nov. 2007.
35. NRAM, Nantero [Online]. Available: http://www.nantero.com/tech.html
36. X. Duan, Y. Huang, and C.M. Lieber, "Nonvolatile memory and programmable logic from molecule-gated nanowires," *Nano Letters*, vol. 2, no. 5, 487–490, 2002.
37. C. Dong, S. Chilstedt, and D. Chen, "FPCNA: A field programmable carbon nanotube array," in *Proceedings of the ACM International Symposium on Field Programmable Gate Arrays*, pp. 161–170, 2009.
38. C. Dong, D. Chen, S. Haruehanroengra, and W. Wang, "3-D nFPGA: A reconfigurable architecture for 3-D CMOS/nanomaterial hybrid digital circuits," *IEEE Transactions on Circuits and Systems I: Regular Papers*, vol. 54, no.11, pp. 2489–2501, 2007.
39. A. DeHon, P. Lincoln, J.E. Savage, "Stochastic assembly of sublithographic nanoscale interfaces," IEEE Transactions on Nanotechnology, vol. 2, no. 3, 165–174, 2003.
40. A. Raychowdhury and K. Roy, "A circuit model for carbon nanotube interconnects: Comparative study with Cu interconnects for scaled technologies," in *Proceedings of the International Conference on Computer-Aided Design*, 2004.
41. S. Eachempati, A. Nieuwoudt, A. Gayasen, N. Vijaykrishnan, and Y. Massoud, "Assessing carbon nanotube bundle interconnect for future FPGA architectures," in *Proceedings of the Design, Automation & Test in Europe Conference*, pp. 1–6, April 2007.
42. A. Gayasen, V. Narayanan, and M.J. Irwin, "Exploring technology alternatives for nano-scale FPGA interconnects" *in Proceedings of Design Automation Conference*, pp. 921–926, June 2005.
43. A. DeHon, "Deterministic addressing of nanoscale devices assembled at sublithographic pitches," *IEEE Transactions on Nanotechnology*, vol. 4, no. 6, pp. 681–687, Nov. 2005.
44. F. Kreupl et al., "Carbon nanotubes for interconnect applications," in *Proceedings of IEEE International Electron Devices Meeting*, 2004.

Nanoscale Application-Specific Integrated Circuits

Csaba Andras Moritz, Pritish Narayanan, and Chi On Chui

1 Fabric Introduction

This chapter provides an overview of the nanoscale application-specific integrated circuit (NASIC)[1] fabric. The NASIC fabric has spawned several research directions by multiple groups. This overview is a snapshot of the thinking, techniques, and some of the results to date. NASIC is targeted as a CMOS-replacement technology. The project encompasses aspects from the physical layer and manufacturing techniques, to devices, circuits, and architectures.

NASICs rely on 2D grids of semiconductor nanowires with computational streaming supported from CMOS. A fabric-centric mindset or integrated approach across devices (i.e., state variables), circuit style, manufacturing techniques, and architectures is followed. This mindset is anchored in the belief that at the nanoscale, the underlying fabric, broadly defined as the state variable in conjunction with the circuit style scalable into large-scale computational systems and its associated manufacturing approach, rather than the device alone, is how significant progress can be made in developing system-level capabilities.

This work is in contrast with efforts that focus on CMOS-replacement devices essentially preserving both CMOS circuit and manufacturing paradigms intact, or directions focusing on the state variables first and then attempting to connect them into a functional structure or circuit. This latter approach might not be feasible: nanoscale devices typically consist of a few atoms and interconnecting such structures post-device formation is a real challenge and might not even be possible.

---

[1] A new derivative of NASICs is MagNASIC/Spin Wave Function (SPWF) that follows magnetic physical phenomena instead of charge-based electronics. 3D NASIC is a NASIC fabric that is based on conventional (CMOS-like) 3D integration combined with nanowire VLSI. A number of other devices than presented here are based on depletion-mode xnwFETs. These and NASIC memories are not discussed in this chapter; relevant papers can be found from the authors' websites.

C.A. Moritz (✉)
Department of Electrical and Computer Engineering, University of Massachusetts, Amherst, MA, USA
e-mail: andras@ecs.umass.edu

N.K. Jha and D. Chen (eds.), *Nanoelectronic Circuit Design*,
DOI 10.1007/978-1-4419-7609-3_7, © Springer Science+Business Media, LLC 2011

Assembling a fabric with self-assembly would require a mindset where the devices are formed at the same time as the main interconnect. While other more conventional approaches might be possible, they would come with a considerable impact on density and system-level performance, potentially losing the benefit of otherwise well-performing devices.

Some of the key design objectives of NASICs include:

- 2D nanowire grid layout to allow partial self-assembly. This is motivated by the fact that self-assembly techniques typically form regular structures.
- Integrated design of device, interconnect and circuit style to accommodate and minimize manufacturing requirements while providing an ultra-dense, high-performance, and low-power solution competitive with or surpassing projected state-of-the-art scaled CMOS at a much lower manufacturing cost.
- Short interconnects that are inherently part of the fabric or means of inter-device interaction based on new physical phenomena, to bring significantly better power efficiency than in CMOS designs.
- Streaming control mechanism with peripheral CMOS support but all logic implemented at the nanoscale. The goal is to minimize device types, eliminate the need for sizing and arbitrary routing, and reduce manufacturing requirements at the nanoscale while providing flexibility in circuit style and achieving better noise margins.
- Reliance on built-in fault masking techniques to mask defects, parameter variation, and noise-related faults. Support in all circuits for 10–15% defect rates (roughly ten orders of magnitude higher rate than in CMOS, e.g., at 90 nm CMOS has 0.4 defects/cm^2 vs. hundreds of millions of defects at the rates assumed in NASICs per cm^2). Support for $3\sigma = 30\%$ parameter variation: this is for compensating for imperfections and device parameter variations in the fabric. This assumption further reduces requirements for precision in manufacturing. The built-in fault masking direction does not exclude using reconfigurable devices in the future; the belief of the group was that the challenges with reconfiguration are still at the material science level. Furthermore, certain types of faults, e.g., due to device parameter variation and noise – will be difficult to overcome even if reconfiguration is made possible with various materials, such as rotaxane [1], amorphous silicon and others [2].
- The simpler manufacturing and fault tolerance should not only reduce costs but also allow more aggressive scaling that can mean higher performance, density, and power efficiency.
- New types of applications, such as neuromorphic architectures resembling brain-like information processing and new types of nanoprocessors focusing on streaming, could directly exploit capabilities in these more regular nanoscale fabrics. Regular architectures and, in general, architectures with minimal feedback paths are better suited to fabrics where arbitrary routing is not supported.

This chapter is organized as follows. First, we describe the core components of the NASIC fabric including the underlying materials, semiconductor nanowires, and cross-nanowire-devices (Sect. 2). This is followed by a discussion on circuit style and methodology for streaming from CMOS, and ways to preserve noise margins during cascading (Sect. 3). We also discuss what this implies for device-level work and describe a methodology for integrated device-circuit exploration at the nanoscale. Additional sections discuss NASIC logic styles (Sect. 4), applications designed for NASICs, such as a nanoprocessor and an image processing architecture (Sect. 5), built-in defect tolerance and parameter variation mitigation techniques (Sect. 6), power and performance implications (Sect. 7), and ongoing manufacturing efforts (Sect. 8).

2   NASIC Building Blocks: Nanowires and xnwFETs

The building blocks of a computational fabric are its materials and active devices. For example, advanced CMOS technologies are built on silicon or silicon-on-insulator (SOI) substrates; with MOSFET devices and copper interconnect. To help readers develop an understanding of the NASIC fabric, this section provides an overview of the fundamental NASIC building blocks: semiconductor nanowires and crossed nanowire field-effect transistors (xnwFETs).

2.1   *Semiconductor Nanowires*

Semiconductor nanowires are nanostructures made of semiconductor material with diameters typically between 2 and 100 nm. Nanowires can be grown to up to a few microns in length and have been fabricated using a variety of materials, including silicon, germanium, zinc oxide, indium phosphide and indium antimonide [3–10].

Semiconductor nanowires are very promising candidates for nanoscale electronics for the following reasons:

- Early studies of the electrical characteristics of nanowires have shown diode-like and field-effect transistor (FET) behavior [10].
- Techniques for growth and assembly of nanowires on to a substrate have been shown in laboratory settings. Scalable assembly is an important criterion for realizing nano-systems. (A discussion of nanowire assembly techniques is presented in Sect. 8 as part of the NASIC manufacturing pathway.)
- It is possible to control the electrical properties, size, composition and doping of nanowires during the growth process [3].
- Techniques, such as ion implantation and silicidation [11], can be used to create high conductivity interconnect regions on nanowires.
- High-density, high-performance integrated systems may be possible since nanowires can be synthesized to ultra-small dimensions.

More information about the synthesis, assembly and properties of nanowires can be found in various review papers and textbooks including [3, 12]. In this chapter, we explore how integrated systems can be built given certain electrical properties and techniques for assembly of nanowires and the benefits of such systems over end-of-the-roadmap CMOS.

2.2   xnwFET Devices

An xnwFET is a transistor consisting of two perpendicular nanowires separated by a dielectric; one nanowire acts as the gate and the other as the channel. Figure 1 shows the schematic of an xnwFET. As with any FET, the electric field/voltage applied at the gate attracts or repels charge carriers in the channel, modulating its conductivity. Prototyping xnwFET behavior has been experimentally demonstrated for silicon and gallium nitride nanowires in [10, 13].

Figure 1 shows a baseline n-type xnwFET with a gate and channel nanowires that have a 10 nm × 10 nm cross-section. While a rectangular cross-section is shown in this figure, it must be noted that both rectangular [14] and cylindrical nanowire [3] structures are possible, depending on the performance target as well as the manufacturing/synthesis process used. In this device, the gate, source, drain, and substrate regions are all doped n^+ (doping concentration is typically 10^{20} dopants/cm^3) to reduce series and contact resistances. The channel is doped p-type and doping concentration choices are discussed below. This xnwFET is an inversion-mode device similar to conventional MOSFETs: applying a positive voltage at the gate terminal attracts electrons into the p-doped channel leading to n-type FET behavior.

In the NASIC design methodology, the electrical characteristics of silicon xnwFET devices are extensively simulated using accurate physics-based 3D simulations of the electrostatics and operations using the Synopsys Sentaurus Device™ [15]. Since any quantum confinement induced threshold voltage (V_{th}) shift [16] is absent in the nanowire diameter of interest, the device electrostatic behavior can be

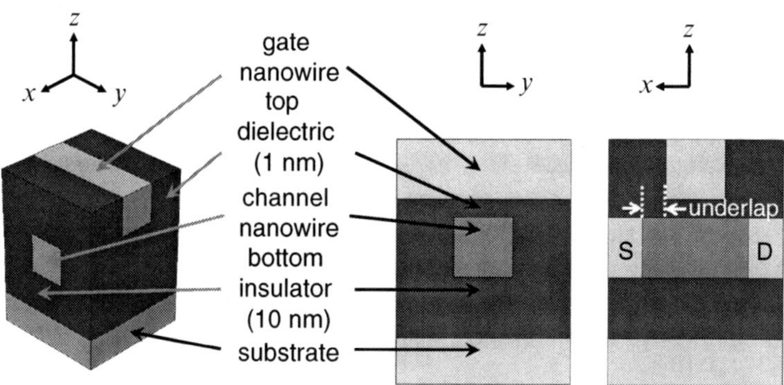

Fig. 1 Crossed nanowire field-effect transistor(xnwFET)

accurately accounted for by the simulator. The drift-diffusion transport models employed need to be calibrated against experimental data to include effects such as carrier scattering due to surface roughness and dielectric/channel interface trapped charges. In the absence of relevant xnwFET characterization data, experimental data from well-characterized nanowire channel FET with similar dimension could instead be chosen [17]. To ensure the representativeness of the calibrated models, the simulated device characteristics are further scaled to confirm similarity with other experimental nanowire FETs. Finally, current–voltage characteristics of the xnwFET device as well as capacitances can be extracted from these simulations. These data can then be incorporated into circuit models and abstracted to higher design levels for validating circuit behavior and evaluating key system-level metrics.

2.2.1 Design Motivation and Nanowire-Level Approaches

Generally speaking, the electric field coupling from the gate nanowire to the channel nanowire in xnwFETs is inherently weak for the following reasons:

- The contact area through the top dielectric at the nanowire cross-point is ultrasmall. Especially for the cylindrical structure, the closest contact is made only at the very center of the cross-point owing to the surface curvature in both gate and channel nanowires.
- The device channel length, i.e., the gate nanowire diameter, is roughly the same as the device channel thickness, i.e., the channel nanowire diameter.

Since poor electrostatic gate-to-channel coupling often compromises the xnwFET integrity as a switching device for circuit implementation (see Sect. 8 for more details), its improvement is mandated through technology-aware explorations of device design and optimization. The individual and combined effects of changing nanowire geometry, sizing, and channel doping concentration are initially analyzed.

The simulated transfer characteristics of both rectangular and cylindrical xnwFET structures are compared in Fig. 2a. A slightly better on-to-off state current ratio (I_{ON}/I_{OFF}) is obtained with the rectangular geometry due to a better gate coupling, as anticipated. This advantage is, however, offset by the technological limitations inherent to the formation and transfer of rectangular nanowires, as discussed in Sect. 8.

Better gate control can alternatively be achieved with an increased gate-to-channel nanowire diameter ratio, as illustrated in Fig. 2b. The use of nanowires with different diameters for NASIC circuits is nonetheless expected to be impractical as xnwFETs in various cascading circuit stages would need to be separately optimized.

A marginal improvement in xnwFET I_{ON}/I_{OFF} can be obtained using a very high channel doping concentration of 10^{19} dopants/cm^3, as shown in Fig. 2c. This heavy doping, unfortunately, leads to a more pronounced random dopant fluctuation issue [18] since only 8–9 dopant atoms would be present in the channel region. Since

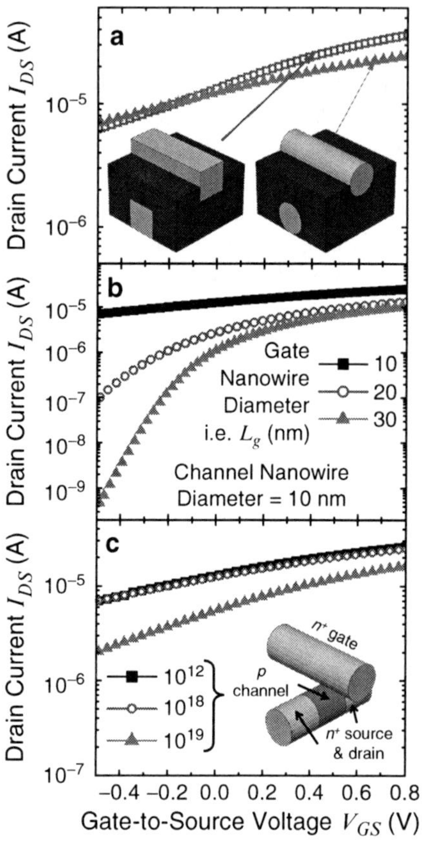

Fig. 2 Simulated xnwFET transfer characteristics with different (**a**) geometry, (**b**) gate nanowire diameter, and (**c**) channel nanowire doping

there resides no dopant atom inside the channel at concentration less than 10^{18} dopants/cm^3, the theoretical concentration is thus "quantized" to be either 0 or above 10^{18} dopants/cm^3 which, in turn, constrains device V_{th} tunability.

In summary, neither the individual nor combined engineering of the above nanowire properties can secure enough gate-to-channel coupling for a sufficient xnwFET $I_{\text{ON}}/I_{\text{OFF}}$ and V_{th}. Device-level measures should, therefore, be considered to meet the NASIC circuit requirements.

2.2.2 xnwFET Device-Level Design Approaches

To be employed in cascading multiple logic circuits, the xnwFETs are required to possess enhancement-mode behavior with a positive V_{th}. The key device design parameters are the $I_{\text{ON}}/I_{\text{OFF}}$ ratio and V_{th}. Since nanowire-level measures (doping, shape, and geometry) do not satisfy the requirements, device-level approaches

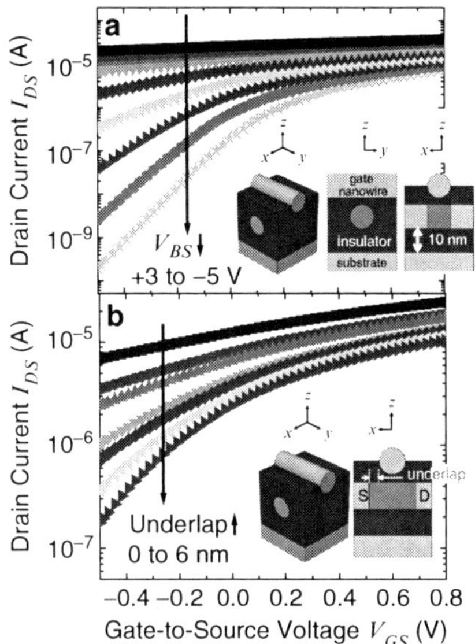

Fig. 3 Simulated xnwFET transfer characteristics with different (**a**) substrate bias and (**b**) source–drain junction underlap distance

should be considered. The two approaches chosen are substrate biasing and source–drain junction underlapping.

A substrate potential bias is applied through the bottom insulator from the wafer backside. As shown in Fig. 3a, a more negative substrate bias (V_{BS}) increases both I_{ON}/I_{OFF} and V_{th} of the xnwFET. This can be attributed to an effective channel doping concentration increase electrostatically induced by the substrate bias. The resultant diminishing of short-channel effects simultaneously improves both key design parameters and the overall device integrity.

As depicted in the inset of Fig. 3b, an underlap region between the gate nanowire edge and the source–drain junction can also be introduced by applying standard spacer technology. In doing so, the effective channel length can be larger than the gate nanowire diameter (or gate length) such that an enhanced channel length-to-thickness ratio becomes apparent. Consequently, diminished short-channel effects are similarly obtained to improve both the I_{ON}/I_{OFF} and V_{th} metrics, as confirmed in Fig. 3b.

This work is currently ongoing yet the combination of both approaches has revealed great promise (Fig. 4). It must be noted that these techniques do not introduce any new manufacturing constraints/challenges. A more detailed discussion of integrated device-circuit explorations that validate NASIC devices and circuit styles will be presented in the next section.

To summarize, in this section, we provided an overview of semiconductor nanowires and xnwFETs. The next section addresses how these basic building blocks are organized into functional networks and circuits for electronics applications.

Fig. 4 I_{ON}/I_{OFF} and V_{th} optimization with substrate bias and junction underlap

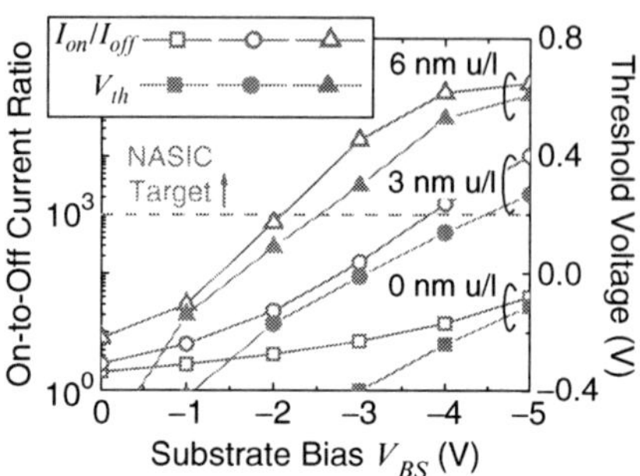

3   NASIC Circuit Styles

An important challenge for a circuit designer at the nanoscale is to identify/invent suitable circuit styles that are feasible with nanoscale devices and the capabilities of non-conventional manufacturing. This is a specific instance of the integrated design approach for nanofabrics where design choices at individual levels are compatible with the fabric as a whole. For example, non-conventional and self-assembly based approaches favor the formation of regular structures, such as parallel arrays and grids with limited ability to customize and interface with these structures. This implies that sizing, arbitrary placement of devices or multilayer complex routing of signals may not be possible given manufacturing restrictions. Consequently, a conventional circuit style such as CMOS, which has many desirable circuit characteristics but places stringent requirements on manufacturing, may not be suitable for nanofabrics.

In general, any circuit style for digital nanofabrics must incorporate all of the following:

- *Combinational circuits* for implementing arbitrary logic functions.
- *Sequential circuits/latching* for temporary storage of signals.
- A *clocking mechanism* for coordinating the flow of data and synchronization with external interfaces.
- *Cascading* of circuits and *signal restoration* for propagation of data through a large-scale design.
- Sufficient *noise margins* to ensure correct functionality.

Furthermore, some circuit characteristics that are desirable for robust system design include:

- *Rail-to-rail voltage swing:* Logic "1s" should be at a voltage level of V and logic "0s" at V_{SS}. This corresponds to the maximum noise margin available.
- No direct paths between V_{DD} and V_{SS}, implying *no static power dissipation.*
- *High input impedance:* There is very little leakage from input to output nodes.
- *Low output impedance:* Output nodes are tied to power supplies through networks of devices and are not floating.

In this section, we discuss circuit styles that have been developed for the NASIC fabric, including static and dynamic styles, as well as the pros and cons of each. These circuits are realized on regular arrays and grids of parallel and perpendicular nanowires, in keeping with the aforementioned restrictions of non-conventional manufacturing. We also present an integrated device-circuit exploration that uses detailed 3D physics-based models of xnwFET devices to validate NASIC circuit styles. It must be noted that circuit style is perhaps the most important factor in determining some key system-level metrics, such as area, operating frequencies, power and energy of the design.

3.1 NASICs with Static Ratioed Logic

Initial NASIC designs employed static ratioed logic and assumed the availability of reconfigurable devices. A typical example of a preliminary NASIC combinational circuit is shown in Fig. 5. It consists of a regular 2D nanowire grid with FETs at certain crosspoints. Thick lines along the periphery are microwires carrying V_{DD}, V_{SS} and signals for programming crosspoints. Thin horizontal lines are p-type nanowires and thin vertical lines are n-type. PFET channels are aligned along the horizontal direction and NFET channels along the vertical. Pull-up and pull-down arrays imprinted on the grid serve as resistive loads to achieve ratioed logic. This functional network of nanowires, and associated peripheral microwire circuitry is called a *tile*.

In Fig. 5 inputs to the tile are received from the top from a previous tile or an external interface. These gate p-type FETs. If all inputs are "0," the output goes to "1," accomplishing NOR functionality. The outputs of this stage gate a diagonal arrangement of n-type FETs; this is a routing stage that "turns the corner" and propagates the signals to the next set of p-type FETs. Outputs from the tile (in this case, carry and sum signals) may then be propagated to other tiles or to external interfaces.

Figure 6 shows a sequential circuit (1-bit flip-flop) implemented on a NASIC tile using static ratioed logic with NOR gates and feedback paths. One key observation from Fig. 6 is that implementing feedback paths and extensive routing on the grid can lead to very poor fabric utilization; incorporating sequential circuits on the grid can thus cause severe area penalties. Therefore, other mechanisms for latching signals may need to be explored.

Static ratioed logic suffers from key design issues that may make it unsuitable for large-scale system design. Firstly, careful sizing of resistive loads is required for correct operation. This may be difficult, given constraints of nano-manufacturing. Large static currents can occur due to direct paths between V_{DD} and V_{SS}. The voltage swing is not rail-to-rail, leading to smaller noise margins. Furthermore, implementing sequential circuits on NASIC tiles was shown to be very area-inefficient. In the next section, we explore dynamic circuit styles for the NASIC fabric that overcome all of the above problems.

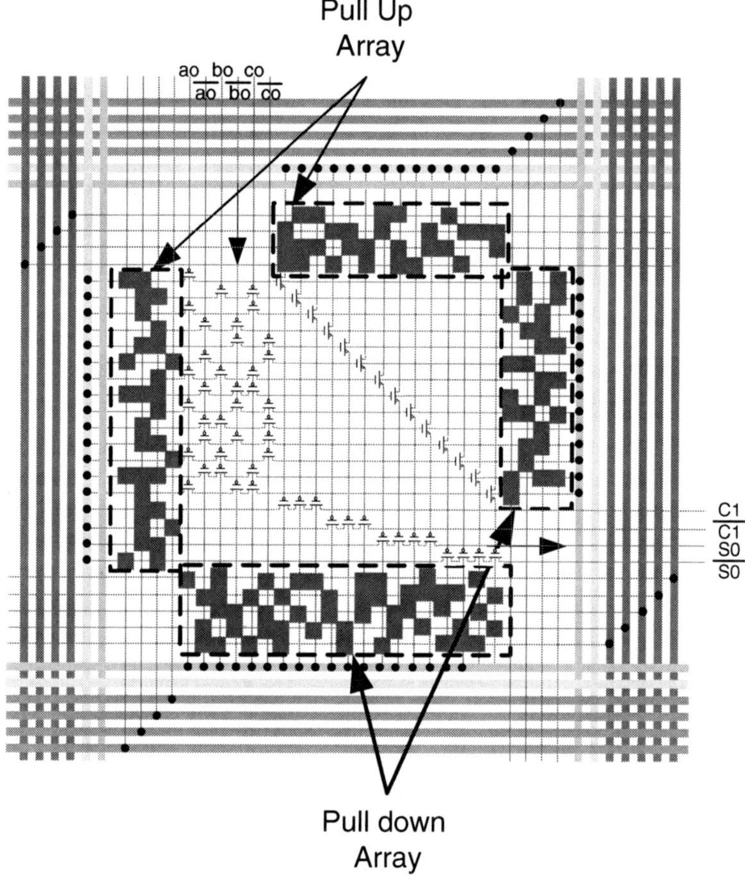

Fig. 5 Initial NASIC design of a 1-bit full adder with static ratioed logic and reconfigurable devices

3.2 NASIC Dynamic Circuits and Timing Schemes

To overcome significant challenges associated with static ratioed logic, dynamic circuit styles were developed for the NASIC fabric. Figure 7 shows a simple NASIC dynamic circuit implemented on a single semiconductor nanowire with xnwFETs. xnwFET source–channel–drain regions are along the length of the nanowire and gates are on perpendicular nanowires. Precharge (*pre*) and evaluate (*eva*) transistors at the top and bottom of the nanowires are used for implementing dynamic logic. The gates for these control transistors are typically microscale wires driven from external reliable CMOS circuitry. The material used for microwires can be tailored based on design requirement: e.g., precharge transistors could be depletion-mode to achieve rail-to-rail voltage swing at the output node.

The output node is precharged to logic "1" by asserting the *pre* signal. Subsequently, the *eva* signal is asserted (and *pre* de-asserted) to evaluate the outputs.

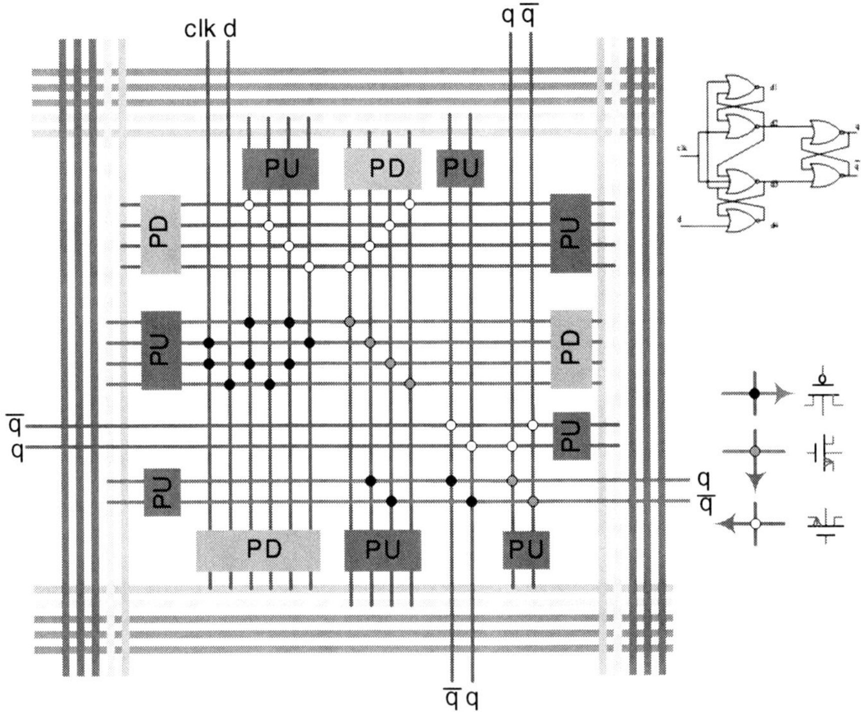

Fig. 6 1-bit flip-flop using NASIC static ratioed logic

In Fig. 7, if all inputs are at logic "1," the output node will be discharged to logic "0" implementing NAND functionality. Other logic functions such as AND, OR and NOR are also possible by suitably altering CMOS control and xnwFET doping type. However, typically NASICs use only NAND and AND functions in the design to implement arbitrary logic, since these require only n-type xnwFETs. Elimination of p-type FETs is advantageous because it considerably eases manufacturing requirements and leads to significant performance improvements, since n-type FETs are typically faster [19].

The dynamic circuit style has many advantages over static ratioed logic. Dynamic circuits are *ratioless*; therefore sizing of resistive loads and pull-up/pull-down arrays are not required. Furthermore, since *pre* and *eva* signals are never asserted simultaneously, static power dissipation does not occur (leakage power may be consumed due to device leakage mechanisms). Rail-to-rail voltage swing may also be achieved by carefully engineering the microwire materials and suitably altering the work function of *pre* and *eva* FETs.

NASIC dynamic style circuits are different from static/dynamic circuits implemented with CMOS – control schemes will be discussed further; the NASIC fabric also incorporates fault tolerance at the circuit level and the control schemes can be engineered to allow more flexibility in the design.

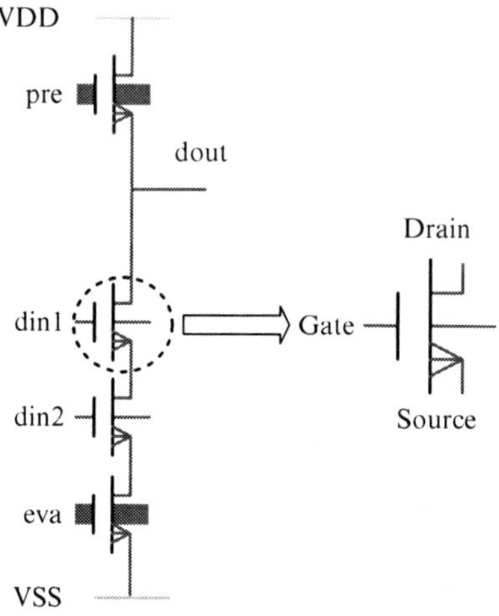

Fig. 7 NASIC dynamic circuit on a single semiconductor nanowire

3.2.1 Sequential Circuits Using Dynamic Circuit Styles

Implementing explicit sequential components, such as latches and flip-flops on regular 2D nanowire grids, is highly area-inefficient, given issues with routing signals and implementing feedback mechanisms. NASIC dynamic circuit styles, therefore, use *implicit* latching of signals on the nanowires themselves without the need for sequential components. This is enabled by using unique control schemes generated from external CMOS.

Figure 8 highlights this concept. It shows a two-stage circuit, with outputs of dynamic circuits in Stage 1 acting as inputs for the Stage 2 dynamic circuit. The associated timing diagram shows data and control signals. First, Stage 1 is precharged and evaluated using *pre1* and *eval* control signals, generating outputs *do1a* and *do1b*. Separate control signals *pre2* and *eva2* are then asserted to evaluate Stage 2, with *do1a* and *do1b* as inputs. During this time, Stage 1 is in a *hold* phase (with *pre1* and *eval* de-asserted), and *do1a* and *do1b* are *implicitly latched* to their evaluated values on the nanowires. The *hold* phase, thus, enables correct cascading of data across dynamic stages. Multiple NASIC dynamic stages can be cascaded together to achieve large-scale designs, such as the NASIC WISP-0 nanoprocessor, with external control schemes to achieve sequential flow of data and pipelining.[2]

Figure 9 shows an example of a 1-bit full-adder implemented in two stages using NASIC dynamic circuit styles and control. Inputs are received on vertical nanowires, which are then NAND-ed together to generate midterms using *pre1* and *eval* signals. The xnwFET channels in this stage are along horizontal nanowires,

---

[2]The use of separate precharge and evaluate signals for successive stages also implies that signal monotonicity issues prevalent in conventional dynamic circuits (that typically require either domino logic, i.e., insertion of a static inverter between dynamic stages or np-CMOS type circuit styles) do not affect NASIC dynamic circuits.

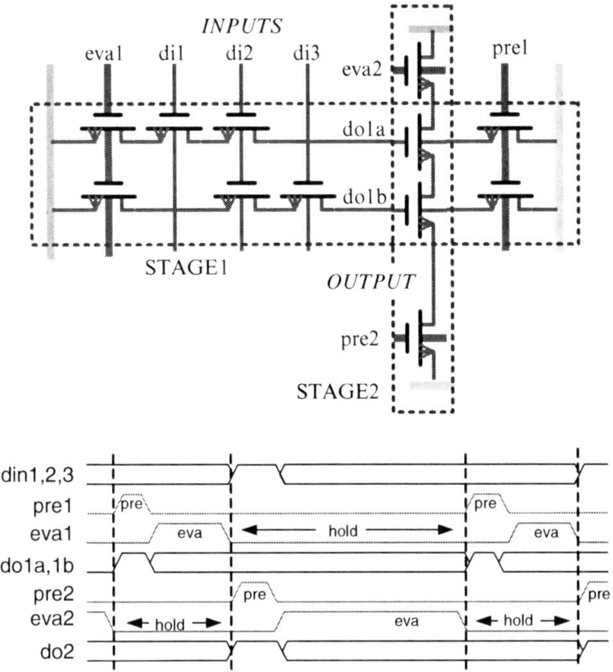

Fig. 8 *Top:* NASIC circuit with two dynamic stages – output of one stage is cascaded to the next. *Bottom:* Control scheme for implicit latching – successive stages are clocked using different precharge and evaluate signals

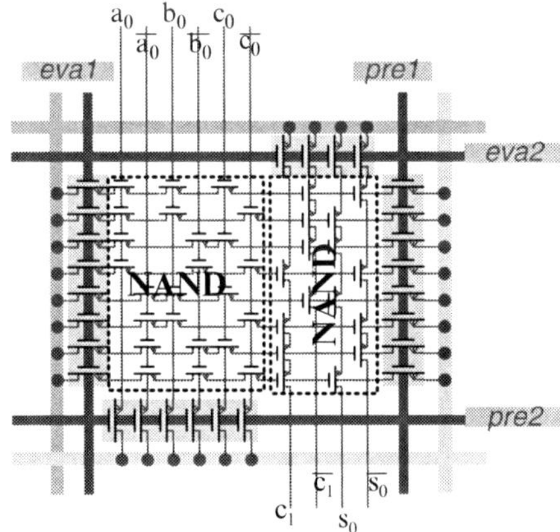

Fig. 9 NASIC 1-bit full-adder using two cascaded dynamic NAND stages

with vertical input gates. The midterms act as inputs for the second NAND stage (whose channels are on the next set of vertical nanowires), and carry and sum outputs are generated along with their complements on vertical nanowires. Subsequent NASIC tiles could then receive these signals as their inputs, achieving large-scale cascaded designs.

3.3   Integrated Device-Circuit Exploration

Nanoscale computing fabrics face challenges at all design levels including manufacturing, devices, circuits, architecture, and fault tolerance. Therefore, design choices at individual levels must be compatible with the fabric as a whole. For example, in addition to having the requisite electrical characteristics, nanodevices should (a) meet circuit requirements and function as expected and (b) not require extensive customization that poses insurmountable challenges to non-conventional manufacturing processes.

In this section, we explore devices and circuits for NASICs in a tightly integrated fashion, with simulations at the circuit level built on accurate 3D device simulations of the electrostatics and operations. Using an integrated approach, co-design of devices and circuits can be accomplished with accurate physics-based device models. Accurate modeling is especially important for NASIC dynamic circuits with single-type FETs [19], where using only n-type devices simplifies manufacturing requirements and improves performance, but may lead to reduced noise margins.

We present the methodology for integrated device-circuit exploration [20], investigate different devices and show how the optimal choice of device parameters such as V_{th}, I_{ON}/I_{OFF} ratios, etc., enables correctly functioning cascaded dynamic circuits. Importantly, we also discuss how these device-level characteristics can be achieved without unrealistic customization requirements on the manufacturing process.

3.3.1   Methodology

Figure 10 summarizes the methodology used for integrated device-circuit explorations. Device-level characterization of xnwFET structures is done using 3D simulations on the Synopsys Sentaurus Device™ simulator [15]. Drain current vs. gate voltage (transfer) characteristics are obtained for drain voltages varying between 0.01 and 1.0 V, which covers the operating range of the devices in the NASIC dynamic circuits. Gate–source and gate–drain capacitances are also extracted as a function of the gate voltage. Regression analysis is carried out on the drain current data, and multivariate polynomial fits (for FET on-region behavior) and exponential fits (for off-region behavior) are extracted using DataFit software.[3] These relationships express the drain current as a function of two independent variables,

---

[3]A note on regression-based vs. analytical modeling: A regression based approach is very generic and can be used to fit arbitrary device characteristics. Coefficients extracted from regression data fits are representative of the device behavior over sweeps of drain-source and gate-source voltages. This is in contrast to conventional in-built models in SPICE for MOSFETs and other devices, which use analytical equations derived from theory and physical parameters, such as channel length and width. The regression coefficients in our approach may not directly correspond to conventional physical parameters. Therefore, different regression fits will need to be extracted for devices with varying geometries, doping, etc.

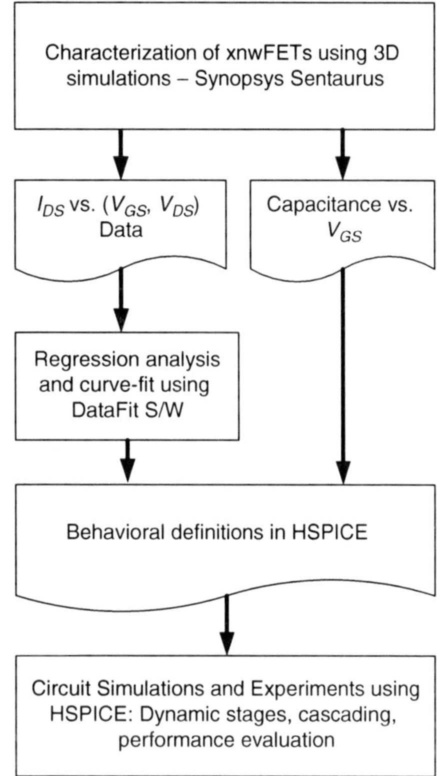

Fig. 10 Methodology for integrated device-circuit explorations

gate–source (V_{GS}) and drain–source (V_{DS}) voltages. These fits are then incorporated into subcircuit definitions for voltage-controlled resistors in HSPICE [21]. Capacitance data from Sentaurus is directly integrated into HSPICE using voltage-controlled capacitance (VCCAP) elements and a piece-wise linear approximation since only small variations in capacitance were observed with V_{GS} increments. The regression fits for currents together with the piecewise linear model for capacitances and subcircuit interconnections define the behavioral model for xnwFET devices used in circuit simulation.

Once behavioral data have been incorporated into HSPICE, multiple experiments are conducted including: DC sweeps of individual devices to verify behavioral models, simulation of a single dynamic stage to verify functionality, and simulation of multiple stages to evaluate the effects of nanowire cascading, charge sharing and diminished noise margins. Devices can also be optimized for circuit-level behavior targeting key metrics, such as performance and power consumption.

As an example, we show three devices that were investigated using this approach. The device parameters are shown in Table 1. By modifying the substrate bias voltage and the underlap, key device parameters, such as V_{th} and I_{ON}/I_{OFF} ratios, were modulated. Current–voltage (I–V) characteristics for the devices are

Table 1 xnwFET devices explored

Device	Gate and channel NWs	N_{ch}	$N_{G/S/D/Sub}$	Ulap (nm)	V_{sub} (V)
#1	10×10 nm^2	10^{19} cm^{-3}	10^{20} cm^{-3}	0	0
#2	10×10 nm^2	10^{19} cm^{-3}	10^{20} cm^{-3}	7	0
#3	10×10 nm^2	10^{19} cm^{-3}	10^{20} cm^{-3}	7	-1

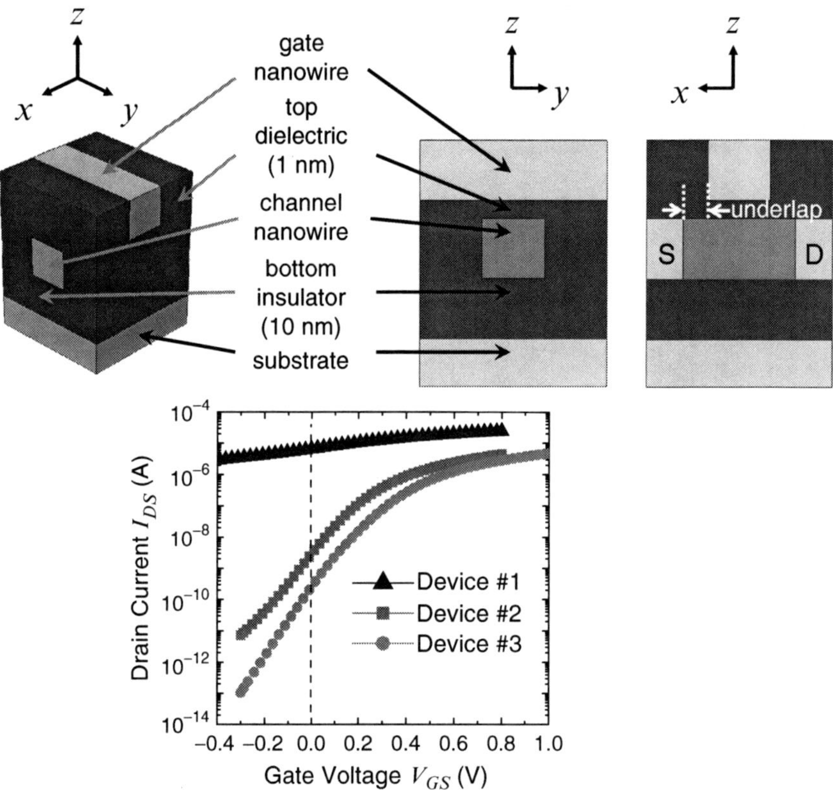

Fig. 11 Simulated transfer characteristics of xnwFETs with parameters listed in Table 1

shown in Fig. 11. Device #1, without any underlap or substrate bias, has poor electrostatic control. Device #2, incorporating underlap, has $I_{ON}/I_{OFF} > 10^3$, but V_{th} is small. Finally, device #3, with both underlap and substrate bias, meets the requirements for correct NASIC functionality.

Using the curve-fit models of the device characteristics, multiple test simulations may be done in HSPICE. DC sweeps of the models verify correct abstraction of device data. Simulations of single NASIC NAND stages using devices #2 and #3 showed them to function correctly.

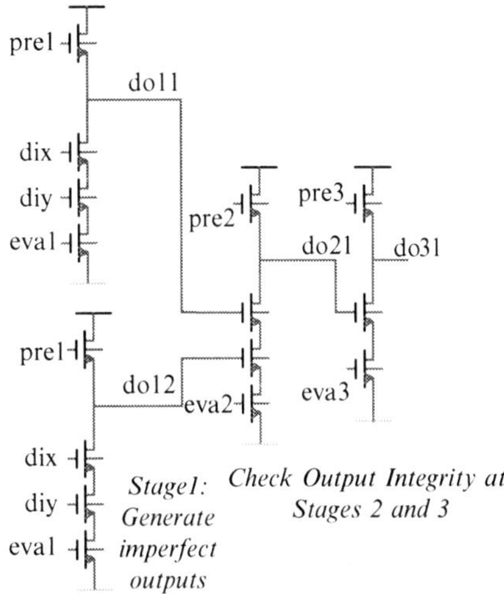

Fig. 12 Circuit for validation of cascading in dynamic NASIC design

For evaluation of dynamic circuit cascading and noise margin issues, a simple three-stage test circuit was used (Fig. 12). In this circuit, Stage 1 generates imperfect outputs and signal integrity was checked at the outputs of Stages 2 and 3. Stage 3 has a single input device and represents a worst-case noise scenario; this stage has the least possible effective resistance and capacitance between the output and V_{SS} nodes. Therefore, it is susceptible to noise on its input node. Separate precharge and evaluate signals were used for the three stages in accordance with NASIC dynamic control concepts.

Figure 13 shows the results of circuit simulation using device #2 and device #3, respectively. From Fig. 13a, we see that a glitch (due to charge-sharing effects) at the output of Stage 2 causes loss of signal integrity at the output of Stage 3. This is because the magnitude of the glitch is larger than the threshold voltage for device #2, causing the input transistor to operate in the linear region. However, in Fig. 13b, the glitch does not affect the Stage 3 output, since its magnitude is smaller than device #3 V_{th}, which ensures that the input transistor stays switched off. Thus, in this example, the integrated device-circuit exploration helps identify and evaluate device characteristics for correct cascading and circuit functionality. Additional experiments, such as detailed evaluation and optimization of key system-level metrics such as performance and power, may be carried out using behavioral models and HSPICE simulations.

A Note on Manufacturing Implications: Manufacturing of nanodevice-based computational systems continues to be very challenging. Therefore, while devices should possess the requisite characteristics to meet circuit requirements and expected functionality, it is equally important that they can be integrated in a manufacturing process without introducing new challenges. For example, while large gate-to-channel ratios (e.g., 20×20 nm^2 gate, 10×10 nm^2 channel) can achieve the required electrostatic control and device characteristics, including V_{th} and current ratio, the inherent problem with these is the dissimilar gate and channel dimensions.

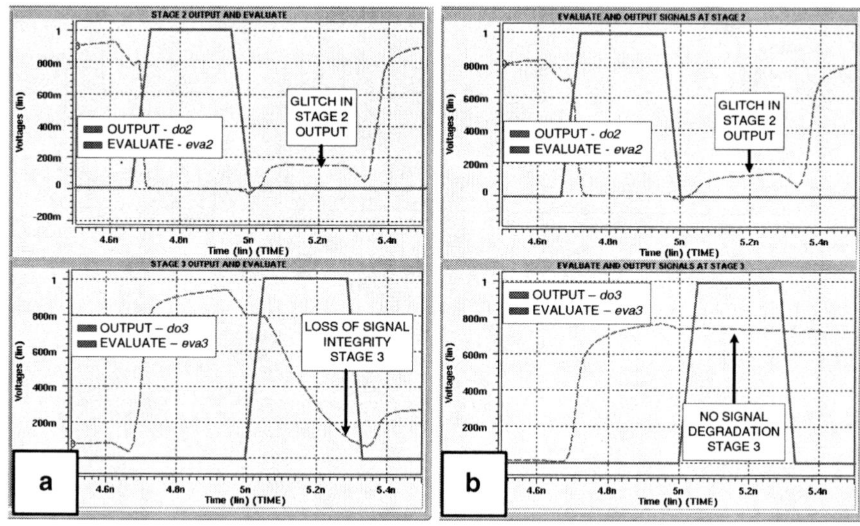

Fig. 13 Results of circuit simulations with xnwFET devices. (**a**) Device #2 with low V_{th} is susceptible to noise and (**b**) device #3 is not affected by glitch

Since the output node of one nanowire acts as gate for the next stage, using 20/10 devices would need asymmetry along the length of the nanowire. This would require varying the radius during growth of nanowires themselves or contacting nanowires of different diameters together after transferring to a substrate. Nanowires with identical gate and channel dimensions do not have these issues, but may suffer from poor electrostatics, as shown in Sect. 2. We, therefore, tune electrical characteristics using techniques such as underlap and substrate biasing. V_{th} tuning using these techniques does not impose any new manufacturing challenges. Biasing in NASIC designs is done for the entire circuit, which is much simpler than biasing individual devices. Spacer requirements for underlap are expected to be comparable to end-of-the-line CMOS technologies.

3.3.2 Control Schemes for the NASIC Fabric

One possible control scheme for the NASIC fabric was shown in Fig. 8. Control schemes may be optimized for higher performance and noise resilience. This does not have any associated cost at the nanoscale in terms of manufacturability since the control schemes are generated by reliable external CMOS circuitry. Figures 14 and 15 show two possible control schemes for the NASIC fabric.

The control scheme in Fig. 14 improves the performance of NASIC designs. In this scheme, *pre2* is overlapped with *eval*, and *pre3* is overlapped with *eva2* leading to a three-phase clock (there are three separate precharge and evaluate signals that repeat every three stages). By overlapping the phases, performance of

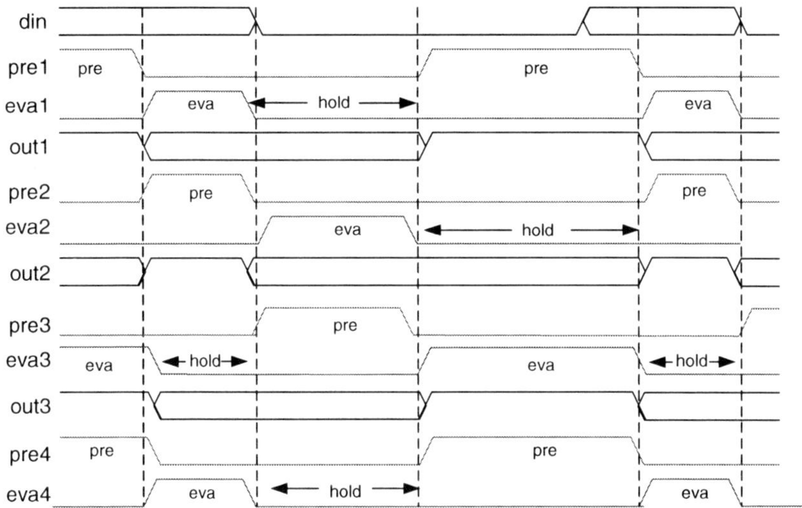

Fig. 14 High-performance three-phase control scheme for NASIC designs

the design can be significantly improved. For example, assuming that the phases in Fig. 8 are of approximately the same duration, up to 33% improvement in performance can be achieved. However, there are two key issues with this scheme:

1. *Feedback circuits may not be possible:* Consider a four-stage feedback path; the output of Stage 4 will be the input of Stage 1. This requires *eva4* to complete before *eva1*, which is not the case with the three-phase control scheme. One possible way to circumvent this issue is to use static portions of the circuits for feedback; however, issues previously discussed for static circuits will be applicable.

2. *Noise issues:* Consider a Stage 3 circuit similar to Fig. 12. While *pre1* is asserted, inputs to Stage 2 go to logic "1," causing charge sharing between Stage 2 output and intermediate nodes. This can cause glitching of the Stage 2 output. Since Stage 3 is evaluating at this time (*eva3* is asserted), the glitch at its input can lead to loss of signal integrity at its output node (this issue is also applicable to the "basic" control scheme shown in Fig. 8).

To overcome issues with a three-phase clocking scheme, a four-phase scheme may be constructed (Fig. 15). In this scheme, in addition to *pre* and *eva*, each dynamic stage goes through two hold phases (labeled *H1* and *H2*). During *H1*, the output is held constant while the next stage is evaluated, similar to the *hold* phase in previous schemes. However, before *pre* is asserted again, an additional *H2* phase is inserted. This ensures that glitches due to precharging do not propagate to subsequent stages. For example, we see that when *eva3* is asserted, Stage 1 is in *H2* and there is no switching activity in this stage. This implies that there is no glitching at the output of Stage 2, whose outputs will be correctly evaluated. When *pre1* is asserted, glitching

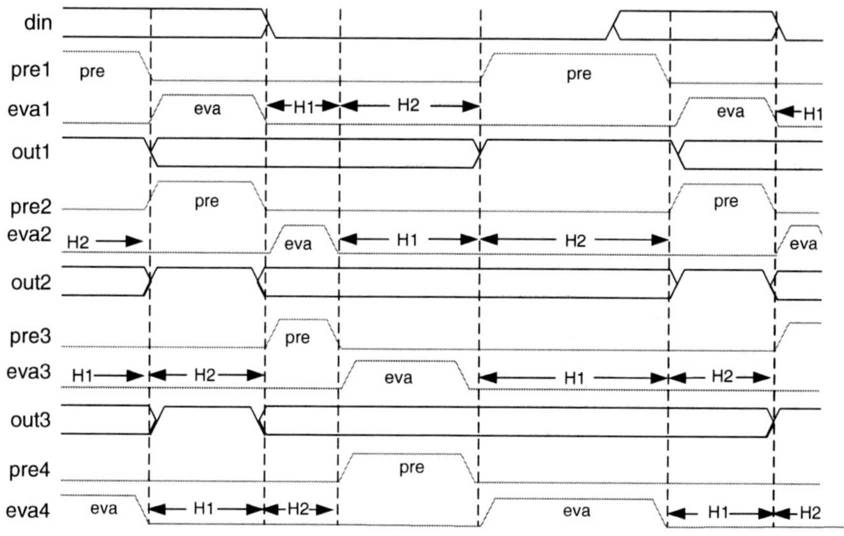

Fig. 15 Four-phase noise-resilient control scheme NASIC

can occur at the Stage 2 output. However, by this time, Stage 3 has completed evaluation (*eva3* is de-asserted) and is not affected by the glitch.

The four-phase clock also implies that feedback paths can be correctly implemented. Since there are four phases and four unique *pre* and *eva* signals, *eva4* completes just ahead of *eva1*, meeting the requirement for feedback.

To summarize, this section dealt extensively with possible NASIC circuit styles and associated control schemes. Dynamic circuits with single-type FETs in conjunction with unique control schemes generated from CMOS are suitable for a regular 2D grid based computational fabric with limited customization requirements. These circuit styles achieve:

- *Combinational Circuits:* NAND gate implementations, which are universal, were shown.
- *Sequential Elements:* Implicit latching on nanowires is achieved by using unique control schemes where successive stages use different *eva* signals.
- *Clocking Mechanism:* The flow of data along multiple NASIC stages is achieved by the control scheme, enabling pipelined designs.
- *Sufficient Noise Margins:* An integrated device-circuit co-design approach was shown, and techniques to tune devices to meet circuit requirements and noise margins were discussed.

Additionally, these dynamic circuits were shown to have *rail-to-rail voltage swings, no static power dissipation* and *high input impedance*. At the time of writing, new control schemes are under investigation to achieve low output impedance that could lead to designs with even better noise resilience.

4 NASIC Logic Styles

NASIC designs typically follow two-level logic styles, such as NAND–NAND. Two-level logic lends itself to relatively straightforward implementation on regular 2D nanowire grids. Figure 16 shows different implementations of a 1-bit full-adder on the NASIC fabric. Figure 16a uses an AND–OR logic implementation with n-type xnwFETs (white boxes) implementing AND logic on the horizontal plane and p-type FETs (black boxes) implementing OR logic on the vertical plane. An improvement over the AND–OR based logic implementation is the NAND–NAND implementation (Fig. 16b), using n-type FETs only. The migration is achieved by suitably altering the external dynamic control scheme. The single-type FET based NAND–NAND scheme eases requirements on manufacturing and significantly improves the performance of NASIC designs by eliminating the slower p-FET devices without any deterioration in density [19].

Simple two-level logic schemes have some inefficiency in terms of generating minterms for complementary signals separately from true values, as well as propagating all complements. Figure 16c shows a heterogeneous two-level (H2L) logic scheme that provides a more efficient logic implementation than simple two-level logic. H2L combines different logic functions into a single stage, e.g., the vertical logic gates in Fig. 16c implement both NAND and AND logic functions. This is labeled as NAND–NAND/AND H2L logic. The availability of multiple logic functions in a single NASIC stage implies efficient logic implementation and higher density compared to simple NAND–NAND schemes. H2L based NASIC processors have been shown to be up to 30% denser than their NAND–NAND based counterparts [22]. H2L logic also significantly improves the yield of NASIC designs. This is due to two factors:

- Simplifying the design – fewer devices are used to implement the same logic function.
- Ability to do nanoscale voting – H2L can be used to design high-yielding NASIC based majority voting blocks.

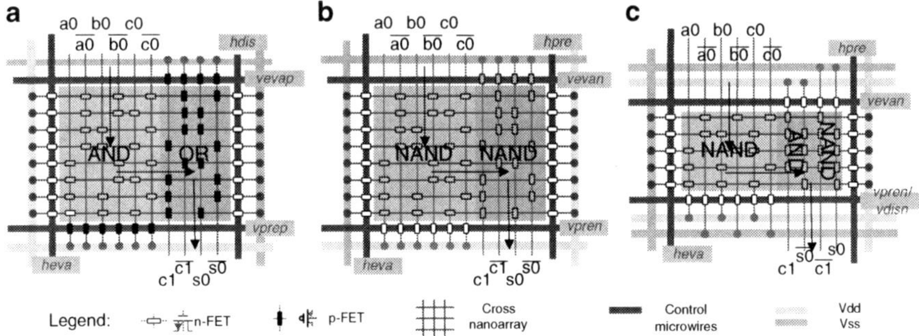

Fig. 16 NASIC logic styles: (a) AND–OR with two types of FETs, (b) NAND–NAND with n-FET-only xnwFETs, and (c) heterogeneous 2-level (H2L) logic

H2L is enabled by modifying the control circuitry and nanowire connections to V_{DD} and V_{SS}. Consequently, H2L schemes for the NASIC fabric do not impose any new challenges from a manufacturability perspective. It must be noted that H2L is a generic logic style that can be applicable to other nanoscale fabrics employing two-level logic schemes. Ongoing work on enhancements to H2L includes three-stage, two-level NASIC design with tiles consisting of one input and two separate output stages generating outputs in a time-multiplexed fashion, improving area utilization and performance.

5 NASIC Architectures

Large-scale computing systems may be designed using the NASIC fabric and the associated framework of building blocks, xnwFET-based circuits, and logic styles discussed in the preceding sections. This section discusses two key architectures for the NASIC fabric: the WISP-0 general-purpose processor and a massively parallel architectural framework with programmable templates for image processing.

5.1 *Wire Streaming Processor*

Wire streaming processor (WISP-0) is a stream processor that implements a five-stage microprocessor pipeline architecture, including *fetch, decode, register file, execute,* and *write back* stages. WISP-0 consists of five nanotiles: program counter (PC), ROM, decoder (DEC), register file (RF), and arithmetic logic unit (ALU). Figure 17 shows the layout. A nanotile is shown as a box surrounded by dashed lines in the figure. In WISP designs, in order to preserve the density advantages of nanodevices, data are streamed through the fabric with minimal control/feedback paths. All hazards are exposed to the compiler. It uses dynamic circuits and pipelining on the wires to eliminate the need for explicit flip-flops and, therefore, improve the density considerably. WISP-0 supports a simple instruction set, including *nop, mov, movi, add* and *multiply* functions. It uses a seven-bit instruction format with three-bit instruction and two-bit source and destination addresses. WISP-0 is used as a design prototype for evaluating key metrics, such as area and performance as well as the impact of various fault-tolerance techniques on chip yield and process variation mitigation. Additional enhancements to this design are ongoing in the NASIC group.

5.1.1 Wisp-0 Program Counter

The WISP-0 program counter is implemented as a four-bit accumulator. Its output is a four-bit address that acts as an input to the ROM. The address is incremented each cycle and fed back using a nano-latch. Figure 18 shows the implementation of

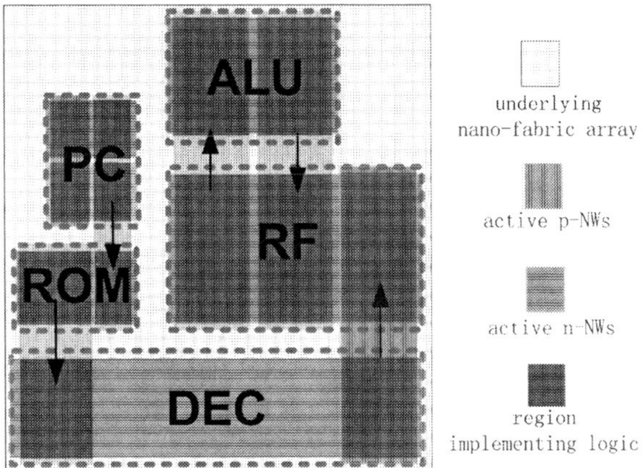

Fig. 17 WISP-0 processor floorplan

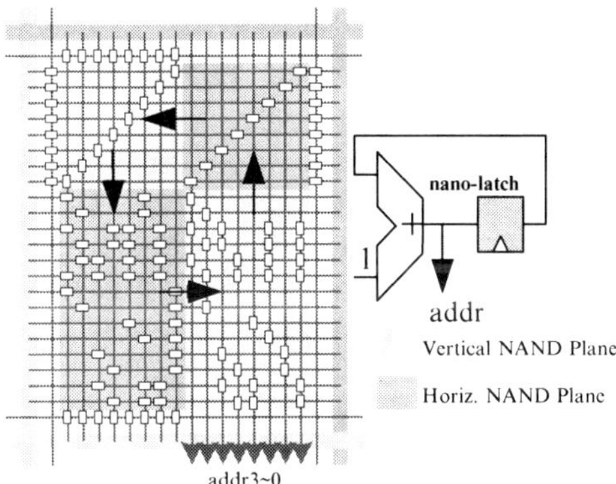

Fig. 18 WISP-0 program
counter

the program counter with a NAND–NAND scheme. Diagonal transistors on the upper two NAND planes implement the nano-latch to delay the output by one cycle and allow the signals to "turn the corner."

5.1.2 Wisp-0 ALU

Figure 19 shows the layout and schematic of the WISP-0 ALU that implements both addition and multiplication functions in a single stage. The arithmetic unit integrates an adder and multiplier together to save area and to ease routing constraints. It takes the inputs (at the bottom) from the register file and produces the write-back result. At the same time, the write-back address is decoded by the 2–4 decoder on the top and is transmitted to the register file along with the result. The result is written to the corresponding register in the next cycle.

More information about WISP-0 can be found in [23, 24].

Fig. 19 WISP-0 ALU

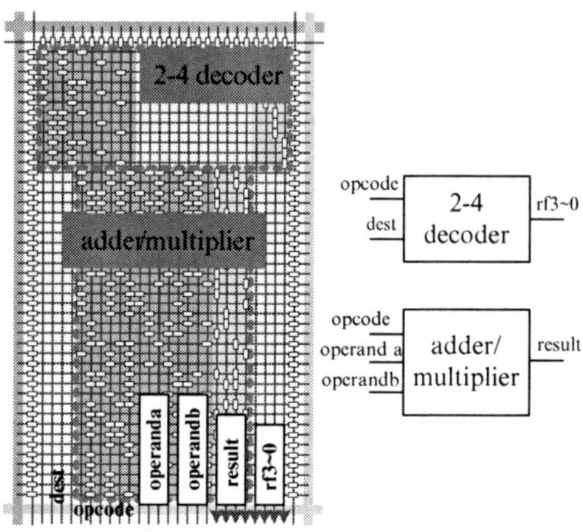

5.2 Nanodevice-Based Programmable Architectures

The processor architecture reviewed in the previous section is composed of dissimilar logic blocks performing a variety of functions interconnected in a prescribed fashion to build nanoscale systems. An alternative architectural style is to use a massively parallel array of identical logic blocks communicating with their neighbors to achieve data processing. Nanodevice-based programmable architectures (NAPA) is an architectural framework for the NASIC fabric using these sorts of collective computation models.

The NAPA design (Fig. 20) is regular and highly parallelized, with a large number of identical functional units or cells performing a given task. Instructions to each cell are transmitted on a limited number of global signals (template rails) controlled from reliable peripheral CMOS circuitry. There is a CMOS input-output controller that serially loads the inputs and reads out outputs. The cells themselves are composed of a small number of nanodevices that perform simple computations and are locally interconnected. The collective behavior of a large number of interconnected cells achieves information processing.

These sorts of collective computation systems may be a "natural fit" for nanoscale implementations because (1) high densities are available for high-level of parallelism, (2) local interconnections with minimal global routing remove the need for complex interconnect structures, and (3) 100% fault-free components are not a requirement because some defective cells in the design may not adversely affect the overall output integrity and allow more aggressive manufacturing processes. However, given the high rates of defects in nanomanufacturing, built-in defect/fault-tolerance techniques will need to be incorporated at various levels of NAPA to ensure that a sufficiently large number of cells are functioning correctly for output integrity.

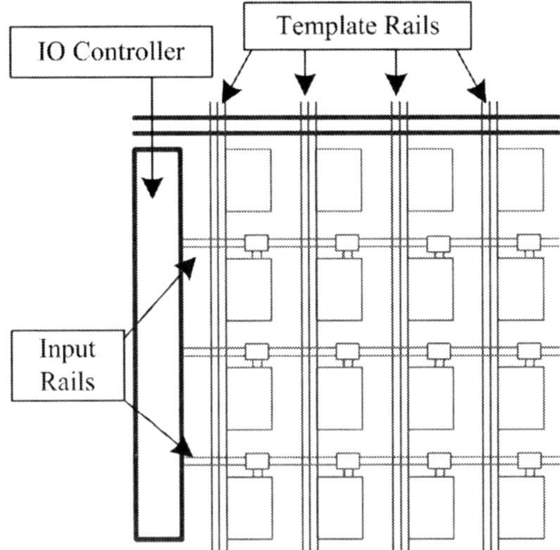

Fig. 20 The NAPA framework

Different applications that use cellular architectures, such as single instruction multiple data (SIMD), digital signal processing (DSP), and cellular nonlinear networks (CNNs) may be mapped to a NAPA design. The architecture and capabilities provided in a single cell determine the application, while the framework for programmability and input–output operations is unchanged. We show one possible implementation based on a computational model used in discrete-time digital CNN [25, 26] for image processing tasks. Multiple image processing tasks are possible by controlling the global instruction signals, which act as the templates for the CNN cells. Hence, the design is fully programmable and can run a variety of processing tasks in a sequence.

In a digital CNN design, each cell processes 1 pixel of an image (the input) and has an associated multi-bit state. A multi-bit template determines the nature of the image processing task. The state evolves as a function of the inputs, the template and signals exchanged with neighbor cells. The output is simply the most significant bit (MSB) of the state variable. The state evolution is given by the relation:

$$x(n + 1) = \sum_{i \in N} a_i y_i(n) + b_i u_i + c$$

where each a_i and b_i are instantaneous template values, $y_i(n)$ and u_i are the outputs and inputs to a particular cell, $x(n + 1)$ is the state and C is a constant term. The term $a_i y_i(n) + b_i u_i$ represents the ith partial sum received from a neighbor. The templates are space-invariant, i.e., the same template values are transmitted to all cells in the design.

Figure 21 shows the layout and circuit-level implementation of a single cell. There is a NASIC tile at the center for generating partial sums and state

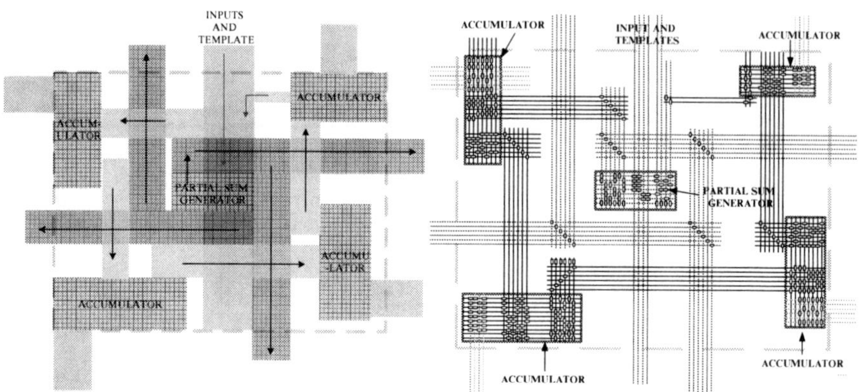

Fig. 21 Floorplan (*left*) and circuit diagram (*right*) of the NAPA cell implementing a digital CNN

accumulator tiles at the corners of each cell. The execution inside a cell progresses through the following stages:

1. *Generation of Partial Sums:* Initially, the control circuitry for the state accumulators is off and the partial sum generator is on. Each cell then generates five partial sums in sequence, one per nearest neighbor, and one for itself. These are transmitted on the broadcast network (Shown by clockwise loop in the left diagram).
2. *State Accumulation:* Once all partial sums are available, state accumulation datapaths are activated (anticlockwise loop). This is enabled by switching off the control circuitry for the partial sum and switching on the control for the state accumulator tiles. During this stage, the broadcast networks may be envisioned to be in an extended hold phase, with no precharge-evaluate operations. Thus, by suitably triggering the dynamic control circuitry, selective activation of datapaths and propagation of signals is achieved on a NASIC fabric without the need for arbitrary routing or additional multiplexers.
3. *Output Calculation:* After all partial sums from neighbors have been accumulated, the new value of the output is calculated and the next partial sum generation cycle begins. After a specified number of iterations, the final output may be read out using the I/O controller.

More information on cellular architectures for NASICs can be found in [27, 28].

6 Built-in Fault Tolerance

Nanoscale manufacturing is expected to have manufacturing defects at many orders of magnitude higher rates than conventional CMOS manufacturing processes. Furthermore, the effects of transient faults due to noise, soft errors, and parameter variations are expected to be significant. Consequently, nanoscale systems need to incorporate some form of fault tolerance to mitigate these effects and obtain functional chips. In this section, we investigate built-in fault-tolerance techniques for the NASIC fabric and evaluate their impact on WISP-0's yield, area and timing. Additionally, WISP-0's density is compared with an equivalent CMOS version developed with state-of-the-art conventional CAD tools and scaled to projected technologies at the end of the ITRS-defined semiconductor roadmap [29].

Two main directions have been proposed to handle defects/faults at nanoscale: reconfiguration and built-in fault tolerance. Most conventional built-in defect/fault-tolerance techniques [30–39], however, are not suitable in nanoscale designs because they were designed for very small defect rates and assume arbitrary layouts for required circuits. Moreover, the circuits used for error correction are often assumed to be defect-free, which cannot be guaranteed in nanoscale fabrics.

Secondly, if reconfigurable devices are available, defective devices might be replaceable after manufacturing. Reconfiguration based approaches [40–43], however, include significant technical challenges: (1) highly complex interfaces are required between micro and nano circuits for extracting defect maps and repro-gramming around defects – this is considered by many researchers a serious manufacturing challenge due to the alignment requirement of a large number of nanowires with programming microwires (MWs); (2) special reconfigurable nano-devices are needed, requiring unique materials with programmable, reversible, and repeatable characteristics; (3) an accurate defect map has to be extracted through a limited number of pins from a fabric with perhaps orders of magnitude more devices than in conventional designs; and (4) reconfiguration based approaches do not protect systems from noise and transient faults due to process variations.

Alternatively, as shown in this section, we can introduce fault tolerance at various granularities, such as fabric, circuit, and architecture levels, to make nanoscale designs functional even in the presence of errors, while carefully manag-ing area tradeoffs. Such built-in fault tolerance could possibly address more than just permanent defects. Faults caused by speed irregularities due to device parame-ter variations, noise, and other transient errors could be potentially masked. Com-pared with reconfiguration-based approaches, this strategy also simplifies the micro-nano interfacing: no access to every crosspoint in the nanoarray is necessary. Furthermore, a defect map is not needed and the devices used do not have to be reconfigurable. Introducing built-in fault tolerance does not impose any new con-straints on manufacturability.

6.1   Techniques for Masking Manufacturing Defects

Permanent defects are mainly caused by the manufacturing process. The small nanowire dimensions combined with the self-assembly process, driven by the promise of cheaper manufacturing, is expected to contribute to high defect rates in nanoscale designs. Examples of permanent defects in NASIC fabrics would include malfunctioning FET devices, broken nanowires, bridging faults between nanowires, and contact problems between controlling MWs and nanowires. For example, in a process that requires ion implantation of segments connecting NASIC FETs for high conductivity, the channels of transistors could be unintentionally implanted and therefore stuck-on.

In NASICs we consider a fairly generic model with both uniform and clustered defects and three main types of permanent defects: nanowires may be broken, the xnwFETs at the crosspoints may be stuck-on (permanently active transistor at crosspoint) or stuck-off (channel is switched off). A stuck-off transistor can also be treated as a broken nanowire. The initial thinking is that the more common defect type is due to stuck-on FETs as a consequence of the ion implantation process used. NASIC fabrics require lithographic masking for implantation steps (to avoid channels at crosspoints where no FETs are placed). Thinking from [44] suggests that we will be able to control the reliability of nanowires fairly well so broken nanowires will be likely less frequent than stuck-on FETs.

6.1.1 Circuit-Level and Structural Redundancy

Figure 22 shows a simple example of a NASIC circuit implementing a NAND–NAND logic function with built-in redundancy: redundant copies of nanowires are added and redundant signals are created and logically merged in the logic planes with the regular signals. As shown in the figure, horizontal nanowires are precharged to "1" and then evaluated. Vertical nanowires are subsequently precharged to "1" and then evaluated. The circuit implements the logic function $o_1 = ab + c$; a' is the redundant copy of a, and so on. Signals a and a' constitute a nanowire pair.

A NASIC design is effectively a connected chain of NAND–NAND (or equivalent) logic planes. Our objective is to mask defects/faults either in the logic stage

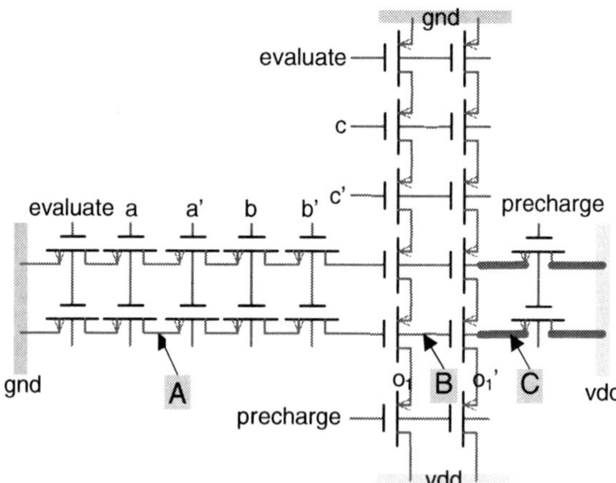

Fig. 22 Simple two-way redundancy

where they occur or the following ones. For example, a break on a horizontal nanowire in the NAND plane (see, e.g., position "A" in the figure) causes the signal on the nanowire to be "1." This is because the nanowire is disconnected from V_{DD}. The faulty "1" signal can, however, be masked by the following logic NAND plane if the corresponding duplicated/redundant nanowire is not defective.

A nanowire break at position "B" can be masked by the NAND plane in the next stage. Similar masking can be achieved for breaks on vertical nanowires. For stuck-on FETs, simple redundancy schemes work well: if one of the two transistors is stuck-on, the redundant copy ensures correct functionality.

6.1.2 Interleaving Nanowires

While the previous technique can mask many types of defects, faults at certain positions are difficult to mask. For example, if there is a break at position "C" in Fig. 22, the bottom horizontal nanowire is disconnected from V_{DD}, preventing precharge. The signal on this nanowire may potentially retain logic "0" from a previous evaluation. Because of NAND logic on the vertical nanowires, the two vertical nanowires would then be set to logic "1." Since both outputs on the vertical nanowires are faulty, the error cannot be masked in the next stage. In Fig. 22, the thicker segments along the horizontal nanowires show the locations at which faults are difficult to mask. We call these segments *hard-to-mask segments*.

For nanotiles with multiple outputs, a particular arrangement of output nanowires and their redundant copies could significantly reduce the size of hard-to-mask segments. Figure 23a presents a design in which each output nanowire and its redundant copy are adjacent to each other. In this arrangement, all segments to

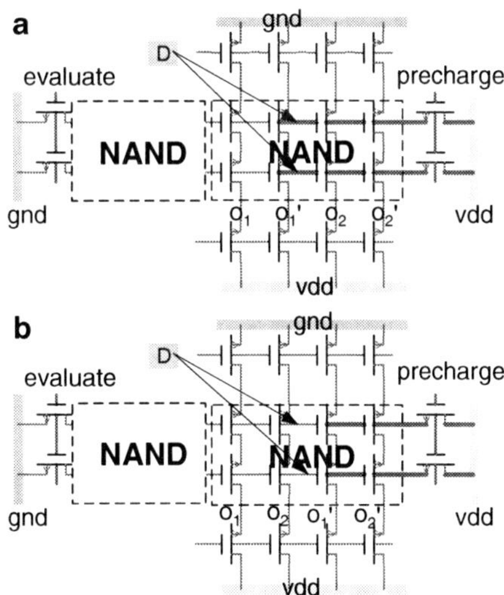

Fig. 23 Interleaving to reduce hard-to-mask regions

the right of the leftmost output nanowire pair (o_1 and o_1' in Fig. 23a) are hard-to-mask. Alternatively, the interleaved version in Fig. 23b shows an arrangement in which the output nanowires and their redundant copies are separated into two groups (o_1 and o_2 form one group; o_1' and o_2' form another group). In this case, the size of the hard-to-mask segments is reduced. In general, the size of hard-to-mask segments can be reduced in larger-scale designs to half, i.e., to half of the region covered by the vertical nanowires plus the segment related to the control FET. This latter region is fixed and for most designs adds a negligible area. Interleaving is also helpful in masking clustered defects because duplicated nanowires are set apart from one another.

6.1.3 Integrated Code-Based Error Masking

A variety of code-based techniques exist for correcting errors in conventional systems, such as Hamming codes [30] or Residue codes [31]. However, most of these require dedicated fault-free encoding and decoding circuitry and may not be directly applicable to nanoscale fabrics with much higher defect rates since correcting circuitry may also have faults. We explore a new integrated code-based error-masking scheme designed for nanoscale logic systems, where codes are directly merged into the fabric at circuit and logic levels. This scheme is built on the principles of Hamming distance and codes, but is achieved without the need for explicit encoding and decoding components.

The Hamming distance between two input codes of the same length is defined as the number of disparate bits. For example, if we consider two input codes "001" and "101," the Hamming distance is 1. Another interpretation of the Hamming distance is the minimum number of bits that need to be flipped for one input code to be indistinguishable from another. In communication channels and memory systems, increasing the distance between input codes (using additional bits) implies that faults (flipped bits) may be detected or corrected.

We apply these basic principles to logic circuits in a novel way. The key innovation is that the positions of transistors in an input logic plane of a NASIC tile can be envisioned to be input codes. This concept is highlighted in Fig. 24. The unshaded region in the input plane of the NASIC tile shows the original input plane of the NASIC 1-bit full-adder. We use the following rule to determine the codes based on FET positions: if there is a FET at the true value of an input signal, we designate the input bit to be "1," if there is a FET at the complementary position, we designate it as "0." For example, the first horizontal row has FETs at a', b', and c'. The corresponding codeword is "000." Similarly, input codes for all the other horizontal nanowires are assigned. These are summarized in the table on the left.

Once the input table is constructed, the Hamming distance may be increased by adding code bits. For example, a distance of 2 may be achieved with a single parity bit. This example shows one possible code achieving a distance of 3 with three additional code bits.

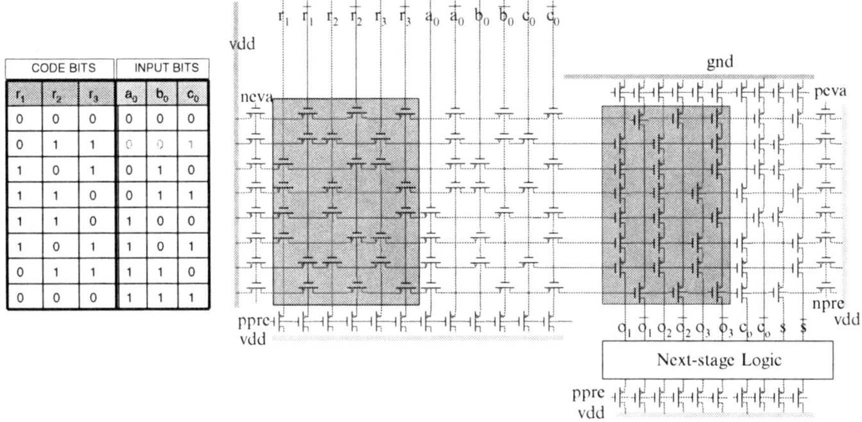

CODE BITS			INPUT BITS		
r_1	r_2	r_3	a_0	b_0	c_0
0	0	0	0	0	0
0	1	1	0	0	1
1	0	1	0	1	0
1	1	0	0	1	1
1	1	0	1	0	0
1	0	1	1	0	1
0	1	1	1	1	0
0	0	0	1	1	1

Fig. 24 Integrated error correction in NASIC designs

Once the code table has been constructed, it may be directly integrated into the logic plane with the positions of xnwFETs on the coding nanowires (left shaded plane) determined by the aforementioned code conversion rule. The coding xnwFETs provide resilience to faults on the input xnwFETs, with at least three errors required to erroneously evaluate a horizontal nanowire to "0" (corresponding to the Hamming distance of 3). The effect of any combination of two or fewer stuck-on defects or broken nanowires (in either the input or code nanowires) will not be propagated to the next stage.

The Hamming distance of 3 was found to be optimal for WISP-0 designs, as will be shown in the results section. The manufacturing process, defect models and yield-area tradeoffs may determine the choice of distance for a particular design.

6.1.4 Voting at the Nanoscale

Majority voting is a well-understood technique for conventional fault tolerance. N independent identical modules generate copies of an output that go to a voting block that determines the correct value of the signal to be a simple majority of the copies. The reliability of the voting scheme (i.e., the probability that a correct output is generated) is determined by the reliability of the individual modules (R_M), the reliability of the voter (R_V), as well as the number of identical copies. This is illustrated in Fig. 25, which plots the reliability of the system against the reliability of the individual module. Different curves correspond to different voter reliabilities and schemes, including triple modular redundancy (TMR – voting on three identical copies) and 6MR (i.e., voting on six copies of the signal). The dotted black line is the break-even point (system reliability = module reliability). Points above the line show regions where voting helps improve the reliability of the system.

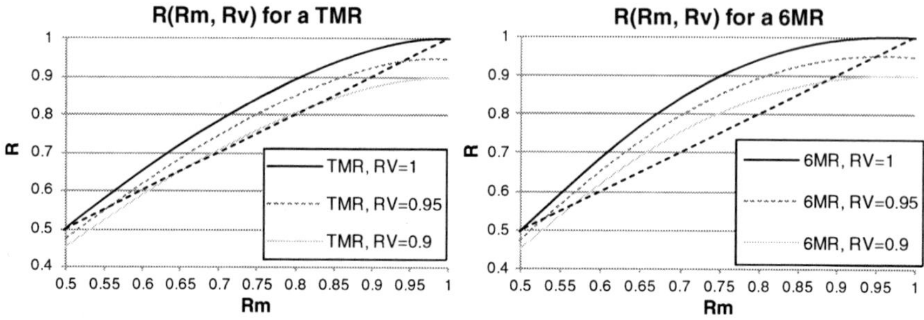

Fig. 25 Reliabilities of voting system vs. module reliability with unreliable voters

Three thin solid lines represent the reliabilities of TMR outputs and three thick lines for the reliabilities of 6MR outputs. From the figure, we can see that if the voting is perfect ($R_V = 1$), it always improves the reliability for $R_M > 0.5$. 6MR is more efficient than TMR but typically with the cost of more components. If the voting circuits could be faulty, voting helps only in a certain range of R_M.

From the figure, there are three possible ways to improve the overall reliability: (a) improve the reliability of each module (R_M), (b) improve the reliability of voting circuits (R_V), and (c) improve the voting logic itself (e.g., 6MR vs. TMR). Based on this discussion, we will show the implementations of nanoscale voting in simple two-level logic (i.e., NAND–NAND) and H2L NAND–AND/NAND logic styles and discuss the benefit of the H2L based implementation.

As mentioned before, complementary signals are necessary in a fabric based on two-level logic. An original signal and its complementary version can provide "dual-rail" redundancy. However, voting on "dual-rail" signals (e.g., a_1 and $\sim a_1$ in Fig. 26) requires signal inversion, which is difficult to achieve with two-level NAND–NAND logic on the layout-constrained 2D grid. Therefore, as shown in Fig. 26, the voting on original signals and complementary signals are separated from each other in a pure NAND–NAND fabric and the "dual-rail" redundancy is effectively unutilized.

With H2L logic, it is possible to produce complementary signals. Given this capability, we can vote only on original outputs and generate the complementary signals in voting circuits using NAND–AND logic only when necessary. As shown in Fig. 27, there is no need to generate complementary output in each module (A_1, A_2, and A_3) for voting. Instead, we generate three more original copies ($a_1{}'$, $a_2{}'$, and $a_3{}'$) for more redundancy and vote on all six signals (6MR). In the voting circuit, the original and complementary outputs are generated for the next stage using H2L NAND–AND/NAND logic.

Note that the area of A_1, A_2 and A_3 in Fig. 27 is actually smaller than in Fig. 26 (since only one set of horizontal midterms is used for output generation) and they provide more redundancy (six copies compared, whereas three copies are in Fig. 26). By using H2L, we improve the voting scheme (use 6MR instead of TMR) as well as the yield of each computing module (R_M). Simulation results in subsequent sections will show that the overall reliability is significantly improved.

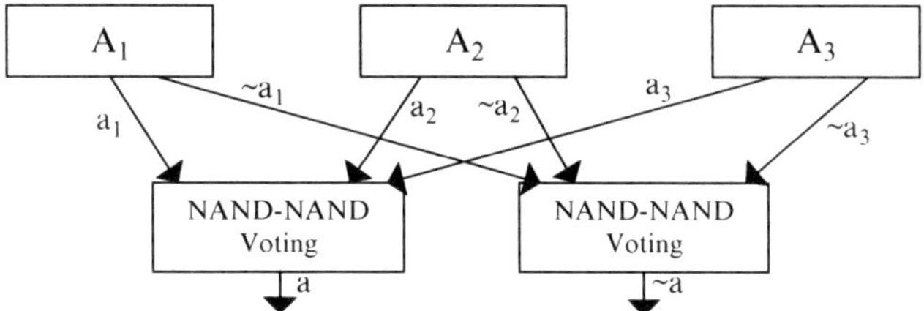

Fig. 26 Nanoscale TMR design in pure NAND–NAND fabric

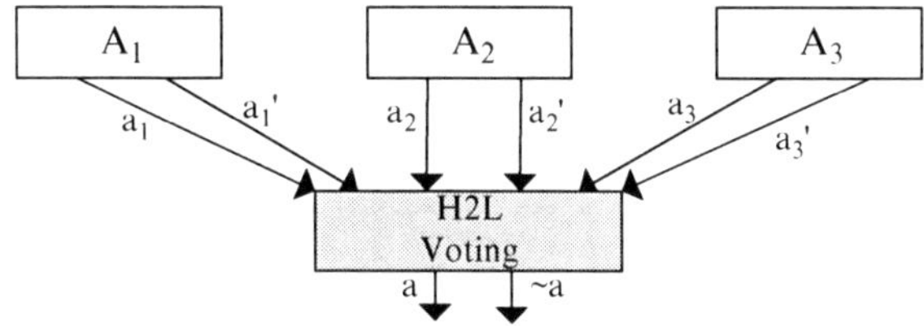

Fig. 27 6MR voting enabled by H2L

6.2 Density Evaluation

One of the key drivers for nanoscale system design is the perceived improvements in density over conventional CMOS technology due to smaller features of nano-materials as well as the smaller "footprint" of nanodevices and interconnect. We compare the density of NASIC designs with various fault tolerance schemes incorporated against an equivalent fault-free CMOS based implementation. We use WISP-0 as a design prototype for this study.

To get a more accurate evaluation on density, we need to take the area overhead of MWs into account. Note that the pitch between MWs in nanoscale WISP-0 also scales down with CMOS technology nodes: a reason why NASIC WISP-0 density changes somewhat with assumptions on MWs. Technology parameters from ITRS 2007 [29] used in the calculations are listed in Table 2.

To get a better sense of what the densities actually mean we normalize the density of nanoscale designs to an equivalent WISP-0 processor synthesized in CMOS. We designed this processor in Verilog, and synthesized it to 180 nm CMOS. We derived the area with the help of the Synopsys Design Compiler. Next, we scaled it to various projected technology nodes based on the predicted parameters by ITRS 2007 [29] assuming area scales down quadratically. For the purpose of this chapter, we assume that the CMOS version of WISP-0 is defect-free with no area overhead for fault tolerance. It is nevertheless expected that even CMOS designs would need redundancy and other techniques to deal with defects, mask irregularities, as well as significant parameter variations.

Density comparisons with CMOS are shown in Fig. 28. The notation used in the graphs is: *w/o Red* stands for WISP-0 without fault-tolerance techniques

Table 2 Parameters for density calculations

Parameter	Values (nm)
NW-pitch	10
NW-width	4

Technology node (ITRS 2007) (nm)	MW pitch (nm)
45	90
32	64
22	44
16	32

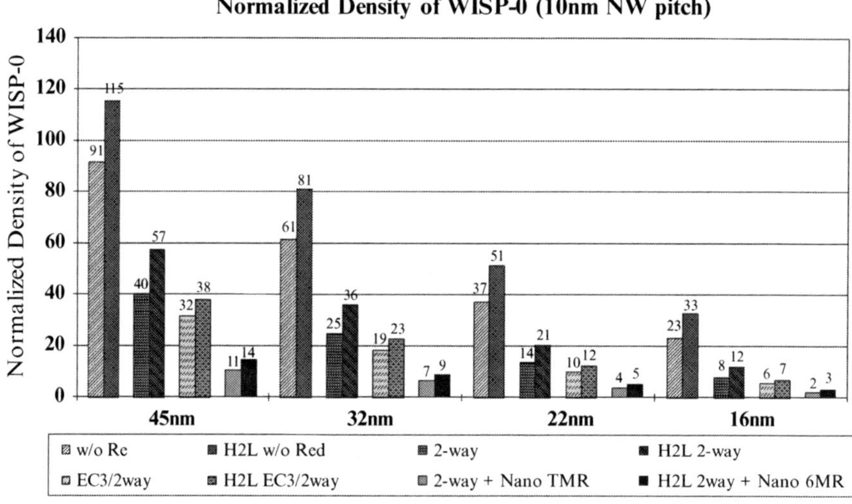

Fig. 28 Density comparison of NASIC WISP-0 with equivalent CMOS implementation

(or baseline); *2-way Red* stands for WISP-0 with two-way redundancy; *2-way Red + TMR/6MR* stands for two-way redundancy plus nanoscale TMR/6MR. While other combinations are possible, we found these to be insightful and representative.

We observe that NASIC designs without redundancy have orders of magnitude density advantages over conventional CMOS design. For example, NASIC H2L WISP-0 without redundancy is 33× denser than an end-of-the-roadmap 16 nm equivalent CMOS implementation. However, with various fault-tolerance techniques, some density advantages are lost. Using 6MR voting schemes with H2L still has 3× density improvement over a fault-free CMOS implementation at 16 nm. This is a somewhat pessimistic number for NASICs because (a) CMOS is assumed to scale perfectly quadratically, (b) no fault tolerance is assumed for CMOS designs, and (c) more aggressive reliable manufacturing processes may be leveraged to achieve NASICs at denser pitches and requiring fewer components with very high levels of fault tolerance.

6.3   Yield Evaluation for WISP-0

The yield impacts of various built-in fault tolerance techniques described in this section were evaluated for NASIC WISP-0 using a logic and circuit-level simulator we developed. Results are summarized in this subsection.

We assumed two types of defects – stuck-on transistors and broken nanowires – in our simulations. Stuck-off transistors are treated as broken nanowires as they have the same effects. From the results (Figs. 29 and 30), we can see that our fault-tolerance technique improves the yield of WISP-0 significantly. The yield of WISP-0 without redundancy is 0 when the defect rate is only 2%. With two-way redundancy, however, the yield of WISP-0 is still more than 20% when 5% transistors are defective. As expected, the error correction technique achieves better yield than two-way redundancy. Compared with a two-way redundancy approach,

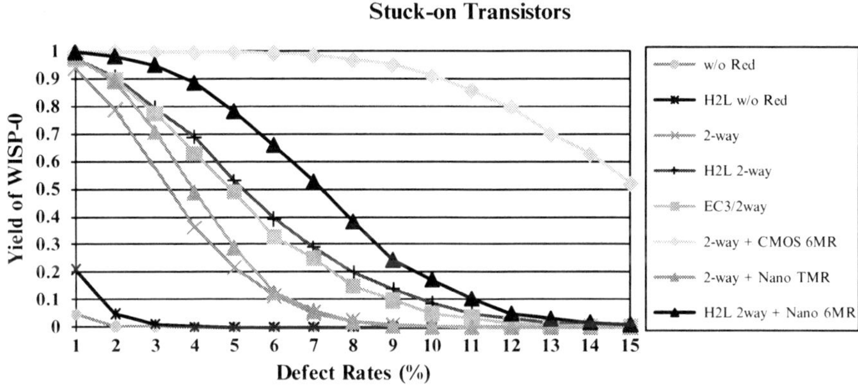

Fig. 29 Yield for WISP-0 with different techniques when only considering defective FETs

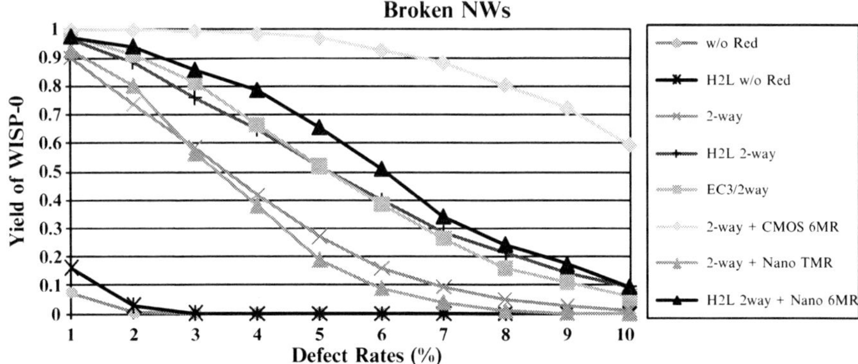

Fig. 30 Yield for WISP-0 with different techniques when only considering broken nanowires

the improvement due to the hybrid approach (*EC3/2-way*) on the yield of WISP-0 is 14% when the defect rate of transistors is at 2%, 129% at 5% defect rate, and 28.6× at 10%. Note that the improvement is greater for higher defect rates. Similar improvements are achieved for broken nanowires.

The H2L logic technique improves the yield considerably. Compared with the AND–OR approach in *2-way Red + TMR* scenario, the improvement of H2L logic on the yield of WISP-0 with *2-way Red + 6MR* is 15% when the defect rate of transistors is 2 and 147% at 5% defect rate. CMOS 6MR refers to using reliable CMOS voters utilizing the dual-rail redundancy from NASIC tiles. As shown in the figures, the CMOS 6MR technique improves the yield of WISP-0 dramatically. This, however, poses some serious constraints: (a) interfacing between nanodevices and CMOS circuitry will be required between NASIC tiles, complicating manufacturability considerably and (b) nanowires will be need to drive large capacitive loads associated with MOSFET gates and parasitics, leading to very poor performance. In these figures, yield values of CMOS 6MR are provided as the upper bound for other voting schemes.

Nanoscale voting eliminates the aforementioned issues imposed by CMOS voting. From the figures, we can see that the nanoscale TMR technique in NAND–NAND NASIC fabric improves the yield when the defect rate is low. However, the improvement vanishes for higher defect rates. For example, if the defect rate for transistors is higher than 7% or the defect rate for broken nanowires is higher than 3%, the nanoscale TMR actually deteriorates the overall yield. This is because with a high defect rate, voting circuits themselves become so unreliable that they negatively impact yield.

With H2L logic, the effectiveness of nanoscale voting is significantly better. In addition to the yield improvement for the logic itself, the nanoscale 6MR technique in the new NASIC fabric consistently improves the yield of WISP-0. Compared with WISP-0 with H2L logic and two-way redundancy, the improvement of nanoscale 6MR technique is 7 and 47%, respectively, when 2 and 5% transistors are defective, respectively. For broken nanowires, similar results are obtained.

It is important to understand the area overhead (or impact on density) of the different fault-tolerance techniques in conjunction with their fault masking ability. We define the yield-density product (YDP) metric to capture these effects. The YDP may be interpreted as an "effective yield" or the number of functional chips obtained per unit wafer size in a parallel manufacturing process.

The YDP results for various defect rates are presented in Figs. 31 and 32, respectively. We can see that the proposed built-in fault-tolerance techniques significantly improve the YDPs of WISP-0. H2L WISP-0 with two-way redundancy achieves the best YDP as the H2L logic technique improves both yield and density. Nanoscale voting techniques (i.e., *Nano TMR* and *Nano 6MR*) improve the yield of WISP-0 with more than 3× area overhead, deteriorating their YDPs. Future NASIC CAD tools can take yield and YDP as different design constraints: if designers care more about yield, nanoscale voting may be necessary. However, if both yield and density are critical, some less area-expensive fault-tolerant schemes, such as two-way redundancy or error correction techniques are desirable.

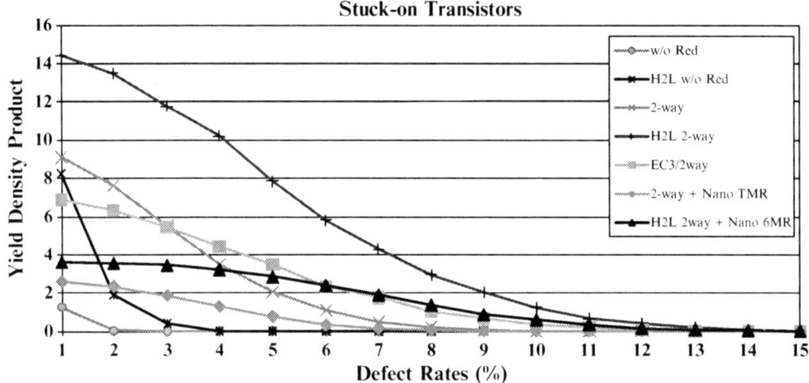

Fig. 31 Yield-density products achieved for WISP-0 when only considering defective FETs

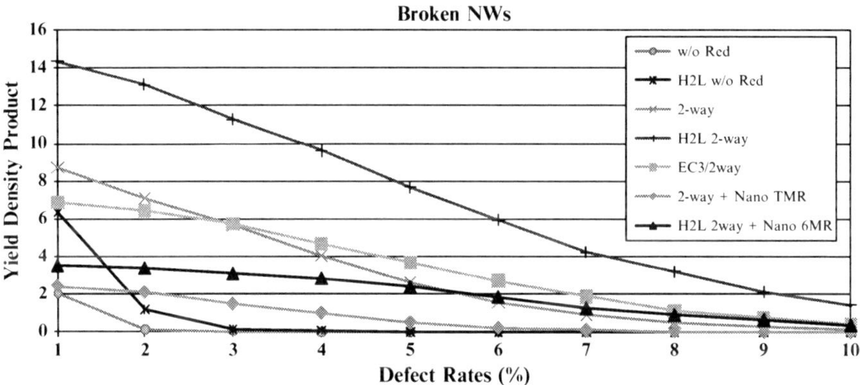

Fig. 32 Yield-density products achieved for WISP-0 when only considering broken nanowires

6.4 Process Variation Mitigation

While it is widely accepted that state-of-the-art CMOS designs need to deal with high levels of parameter variation, the defect rates expected are still fairly low. For example, at 90 nm, the expected defect rate is only 0.4 defects/cm^2. Process variation is often dealt with independently from defect tolerance and can be even handled at the architecture level [45–48] In contrast, nanoscale fabrics based on self-assembly manufacturing processes are expected to have much higher defect rates (in NASICs, we assume ten orders of magnitude higher or hundreds of millions to billions of defective devices per cm^2) in conjunction with high levels of parameter variation. These high defect rates require a layered approach for fault tolerance and typically involve incorporating carefully targeted redundancy at multiple system levels, e.g., structure, circuit, and architecture levels, as has been shown in the previous section. Given that this redundancy is already built-in for yield purposes, it is imperative to understand whether it could be exploited to mask delay faults caused by process variation. We can then develop redundancy-based techniques that are more tailored towards process variation than defect tolerance, e.g., by trading off yield for higher performance, or the opposite, depending on design requirements. We can also apply redundancy non-uniformly, taking into consideration defect and process variation masking needs in specific circuits. For example, it might be more critical to have higher resilience against variations in circuit blocks that are part of the critical path, than in blocks that have plenty of timing slack. In non-critical regions, we might be able to apply techniques focusing on improving yield, without sacrificing performance.

6.4.1 Initial Study on Impact of Redundancy for Masking Delay Faults

To see whether redundancy can mitigate the impact of process variation, we run Monte Carlo simulations of a WISP-0 NASIC streaming nanoprocessor design based on the NASIC fabric with built-in defect tolerance. In our simulations, we consider five parameters that may vary: wire width, metalized wire resistivity, contact resistance, pitch, and gate doping. The wire width modifies both the gate width and length, as NASIC xnwFETs are at the crosspoints between nanowires. Gate doping varies the FET equivalent ON resistance [49]. We assume variations of up to $3\sigma = 30\%$ for all of these parameters. We have explored the frequency distribution of a design without redundancy and after applying two techniques that were developed for defect tolerance: two-way redundancy and majority voting. Two-way redundancy is based on duplicating every nanowire and device. For majority voting, multiple copies of a logic block are created and a voter determines the block output based on the majority of the individual outputs.

Representative simulation results are shown in Fig. 33. All speeds are normalized to the speed at which that type of circuit would run with zero process variation to allow comparisons of the distributions despite differences in their mean frequency. The figure shows that the normalized frequency for circuits using redundancy tends to be significantly higher than without redundancy. In other words, introducing redundancy greatly increases the percentage of circuits that will operate at or above the nominal frequency in the presence of process variations.

Furthermore, when majority voting is used, the variation in frequency is barely half that of the non-redundant and two-way redundant circuits. These results attest to the potential for redundancy to counter the effects of process variation in addition to introducing defect tolerance and motivated us to consider these techniques in more detail and investigate specific approaches to trade off yield and performance. A number of techniques we are currently actively pursuing are presented next.

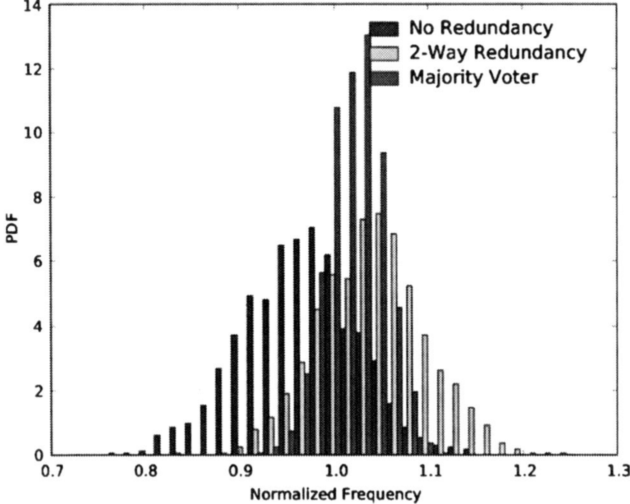

Fig. 33 Frequency distribution of simulated NASIC WISP-0 processor

6.4.2 Design of Fast-Track Techniques

The key idea we are actively exploring is based on inserting nanoscale voters at key architectural points in a design but also unbalancing the redundancy in each of the voter input blocks. The intuition behind this technique is to have some inputs (in some of the blocks) arrive faster than others to address the increased delay due to parameter variation and the delay due to the added redundancy itself. In addition, we leverage the property of NASIC circuits that logic "0" faults are much less likely than logic "1" related faults: therefore, we bias the voters towards logic "0." Other types of nanoscale fabrics might require a different biasing logic level. In NASICs, faulty "0s" are much less likely, as their occurrence requires a higher degree of faults on nanowires or devices in NASICs.

The combination of unbalanced input blocks and biasing defines a variety of new techniques depending on the input configuration, redundancy levels for each input, voting type, and biasing applied. The biasing towards "0" also helps mask the impact of the lesser redundancy inputs on overall yield and helps preserve yield. If a lesser redundancy block outputs a "0," there might be no need to wait for slower outputs from higher redundancy blocks. By biasing towards "0," the faulty logic "1s" are more easily overridden by the correctly operating circuits, allowing the system as a whole to be run at a higher frequency. We have explored many combinations of redundancy configurations for both defects and process variations and fully implemented in the NASIC processor simulator.

For example, along these ideas, a so-called FastTrack voting scheme example and notation are shown in Fig. 34. Note that the redundancy applied to input blocks to the voter is unbalanced. The figure implies a voter with three blocks, two having

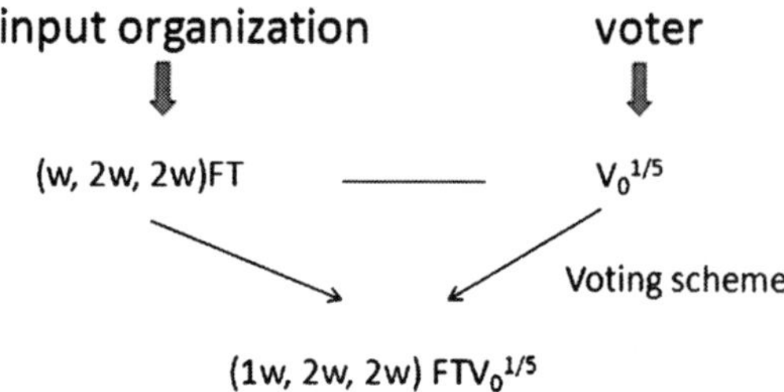

Fig. 34 FastTrack redundancy schemes and associated notation

two-way redundancy, and one with no redundancy. Furthermore, the voter is biased towards "0" (see subscript on FTV): if 1 out of 5 inputs coming from the three blocks are "0" the output is voted as logic "0." This is in contrast to a regular majority voter wherein three out of five inputs would need to be "0" to vote logic "0."

Similarly, a plurality voter would require a plurality of "0s" to output "0." Plurality and majority voters are in essence instances of biased voters. The combination of input organization and voter type results in a voting scheme denoted, e.g., as $(1w, 2w, 2w)FTV_0^{1/5}$. This notation can be generalized to voting schemes with unbalanced redundancy in their input blocks and biasing, in general. If there is no fast tracking, the FTV notation is replaced with V. If there is no biasing of the output, the subscript and superscript after the voter type can be omitted. In the next section, the experimental setup and method of evaluation are outlined. This is followed by initial results for a variety of voting schemes based on the above concepts.

6.4.3 Method of Evaluation

Delay calculations are initially done using RC-equivalent circuits. A more accurate approach would be to incorporate the device physical model and include changes in actual characteristics due to manufacturing imperfections. This is a direction that is currently pursued but more challenging since for every variation assumed, a new 3D model for the devices is extracted from the device simulation.

To this end, we model nanowires with each transistor being changed into ON (low resistance) and OFF (high resistance) modes by its gate input. There is also the resistance of the nanowire interconnect and contact resistance with the CMOS wires that drive the gates. Capacitance sources include inter-wire and junctions between the wires where transistors are formed. The values of resistive and capacitive elements are calculated using their geometry, together with several parameters which vary between devices. For our initial experiments, the nominal value of the transistor length evaluated is 4 nm and the width is 4 nm, the square aspect ratio being due to the crossed-nanowire devices. The width of the nanowire depends, for example, on the size of the catalyst nanoparticles used as seeds for vapor–liquid–solid (VLS) nanowire growth. Variations in nanoparticle sizes, therefore, directly correlate with variations in nanowire width. The standard deviation of wire widths has been shown to be around 10% of the mean [50]. In addition, the width of the nanowire is assumed to be uniform along its length.

The transistor ON resistance is also varied by the gate geometry – i.e., length and width – that are determined by the width of the nanowires. We do not use such traditional CMOS parameters as V_{th} and L_{eff}. Instead, we abstract variations in physical parameters into a variation in transistor resistance R_{ON}. Similarly, we calculate the capacitances based on geometries of the nanowires. This allows us to make predictions of delay such that they are anchored in experimental work and geometric variation.

All device parameters are taken from the literature on manufacturing of nano-scale devices and our own devices. n-Type xnwFETs are used in these calculations. The n-type devices are assumed to be silicon nanowires lightly doped with phosphorus. The nominal ON resistance for the specified geometries for n-type devices (R_{ON}) has been calculated to be 3.75 Ω based on experimental work [3]. The overall standard deviation of transistor ON resistance has been found to be $\sim$20% [51], including variation of gate geometry. After removing the variation in gate geometry, this points to a variation of 10% in transistor ON resistance for a square transistor. The nominal contact resistance with CMOS wires delivering V_{DD} and GND has been found to be $\sim$10 kΩ [52].

Interconnect variation is modeled geometrically. The diameter, resistivity, and contact resistance of each nanowire vary independently. Interconnect is assumed to be obtained by transforming silicon nanowires into nickel silicide (NiSi) via silicidation. The resistivity of the resulting nanowires after thermal annealing averages 9.5 $\mu\Omega$ cm [11]. Nonuniform metallization can lead to variations in the resistivity of interconnect.

The capacitance is calculated geometrically using the standard expressions for cylindrical wires. These are calculated using the wire diameter and pitch together with circuit layout. The dielectric is SiO_2 with a dielectric constant of 3.9. The nominal value for parallel nanowire capacitance is calculated to be 53.49 pF/m and for junction overlap capacitance, it is 0.602 aF. Variations in the spacing of nanowires, due to self-assembly based techniques for creating parallel arrays, will lead to variations in parallel nanowire capacitance. The thickness of the dielectric layer between wires (nominal = 5 nm) is determined by the inter-wire spacing, as the space between them is filled with SiO_2 during manufacturing. These parameters are summarized in Table 3.

Table 3 Parameters used for timing simulation

Parameter	Nominal value	Standard deviation
Wire resistivity of NiSi (ρ_{NiSi})	9.5 $\mu\Omega$ cm	10%
Wire diameter (d)	4 nm	10%
Wire pitch	10 nm	10%
Contact resistance	10 Ω	10%
Transistor ON resistance (R_{ON})	3.75 kΩ	10%
Oxide dielectric constant (ε_r)	3.9	None
Oxide dielectric thickness (t_{ox})	5 nm	10%

In order to evaluate the effects of parameter variation on the WISP-0 processor, we built a simulator that incorporates physical design parameters and is capable of handling both parameter variation and device faults. The simulator uses Monte Carlo techniques to capture statistical distributions. It takes as input a WISP-0 circuit design and a set of manufacturing parameters as described above, including the amount of variation to apply. The simulator based on the variation model given as input varies the characteristics of each nanowire, xnwFETs, and so forth, independently.

Additionally, it takes settings for defects. The defect model consists of a set of defect types and probabilities for each, plus a description of any clustering behavior [23]. For instance, the defect model for a given test might be "5% probability of each transistor being stuck-on, with uniform distribution of defects."

The simulator then generates a series of designs based on these parameters, with the values of each individual circuit element changing based on the parameters defined. The speed is then determined by simulating the WISP-0 design and measuring whether it generates the correct output, searching to find the fastest speed at which the correct output is generated by that test WISP-0 nanoprocessor. If the correct output is not generated at any speed due to defects, then the nanoprocessor is noted as faulty and no speed is recorded for that sample. This is repeated many times and each maximum speed is recorded, giving an overall statistical distribution of operating speeds.

It may seem that this could be more quickly done by simply measuring the speed of each gate and taking the speed of the slowest gate as the speed of the nanoprocessor in toto, but this would be inaccurate. This is because in many cases, the fault caused by a gate working incorrectly will be masked by the built-in fault tolerance. Therefore, the only accurate way to determine the speed is through simulation of the complete nanoprocessor including all its circuits. Additionally, the nominal frequency of the design, equal to the maximum frequency of the design with all parameters set to their nominal (i.e., zero variation) values, is measured.

6.4.4 Fast Track Results

This section shows initial results with the FastTrack techniques for WISP-0. Three sets of graphs capture the effectiveness of the techniques by plotting effective yield, normalized performance, and effective-yield normalized-performance products. The effective yield is defined as the yield per unit area. The performance is normalized to the slowest result. The third metric is introduced to provide an initial idea for how the techniques work for both performance and yield given the same area overhead.

The results (Figs. 35–37) show that the new FastTrack techniques do best in improving performance but impact yield somewhat, especially at higher defect rates. There is no one single technique that performs best overall, considering various defect rates, when both yield and performance are important. The scheme $(3w, 2w, w)\text{FTV}_0^{2/6}$ performs best in terms of performance and does well up to 6%

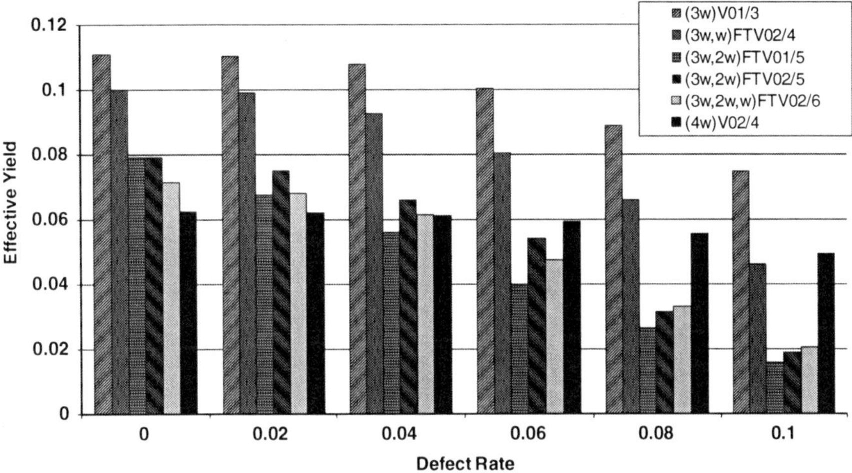

Fig. 35 Effective yield comparison of techniques used for parameter variation applied on WISP-0 processor

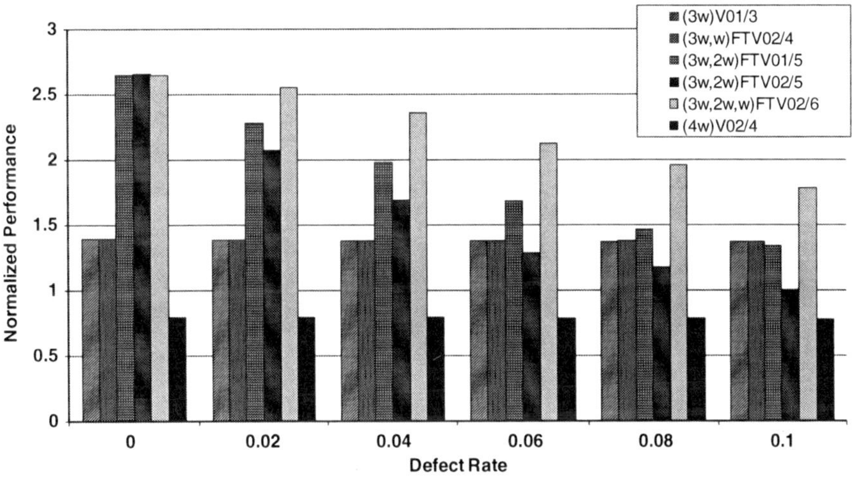

Fig. 36 Normalized performance of techniques used for parameter variation applied on WISP-0 processor

defect rates even in terms of yield. As expected, the schemes having the highest redundancy levels, such as $(4w)V_0^{2/4}$, do best at very high defect rates. In these experiments, the fault-tolerance techniques have been applied uniformly across all circuits, but voting has been added only after the ROM, DECODER, and ALU modules in WISP-0. However, a different balance between fast tracking and redundancy can be used, depending on the criticality of the path in practice. For more details, readers can refer to [53, 54].

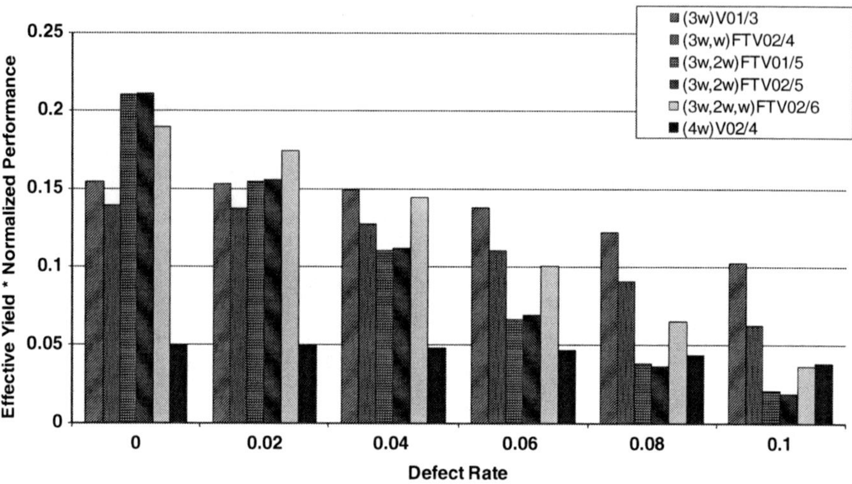

Fig. 37 Effective yield × normalized performance of parameter variation mitigation techniques for WISP-0 processor

Another important result is that the FastTrack schemes can be used also for a performance play. If the parameter variation is less of an issue, but a high level of redundancy is needed for yield purposes, FastTrack can mitigate the performance impact of the high redundancy blocks. While FastTrack voters add to the overall latency of the design, they do not significantly affect the throughput. For example, in a processor design, they would increase the overall latency of the pipeline somewhat. H2L allows voting to be implemented with very few NASIC gate delays.

7 Discussion on Performance and Power

Performance and power consumption in NASICs designs depend on a number of factors, such as the underlying semiconductor materials, device structures, manufacturing modules, amount of redundancy, logic and circuit schemes, as well as optimizations. Ongoing effort is to derive projections similar to the ITRS roadmap depending on these factors. A brief review of the potential benefits and tradeoffs compared to end-of-the-line projected CMOS is discussed below.

Fundamentally, xnwFET devices should be able to attain a per-device performance comparable to or better than in CMOS devices of similar dimension. The better performance is projected to be a result of more aggressive scaling. As mentioned, the devices for NASICs with various materials and structures are currently in the process of being optimized: a key challenge has been to increase the I_{ON} current. NASICs are based on a dynamic style of evaluations that has been shown to be faster than static CMOS MOSFETs at the same feature sizes. Without redundancy, the NASIC circuits have fewer devices per gate although ultimately at the system level, the number of devices also depends on the effectiveness of the logic style in the fabric. In this direction, the NASIC H2L and other emerging NASIC logic styles based on three-stage tiles overcome some of the limitations of traditional two-level logic (e.g., there is no need to carry complementary signals across all tiles anymore) commonly adopted for 2D layouts. NASIC redundancy techniques have been simulated and the results show a close to linear degradation of performance per tile for higher degrees of redundancy rather than a perhaps expected quadratic degradation, i.e., two-way redundancy would impact performance by roughly $2\times$. The performance impacts could also be masked somewhat by the FastTrack techniques, as described in previous sections.

NASICs have no static power: no direct current path exists between V_{DD} and V_{SS}. Conventional wisdom would dictate that a high I_{ON}/I_{OFF} ratio is necessary to avoid leakage from becoming an issue. However, in NASICs, the control FETs are microwire-based and their work function can be tailored. This allows these devices to act like conventional ground and V_{DD}-gating FETs, strongly limiting leakage during precharge and hold phases. The leakage in between the intermediate xnwFET nodes primarily contributes to charge redistribution and has been found to not be a problem. In our simulation, we found that an I_{ON}/I_{OFF} ratio of 1,000 might be sufficient to control leakage if the control FETs are optimized. Active power is related to precharging and discharging of the nanowires.

There are many optimizations possible to reduce it. This includes reducing unnecessary precharging events, controlling charge sharing between intermediate nodes on nanowires, optimizing toward logic "1" outputs vs. logic "0," etc. Note that when a nanowire evaluates to logic "1," there is no discharge and the power consumed is primarily related to leakage/charge redistribution to intermediate nodes and possibly leakage to V_{SS} (that, as has been discussed above, is mitigated with the microwire-based FETs).

As an initial point of comparison between NASICs and CMOS for the WISP-0 processor, a preliminary evaluation has been completed. Assuming a 0.7 V V_{DD}, an I_{ON}/I_{OFF} ratio of 1,000, and a contact resistance of 10 kΩ, the NASIC WISP-0 was found to achieve $\sim 7\times$ better power-per-performance than in a 16-nm CMOS WISP-0. The CMOS version assumes a fully synthesized chip from Verilog using commercial CAD tools for synthesis and place-and-route. It assumes a somewhat optimistic performance scaling between technology nodes and does not consider the impact of interconnect delays. These results are thus preliminary and mainly indicative of the NASIC fabric's potential, since they use simple RC models that are not based on detailed device-level simulations of xnwFET devices.

Our current work with optimizing and using xnwFETs based on accurate physics models and manufacturing imperfections as well as experimental validation, and efforts to build tools allowing a plug-and-play evaluation for various device structures we develop, will provide a more realistic comparison. These efforts are ongoing and hopefully will be reported in the next few years to the research community.

8 Manufacturing

Reliable manufacturing of large-scale nanodevice based systems continues to be very challenging. Self-assembly based approaches, while essential for the synthesis and scalable assembly of nano-materials and structures at very small dimensions, lack the specificity and long-range control shown by conventional photolithography. Other non-conventional approaches, such as electron-beam lithography, provide the necessary precision and control and are pivotal in characterization studies; but these are not scalable to large-scale systems. Examples of small nanoscale prototypes include a carbon nanotube FET based ring oscillator [55] and an XOR gate using superlattice nanowire pattern transfer technique (SNAP) based semiconductor nanowires and electron-beam lithography [14].

A manufacturing pathway for a nanodevice based computing system needs to achieve three important criteria:

- *Scalability:* Large-scale simultaneous assembly of nanostructures/devices on a substrate must be possible.
- *Interconnect:* Nanodevices, once assembled, must be interconnected in a prescribed fashion for signal propagation and achieving requisite circuit functionality.
- *Interfacing:* The nanosystem must effectively communicate with the external world.

In this chapter, we explore a manufacturing pathway for NASICs [56] that realizes the fabric as a whole, including devices, interconnect and interfacing. This pathway employs self-assembly based approaches for scalable assembly of semiconductor nanowires, and conventional lithography-based techniques for parallel and specific functionalization of nanodevices and interconnects. While individual steps have been demonstrated in laboratory settings, challenges exist in terms of meeting specific fabric requirements and integration of disparate process steps.

8.1 Fabric Choices Targeting Manufacturability

Before delving into the details of the manufacturing pathway, it is instructive to look at certain NASIC fabric decisions that significantly mitigate requirements on manufacturing. Choices have been made at the device, circuit, and architecture levels targeting feasible manufacturability while carefully managing design constraints. This is in direct contrast to other technologies, such as CMOS, which optimize designs for performance and area, but place stringent requirements on the manufacturing process.

- NASIC designs use regular semiconductor nanowire crossbars without any requirement for arbitrary sizing, placement or doping. Regular nanostructures with limited customization are more easily realizable with unconventional nanofabrication approaches.

- NASIC circuits require only one type of xnwFET in logic portions of the design.
- Local interconnection between individual devices as well as between adjacent crossbars is achieved entirely on nanowires; interconnection of devices does not introduce new manufacturing requirements.
- NASICs use dynamic circuit styles with implicit latching on nanowires. Implicit latching reduces the need for complex latch/flip-flop components that require local feedback.
- Tuning xnwFET devices to meet circuit requirements is done in a "fabric-friendly" fashion; techniques such as gate underlap and substrate biasing do not impose new manufacturing constraints.
- NASICs use built-in fault-tolerance techniques to protect against manufacturing defects and timing faults caused by process variation. Built-in fault-tolerance techniques do not need reconfigurable devices, extraction of defect maps, or complex micro-nano interfacing, as required by reconfiguration based fabrics. All fault tolerance is added at the nanoscale and made part of the design.

These fabric choices reduce manufacturing requirements down to two key issues: assembling nanowire grids onto a substrate and defining the positions of xnwFETs and interconnect. The latter step, also called *functionalization*, is the price paid for a manufacturing-time customization. The manufacturing pathway and associated challenges are discussed in the next subsection. Note that by adjusting the nanowire pitch, any manufacturing issue can be managed, but the goal is to achieve the smallest possible pitch.

8.2 *Manufacturing Pathway*

Key steps in the NASIC manufacturing pathway are shown in Fig. 38. Part (a) shows a NASIC 1-bit full-adder circuit. Horizontal nanowires are grown and aligned on a substrate, as shown in part (b). In general, nanowire alignment can be *in-situ* or *ex-situ*. *In-situ* refers to techniques where nanowires are aligned in parallel arrays during the synthesis phase itself. On the other hand, ex situ refers to techniques where nanowire synthesis and alignment are carried out separately. Currently, there are research groups pursuing both in situ and *ex-situ* assembly.

Lithographic contacts for V_{DD} and GND as well as some control signals are created (b). A photolithography step is used to protect regions where transistors will be formed while creating high conductivity regions, using ion implantation, elsewhere (c, d). Ion implantation creates $n^+/p/n^+$ regions along the nanowires, which, under suitable electrical fields, act as inversion-mode source/channel/drain regions.

The gate dielectric layer is then deposited (or oxide is grown) (e) followed by alignment of vertical nanowires. The above steps are now repeated for the vertical nanowire layer (f–h). During ion implantation on vertical nanowires (H), channels along horizontal nanowires are self-aligned against the vertical gates.

Key individual steps and challenges are discussed in detail in the following subsections.

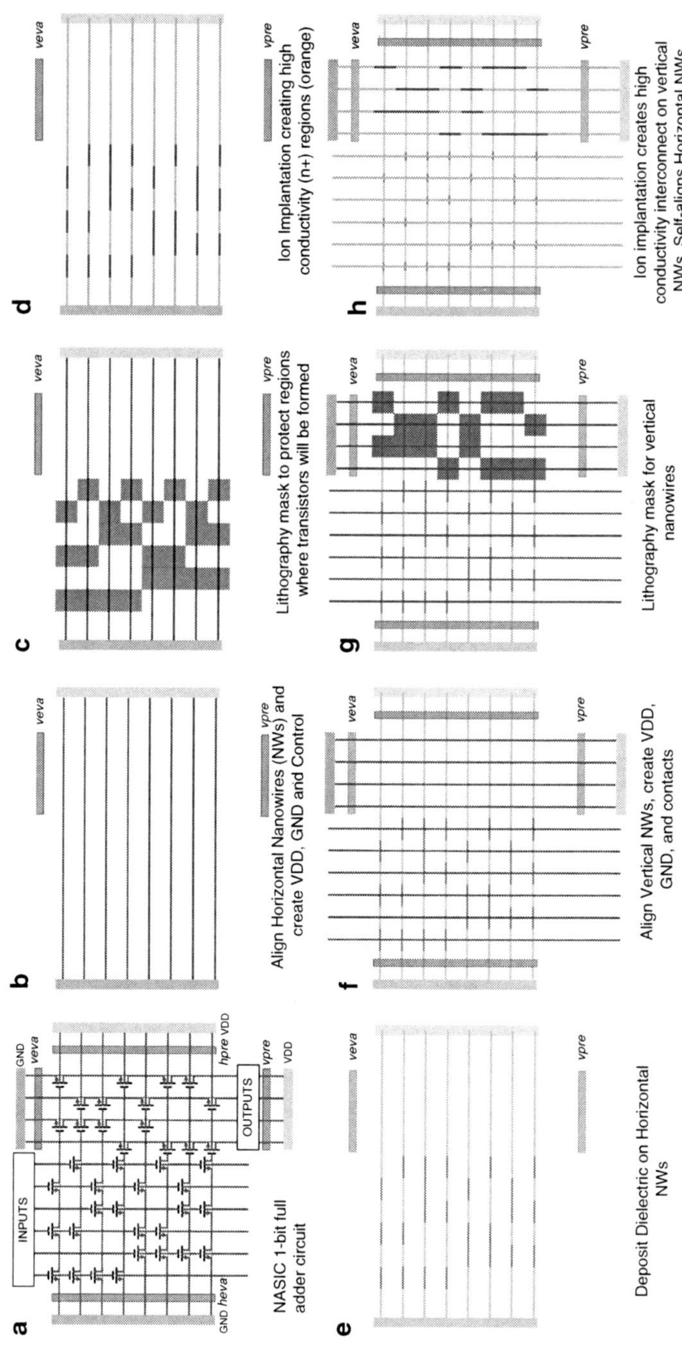

Fig. 38 NASIC manufacturing pathway

8.2.1 Nanowire Growth and Alignment

The ideal technique to form aligned nanowire arrays should guarantee an intrinsic and concurrent control over three key parameters: (a) the number of nanowires, (b) the internanowire pitch, and (c) the nanowire diameter within the array. The state-of-the-art semiconductor nanowire array formation with alignment techniques can be broadly classified into the following three categories:

In situ Aligned Growth: In situ techniques refer to techniques where cylindrical nanowires are synthesized directly on a substrate in an aligned fashion and do not require a separate transfer step. Examples of these techniques include gas flow guiding, electric field guiding, substrate or template guiding [57–61]. While arrays containing more of less parallel nanowires have been demonstrated using these techniques, the control of pitch and diameter depends on catalyst engineering, i.e., the ability to form a nanoscale periodic array of catalyst nanoparticles as a dotted straight line, as well as catalyst material compatibility with the substrate.

Ex situ Growth and Alignment: In ex situ growth and alignment, cylindrical nanowires are grown using techniques, such as VLS growth [3], and then aligned onto a substrate in a subsequent process step. Examples include Langmuir–Blodgett technique [62, 63]. fluidic alignment [64], organic self-assembly [65, 66], and contact printing [67]. A wide choice of material and synthesis techniques is available with ex situ growth and alignment. Narrow distribution of nanowire diameters is possible since nanowires can be purified post-synthesis. However, the key issue is substrate patterning to align nanowires during the transfer step.

Nanolithography-Based Pattern and Etch: In these approaches, a semiconductor material layer, preformed on the target substrate surface, is first patterned by nanolithography and then anisotropically etched to create a periodic rectangular nanowire array. Examples include SNAP [14] and nanoimprint lithography [68]. These techniques demonstrate excellent pitch and diameter control, but face some key issues: choice of material is limited to silicon; also, only one layer of nanowires can be formed using this technique. The surface of those nanowires is usually somewhat damaged too.

Although many pioneering contributions have been made by the techniques discussed above, there remain several areas for improvement. For instance, some of them rely on the advanced and expensive fabrication processes, such as electron-beam lithography and superlattice epitaxy, which have not been used in any mainstream production facilities. In addition, many chemical self-assembly methods could only cover a localized wafer footprint.

Even if these nanotechnologies might possibly be scaled up to cover the entire wafer, wafer-scale precision control, as demanded by low-cost and high-yield manufacturing, would be deemed very difficult if not impossible. The most promising pathway, among others, is the hybrid top-down/bottom-up directed self-assembly (DSA) approach that is, however, limited by its slow throughput, inaccurate registration, and high defect density.

To address the challenges outlined above in controlling the three key parameters, a novel fabrication process combining top-down lithography and crystallographic etching is being considered that comprises a three-step formation strategy:

- The creation of an intrinsically controlled nanoscale periodic line array on the substrate surface.
- The selective attachment of the 1D nanostructures only onto the lines within the array.
- The transfer of the aligned nanostructure array to different substrates.

In the first step, a nanoscale line array on the substrate surface is created with a controlled number of lines, pitch, and linewidth, as illustrated in Fig. 39. The number of lines and pitch can be defined by conventional lithography (Fig. 39a), that arguably offers the ultimate intrinsic control and requires a minimal alteration of the existing layout design infrastructure. Even though state-of-the-art lithography [29], such as extreme ultraviolet (EUV) lithography, might be capable of meeting the pitch requirement, a sub-lithographic pitch could indeed be achieved using the spacer double-patterning technique, as shown in Fig. 40. Besides being a cost-effective way to manufacture the necessary nanoscale pitch from a coarse pitch line array, this approach also permits an alignment to the pre-existing features on the substrate as a side benefit if necessary.

The nanoscale and usually sub-lithographic linewidth can be controlled by using crystallographic etching (Fig. 39b) to be larger than, similar to, or smaller than the target nanowire diameter. Contrary to a generic isotropic wet etching, this novel use of crystallographic etching bears several advantages. First, it offers a very good control over the resultant silicon linewidth, as determined by the etching time of the

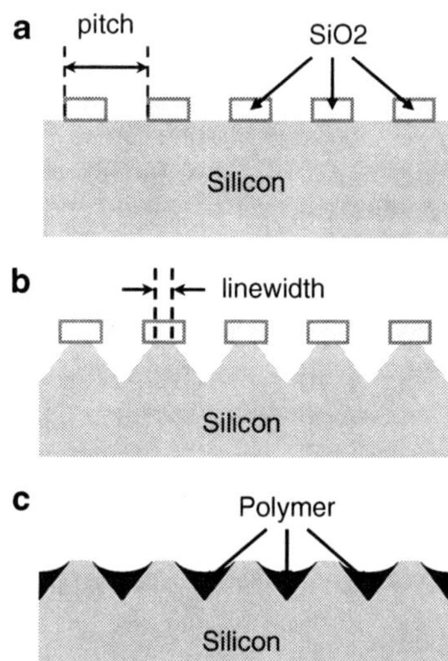

Fig. 39 Formation of a nanoscale line array with periodically dissimilar surface properties

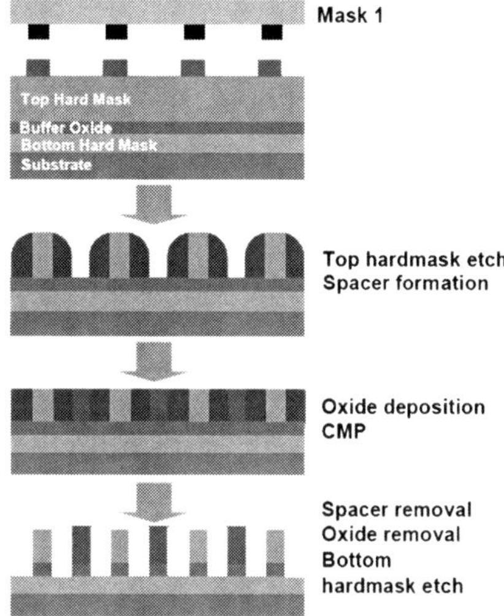

Fig. 40 Schematic process flow for spacer double patterning

slowest-etching crystal plane. The second one is the creation of precise and repeatable sidewall profiles (Fig. 39b) for subsequent filling of the inter-space. For example, a material with property different from silicon, such as polymer, can be used to fill the inter-space to form a resultant surface with periodically dissimilar properties (Fig. 39c).

The second step of the formation strategy involves the selective attachment of nanowires only onto the exposed Si line array on the substrate surface, i.e., onto the ridge of the Si triangles (Fig. 39c), but not onto the inter-space polymer (Fig. 41a). This can be achieved by manipulating the surface chemistry of the nanoscale substrate lines as well as that of the to-be-assembled nanowires. In other words, the nanowire surface should be attractive to the substrate lines, but repulsive to the surface of the inter-space polymer or another nanowire. This desired selectivity could come from either a hydrophobic–hydrophilic interaction or a coulombic interaction. After which, the inter-space polymer would be removed without meddling with the assembled nanowire array (Fig. 41b). This crucial step could in fact serve as a lift-off process to remove any nanowire that might have randomly adsorbed on the polymer surface, a process that can minimize the resultant array defectivity and, thus, greatly enhance the manufacturing yield.

The last step is the transfer of the aligned nanowire array onto the target substrate surface (Fig. 42), which is readily accomplished with either the wafer manufacturing bonding process or the established stamping transfer process [68–70] after slight modifications. The key requirement in this step is an accurate alignment between the Si substrate and target substrate, which could fortunately be fulfilled using a crystallographic alignment mark etched into the Si substrate. This work is currently ongoing, yet preliminary results (Figs. 43 and 44) have revealed great promise of this novel technology.

                                                   C.A. Moritz et al.

Fig. 41 Formation of the aligned nanowire array at the ridge of the triangular features

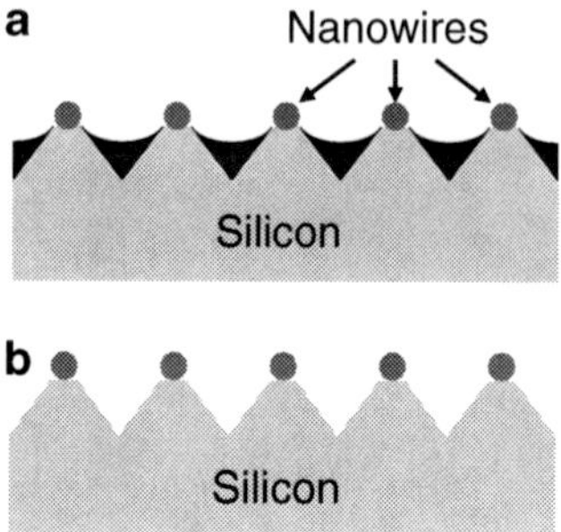

Fig. 42 Transfer of the aligned nanowire array onto the target substrate surface

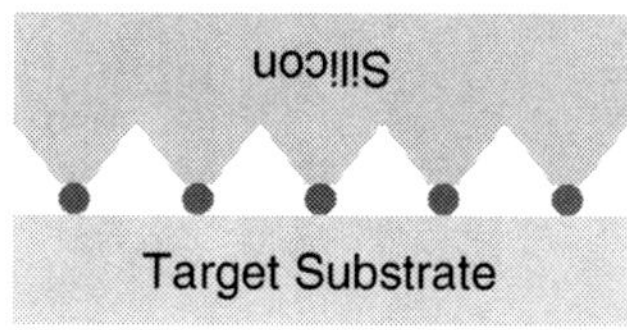

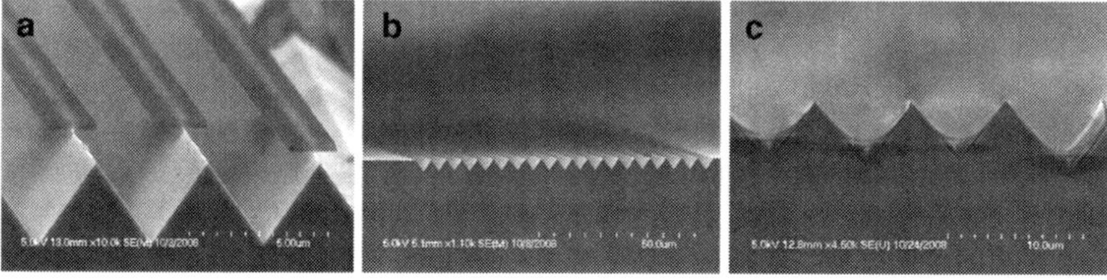

Fig. 43 SEM images captured after (**a**) crystallographic etching (Fig. 39b), (**b**) SiO$_2$ strip removal, and (**c**) polymer coating (Fig. 39c), during the formation of a surface with periodic nanoscale silicon ridges in between polymer

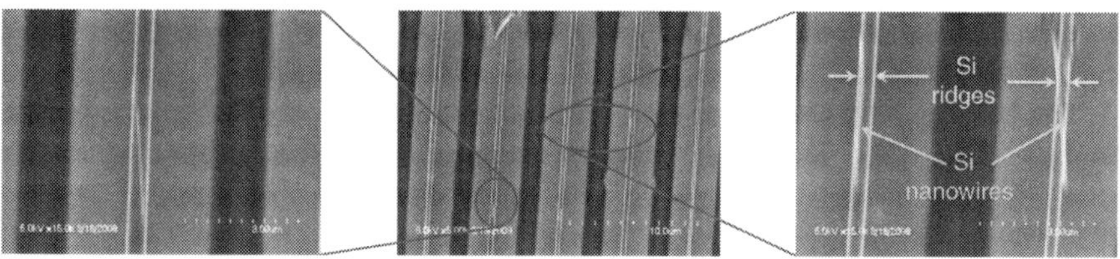

Fig. 44 Preliminary results demonstrating the silicon nanowire selective attachment onto the silicon ridge array (Fig. 41b). The center SEM image is a low-magnification top view above the ridge array. The *left* and *right* images are the zoomed-in views unveiling the assembled silicon nanowires.

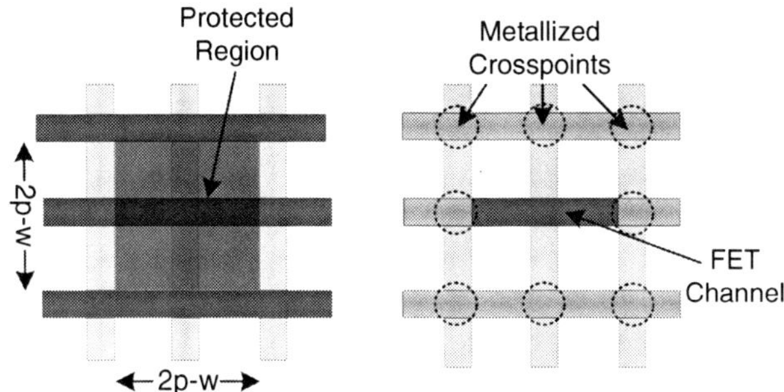

Fig. 45 Lithography requirements for NASIC grid functionalization

8.2.2 Nanowire Grid Functionalization

Nanowire grid functionalization refers to defining the positions of xnwFET devices and interconnects on nanowire arrays assembled on a substrate. In the NASIC manufacturing pathway, this is achieved by using ion implantation in conjunction with photolithography masks; Fig. 38c–d and g–h. The masks protect the native semiconductor materials on the nanowires to form the xnwFET channels in certain regions, and ion implantation creates a high-conductivity interconnect as well as the gate material for the nanowires. Thus, devices and interconnect for the nanowires are formed in a parallel process step.

The key consideration for nanowire grid functionalization is the minimum feature size required for photolithographic masks. This value is shown to be $(2 \times \text{pitch} - \text{width})$ squares (Fig. 45). For example, a nanowire grid with 20 nm pitch and 10 nm width would require 30×30 nm^2.

It is important to note that the lithographic requirements for NASICs are much simpler than a CMOS process of comparable minimum feature size. Masks are used only to protect semiconducting channel regions and not for creating complex patterns on a substrate. Consequently, precise shaping and sharp edges are not necessary. Furthermore, built-in fault-tolerance techniques can tolerate faults due to improperly formed devices and interconnect. Also, fewer masks imply much smaller manufacturing costs for NASIC designs. More in-depth treatment of the NASIC manufacturing techniques can be accessed from the authors' websites.

9 Summary and Future Work

NASIC is a new technology targeting CMOS replacement. All fabric aspects have been actively researched, including devices, manufacturing, circuit and logic styles, and architectures. A fabric-centric mindset or integrated approach across devices, circuit style, manufacturing techniques and architectures is followed. This mindset is anchored in a belief that at the nanoscale, the underlying fabric, broadly defined as the state variable in conjunction with the circuit style scalable into large-scale computational systems and its associated manufacturing approach, rather than the device alone, is how significant progress can be made to obtain system-level capabilities.

The work to date has validated correct fabric operation across both xnwFET devices and circuits, developed circuit structures and fault-tolerance techniques, as well as architectures for processors and image processing. Ongoing efforts are in optimizing devices and developing key manufacturing modules based on both ex situ and in situ approaches. At the circuit level, the current focus is on integrating fault tolerance with support for defects and other fault types caused by process variation and noise, and performance optimizations. Furthermore, there is a concentrated effort to optimize the fabric across devices and circuits for minimizing power consumption. Larger processor designs based on NASICs are in progress.

Exercises 1 NASIC Device Design

1. This question aims to illustrate the effect of different nanowire geometries on silicon xnwFET switching behavior. Using the device structure shown in Fig. 1, the following assumptions are made:

   - Gate nanowire width = channel nanowire width = 10 nm
   - Silicon dioxide (SiO_2) top dielectric thickness = 1 nm
   - Gate, source, and drain n-type doping concentration = 10^{20} cm^{-3}
   - Channel p-type doping concentration = 10^{19} cm^{-3}
   - Source–drain junction underlap length = 0 nm

   Further, ignore the presence of the bottom insulator and substrate, SiO_2 fixed and trapped charges, and silicon–SiO_2 interface trap charge.

   (a) Calculate the theoretical threshold voltage (V_{th}) and ideal subthreshold swing (SS) for a rectangular xnwFET using the following equations. (*Hint*: Convert all the nm units into cm in your calculations.)

   $$C_{ox} = \frac{3.45 \times 10^{-13}}{t_{SiO_2}} \quad (\text{F/cm}^{-2})$$

   $$\phi_{F(gate,ch)} = 0.026 \times \ln\left(\frac{N_{gate,ch}}{10^{10}}\right) \quad (\text{V})$$

   $$V_{th} = -\phi_{F(gate)} + \phi_{F(ch)} + \frac{\sqrt{6.68 \times 10^{-31} \times N_{ch} \times \phi_{F(ch)}}}{C_{ox}} (\text{V})$$

$$SS = 0.060 \times \left(1 + \frac{1}{C_{ox}}\left(\sqrt{\frac{4.18 \times 10^{-32} \times N_{ch}}{\phi_{F(ch)}}}\right)\right) \quad (\text{V/dec})$$

    (b) If the rectangular *channel* nanowire in (a) is replaced by a cylindrical one
        (i) Repeat (a) for the middle of the channel nanowire on the z–y plane
       (ii) Repeat (a) for the edge of the channel nanowire on the z–y plane
      (iii) Compare these results with (a) and comment on the subthreshold behavior of the two devices.
   (c) If the rectangular *gate* nanowire in (b) is also replaced by a cylindrical one
        (i) Repeat (a) for the middle of the channel nanowire on the z–y plane and at the middle of the gate nanowire on the z–x plane
       (ii) Repeat (a) for the middle of the channel nanowire on the z–y plane and at the edge of the gate nanowire on the z–x plane
      (iii) Repeat (a) for the edge of the channel nanowire on the z–y plane and at the edge of the gate nanowire on the z–x plane
      (iv) Compare these results with (a) and (b), and comment on the subthreshold behavior of the two devices.

2. This question aims to illustrate the effect of different channel p-type doping on the rectangular silicon xnwFET threshold voltage in meeting the NASIC circuit requirements. Using the device structure shown in Fig. 1 and applying the same set of assumptions made in Exercise 1 except the channel doping concentration.
   (a) Plot the channel doping concentration (in log scale) vs. the number of p-type dopant atoms (in linear scale from 0 to 50 atoms).
   (b) Based on (a) and using the equations given in Exercise 1, calculate the xnwFET threshold voltage (in linear scale) vs. the channel doping concentration (in log scale from 10^{12} to 10^{20} cm^{-3}). Comment on V_{th} tunability by channel doping concentration.

Exercise 2 NASIC Dynamic Circuits

3. Given a three-input NASIC NAND working in the clocking scheme described in Fig. 8, assume the following:

- Resistance between drain and source when a transistor opens is constant (R).
- Capacitances at the output and internal nodes are also constant (C_{out} and C_{int}).

Calculate the power consumption of the NAND gate for the worst case. Consider the following cases:

   (a) 1/frequency is small enough for all nodes to be charged fully.
   (b) 2/frequency is small enough for the output node to be charged fully, but not for the internal nodes.
   (c) Generalize your expressions for an N-input NASIC NAND gate.

Exercise 3 NASIC Logic Design

4. Consider the design of an instruction ROM for a processor. The input to this logic block is the program counter value, and the output is an instruction. Design an instruction ROM on a single NASIC tile with NAND–NAND logic. Use the following instructions and format:

nop
add r1, r2 (add contents of r1 and r2 and store it in r2)
mov r2, r3
mul r1, r0
Instruction format: three-bit opcode/two-bit source register/two-bit source and destination registers
Opcodes: nop 000, *mov* 001, *add* 011, *mul* 100
(Hint: Remember that NASICs follow two-level logic with propagation of true and complementary signals.)

Exercise 4 Area Estimation

5. Consider a NASIC design with 10 nm nanowire pitch and 32 nm microwire pitch (corresponding to the 16 nm ITRS node).
    (a) What is the area of a single tile with 8 input nanowires, 12 midterms, and 5 outputs?
    (b) Consider this tile has 60% crosspoint utilization (i.e., 60% of the crosspoints have FETs). If many such tiles were implemented on a single CMOS die, how many devices/die may be achieved? How does this compare to state-of-the-art and projected end-of-the-line CMOS?

Exercise 5 NASIC Built-in Fault Tolerance

6. Consider a four-tile design with tile sizes in the format (inputs × minterms-× outputs).

    - Tile 1: 6 × 7 × 8
    - Tile 2: 8 × 12 × 6
    - Tile 3: 6 × 10 × 4
    - Tile 4: 4 × 4 × 6

    (Inputs and outputs include true and complementary signals.)

    (a) Given a nanowire pitch of 10 nm and microwire pitch of 32 nm, what is the area of this design?
    (b) *n*-Way redundancy: what is the area of the design with two-way and three-way redundancy? (Remember that both horizontal and vertical nanowires will be replicated.)

(c) Integrated error masking with Hamming codes: Additional code bits can be incorporated on vertical nanowires using Hamming codes to increase distance and mask faults. A single parity bit can increase the distance to 2; additional bits will be required for more distance. What will be the area of the design incorporating a parity bit for vertical nanowires and two-way redundancy for horizontal nanowires?

(d) Consider the following yield values at different defect rates for the design incorporating the techniques in parts (b) and (c).

Defect rate	2-Way	3-Way	Parity
0.01	0.951	0.999	0.912
0.02	0.779	0.993	0.718
0.03	0.579	0.979	0.526
0.04	0.378	0.951	0.315
0.05	0.242	0.911	0.172
0.06	0.114	0.82	0.084
0.07	0.058	0.74	0.045
0.08	0.025	0.647	0.014
0.09	0.011	0.55	0.002
0.1	0.005	0.431	0.003

Construct yield and YDP graphs for the data. The YDP (also called effective yield) is a measure of yield-area tradeoffs. It represents the number of correctly functioning chips that may be obtained per unit area. Note from your graphs that (a) different levels of fault tolerance may be applicable at different defect rates and (b) a technique with the highest yield may not have the best YDP for a particular defect rate, implying that a technique with lower yield may still produce more functioning chips per unit area.

Acknowledgments The authors acknowledge support from the Focus Center Research Program (FCRP) – Center on Functional Engineered Nano Architectonics (FENA). This work was also supported by the Center for Hierarchical Manufacturing (CHM/NSEC 0531171) University of Massachusetts at Amherst and NSF awards 0508382, 0541066 and 0915612.

There are several researchers who are actively contributing to various aspects of NASICs, in addition to the authors. Prof. Anderson at UMass is developing information-theoretical models for projecting the ultimate capabilities of NASICs. In addition to the efforts based on the promising ex situ techniques presented, experimental techniques are being investigated for in situ self-assembly-based NASIC fabric formation by Profs. Mihri and Cengiz Ozkan at UCR, as well as investigators in the UMass CHM Nanotechnology Center.

Profs. Pottier, Dezan, and Lagadec and their groups at Universite Occidentale in Bretagne, France, collaborate in developing CAD tools. Profs. Koren and Krishna at Umass are collaborating in devising techniques that allow efficient fault masking in NASICs.

The authors appreciate the valuable feedback and support at various phases of the project from Dr. Avouris, IBM; Dr. Kos Galatsis, UCLA; Profs. Kostya Likharev, Stony Brook University; Mark Tuominen, UMass; James Watkins, UMass; Richard Kiehl, UC Davis; Kang Wang, UCLA; and many others.

This project would have not been possible without the effort of current and former graduate students. Key contributors include Dr. Teng Wang, currently at Qualcomm; Dr. Yao Guo, currently at Peking University; Dr. Mahmoud Bennaser, currently at Kuwait University; Dr. Kyeon–Sik Shin, currently a post-doc at UCLA; Michael Leuchtenburg, Prachi Joshi, Lin Zhang, Jorge Kina, Pavan Panchakapeshan, Rahul Kulkarni, Mostafizur Rahman, Prasad Shabadi, Kyongwon Park, and Trong Tong Van.

References

1. C. P. Collier, E. W. Wong, M. Belohradský, F. M. Raymo, J. F. Stoddart, P. J. Kuekes, R. S. Williams, and J. R. Heath, Electronically configurable molecular-based logic gates, *Science*, vol. 285, pp. 391–394, 1999.
2. K. Galatsis, K. Wang, Y. Botros, Y. Yang, Y-H. Xie, J. Stoddart, R. Kaner, C. Ozhan, J. Liu, M. Ozkan, C. Zhou, and K. W. Kim, Emerging memory devices, *IEEE Circ. Dev. Mag.*, vol. 22, pp. 12–21, 2006.
3. W. Lu and C. M. Lieber, Semiconductor nanowires, *J. Phys. D Appl. Phys.*, vol. 39, pp. R387–R406, 2006.
4. Y. Cui, X. Duan, J. Hu, and C. M. Lieber, Doping and electrical transport in silicon nanowires, *J. Phys. Chem. B*, vol. 104, pp. 5213–5216, 2000.
5. A. B. Greytak, L. J. Lauhon, M. S. Gudiksen, and C. M. Lieber, Growth and transport properties of complementary germanium nanowire field-effect transistors, *Appl. Phys. Lett.*, vol. 84, pp. 4176–4178, 2004.
6. J. Xiang, W. Lu, Y. Hu, Y. Wu, H. Yan, and C. M. Lieber, Ge/Si nanowire heterostructures as high-performance field-effect transistor, *Nature*, vol. 441, pp. 489–493, 2006.
7. M. I. Khan, X. Wang, K. N. Bozhilov, and C. S. Ozkan, Templated fabrication of InSb nanowires for nanoelectronics, *J. Nanomater.*, vol. 2008, pp. 1–5, 2008.
8. Y. Li, G. W. Meng, L. D. Zhang, and F. Phillipp, Ordered semiconductor ZnO nanowire arrays and their photoluminescence properties, *Appl. Phys. Lett.*, vol. 76, pp. 2011–2013, 2000.
9. X. Duan, Y. Huang, Y. Cui, J. Wang, and C. M. Lieber, Indium phosphide nanowires as building blocks for nanoscale electronic and optoelectronic devices, *Nature*, vol. 409, pp. 66–69, 2001.
10. Y. Huang, X. Duan, Y. Cui, L. J. Lauhon, K-Y. Kim, and C. M. Lieber, Logic gates and computation from assembled nanowire building blocks, *Science*, vol. 294, no. 5545, pp. 1313–1317, 2001.
11. Y. Wu, J. Xiang, C. Yang, W. Lu, and C. M. Lieber, Single-crystal metallic nanowires and metal/semiconductor nanowire heterostructures, *Nature*, vol. 430, pp. 61–65, 2004.
12. *Semiconductor Nanowires: Fabrication, Physical Properties and Applications*, Warrendale, PA: Materials Research Society, 2006.
13. Z. Zhong, D. Wang, Y. Cui, M. W. Bockrath, and C. M. Lieber, Nanowire crossbar arrays as address decoders for integrated nanosystems, *Science*, vol. 302, pp. 1377–1379, 2003.
14. D. Wang, B. Sheriff, M. McAlpine, and J. Heath, Development of ultra-high density silicon nanowire arrays for electronics applications, *Nano Res.*, vol. 1, pp. 9–21, 2008.
15. *Sentaurus Device User Guide*, Synopsys, Inc., 2007.
16. A. Marchi, E. Gnani, S. Reggiani, M. Rudan, and G. Baccarani, Investigating the performance limits of silicon-nanowire and carbon-nanotube FETs, *Solid State Electron*, vol. 50, pp. 78–85, 2006.
17. S. D. Suk, M. Li, Y. Y. Yeoh, K. H. Yeo, K. H. Cho, I. K. Ku, H. Cho, W-J. Jang, D-W. Kim, D. Park, and W-S. Lee, Investigation of nanowire size dependency on TSNWFET, in *Proc. IEEE Electron Devices Meeting*, pp. 891–894, 2007.
18. P. Stolk, F. Widdershoven, and D. Klaassen, Modeling statistical dopant fluctuations in MOS transistors, *IEEE Trans. Electron Devices*, vol. 45, pp. 1960–1971, 1998.
19. P. Narayanan, M. Leuchtenburg, T. Wang, and C. A. Moritz, CMOS control enabled single-type FET NASIC, in *Proc. IEEE Int. Symp. on VLSI*, April 2008.
20. P. Narayanan, K. W. Park, C. O. Chui, and C. A. Moritz, Validating cascading of crossbar circuits with an integrated device-circuit exploration, in *Proc. IEEE/ACM Symposium on Nanoscale Architectures*, San Francisco, CA, 2009.
21. *HSPICE User's Manual*, Campbell, CA: Meta-Software, Inc., 1992.
22. T. Wang, P. Narayanan, and C. A. Moritz, Heterogeneous 2-level logic and its density and fault tolerance implications in nanoscale fabrics, *IEEE Trans. Nanotechnol.*, vol. 8, no. 1, pp. 22–30, 2009.

23. C. A. Moritz, T. Wang, P. Narayanan, M. Leuchtenburg, Y. Guo, C. Dezan, and M. Bennaser, Fault-tolerant nanoscale processors on semiconductor nanowire grids, *IEEE Trans. Circ. Syst. I Special Issue on Nanoelectronic Circuits and Nanoarchitectures*, vol. 54, no. 11, pp. 2422–2437, 2007.

24. T. Wang, P. Narayanan, and C. A. Moritz, Combining 2-level logic families in grid-based nanoscale fabrics, in *Proc. IEEE/ACM Symposium on Nanoscale Architectures*, San Jose, CA, Oct. 2007.

25. M. Sindhwani, T. Srikanthan, and K. V. Asari, VLSI efficient discrete-time cellular neural network processor, *IEEE Proc. Circuits Devices Syst.*, vol. 149, pp. 167–171, 2002.

26. L. O. Chua and L. Yang, Cellular neural networks: theory, *IEEE Trans. Circuits Syst. I*, vol. 35, pp. 1257–1272, 1988.

27. P. Narayanan, T. Wang, M. Leuchtenburg, and C. A. Moritz, Comparison of analog and digital nanoscale systems: issues for the nano-architect, in *Proc. IEEE Nanoelectronics Conference*, March 2008.

28. P. Narayanan, T. Wang, and C. A. Moritz, Programmable cellular architectures at the nanoscale, unpublished manuscript.

29. *International Technology Roadmap for Semiconductors*, 2007 edition. Available online at http://public.itrs.net/

30. A. A. Bruen and M. A. Forcinito, *Cryptography, Information Theory, and Error-Correction*, New York: Wiley-Interscience, 2005.

31. I. L. Sayers and D. J. Kinniment, Low-cost residue codes and their applications to self-checking VLSI systems, *IEEE Proc.*, vol. 132, Pt. E, no. 4, 1985.

32. H. Krishna and J. D. Sun, On theory and fast algorithms for error correction in residue number system product codes, *IEEE Trans. Comput.*, vol. 42, no. 7, 1993.

33. W. H. Pierce, *Interconnection Structure for Redundant Logic, Failure-Tolerant Computer Design*, New York: Academic Press, 1965.

34. R. E. Lyions and W. Vanderkulk, The use of triple modular redundancy to improve computer reliability, *IBM J. Res. Develop.* vol. 6, no. 2, 1962.

35. R. C. Rose and D. K. Ray-Chaudhuri, On a class of error-correcting binary group codes, *Info. Control*, vol. 3, pp. 68–79, March 1960.

36. A. Hocquengham, Codes correcteurs d'erreurs, *Chiffre*, vol. 2, pp. 147–156, 1959.

37. T. R. N. Rao, *Error Coding for Arithmetic Processor*, New York: Academic Press, 1974.

38. B. W. Johnson, *Design and Analysis of Fault-Tolerant Digital Systems*, Reading, MA: Addison-Wesley, 1989.

39. D. B. Armstrong, A general method of applying error correction to synchronous digital systems, *Bell Syst. Tech. J.*, vol. 40, pp. 477–593, 1961.

40. A. DeHon. Nanowire-based programmable architectures, *ACM J. Emerg. Tech.. Comput. Syst.*, vol. 1, pp. 109–162, 2005.

41. K. K. Likharev and D. B. Strukov, CMOL: devices, circuits, and architectures. Introducing molecular electronics, G. F. G. Cuniberti and K. Richter (eds.), 2005.

42. D. B. Strukov and K. K. Likharev, Reconfigurable hybrid CMOS devices for image processing, *IEEE Trans. Nanotechnol*, vol. 6, pp. 696–710, 2007.

43. G. S. Snider and R. S. Williams, Nano/CMOS architectures using a field-programmable nanowire interconnect, *Nanotechnology*, vol. 18, pp. 1–11, 2007.

44. Y. Li, F. Qian, J. Xiang, and C. M. Lieber, Nanowire electronic and optoelectronic devices, *Mater. Today*, vol. 9, pp. 18–27, 2006.

45. M. Bennaser, Y. Guo, and C. A. Moritz, Designing memory subsystems resilient to process variations, in *Proc. Annual Symposium on VLSI*, pp. 357–363, Porto Alegre, Brazil, March 2007.

46. D. Burnett, K. Erington, C. Subramanian, and K. Baker, Implications of fundamental threshold voltage variations for high-density SRAM and logic circuits, in *Proc. Symposium on VLSI Technology*, pp. 14–15, June 1994.

47. E. Humenay, D. Tarjan, and K. Skadron, Impact of parameter variations on multi-core chips, in *Proc. Workshop on Architecture Support for Gigascale Integration*, June 2006.

48. A. Agarwal, B. Paul, H. Mahmoodi, A. Datta, and K. Roy, A process-tolerant cache architecture for improved yield in nanoscale technologies, *IEEE Trans. VLSI Syst.*, vol. 13, no. 1, pp. 27–38, Jan. 2005.
49. X. Tang, V. K. De, and J. D. Meindl, Intrinsic MOSFET parameter fluctuations due to random dopant placement, *IEEE Trans. VLSI Syst.*, vol. 5, no. 4, pp. 369–376, Dec. 1997.
50. E. Garnett, W. Liang, and P. Yang, Growth and electrical characteristics of platinum-nanoparticle-catalyzed silicon nanowires, *Adv. Mater.*, vol. 19, pp. 2946–2950, 2007.
51. Y. Cui, Z. Zhong, D. Wang, W. U. Wang, and C. M. Lieber, High-performance silicon nanowire field effect transistors, *Nano Lett.*, vol. 3, no. 2, pp. 149–152, 2003.
52. J. Kim, D. H. Shin, E. Lee, and C. Han, Electrical characteristics of singly and doubly connected Ni Silicide nanowire grown by plasma-enhanced chemical vapor deposition, *Appl. Phys. Lett.*, vol. 90, p. 253103, 2007.
53. M. Leuchtenburg, P. Narayanan, T. Wang, and C. A. Moritz, Process variation and 2-way redundancy in grid-based nanoscale processors, in *Proc. IEEE Conference on Nanotechnology*, Genoa, Italy, 2009.
54. P. Joshi, M. Leuchtenburg, P. Narayanan, and C. A. Moritz, Variation mitigation versus defect tolerance at the nanoscale, in *Proc. Nanoelectronic Devices for Defense & Security Conference*, Fort Lauderdale, FL, 2009.
55. Z. Chen et al., An integrated logic circuit assembled on a single carbon nanotube, *Science*, vol. 311, p. 1735, 2006.
56. P. Narayanan, K. W. Park, C. O. Chui, and C. A. Moritz, Manufacturing pathway and associated challenges for nanoscale computational systems, in *Proc. IEEE Conference on Nanotechnology*, Genoa, Italy, 2009.
57. R. He, D. Gao, R. Fan, A. I. Hochbaum, C. Carraro, R. Maboudian, and P. Yang, Si nanowire bridges in microtrenches: integration of growth into device fabrication, *Adv. Mater.*, vol. 17, pp. 2098–2102, 2005.
58. Y. Shan and S. J. Fonash, Self-assembling silicon nanowires for device applications using the nanochannel-guided 'grow-in-Place' approach, *ACS Nano*, vol. 2, pp. 429–434, 2008.
59. A. Ural, Y. Li, and H. Dai, Electric-field-aligned growth of single-walled carbon nanotubes on surfaces, *Appl. Phys. Lett.*, vol. 81, pp. 3464–3466, 2002.
60. O. Englander, D. Christensen, J. Kim, L. Lin, and S. J. S. Morris, Electric-field assisted growth and self-assembly of intrinsic silicon nanowires, *Nano Lett.*, vol. 5, pp. 705–708, 2005.
61. Y-T. Liu, X-M. Xie, Y-F. Gao, Q-P. Feng, L-R. Guo, X-H. Wang, and X-Y. Ye, Gas flow directed assembly of carbon nanotubes into horizontal arrays, *Mater. Lett.*, vol. 61, pp. 334–338, 2007.
62. X. Chen, M. Hirtz, H. Fuchs, and L. Chi, Fabrication of gradient mesostructures by Langmuir–Blodgett rotating transfer, *Langmuir*, vol. 23, pp. 2280–2283, 2007.
63. D. Whang, S. Jin, and C. M. Lieber, Nanolithography using hierarchically assembled nanowire masks, *Nano Lett.*, vol. 3, pp. 951–954, 2003.
64. X. Xiong, L. Jaberansari, M. G. Hahm, A. Busnaina, and Y. J. Jung, Building highly organized single-walled carbon-nanotube networks using template-guided fluidic assembly, *Small*, vol. 3, pp. 2006–2010, 2007.
65. K. Heo, E. Cho, J-E. Yang, M-H. Kim, M. Lee, B. Y. Lee, S. G. Kwon, M-S. Lee, M-H. Jo, H-J. Choi, T. Hyeon, and S. Hong, Large-scale assembly of silicon nanowire network-based devices using conventional microfabrication facilities, *Nano Lett.*, vol. 8, pp. 4523–4527, 2008.
66. B. J. Jordan, Y. Ofir, D. Patra, S. T. Caldwell, A. Kennedy, S. Joubanian, G. Rabani, G. Cooke, and V. M. Rotello, Controlled self-assembly of organic nanowires and platelets using dipolar and hydrogen-bonding interactions, *Small*, vol. 4, pp. 2074–2078, 2008.
67. A. Javey, S. Nam, R. S. Friedman, H. Yan, and C. M. Lieber, Layer-by-layer assembly of nanowires for three-dimensional, multifunctional electronics, *Nano Lett.*, vol. 7, pp. 773–777, 2007.

68. T. Martensson, P. Carlberg, M. Borgstrom, L. Montelius, W. Seifert, and L. Samuelson, Nanowire arrays defined by nanoimprint lithography, *Nano Lett.*, vol. 4, pp. 699–702, 2004.
69. S. J. Kang, C. Kocabas, H-S. Kim, Q. Cao, M. A. Meitl, D-Y. Khang, and J. A. Rogers, Printed multilayer superstructures of aligned single-walled carbon nanotubes for electronic applications, *Nano Lett.*, vol. 7, pp. 3343–3348, 2007.
70. M. C. McAlpine, H. Ahmad, D. Wang, and J. R. Heath, Highly ordered nanowire arrays on plastic substrates for ultrasensitive flexible chemical sensors, *Nat. Mater.*, vol.6, pp. 379–384, 2007.

Imperfection-Immune Carbon Nanotube VLSI Circuits

Nishant Patil, Albert Lin, Jie Zhang, Hai Wei, H.-S. Philip Wong, and Subhasish Mitra

Abstract Carbon Nanotube Field Effect Transistors (CNFETs), consisting of semiconducting single walled Carbon Nanotubes (CNTs), show great promise as extensions to silicon CMOS. While there has been significant progress at a single-device level, a major gap exists between such results and their transformation into VLSI CNFET technologies. Major CNFET technology challenges include mis-positioned CNTs, metallic CNTs, and wafer-scale integration. This work presents design and processing techniques to overcome these challenges. Experimental results demonstrate the effectiveness of the presented techniques.

Mis-positioned CNTs can result in incorrect logic functionality of CNFET circuits. A new layout design technique produces CNFET circuits implementing arbitrary logic functions that are immune to a large number of mis-positioned CNTs. This technique is significantly more efficient compared to traditional defect- and fault-tolerance. Furthermore, it is VLSI-compatible and does not require changes to existing VLSI design and manufacturing flows.

A CNT can be semiconducting or metallic depending upon the arrangement of carbon atoms. Typical CNT synthesis techniques yield ~33% metallic CNTs. Metallic CNTs create source-drain shorts in CNFETs resulting in excessive leakage ($I_{on}/I_{off} < 5$) and highly degraded noise margins. A new technique, VLSI-compatible Metallic-CNT Removal (VMR), overcomes metallic CNT challenges by combining layout design with CNFET processing. VMR produces CNFET circuits with I_{on}/I_{off} in the range of 10^3-10^5, and overcomes the limitations of existing metallic-CNT removal techniques.

We also present the first experimental demonstration of VLSI-compatible CNFET combinational circuits (e.g., computational elements such as half-adder sum-generators) and storage circuits (e.g., sequential elements such as D-latches) that are immune to inherent CNT imperfections. These experimentally-demonstrated circuits form essential building blocks for large-scale digital computing systems.

N. Patil (✉)

Department of Electrical Engineering, Stanford University, Stanford, CA 94305, USA

e-mail: nppatil@stanford.edu

N.K. Jha and D. Chen (eds.), *Nanoelectronic Circuit Design*,

DOI 10.1007/978-1-4419-7609-3_8, © Springer Science+Business Media, LLC 2011

1 Introduction

Energy efficiency is the most significant challenge for continued integration of systems according to Moore's law – which is the principal driver behind the semiconductor industry. *Carbon nanotubes* (CNTs) provide a very promising path towards solving this outstanding challenge. CNTs are cylindrical nanostructures of carbon with exceptional electrical, thermal, and mechanical properties [36]. CNTs with a single shell of carbon atoms are called *single-walled carbon nanotubes* (SWNTs) and have a diameter between 0.5 and 3 nm. In this chapter, all CNTs are SWNTs. CNTs are grown through a process of chemical synthesis [7].

Using CNTs, we can create *carbon nanotube field-effect transistors* (CNFETs), as shown in Fig. 1. Multiple semiconducting CNTs are grown or deposited on an insulating substrate. The regions of the CNTs under the gate act as the channel region of the CNFET. The conductivity of the channel region is controlled by the voltage applied to the gate similar to the case for silicon MOSFETs. The regions of the CNTs outside the gate are heavily doped and act as the source and drain regions of the CNFET.

Figure 2 shows the layout of a CNFET inverter. The gate, source and drain contacts, and interconnects are defined by conventional lithography. The distances between gates and contacts are limited by the lithographic feature size (lithographic pitch in Fig. 2: 64 nm for the 32-nm technology node). Since CNTs are grown using chemical synthesis, the inter-CNT distance (sub-lithographic pitch in Fig. 2) is not limited by lithography.

CNFETs are excellent candidates for building highly energy-efficient next-generation integrated systems. A "perfect" CNFET technology is predicted to be $5\times$ faster than silicon transistors, while consuming the same power [38]. However, several significant challenges must be overcome before such benefits can be experimentally realized. Major sources of imperfections inherent in CNFET circuits [9, 30, 42] are (1) *mis-positioned CNTs*, (2) *metallic CNTs* (*m-CNTs*), and (3) *CNT density variations*.

In spite of several unanswered questions about CNFETs, three major trends are clear:

1. Future VLSI systems cannot rely solely on current chemical synthesis for guaranteed perfect devices. Mis-positioned CNTs can result in incorrect logic

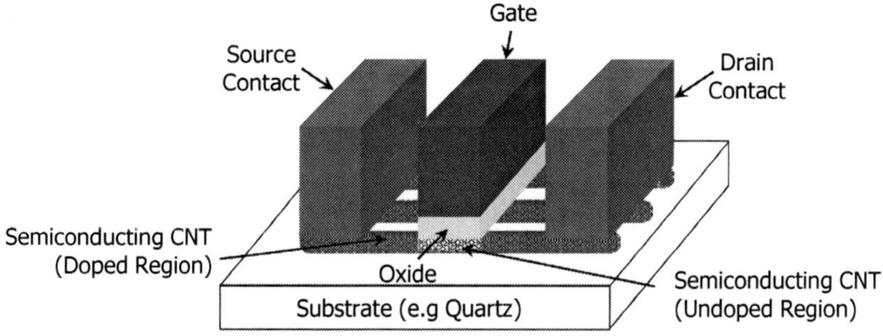

Fig. 1 Carbon-nanotube field-effect transistor (CNFET)

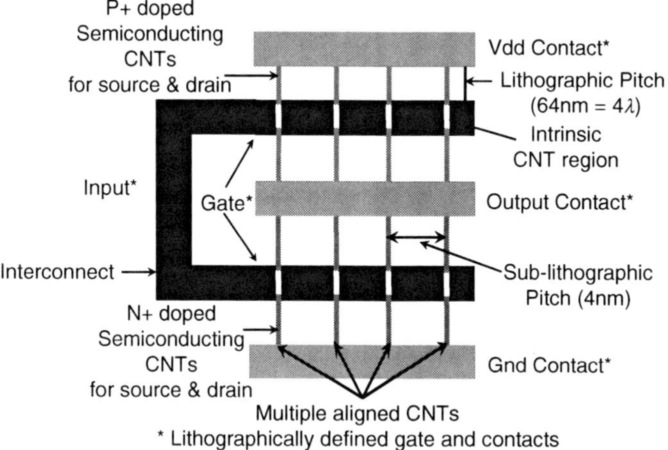

Fig. 2 CNFET inverter

functionality. Similarly, no known CNT growth technique guarantees 0% m-CNTs. m-CNTs create source-drain shorts resulting in excessive leakage, severely degraded noise margins, and delay variations. Also, the best CNT growth techniques have CNT density variations.

2. Expensive defect- and fault-tolerance techniques will not enable wide adoption of CNFET-based circuits. Expensive techniques, in fact, will create major roadblocks to the adoption of CNFET technologies.

3. New design techniques must be compatible with VLSI processing and have minimal impact on existing VLSI design flows. Investments made in VLSI design infrastructure are too large to be ignored. For example, techniques that rely on separate customization of every chip can be prohibitively expensive if not designed carefully.

New design techniques, together with advances in CNT processing, must be employed to create CNFET circuits that are immune to inherent CNT imperfections. In this chapter, we address the topic of reliable and robust CNFET circuits in the presence of CNT imperfections. First, we describe a robust *mis-positioned-CNT-immune* logic design technique that generates layouts for CNFET circuits that function correctly in the presence of a large number of mis-positioned CNTs. Next, we describe two techniques for creating CNFET circuits immune to metallic CNTs – *asymmetrically correlated CNT technology* (ACCNT) *and VLSI-compatible metallic-CNT removal* (VMR). Lastly, we analyze the reliability of CNFET circuits in the presence of metallic-CNT removal and CNT density variations.

2 Mis-Positioned-CNT-Immune Logic Design

A *mis-positioned CNT* is a CNT that passes through a layout region where a CNT was not intended to pass. This may be caused due to misalignment or lack of control of correct positioning of CNTs during CNT growth. A large fraction (99.5%) of the CNTs grown on single-crystal quartz substrates grow aligned, i.e., straight and parallel to each other [15]. Even for CNTs on quartz substrates, a non-negligible fraction of CNTs are misaligned. It is very difficult to correctly position and align *all* the CNTs for all CNFETs for VLSI. Mis-positioned CNTs can cause shorts (Fig. 3a) and incorrect logic functionality in CNFET logic circuits (Fig. 3b). Using design principles described in [25, 27], we experimentally demonstrated in [26, 27] circuits that are immune to a large number of mis-positioned CNTs. We refer to these circuits as *mis-positioned-CNT-immune* circuits. This is accomplished by etching CNTs from regions pre-defined during layout design, so that any mis-positioned CNT cannot result in incorrect logic functionality. This technique is VLSI-compatible since it does not require any die-specific customization or test and reconfiguration. *All* logic cells have CNTs removed in the predefined layout regions using oxygen plasma [26, 27]. This is feasible since the penalties associated with this technique are small.

Figure 3 Incorrect logic functionality caused by misaligned and mis-positioned CNTs.

Given the layout of a circuit implementing a logic function, we can determine whether one or more misaligned or mis-positioned CNTs can result in incorrect logic function implementation. We will use a graph abstraction of the layout for this purpose. Figure 4 shows two possible structures for a NAND circuit (the metal interconnect connecting the PFET and the NFET inputs are not shown for clarity). These structures will be used to illustrate our technique. The two gates in the pull-up network of Fig. 4a (corresponding to inputs A and B) are not at the same horizontal level in order to create a compact layout in the horizontal direction. This also reduces the gate capacitance for the gates in the pull-down network.

In Fig. 4b, the two gates in the pull-up network are at the same horizontal level separated by an etched region. The regions outside the cell boundaries are devoid of CNTs (removed by etching of CNTs). In the ideal case, all CNTs should grow on the substrate in one direction from one contact to another, only passing under the

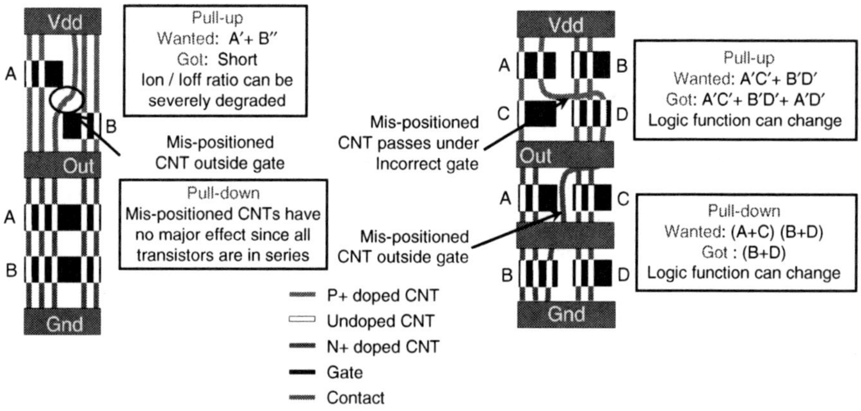

Fig. 3 Incorrect logic functionality caused by misaligned and mis-positioned CNTs

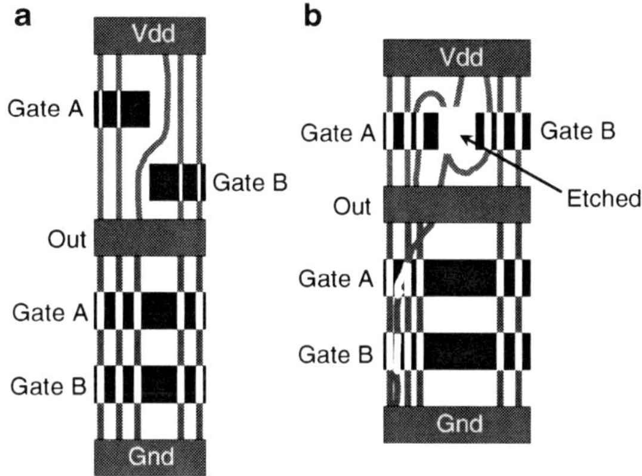

Fig. 4 (**a**) Mis-positioned-CNT-vulnerable NAND. (**b**) Mis-positioned-CNT-immune NAND

gates. However, in reality, not all CNTs will be perfectly aligned and positioned. Consider the mis-positioned CNT in Fig. 4a, which has all doped regions because it is misaligned. This CNT will cause a short between the Vdd and Output nodes. On the other hand, any misaligned or mis-positioned CNT in Fig. 4b will not cause a malfunction. This is because any CNT in the pull-up network of Fig. 4b either passes under the gates corresponding to inputs A or B, or passes through the etched region between the two gates, in which case it will not conduct. Given sufficient CNT density (> 100 CNTs/μm), [1]the chance that no CNT passes through a given gate (e.g., gate A or gate B) is very small.

The first step in our automated analysis is to divide the cell into pull-up and pull-down regions. We will analyze each of these regions separately. Each region is decomposed into a finely divided *square grid*. The dimension of each side of a square in this grid is equal to the smallest lithography feature size. Each square in this grid has a *label*: contact (C), doped (D), etched (E), or gate (G). For a gate, we also include the input variable associated with that gate (e.g., G_A in Fig. 4). For a contact, we label a Vdd contact as C_V, ground contact as C_G, output contact as C_O, and any other intermediate contact as C. Figure 4 shows the grid decomposition of the NAND cell layouts in Fig. 3.

We create a graph where each square in the grid is a node in the graph. The label associated with a node is the same as that associated with the corresponding square as described in Table 1. There is an edge between two nodes in the graph if and only if the corresponding squares in the grid are adjacent, i.e., the squares corresponding to those two nodes have a boundary or a vertex in common. Two nodes with an edge between them are referred to as *neighboring nodes*.

Any CNT can be represented by a path from a contact node to another contact node in the graph extracted from the layout. A CNT in a given square in the grid decomposition of the layout can grow into any of its adjacent squares in the grid.

Table 1 Labels associated with nodes in the graph

Node type	Node label	Boolean function
Gate with input variable A	G_A	A
Gate with input variable A'	$G_{A'}$	A'
Doped region	D	1
Etched region	E	0
Vdd contact	C_V	1
Gnd contact	C_G	1
Output contact	C_O	1
Any intermediate contact	C	1

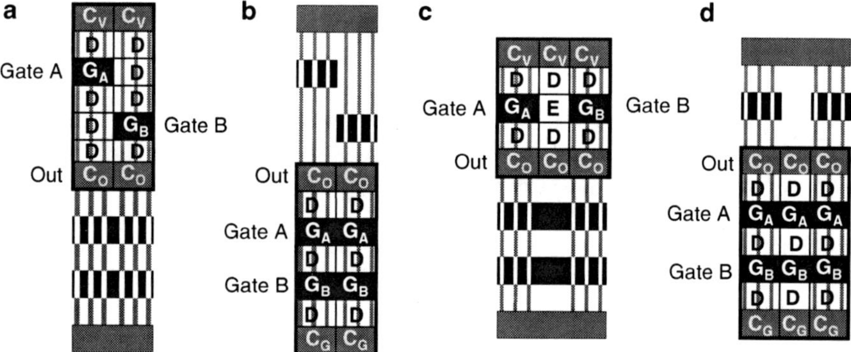

Fig. 5 Grid decomposition of NAND cell layouts in Fig. 4

Any CNT can be represented by a path in this graph since the edges in the graph account for all possible growth directions of CNTs.

Figure 5 shows the graphs corresponding to the layouts in Fig. 4. To reduce the number of nodes and edges in the graph, we can combine neighboring nodes with the same label in Fig. 4 into a single node. The set of neighbors of this new combined node comprises the union of the neighbors of its constituent nodes. Figure 6 shows the reduced versions of the graphs in Fig. 5. Since we preserve the neighbor list of the graph nodes, we can work on the reduced graph to determine whether a given layout is immune to misaligned and mis-positioned CNTs.

Each node in the graph has an associated Boolean function as defined in Table 1. For example, gate node G_A has the associated function A since the CNTs under that gate conduct when gate A is turned on (for the PFET pull-up, CNTs under the gate conduct when A is low, while for the NFET pull-down, CNTs under the gate conduct when A is high). For a node with label D, the associated function is 1 since the doped region of the CNT always conducts. For an undoped region, the corresponding Boolean function is 0 since an undoped region of the CNT does not conduct. For an etched region, the corresponding Boolean function is 0 since the CNT portion in the etched region is removed and does not conduct.

To determine whether the pull-up (pull-down) network implements the correct function in the presence of misaligned and mis-positioned CNTs, we need to

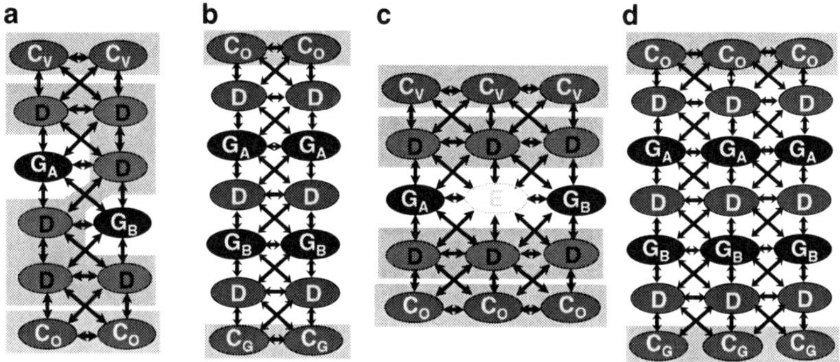

Fig. 6 Graph representations of NAND cell layouts in Fig. 4

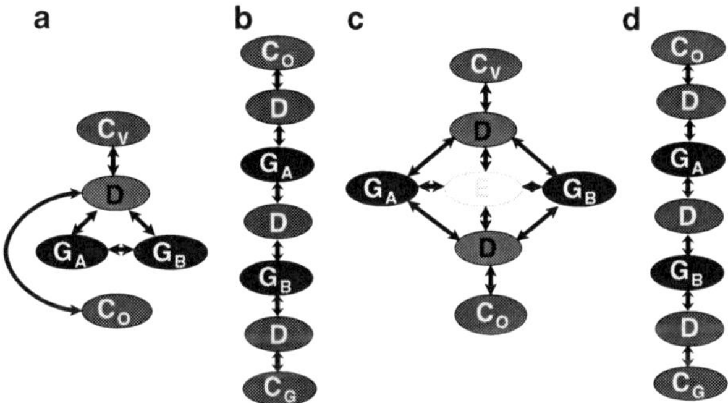

Fig. 7 Reduced graphs for NAND cell layouts in Fig. 4

traverse all possible paths between the Vdd contact (Gnd contact) node and the output contact in the corresponding graph. For the pull-down network, we compare against the complement of the function implemented by the standard logic cell. In the case of the pull-up network, we apply De Morgan's law and then complement the variables. This step simplifies the analysis (Fig. 7).

Let us first consider paths with no loops. The *Boolean function associated with each path* is obtained by ANDing the Boolean functions associated with the nodes along the path. This Boolean function of a path represents the switch-level function implemented by a CNT traversing that path. For example, in Fig. 7c, the path C_v–D–G_A–D–C_O has the Boolean function A and the path C_v–D–E–D–C_O has the Boolean function 0. In Fig. 7c, the path C_v–D–C_O has the Boolean function 1. For paths with loops, we traverse the loops only once, since the Boolean expression associated with the path will not change with multiple traversals of the same loops [6].

The OR of the Boolean functions associated with all paths must be identical to the intended function (function without mis-positioned CNTs) for the circuit to be immune to mis-positioned CNTs. This is because the paths represent all possible CNT mis-positioning scenarios. In our implementation, we used standard equivalence checking techniques for this purpose [2].

We show a few examples of this graph traversal in Fig. 8. Note that the pull-down network is compared to (A & B), which is the complement of the NAND function while the pull-up network is compared to (A OR B), which is the function obtained by applying De Morgan's law, and then complementing the inputs. The functionality of pull-down networks in both mis-positioned-CNT-vulnerable and mis-positioned-CNT-immune NAND cell designs is not affected by mis-positioned CNTs. The reader can verify that the OR of the Boolean functions of all paths in the graphs of Fig. 7b, d are identical to the intended function implemented by the pull-down network of a NAND cell. These graphs have special properties and we refer to them as *straight line graphs*.

A mis-positioned CNT in the pull-up network of the mis-positioned-CNT-vulnerable cell may create a short between Vdd and Output because the Boolean function of the path C_V-D-C_O in Fig. 7a is 1 (which includes all possible minterms). However, the pull-up network of the mis-positioned-CNT-immune NAND is

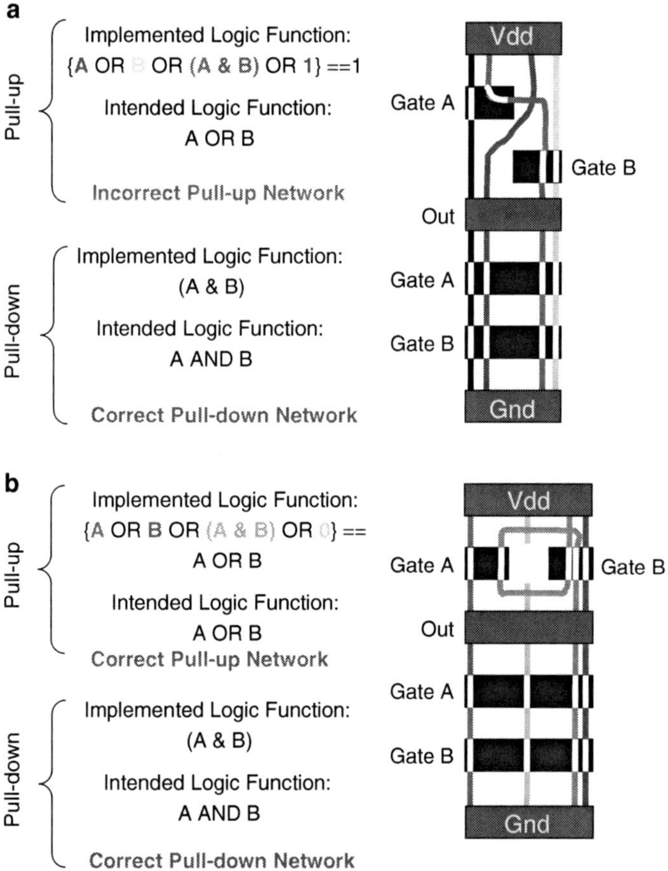

Fig. 8 Traversal of NAND cell graphs in Fig. 7

immune to mis-positioned CNTs because the Boolean function of any path cannot include logic terms that are not present in the original function.

In summary, the algorithm to determine mis-positioned CNT immunity for a given layout comprises the following steps:

1. Perform a grid decomposition of the cell layout.
2. Extract a graph representation of the layout.
3. Perform a graph reduction to obtain the reduced graph.
4. Traverse all paths in the graph and OR Boolean functions corresponding to paths.
5. Determine whether the ORed Boolean function of all paths equals the intended function.
6. Layout is mis-positioned-CNT immune if and only if the OR of the Boolean functions of all paths in the graph is equal to the intended function.

We can implement mis-positioned-CNT-immune layouts of the pull-up and pull-down networks for a standard logic cell. The inputs to our algorithm are the pull-up and pull-down networks represented as a combination of series and parallel transistor networks. We first discuss the case where the pull-up and pull-down networks are expressed in the sum-of-products (SOP) or product-of-sums (POS) form.

For a network specified in the SOP form, the mis-positioned-CNT-immune layout implementation is shown in Fig. 9. Any path in the corresponding graph between the Vdd/Gnd contact and the Output contact must pass through all the gates corresponding to a product term or through an etched region. Hence, the Boolean function corresponding to that path cannot include minterms that are not present in the SOP representation of the network. Hence, the overall layout in Fig. 9 is immune to mis-positioned CNTs. The CNTs can be removed from the lithographically defined etched regions using oxygen plasma [26].

For a network specified in POS form, the mis-positioned-CNT-immune layout implementation is shown in Fig. 10. This structure is immune to mis-positioned CNTs since each sum term is immune to mis-positioned CNTs (by the same

Fig. 9 Mis-positioned-CNT-immune layout for a network specified in the SOP form

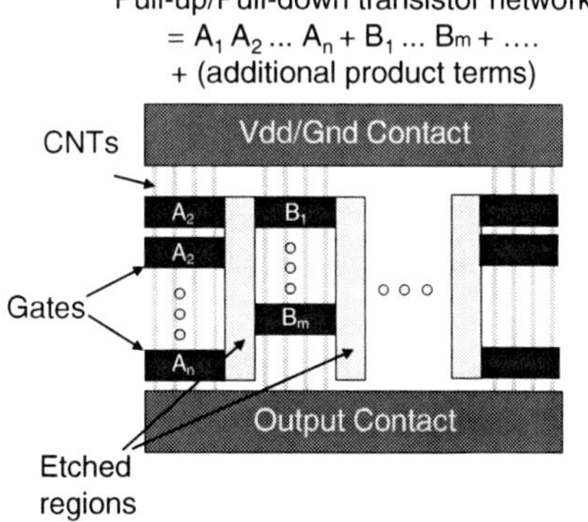

argument as the SOP case). Any CNT that spans multiple sum terms must pass through a contact and this does not change the function implemented by the cell.

An arbitrary representation of the network may not be in either SOP or POS forms. In that case, each sum term is implemented with parallel CNFETs with gates at the same horizontal level and etched regions between the gates (similar to Figs. 9 and 10). Each product term is implemented as a series network of CNFETs. For example, consider the network represented by the expression

A + (B + C) (D + E). We first implement the term (B + C) (D + E) similar to Fig. 10. Next, we implement A in parallel with this structure and separated by an etched region similar to Fig. 9. Figure 11 illustrates the fabrication steps to fabricate the transistor network in Fig. 12. We use additional metal layers to route the inputs to the logic function as is done in CMOS standard cells.

The mis-positioned-CNT-immune design technique works for any arbitrary function with any number of inputs, as shown in Figs. 9, 10, and 12. The inputs can be routed in the standard cell using multiple metal layers after the etched regions are created. The use of multiple metal layers for routing in standard cells is common practice in CMOS. Other nanoscale standard cell design techniques described in [1] also use metal 1, gate and metal 2 wiring within standard cells.

Fig. 10 Mis-positioned-CNT-immune layout for a network specified in the POS form

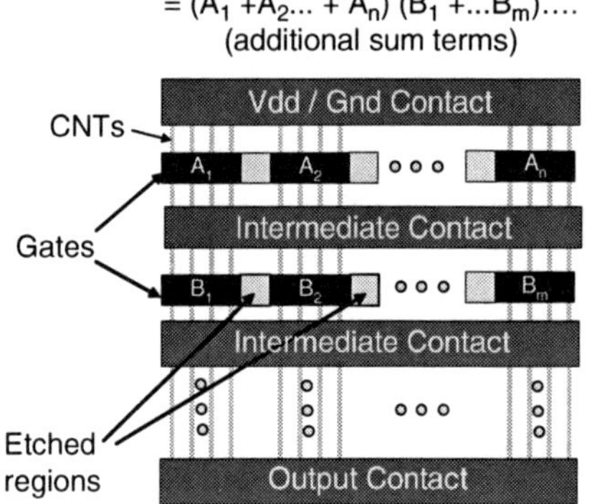

Fig. 11 Mis-positioned-CNT-immune layout for network represented by the function A + (B + C) (D + E)

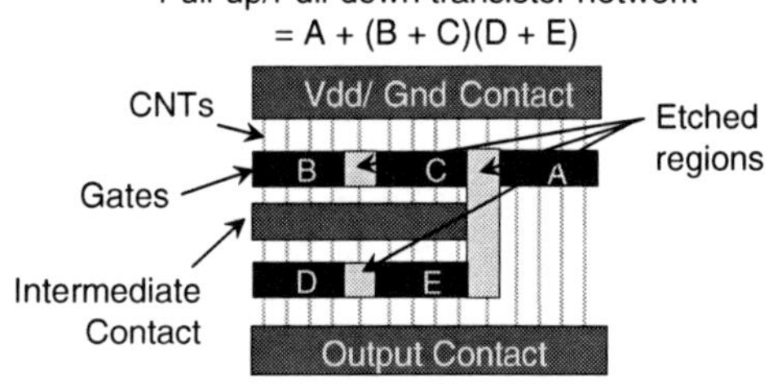

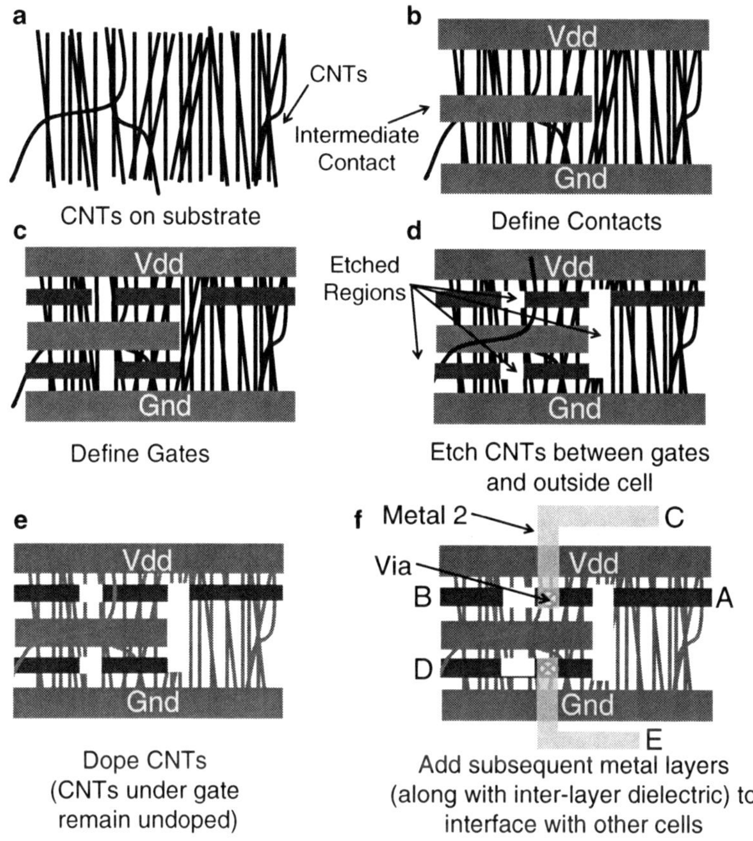

Fig. 12 Steps to create misaligned-and-mis-positioned CNT immune layout of Fig. 11

To make an arbitrary CMOS layout misaligned-and-mis-positioned-CNT-immune, we need to have etched regions in the CNFET layout outside the CMOS diffusion and active regions of the transistors to render the CNTs in those regions nonconductive. This represents a sufficient but not necessary condition for mis-positioned-CNT-immune design.

The worst-case (for minimum-sized CNFET logic cells) energy, delay and area penalties of mis-positioned-CNT-immune logic cells are 18, 13, and 21%, respectively, compared to designs that are not mis-positioned-CNT-immune [25, 27]. These penalties decrease for non-minimum-sized standard cells. These penalties are significantly lower than traditional defect-and-fault-tolerance techniques [8, 14, 34, 35, 37]. Furthermore, this design technique is VLSI-compatible since it does not require changes to the existing VLSI design flow [3]. There is no need to test or reconfigure for mis-positioned CNTs (logic cells are designed to be immune to mis-positioned CNTs by construction [25, 27]).

Mis-positioned CNTs impose a major barrier to practical implementations of CNFET-based logic circuits. It is possible to overcome this barrier by designing

CNFET-based logic circuits that are inherently immune to mis-positioned CNTs. With sufficiently high CNT density, along with the use of mis-positioned-CNT-immune design, CNT alignment and positioning is no longer a problem.

3 Metallic-CNT-Immune CNFET Circuits

The existence of m-CNTs poses another major barrier for CNFET-based digital VLSI logic circuits. Depending upon the arrangement of carbon atoms in a CNT (i.e., *chirality*), a CNT can be *semiconducting* (with large bandgap) or *metallic* (with zero or almost zero bandgap). Unlike semiconducting CNTs (s-CNTs), the conductivity of m-CNT cannot be controlled by the gate. This creates source-drain shorts in CNFETs.

No existing CNT growth technique can grow exclusively s-CNTs. Preferential s-CNT growth techniques yields 90–96% s-CNTs [12, 17]. This results in an ensemble of m-CNTs and s-CNTs in CNFETs. CNFETs based on such an ensemble pose major challenges in CNFET digital design [40, 41]: (1) excessive leakage: I_{on}/I_{off} can be $1,000\times$ worse compared to state-of-the-art Si-CMOS; (2) severely degraded noise margins; (3) increased delay variations.

The best way to solve the m-CNT challenge is to grow or deposit s-CNTs exclusively or predominantly (>99.99%) to start with—this is an extremely difficult problem [13, 16, 17, 33]. Hence, we must either remove m-CNTs after growth or design CNFET circuits tolerant to m-CNTs.

3.1 ACCNT: Asymmetrically Correlated CNT Technology

ACCNT is a design methodology that uses asymmetrically correlated CNTs to achieve metallic-CNT tolerance [19]. Even in the presence of m-CNTs, ACCNT CNFETs attain a high on-off ratio as would a s-CNT-only CNFET (i.e., CNFET with only s-CNTs), and ACCNT circuits function correctly as would circuits made entirely from s-CNTs.

CNFETs fabricated on aligned CNTs have an inherent correlation among the CNFETs; furthermore, the correlation is asymmetric. For example, consider two CNFETs fabricated vertically adjacent to each other (CNFET1 and CNFET2 in Fig. 13). If we assume there is no variation along the length of the CNT, then these two CNFETs essentially consist of the same CNTs (different segments of the same CNTs). Hence, for example, we can conclude that if CNFET1 contains three m-CNTs and 6 s-CNTs, then CNFET2 also contains exactly three m-CNTs and six s-CNTs. Thus, CNFETs fabricated along the same vertical column are highly correlated.

Next, consider two CNFETs fabricated horizontally side-by-side (CNFET1 and CNFET3). These two CNFETs consist of different CNTs in their transistor channel region. Thus, we cannot say much about CNFET3 even if we know the metallic vs.

Fig. 13 Illustration of the asymmetric correlations inherent in CNFETs fabricated on aligned CNTs. (*Top*) SEM of aligned CNTs grown from striped catalysts. (*Bottom*) Conceptual cartoon illustrating that CNFETs fabricated along the same column consist of the same CNTs and are highly correlated ($\sim$identical); CNTs fabricated along the same row consist of different CNTs and are highly uncorrelated ($\sim$independent)

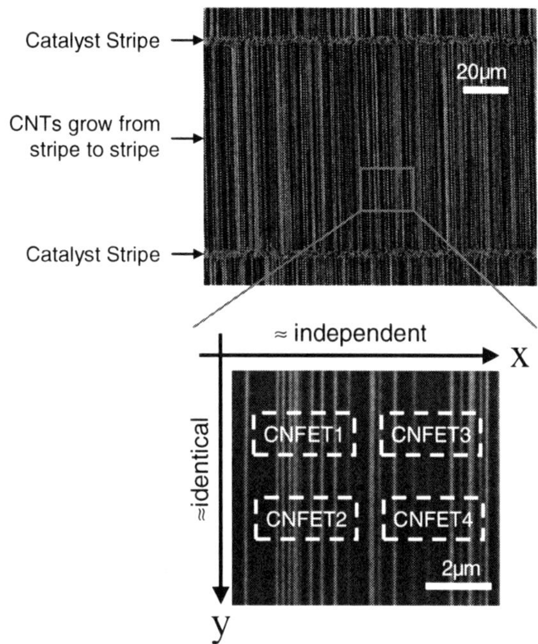

semiconducting makeup of CNFET1. That is, CNFETs fabricated along the same row are relatively uncorrelated.

For ease of analysis, we make the following simplifying assumptions regarding the asymmetric correlations:

(A1). CNFETs along the x-direction are independent (thus the x-direction correlation is 0)

(A2). CNFETs along the y-direction are identical (thus the y-direction correlation is 1)

As a baseline, first consider a conventionally laid-out CNFET with 4 CNTs. s-CNTs have high on-off ratios of up to $\sim10^6$, while m-CNTs have an on/off ratio of 1. Thus, assuming the s-CNT on-current is the same as the m-CNT current, if just one of the 4 CNTs in this CNFET were an m-CNT, the I_{on}/I_{off} of the CNFET would degrade to ~4. On the other hand, if all 4 CNTs were s-CNTs, the CNFET would have an overall device on/off ratio of $\sim10^6$. Consequently, in order for the conventional CNFET to exhibit semiconducting device characteristics with high I_{on}/I_{off} ratio, all the CNTs need to be s-CNTs.

Define p_{semi} as the probability a given CNT is semiconducting (Table 2). Then, the probability ($P_{OverallSemi}$) that a conventional CNFET device exhibits overall semiconducting behavior is

$$P_{OverallSemi} = \left(p_{semi}\right)^{N_{CNT}} \qquad (1)$$

Table 2 Construction of the ACCNT CNFET

	Schematic	Probability of Overall Semiconducting Characteristics
Carbon Nanotube (CNT)		p_{semi}
CNFET with N_{CNT} CNTs forming the channel		$= \left(p_{semi} \right)^{N_{CNT}}$
ACCNT Row with m_{cols} CNFETs (all independent) in series	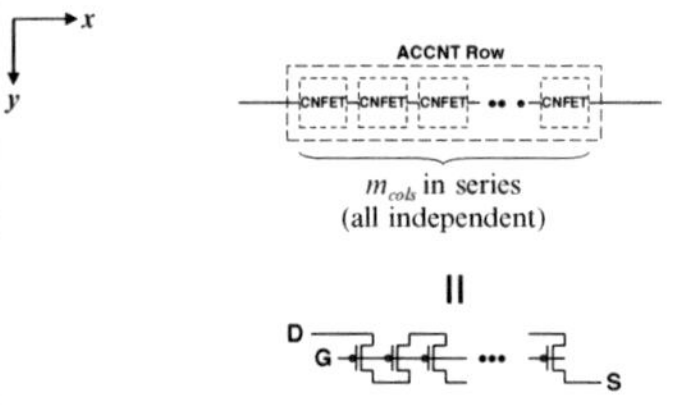 ACCNT Row — CNFET–CNFET–CNFET– ••• –CNFET — m_{cols} in series (all independent) ‖	$P_{ACCNT\ Row}$ $= Prob \left(\begin{array}{c} \text{at least one CNFET} \\ \text{is semiconducting} \end{array} \right)$ $= 1 - Prob \left(\begin{array}{c} \text{all of the CNFETs} \\ \text{are not semiconducting} \end{array} \right)$ $= 1 - \left(1 - P_{CNFET} \right)^{m_{cols}}$ $= 1 - \left(1 - p_{semi}^{\ N_{CNT}} \right)^{m_{cols}}$
ACCNT CNFET with n_{rows} ACCNT Rows (all identical) in parallel	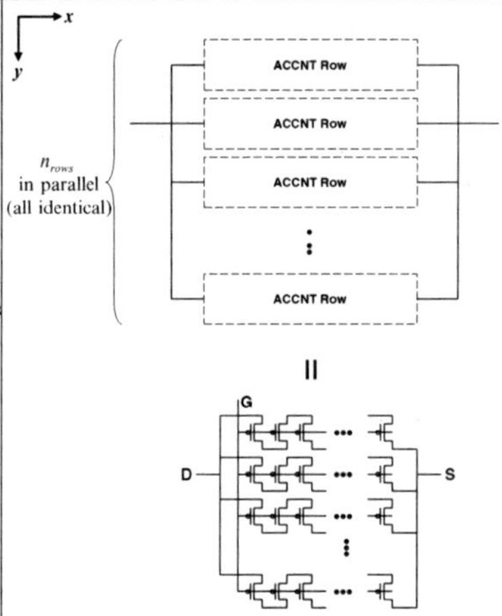 n_{rows} in parallel (all identical); ACCNT Row ⋮ ACCNT Row ‖	$P_{ACCNTCNFET} = \left(P_{ACCNTRow} \right)^1$ $= \left(1 - \overline{1 - p_{semi}^{\ N_{CNT}}} \right)^{m_{cols}}$ *In the first equality: if the rows were all independent, the exponent would be n_{rows}; but since the rows are all identical, the exponent is reduced to 1.

The second column illustrates how the ACCNT CNFET is constructed from conventional CNFETs. The third column summarizes the probability ($P_{OverallSemi}$) of obtaining overall semiconducting characteristics with high on/off ratio. (For clarity, $P_{OverallSemi}$ for a CNFET, ACCNT Row, and ACCNT CNFET is labeled as P_{CNFET}, $P_{ACCNT\ Row}$, and $P_{ACCNT\ CNFET}$, respectively.)

where N_{CNT} is the number of CNTs in the device (Table 2.1). For the conventional CNFET, $N_{CNT} = 4$ and $p_{semi} \approx 2/3$; so $P_{OverallSemi}$ for such a CNFET is 19.8%. Considering there may be a thousand to a billion transistors per chip, a device yield (in terms of obtaining semiconducting device characteristics) of 19.8% would not acceptable for VLSI applications. However, ACCNT CNFETs can improve $P_{OverallSemi}$ arbitrarily close to 100%, which enables VLSI applications of CNFET circuits.

The concept here is similar to the well-established methods of combating device shorts through redundant series connections [22, 23]; however, an important distinction must be made here. The ACCNT design notes the asymmetric correlation intrinsic to aligned CNTs and utilizes it to achieve *both* metallic-CNT tolerance and undiminished current drive.

Specifically, to achieve metallic-CNT tolerance, CNFETs are connected in series along the x direction (independent direction). The gates of all CNFETs are connected together, so that the overall series connection acts like one larger CNFET device; we call this an "ACCNT Row." Schematics of the ACCNT construction are shown in Table 2. The ACCNT Row will exhibit semiconducting device characteristics if and only if there exists at least one semiconducting-CNT-only CNFET in the series connection. Thus, a series link of m_{cols} CNFETs will have a probability of exhibiting semiconducting characteristics as follows:

$$P_{\mathrm{OverallSemi}} = 1 - \left(1 - p_{\mathrm{semi}}{}^{N_{\mathrm{CNT}}}\right)^{m_{\mathrm{cols}}} \tag{2}$$

Note that this is only true if the devices are statistically independent of each other; thus, the ACCNT design employs series connections along the x direction in which the CNFETs are independent (vs. the y-direction in which the devices are identical). From (2), as m_{cols} increases, $P_{\mathrm{OverallSemi}}$ improves and asymptotically approaches 100%; thus $P_{\mathrm{OverallSemi}}$ can be engineered arbitrarily close to 100% by an appropriate choice of the design parameter m_{cols}.

However, by connecting m_{cols} CNFETs in series in an ACCNT Row, the total series resistance increases by a factor of m_{cols} and thus decreases the total drive current by approximately a factor of m_{cols}. This is problematic, since most circuit applications require a minimum available current drive to achieve correct functionality and speed. However, the widths of the CNFET cannot simply be increased for increased current, as this would effectively increase N_{CNT} and severely degrade $P_{\mathrm{OverallSemi}}$.

Instead, ACCNT improves current drive by exploiting the correlation in the y-direction to connect identical ACCNT Rows in parallel (Table 2). The ACCNT Rows are aligned such that the first CNFET in each ACCNT Row are aligned in the y-direction, and are thus all identical; the second CNFET in each ACCNT Row are aligned, and are thus all identical; so forth. By connecting n_{rows} of these y-tiled ACCNT Rows, the current drive is then increased by a factor of n_{rows}. It is important to note that $P_{\mathrm{OverallSemi}}$ remains unchanged, (2), since the rows are all identical (perfectly dependent) under assumption (A2). That is, there are only two possibilities: either all rows exhibit semiconducting characteristics or all rows exhibit non-semiconducting characteristics; thus, the probability of all rows being semiconducting is the same as the probability of a single row being semiconducting. Consequently, $P_{\mathrm{OverallSemi}}$ remains unchanged. In summary, the ACCNT design restores lost current drive by connecting ACCNT Rows (or more generally, CNFETs) in parallel along the y-direction.

Using ACCNT, a high $I_{\mathrm{on}}/I_{\mathrm{off}}$ ratio is achieved, similar to that of CNFETs with s-CNTs only, despite the presence of m-CNTs (Fig. 14a). As a result, the issues of noise margin degradation caused by m-CNTs are fixed. Using ACCNT, we successfully demonstrated ACCNT CNFETs and inverters, at wafer-scale, that satisfy the requirements of digital logic (Fig. 14b).

To summarize, there are three main design parameters in the ACCNT design – n_{rows}, m_{cols}, and N_{CNT}. Equation (2) describes the ACCNT design and Table 3 outlines the ACCNT design construction. As an example, to obtain $P_{\text{OverallSem}i} = 99.99\%$ with $p_{\text{semi}} = 2/3$, m_{cols} must be at least 9 for $N_{\text{CNT}} = 1$ and at least 41 for $N_{\text{CNT}} = 4$. n_{rows} can be set depending on the current drive requirements; for example, to exactly preserve current drive as compared to the conventional CNFET, set $n_{\text{rows}} = m_{\text{cols}}$.

ACCNT, being a purely circuit design solution, will in some scenarios incur expensive design costs. In particular, when the percentage of m-CNTs is very large (>10%), the ACCNT overhead can become quite large. This also translates into significant delay and power overheads (e.g., from larger device capacitances and longer interconnects). A solution employing a combination of design and processing techniques could potentially fix the m-CNT problem at substantially less cost. We next describe such a solution, VMR (Sect. 3.2), which combines layout design with processing to overcome these challenges. Also, the new concept of CNT correlation, as illustrated in ACCNT, can also be effectively used as a unique optimization variable during layout design of CNFET circuits (Sect. 8.3).

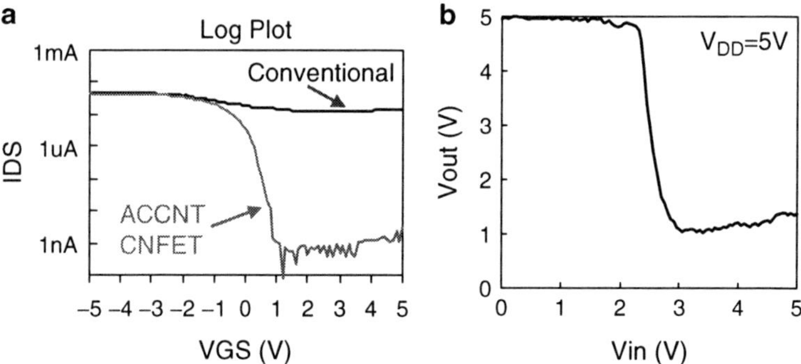

Fig. 14 ACCNT CNFET and ACCNT inverter. (**a**) ACCNT CNFET exhibits high $I_{\text{on}}/I_{\text{off}}$. (**b**) ACCNT inverter results

Table 3 Minimum average number of CNTs per CNFET (N_{min}) for CNFET failure probability (p_{F}) = 10^{-8} with uniform CNT density and in the presence of CNT density variation

p_{m}	33%	33%	10%	10%
p_{rs}	16%	0%	16%	0%
p_{f}	43.7%	33%	24.4%	10%
N_{min} (uniform density)	23	17	13	8
N_{min} (with density variation)	29	24	20	18

3.2 VMR: VLSI-Compatible Metallic-CNT Removal

m-CNTs can be removed after growth using sorting [13, 16], selective chemical etching [39] or single-device electrical breakdown (SDB) [5, 18]. For CNFET-based VLSI circuits, 99.99% of all m-CNTs must be removed (Fig. 15a). Existing techniques for m-CNT removal are either not VLSI-compatible or do not remove enough m-CNTs. Some also impose restrictions such as radial CNT alignment [16] or narrow CNT diameter distributions [39]. SDB achieves $\sim$100% m-CNT removal [5, 18], but suffers from the following major VLSI challenges:

1. SDB requires top-gate oxide of CNFETs to be thick enough to withstand the high voltage required for m-CNT breakdown (5–10 V [32]) resulting in reduced circuit performance.
2. It is impractical to contact gate, source, and drain of each CNFET individually in gigascale ICs.
3. m-CNT fragments can produce incorrect logic functionality because internal contacts cannot be accessed for gigascale ICs (Fig. 14b).

We describe a new technique called VMR (VLSI-compatible metallic CNT removal) that overcomes the VLSI challenges of existing m-CNT removal techniques [31]. VMR combines layout design with CNFET processing to remove m-CNTs in CNFET circuits. VMR consists of the following steps:

1. Aligned CNTs are grown on single-crystal quartz wafers and then transferred to silicon wafers with 100-nm thick silicon dioxide [26, 29]. A special layout called VMR structure is fabricated (Figs. 16 and 18a). The *VMR structure* consists of inter-digitated electrodes at minimum metal pitch.
2. Next, VLSI-compatible m-CNT electrical breakdown is performed by applying high voltage across m-CNTs all at once using the VMR structure (Fig. 17). The back-gate turns off the s-CNTs.
3. Unwanted sections of the VMR structure are removed and CNTs are etched from predefined layout regions corresponding to the final intended design (using

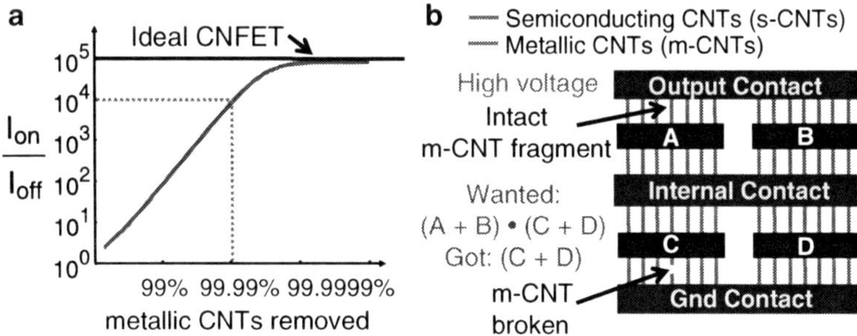

Fig. 15 (a) Simulation results for I_{on}/I_{off} vs. %m-CNTs removed. Initial m-CNT percentage = 33%. 99.99% m-CNT removal required for $I_{on}/I_{off} = 10^4$. (b) m-CNT fragments left behind after SDB cause incorrect logic functionality because internal contacts cannot be accessed for large ICs

Fig. 16 (**a**) Top view and (**b**) cross-sectional view of VMR structure consisting of six inter-digitated VMR electrodes. (**c**) SEM Image of VMR structure (*top view*). Width of VMR electrodes = 50 μm. Back-gate oxide (SiO₂) thickness = 100 nm

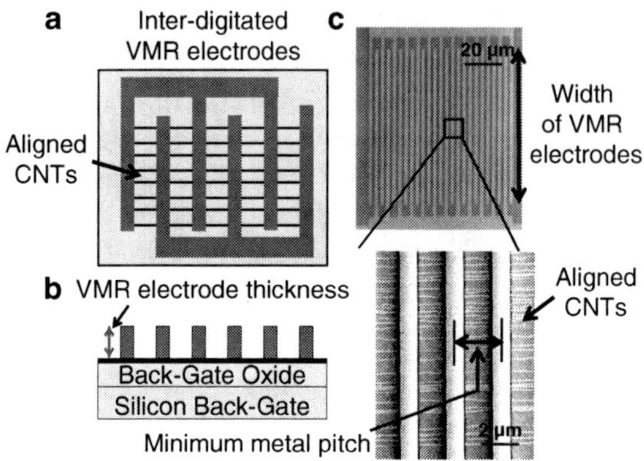

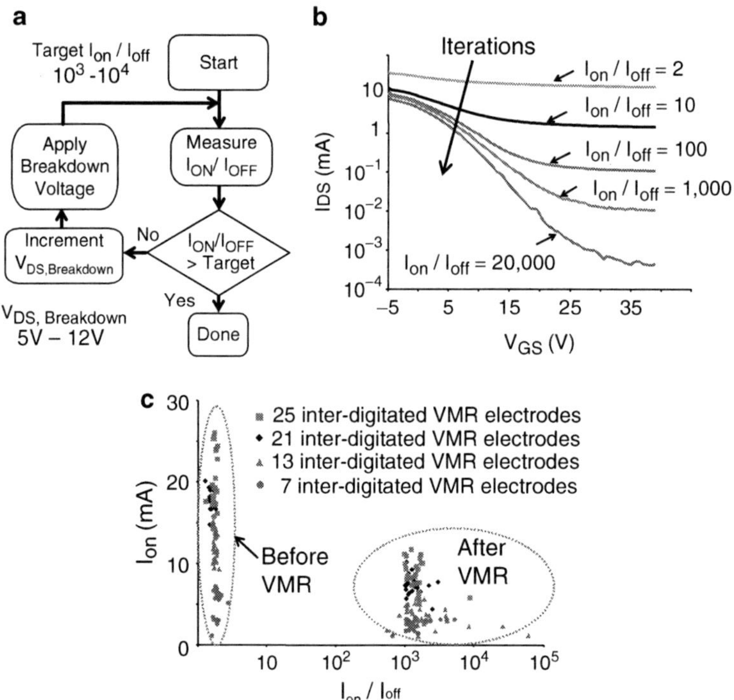

Fig. 17 (**a**) Electrical breakdown of m-CNTs using a back-gated VMR structure. The back-gate is used to turn off s-CNTs. (**b**) Typical current-voltage (I_{DS}–V_{GS}) characteristics for a VMR structure (Fig. 16b) at different points during the m-CNT breakdown process. $V_{DS} = -1$ V. (**c**) I_{on} vs. I_{on}/I_{off} for 104 VMR structures with 7, 13, 21, 25 interdigitated VMR electrodes. VMR electrode width = 50 μm. Average s-CNT current yield, defined as $(I_{on}-I_{off})_{After\_VMR} / (I_{on}-I_{off})_{Before\_VMR}$, is 95%. I_{on} is measured at $V_{GS} = -5$V. I_{off} is measured at $V_{GS} = 40$ V. High back-gate voltage is needed because of the thick back-gate oxide (100 nm SiO₂)

    mis-positioned-CNT-immune layouts [25, 26, 27]) (Fig. 18b). VMR electrode sections that will form CNFET contacts in the final design are not removed.

4. Fabrication of the final intended circuit is completed by patterning top-gates with thin-oxide, and interconnects that span multiple metal layers (Fig. 18c).

The VMR structure is independent of the final intended design. However, *any arbitrary final design* (e.g., Fig. 18d) *can be created* by etching out parts of the VMR structure as long as the following modification is incorporated in logic library cell layouts: an additional contact (at minimum metal pitch) is added for series-connected CNFETs for logic library cells (Fig. 19). Regions of the VMR structure to be etched out are pre-defined during layout design and no die-specific customization is required. The additional contacts incur <2% area penalty and 3% delay penalty at the chip level, based on synthesis results of large designs (OpenRISC processor, Ethernet controller [24]).

VMR retains the benefits of SDB while overcoming VLSI challenges faced by SDB:

1. VMR does not require any top-gate because the back-gate is used for m-CNT breakdown. This enables thin top-gate oxide to be fabricated for high performance

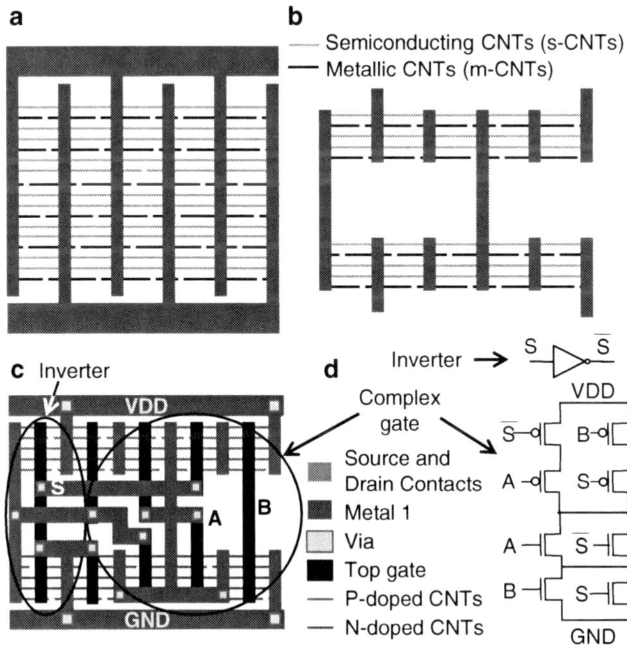

Fig. 18 VMR example for an arbitrary final design. (**a**) Fabricate VMR electrodes (Fig. 16) and perform electrical breakdown of m-CNTs using back-gate to turn-off s-CNTs (Fig. 17). (**b**) Etch CNTs in predefined layout regions using the mis-positioned-CNT-immune layout design technique. Etch unneeded sections of VMR electrodes to create individual CNFET contacts. VMR electrode sections that will form CNFET contacts in the final circuit are not removed. (**c**) Define top-gates with thin top-gate oxide, perform CNT doping, and fabricate subsequent metal layers to create final circuit. (**d**). Schematic of final design fabricated using VMR

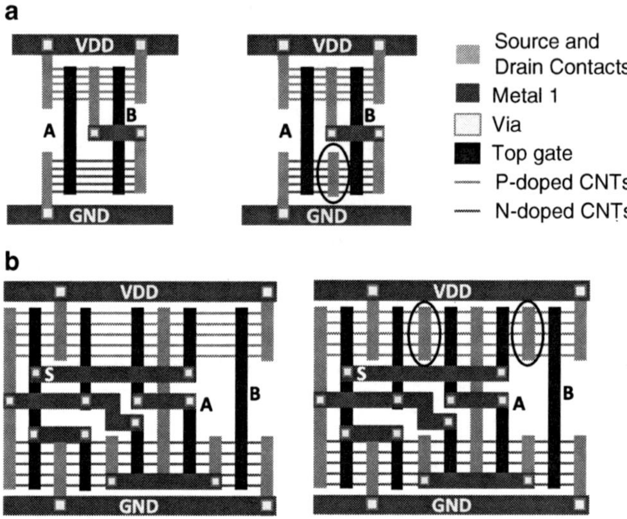

Fig. 19 Layout modifications needed for VMR. (**a**) NAND layout with added contacts (marked with *black circle*). (**b**) Multiplexer layout with added contacts (marked with *black circles*)

CNFETs *after* m-CNT breakdown. The effect of the back-gate capacitance during normal circuit operation can be minimized by using a thick back-gate oxide since the back-gate is not needed during normal circuit operation.

2. VMR does not require any mechanism to contact each CNFET separately.
3. The problem of m-CNT fragments is overcome since all internal contacts can be accessed using the VMR structure (parts of which are etched to create the final intended design).

A high I_{on}/I_{off} ratio of 10^3–10^5 is achieved for various circuit elements, e.g., NAND, multiplexer in Fig. 19, fabricated using VMR.

Figure 20 shows an experimental demonstration of a cascaded CNFET circuit implementing a half-adder sum generator (XNOR) function fabricated using VMR. Such logic circuits are immune to both mis-positioned and metallic CNTs. In this circuit, the outputs of CNFET gates drive other CNFET gates. Voltage swing is reduced due to the use of only p-type CNFETs (similar to all-PMOS logic) [21]. Integration of CNT doping to fabricate *n*-type CNFETs can improve the voltage swing. These cascaded CNFET circuits, fabricated using VMR and wafer-scale processing techniques [26, 29], satisfy requirements of digital logic (gain and cascadability) [21] in a VLSI-compatible manner.

VMR electrodes require careful design to ensure that voltage drops along electrodes (Fig. 21b, c) do not prevent m-CNT removal (m-CNT breakdown requires a certain voltage drop across m-CNTs [32]). Hence, there exists a constraint on the widths of VMR electrodes (Fig. 22d). However, VMR can be readily scaled by connecting multiple VMR electrodes in parallel (Figs. 16 and 17). Preferential s-CNT growth [12, 17, 33] or CNT enrichment [13, 16] can further improve the scalability of VMR (Fig. 13d) and reduce VMR current requirements.

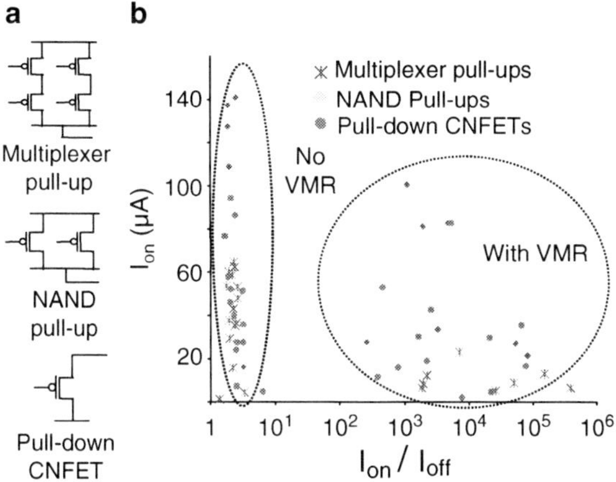

Fig. 20 (**a**) Circuit elements used in final CNFET circuits: Multiplexer pull-up, NAND pull-up and pull-down CNFET. (**b**) Scatter plot of I_{on} vs. I_{on}/I_{off} for the circuit elements fabricated with and without VMR. All CNFETs have width = 50 μm and length = 1 μm. I_{on} for circuit elements without VMR does not include m-CNT current (I_{off}). The total number of circuit elements measured = 81. All CNFET circuit elements fabricated using VMR show high I_{on}/I_{off} (10^3–10^5). I_{on} decreases after VMR because some s-CNTs are removed during VMR (see Fig. 17c)

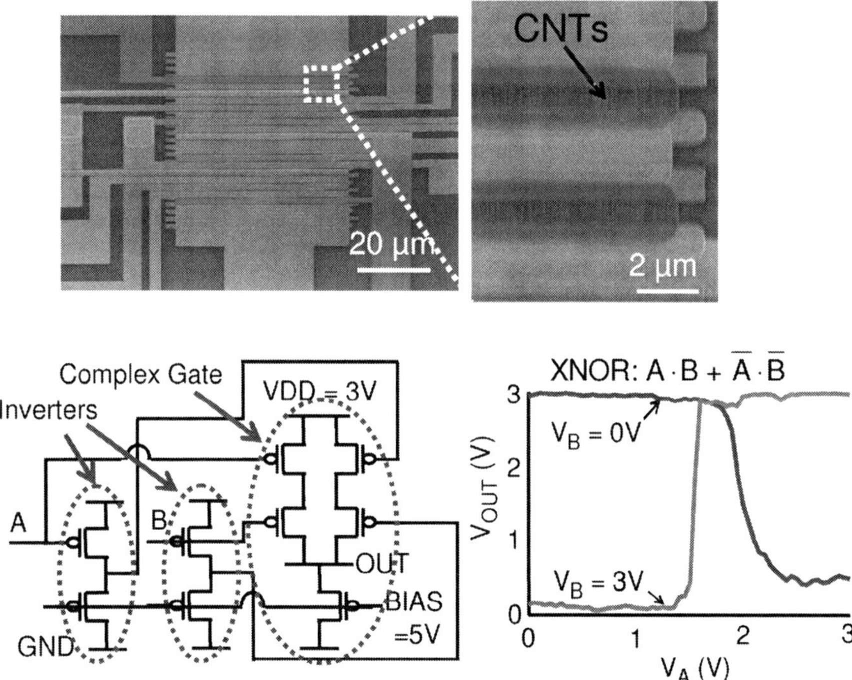

Fig. 21 Experimental demonstration of multistage cascaded CNFET logic circuit implementing the equivalent of a half-adder sum generator function (XNOR) fabricated in a VLSI-compatible manner. All biases for the pull-down CNFETs in the circuit are connected together similar to all-PMOS logic circuits [21]. Gains = 12 and 28 when V_B = 0 V and 3 V, respectively

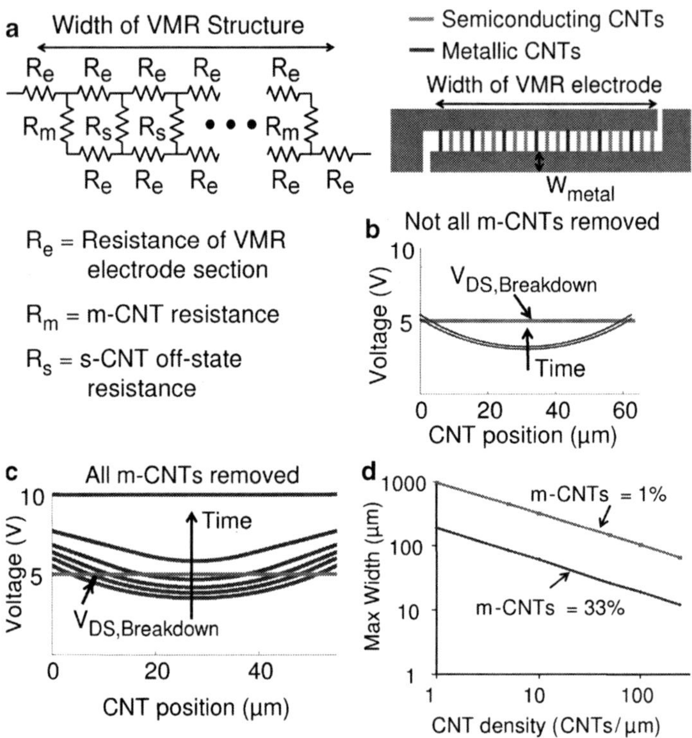

Fig. 22 VMR structure simulation results. (**a**) Resistor network and schematic representing VMR structure with palladium VMR electrodes. Width of palladium VMR electrodes (W_{metal}) = 100 nm, Thickness of palladium VMR electrodes = 100 nm. Resistance of electrode section (R_e) depends on CNT density. Resistance of m-CNT (R_m) = 50 kΩ [32]. Resistance of s-CNT (off-state) R_s = 50 MΩ. CNT density = 10 CNTs/μm. (**b**) Voltage profile simulation of a VMR electrode with 10 V applied at the two ends. Width of VMR electrode = 62 μm. Some m-CNTs are below the breakdown threshold ($V_{DS,Breakdown}$) of m-CNTs = 5 V for CNT length = 1 μm [32] because of voltage drops along VMR electrodes. Hence, not all m-CNTs are removed. (**c**) Voltage profile simulation of a VMR electrode with 10 V applied at the two ends. Width of VMR electrode = 55 μm. All m-CNTs in VMR structure are removed. (**d**) Maximum width of VMR electrodes. Maximum width decreases with higher CNT density due to increased voltage drop through electrodes.

4 Probabilistic Analysis of CNFET Circuits

CNTs are grown using chemical synthesis. As a result, it is extremely difficult to ensure that all CNTs are s-CNTs and with uniform CNT density. Density variations in CNT growth coupled with m-CNTs and m-CNT removal can compromise reliability of CNFET circuits. We demonstrate how the reliability of CNFET circuits can be significantly improved by taking advantage of CNT correlation in aligned CNT growth (mentioned earlier in Section 2 in the context of ACCNT CNFETs).

An important failure case of a CNFET is when there is no s-CNT left in the CNFET [40]. We assume that the probability of any CNT being an m-CNT (s-CNT) is p_m (p_s), independent of the types of any of its neighboring CNTs, with $p_m + p_s = 1$.

In the case of *ideal removal* of m-CNTs (all m-CNTs are removed without removing any s-CNTs), each CNT has a probability $p_f = p_m$ of not being an s-CNT. In situations when there is inadvertent removal of s-CNTs, this probability increases to

$$p_f = p_m + p_s p_{rs},\tag{3}$$

where p_{rs} is the probability that an s-CNT will be inadvertently removed. We generalize the above discussion by considering p_f as the failure probability for each CNT. Then, for a CNFET with N independent CNTs (before any removal), the failure probability for the CNFET (p_F) is given by

$$p_F = p_f^N.\tag{4}$$

Equation (4) shows that the failure probability of a CNFET exponentially decreases with the number of CNTs. When CNT density variation is taken into account, p_F is derived below based on the notion of conditional probability

$$p_F = \sum_{N_i} p_f^{N_i} \text{Prob}[N(W) = N_i],\tag{5}$$

where W is the width of the CNFET.

For a given CNFET width, CNT density variation (or equivalently variation in $N(W)$) always results in a higher probability of failure (p_F) compared to the case with uniform CNT density. To prove this, the arithmetic mean-geometric mean inequality can be applied to (5), which gives

$$\begin{aligned}
p_F &= \sum_{N_i} p_f^{N_i} \text{Prob}[N(W) = N_i] \\
&\geq \left[\prod_{N_i} \left(p_f^{N_i} \right)^{\text{Prob}[N(W)=N_i]} \right]^{1/\sum_{N_i} \text{Prob}[N(W)=N_i]} = p_f^{\mu[N(W)]}.
\end{aligned}\tag{6}$$

The right-hand side of inequality in (6) is the failure probability of the uniform density case.

To quantify such degradation in p_F due to CNT density variation, we can define N_{min} as the minimum average CNT count $\mu(N(W))$ for a target value of failure probability (p_F). Such a definition of N_{min} corresponds to the minimum CNFET width circuit designers must satisfy for a certain average inter-CNT spacing of the given technology. Table 3 shows the values of N_{min} with varying p_m and p_{rs} for uniform CNT density as well as in the presence of CNT density variation [42].

As shown in the table, CNT density variation significantly increases N_{min} for all cases, especially for small values of CNT failure probability (p_f).

We now describe a way to utilize the correction in aligned CNT growth (discussed in Sect. 2.1) to improve the reliability of CNFET circuits. We consider a nonideal m-CNT removal process which removes m-CNTs with probability p_{rm} and inadvertently removes s-CNTs with probability p_{rs}. Consider a pair of cross-coupled inverters (Fig. 23) each with a p-type CNFET (*PFET*) and an n-type CNFET (*NFET*). Static noise margin (*SNM*), defined as the maximum nested square between the normal and mirrored voltage transfer curves (*VTCs*) for the two inverters [20], is used as a metric for the robustness of the cross-coupled inverters (Fig. 23b). When m-CNTs are present, the inverters do not give a full rail-to-rail output, which reduces the gain as well as the noise margin (Fig. 23c).

Suppose that SNM_R is the required static noise margin that such cross-coupled inverters must satisfy. Given the CNT density distribution, CNT processing parameters (p_{rm}, p_{rs}) and the width of the CNFETs (W), we can calculate the probability that a gate will fail to satisfy this SNM requirement. We refer to this as *PNMV*, or *probability of noise margin violation*. We show that the layout of such cross-coupled inverters can be optimized to reduce the PNMV by up to three orders of magnitude. Five different layout styles are studied, as shown in Fig. 24. The different styles have different degrees of correlation among the four CNFETs comprising the cross-coupled inverters. The symmetries in VTCs caused by such a correlation are also shown in Fig. 24. Style 1 (Fig. 24a) has perfect correlation among the drive strengths of all four CNFETs. Style 2 (Fig. 24b) has perfect correlation between the PFET and NFET of each inverter, while the CNFETs in the different inverters are uncorrelated. Style 3 (Fig. 24c) has perfect correlation between the PFET of one inverter and the PFET of the second inverter and similarly for the NFETs. Style 4 (Fig. 24d) has perfect correlation between the PFET of one inverter and the NFET of the other and vice versa. Style 5 (Fig. 24e) has completely uncorrelated CNTs. A CNFET SPICE model [10, 11] with a 32-nm CNFET technology is used to simulate the VTCs. We assume a uniform CNT diameter of 1.5 nm and $p_m = 1/3$. SNM_R is assumed to be Vdd/4 and $p_{rs} = 16\%$.

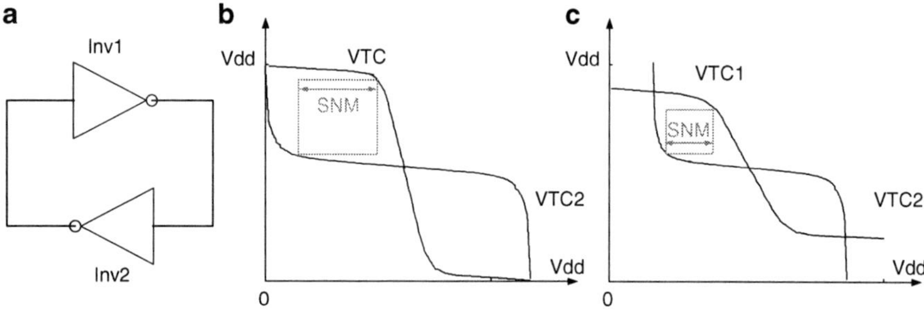

Fig. 23 (**a**) Cross-coupled CNFET inverters (**b**) with voltage transfer curves without m-CNTs and (**c**) with m-CNTs

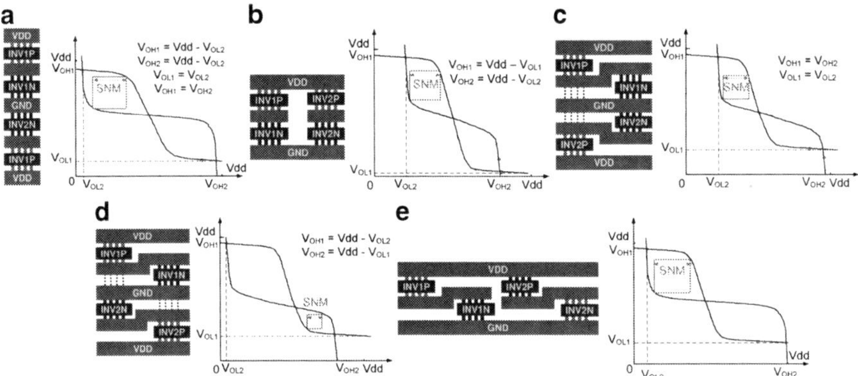

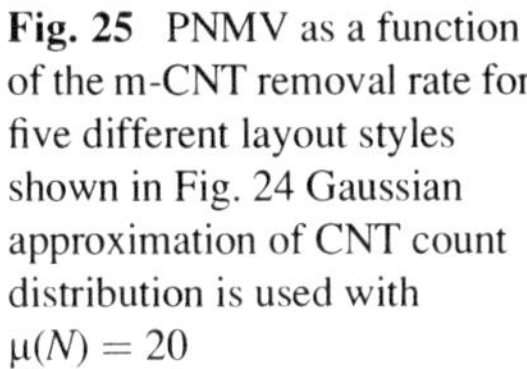

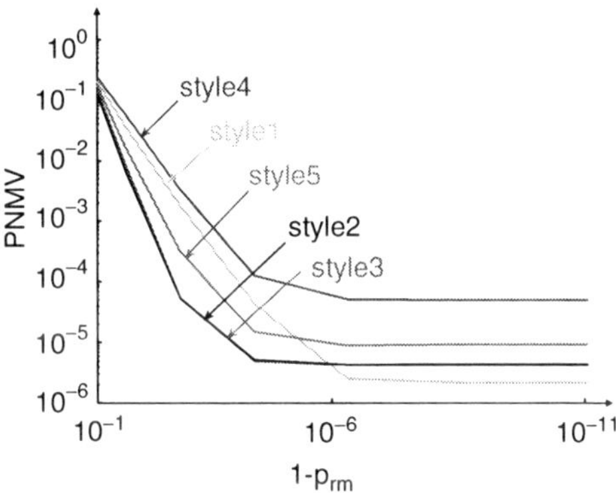

Fig. 24 CNFET cross-coupled inverter layout styles 1–5 (**a** through **e**) with varying degrees of CNT correlation

Fig. 25 PNMV as a function of the m-CNT removal rate for five different layout styles shown in Fig. 24 Gaussian approximation of CNT count distribution is used with $\mu(N) = 20$

Figure 25 shows the PNMV for the five layout styles as a function of the survival probability of m-CNTs ($1-p_{\text{rm}}$). Symmetry in the VTCs of the two CNFET inverters, introduced by correlation among CNFETs, plays an important role in determining the PNMV. Layout styles 1, 2, and 3 ensure symmetry between the two "eye-openings" of the VTC curves and, therefore, have low PNMV values. Style 4 has the highest PNMV for all values of p_{rm} because of the anti-symmetry in the VTC curves (Fig. 24d). Style 5 has no inherent symmetry because all the CNFETs are uncorrelated and has intermediate values of PNMV.

When the removal probability of m-CNTs is very high ($1 - p_{\text{rm}} < 10^{-6}$), there are negligible number of surviving m-CNTs. In this limiting case, failure of this circuit is due to no CNTs left in the CNFETs because of either inadvertent removal of s-CNTs or CNT density variation. The PNMV value in this case for layout styles 1, 2, 3, and 5 (style 4 is an exception because of its anti-symmetry) can be given by

$$\text{PNMV} = 1 - (1 - p_{\text{F}})^k \approx k p_{\text{F}}, \tag{7}$$

where k is number of uncorrelated CNFETs in the circuit and p_F is the failure probability for each one of them, calculated using (6). Style 1 is has the lowest PNMV because of perfect correlation between the four CNFETs ($k = 1$). Styles 2 and 3 ($k = 2$) and style 5 ($k = 4$) have higher values of PNMV.

For lower probabilities of m-CNT removal ($1-p_{rm} > 10^{-6}$), the presence of m-CNTs can no longer be ignored and becomes the dominant factor in PNMV. In this case, style 1 does not perform as well as styles 2, 3, and 5, since the presence of m-CNTs will cause equal degradation of all four correlated CNFETs in style 1, while such an equal degradation is unlikely in the case of uncorrelated CNFETs in styles 2, 3, and 5.

5 Conclusion

We have the following messages for researchers interested in digital system design in emerging nanotechnologies such as CNFETs:

1. Imperfection-immune design is required.
2. It is extremely important to optimize *across* processing and design. Otherwise, we will end up with solutions that are either impractical or highly inefficient.
3. Traditional defect- and fault-tolerance techniques must be applied very carefully. Otherwise, the benefits of such emerging nanotechnologies can significantly reduce.
4. Design techniques must be compatible with VLSI processing and design flows.

Imperfection-immune design techniques and experimental results, presented in this chapter, are based on these principles, and may pave the way for VLSI CNFET technologies.

Exercises

1. Draw the mis-positioned-CNT-immune layouts for the following functions:

   (a) $A + (B + C) \cdot D$
   (b) $A \cdot (B + C) + D$

2. Draw the graph representations for the above layouts.

3. ACCNT is a metallic-CNT tolerant technology that takes advantage of asymmetric correlations inherent in CNFETs fabricated on aligned CNTs. List two benefits of ACCNT. List two costs of ACCNT.

4. Design an ACCNT CNFET: You are given that the CNT material has a density of 100 CNTs/μm and that the semiconducting-CNT percentage is 99%. Also, you are given that each CNT contributes an approximate on-current of 10 μA.

Lastly, your lithography technology has a minimum feature size of 22 nm. Your design must achieve a metallic-CNT tolerance of 99.9999999% (i.e., less than 1 in a billion yield loss) and an on-current of approximately 20 µA.

(c) What is your choice of n_{rows}, m_{cols}, *and* N_{CNT}?
(d) Now assume that the semiconducting-CNT percentage has improved from 99 to 99.9%. How do your design variables change?
(e) Comparing your designs in (a) and (b), what is the approximate percent-area savings if the semiconducting-CNT percentage is improved from 99 to 99.9%?

5. John Doolittle is attempting to create an ACCNT design, but instead incorrectly exploits the asymmetric correlations. In his erroneous design, he connects correlated CNFETs in series to form a row and connects statistically independent rows (CNFETs) in parallel (versus connecting statistically independent CNFETs in series and correlated rows in parallel). Derive the expression for the metallic-CNT tolerance, $P_{\mathrm{OverallSemi}}$ of Doolittle's creation in terms of p_{semi}, n_{rows}, m_{cols}, and N_{CNT}?

6. In probability theory, the famous Jensen's inequality can be stated as follows: if X is a random variable and f is a continuous, convex function, then

$$f[E(X)] \leq E[f(X)],$$

where $E()$ is the expectation of a random variable. Use Jensen's inequality to prove (4) in Chapter "Graphene Transistors and Circuits." Can you give a general interpretation of Jensen's inequality?

7. Suppose that that $N(W)$ follows a Poisson distribution with parameter $\lambda = \alpha W$ (i.e., assume the average of the CNTs is proportional to the width of the CNFET). Derive an analytical expression for p_{F} based on (10.3) in Chapter "Graphene Transistors and Circuits". If the required $p_{\mathrm{F}} = R$, what is expression for N_{min}?

8. Suppose in a certain CNFET fabrication process, each individual s-CNT has $I_{\mathrm{on}}/I_{\mathrm{off}} = 10^6$, and each m-CNT has $I_{\mathrm{on}}/I_{\mathrm{off}} = 1$. You are also told that the on current (I_{on}) of an s-CNT and the on current of an m-CNT are approximately the same. The probabilistic behavior of this process (including a m-CNT removal step) can be described using the parameters p_{m}, p_{rm}, *and* p_{rs} as defined in Sect. 10.3

(f) Give an expression for the ratio of average I_{on} to average I_{off}, i.e., $\mu(I_{\mathrm{on}})/\mu(I_{\mathrm{off}})$.
(g) If $p_{\mathrm{m}} = 33\%$ is fixed, what is the minimum possible p_{rm} in the m-CNT removal step, such that the $\mu(I_{\mathrm{on}})/\mu(I_{\mathrm{off}})$ given in part (a) is above 10^4?

References

1. D. Atienza, S.K. Bobba, M. Poli, G. De Micheli, and L. Benini, "System-level design for nano-electronics," in *Proc. IEEE International Conference in Electronics, Circuits and Systems*, pp. 747–751, 2007.
2. I. Beer, et al., "RuleBase: An industry-oriented formal verification tool," in *Proc. Design Automation Conf.*, pp. 655–660, 1996.
3. S. Bobba, et al., "Design of compact imperfection-immune CNFET layouts for standard-cell-based logic synthesis," in *Proc. Design Automation and Test in Europe*, pp. 616–621, 2009.
4. P.G. Collins, M.S. Arnold, and P. Avouris, "Engineering carbon nanotubes and nanotube circuits using electrical breakdown," *Science*, vol. 292, pp. 706–709, 2001.
5. T.H. Cormen, C.E. Leiserson, R.L. Rivest, and C. Stein, *Introduction to Algorithms*, Cambridge, MA: MIT, 1990.
6. H. Dai, "Carbon nanotubes, from synthesis to integration and properties" *Accounts of Chemical Research*, vol. 35, 1035–1044, 2002.
7. A. DeHon, and H. Naeimi, "Seven strategies for tolerating highly defective fabrication," *IEEE Design and Test of Computers*, vol. 22, no. 4, pp. 306–315, 2005.
8. J. Deng, N. Patil, K. Ryu, A. Badmaev, C. Zhou, S. Mitra, and H.-S.P. Wong, "Carbon nanotube transistor circuits: Circuit-level performance benchmarking and design options for living with imperfections," in *Proc. Intl. Solid State Circuits Conf.*, pp. 70–588, 2007.
9. J. Deng, and H.-S.P. Wong, "A compact SPICE model for carbon nanotube field effect transistors including non-idealities and its application. Part I: Model of the intrinsic channel region," *IEEE Transactions on Electron Devices*, vol. 54, no. 12, pp. 3186–3194, 2007.
10. J. Deng, and H.-S.P. Wong, "A compact SPICE model for carbon nanotube field effect transistors including non-idealities and its application. Part II: Full device model and circuit performance benchmarking," *IEEE Transactions on Electron Devices*, vol. 54, no. 12, pp. 3195–3205, 2007.
11. L. Ding, et al.,"Selective growth of well-aligned semiconducting single-walled carbon nano-tubes," *Nano Letters*, vol. 9, no. 2, pp. 800–805, 2009.
12. M. Engel, et al., "Thin film nanotube transistors based on self-assembled, aligned, semicon-ducting carbon nanotube arrays," *ACS Nano*, vol. 2, no. 12, pp. 2445–2452, 2008.
13. S.C. Goldstein, and M. Budiu, "NanoFabrics: Spatial computing using molecular electronics," in *Proc. Intl. Symp. Computer Architecture*, pp. 178–191, 2001.
14. S.J. Kang, et al., "High-performance electronics using dense, perfectly aligned arrays of single-walled carbon nanotubes," *Nature Nanotechnology*, vol. 2, pp. 230–236, 2007.
15. M. LeMieux, et al., "Self-sorted, aligned nanotube networks for thin-film transistors," *Science*, vol. 321, pp. 101–104, 2008.
16. Y. Li, et al., "Preferential growth of semiconducting single-walled carbon nanotubes by a plasma nnhanced CVD method," *Nano Letters*, vol. 4, pp. 317–321, 2004.
17. A. Lin, et al., "Threshold voltage and on-off ratio tuning for multiple-tube carbon nanotube FETs," *IEEE Transactions on Nanotechnology*, vol. 8, no. 1, pp. 4–9, 2009.
18. A. Lin, et al., "A metallic-CNT-tolerant carbon nanotube technology using asymmetrically-correlated CNTs (ACCNT)," in *Proc. Symposium on VLSI Technology*, pp. 182–183, 2009.
19. J. Lohstroh, E. Seevinck, and J.D. Groot, "Worst-case static noise margin criteria for logic circuits and their mathematical equivalence," *Journal of Solid-State Circuits*, vol. 18, no.6, pp. 803–806, 1983.
20. E.J. McCluskey, *Logic Design Principles*, Englewood Cliffs, NJ: Prentice-Hall, 1986.
21. E.F. Moore, and C.E. Shannon, "Reliable circuits using less reliable relays. Part I," *Journal of the Franklin Institute*, vol. 262, no. 3, pp. 191–208, 1956.
22. E.F. Moore, and C.E. Shannon, "Reliable circuits using less reliable relays. Part 2," *Journal of the Franklin Institute*, vol. 262, no. 4, pp. 281–297, 1956.
23. http://www.opencores.org.

24. N. Patil, J. Deng, H.-S. P. Wong, and S. Mitra. "Automated design of misaligned-carbon-nanotube-immune circuits," in *Proc. Design Automation Conference*, pp. 958–961, 2007.
25. N. Patil, et al., "Integrated wafer-scale growth and transfer of directional carbon nanotubes and misaligned-carbon-nanotube-immune logic structures," in *Proc. Symposium on VLSI Technology*, pp. 205–206, 2008.
26. N. Patil, et al., "Design methods for misaligned and mis-positioned carbon-nanotube-immune circuits," *IEEE Transactions on Computer-Aided Design of Integrated Circuits and Systems*, pp. 1725–1736, 2008.
27. N. Patil, et al., "Circuit-level performance benchmarking and scalability of carbon nanotube transistor circuits," *IEEE Transactions on Nanotechnology*, vol. 8, no.1, pp.37–45, 2009.
28. N. Patil, et al., "Wafer-scale growth and transfer of aligned single-walled carbon nanotubes," *IEEE Transactions on Nanotechnology*, vol. 8, no. 4, pp. 498–504, 2009.
29. N. Patil, et al., "Digital VLSI logic technology using carbon nanotube FETs: Frequently asked questions," in *Proc. Design Automation Conference*, pp. 304–309, 2009.
30. N. Patil, et al., "VMR: VLSI-compatible metallic carbon nanotube removal for imperfection-immune cascaded multi-stage digital logic circuits using carbon nanotube FETs," in *Proc. Int. Electron Devices Meeting*, pp. 573–576, 2009.
31. E. Pop, "The role of electrical and thermal contact resistance for Joule breakdown of single-wall carbon nanotubes," *Nanotechnology*, vol. 19, 295202, 2008.
32. L. Qu, D. Feng, and L. Dai, "Preferential syntheses of semiconducting vertically aligned single-walled carbon nanotubes for direct use in FETs," *Nano Letters*, vol. 8, no. 9, pp. 2682–2687, 2008.
33. R.M. Rad, and M. Tehranipoor, "A hybrid FPGA using nanoscale cluster and CMOS scale routing," in *Proc. Design Automation Conference*, pp. 727–730, 2006.
34. W. Rao, A. Orailoglu, and R. Karri, "Fault-tolerant nanoelectronic processor architectures," in *Proc. Asia South Pacific Design Automation Conf.*, pp. 311–316, 2005.
35. R. Saito, G. Dresselhaus, and M. Dresselhaus, *Physical Properties of Carbon Nanotubes*, London: Imperial College, 1998.
36. M.B. Tahoori, "Application-independent defect-tolerance of reconfigurable nano-architectures," *ACM Journal Emerging Technologies in Computing*, vol. 2, pp. 197–218, 2006.
37. L. Wei, D.J. Frank, L. Chang, and H.-S.P. Wong, "A non-iterative compact model for carbon nanotube FETs incorporating source exhaustion effects," in *Proceedings of the International Electron Devices Meeting*, pp. 917–920, 2009.
38. G. Zhang, et al., "Selective etching of metallic carbon nanotubes by gas-phase reaction," *Science*, vol. 314, pp. 974–979, 2006.
39. J. Zhang, N. Patil, and S. Mitra, "Design guidelines for metallic-carbon-nanotube-tolerant digital logic circuits," in *Proc. Design Automation and Test in Europe*, pp. 1009–1014, 2008.
40. J. Zhang, N. Patil, and S. Mitra, "Probabilistic analysis and design of metallic-carbon-nanotube-tolerant digital logic circuits," *IEEE Transactions on Computer-Aided Design of Integrated Circuits and Systems*, vol. 38, no. 9, pp. 1307–1320, 2009.
41. J. Zhang, N. Patil, A. Hazeghi, and S. Mitra, "Carbon nanotube circuits in the presence of carbon nanotube density variations," in *Proc. Design Automation Conference*, pp. 71–76, 2009.

FPCNA: A Carbon Nanotube-Based Programmable Architecture

Chen Dong, Scott Chilstedt, and Deming Chen

Abstract In the hunt to find a replacement for CMOS, material scientists are developing a wide range of nanomaterials and nanomaterial-based devices that offer significant performance improvements. One example is the carbon nanotube field-effect transistor (CNFET), which replaces the traditional silicon channel with an array of semiconducting carbon nanotubes (CNTs). Due to the increased variation and defects in nanometer-scale fabrication and the regular nature of bottom-up self-assembly, field-programmable devices are a promising initial application for such technologies. In this chapter, we detail the design and evaluation of a nanomaterial-based architecture called FPCNA (field-programmable carbon nanotube array). Nanomaterial-based devices and circuit building blocks are developed and characterized, including a lookup table created entirely from continuous CNT ribbons. To determine the performance of these building blocks, variation-aware physical design tools are used, with statistical static timing analysis (SSTA) that can handle both Gaussian and non-Gaussian random variables. When the FPCNA architecture is evaluated using this computer-aided design (CAD) flow, a $2.75\times$ performance improvement is seen over an equivalent CMOS FPGA at a 95% yield. In addition, FPCNA offers a $5.07\times$ footprint reduction compared with a baseline FPGA.

Keywords Discretized SSTA · FPCNA · FPGA · Nanoelectronics · Variation-aware CAD

D. Chen (✉)
Department of Electrical and Computer Engineering, University of Illinois, at Urbana-Champaign, IL, USA
e-mail: dchen@illinois.edu

N.K. Jha and D. Chen (eds.), *Nanoelectronic Circuit Design*, DOI 10.1007/978-1-4419-7609-3_9, © Springer Science+Business Media, LLC 2011

1   Introduction

At and below the 22 nm process node, the conventional top–down manufacturing process faces serious challenges due to physical device limitations and rising production costs. Shifting to a bottom-up approach, in which self-assembled nanoscale building blocks, such as CNTs, are combined to create integrated functional devices, offers the potential to overcome these challenges and revolutionize electronic system fabrication.

The nature of the chemical synthesis process allows smaller feature sizes to be defined, but offers less control over individual device location. This leads to very regular structures, making it ideal for creating the repetitive architectures found in field-programmable gate arrays (s). In addition, the programmability of FPGAs allows reconfiguration around the large number of fabrication defects inherent in nanoscale processes, which helps to provide the high level of fault tolerance needed for correct nanocircuit operation.

In this chapter, we review a CNT-based FPGA architecture called . We detail the building blocks of FPCNA, including the CNT lookup table (LUT), which makes up its programmable logic. We also describe a high-density routing architecture using a recently developed nanoswitch device. In these designs, we make special considerations to mitigate the negative effects of nanospecific process variations. We characterize the components considering both these variations and circuit-level delay variations.

To evaluate the performance of this architecture, a typical FPGA design flow is used, and variation-aware placement and routing algorithms are developed. These algorithms are enhanced from the popular physical design tool VPR [1] and use SSTA to improve the performance yield. The SSTA is performed with both normal and non-Gaussian variation models. The results show that FPCNA offers significant performance and density gains compared to the conventional CMOS FPGA, demonstrating the potential for use of CNT devices in next-generation FPGAs.

This chapter is organized as follows: In Sect. 2, we review the related works in nano-FPGA architecture. We provide an overview of the design flow in Sect. 3. Section 4 introduces the nanoelectronic devices used in the designs. In Sect. 5, we present the FPCNA architecture. Section 6 discusses the fabrication of CNT-based LUT. We present circuit-level characterization results in Sect. 7. These results guide the CAD flow, which is detailed in Sect. 8. In Sect. 9, we present experimental results that show the advantages of FPCNA over a conventional CMOS FPGA, and Sect. 10 concludes the chapter.

2   Related Work

There have been a number of nanomaterial-based programmable architectures proposed in the literature. In an early work [2], Goldstein and Budiu presented an island-style fabric in which clusters of nanoblocks and switch blocks are interconnected in an array structure. Each nanoblock consists of a grid of nano-wires, which can be configured to implement a three-bit input to three-bit output Boolean function. Routing channels are created between clustered nanoblocks to provide low-latency communication over longer distances. A comprehensive PLA-based architecture, known as NanoPLA, was presented by DeHon in [3]. This architecture builds logic from crossed sets of parallel semiconducting nanowires. Axial doping is used to address the individual nanowires, and OR-plane crossbars are programmed by applying a voltage across a pair of crossed nanowires. Nano-wire field-effect transistor (FET) restoring units are used at the output of the programmable OR-planes to restore the output signals and provide inversion. Snider et al. proposed a CMOS-like logic structure based on nanoscale FETs in [4], where nanowire crossbars were doped to create arrays of either n-type and p-type semiconductors. These crossbars are made from horizontal metallic wires and vertical semiconducting wires and can be connected using undoped program-mable switch arrays to create AND-OR-INVERT functions.

CMOS-nano hybrid FPGAs have also been explored. In [5], an architecture was presented by Gayasen et al. that combines CMOS logic blocks/clusters with different types of nanowire routing elements and programmable molecular switches. The results show that this architecture could reduce chip area by up to 70% compared to a traditional CMOS FPGA architecture scaled to 22 nm. Using an opposite approach, Rad et al. presented a FPGA with nanowire logic clusters in [6], where the intercluster routing remains in CMOS. Similar to Gayasen et al. up to 75% area reduction is seen, with performance comparable to a traditional FPGA. In [7], an innovative cell-based architecture was introduced by Strukov and Likharev. This architecture, called CMOL, uses specially doped silicon pins on the upper layers of a CMOS stack to provide contacts to a nanowire array interconnect layer. Logic functions are implemented by a combination of CMOS inverter arrays and nanowire OR-planes based on molecular-switches. Interconnect signals can also be routed through the nanowires by connecting specific crosspoints. A generalized CMOL architecture, called FPNI, was proposed by Snider et al. in [8]. Unlike CMOL's inverter array architecture, the logic in FPNI is implemented in CMOS arrays of NAND/AND functions paired with buffers and flip-flops, and the nanowires are used only for routing. FPNI also improved upon the manufacturability of CMOL by requiring less alignment accuracy between the CMOS and nanowire layers. In [9], Dong et al. proposed a 3D nano-FPGA architecture that distributes the components of a 2D FPGA into vertically stacked CMOS and nanomaterial layers. To connect the layers, vias made from CNT bundles were used to transmit signals and dissipate heat. Researchers have also proposed using carbon nanotube-based memory (i.e. NRAM [10–12]) as block storage for FPGA configuration data [13], interconnect switch memory [9], and FPGA LUT memory [14].

While many of the aforementioned studies demonstrate the use of nanowire crossbars for logic and interconnect, they do not explore the possibility of using carbon-based nanoelectronics to implement FPGA logic. In addition, the previous designs were not evaluated using variation-aware CAD tools to predict the performance yield under the large variations in nanoscale fabrication.

3   Design Flow

The development of a nanomaterial-based programmable device involves a number of design steps. To show the relationships between these steps, we illustrate the overall flow in Fig. 1. In this figure, the dashed box shows how the nanocircuit design flow bridges the development of nanoelectronic devices with the fabrication of nanocircuits.

Using the latest advances in nanoelectronic device research, low-level circuit elements, such as logic and memory, are designed, and these elements are assembled into high-level architectures, such as programmable logic tiles and interconnect structures. Circuit characterization is then used to model the behavior of these structures, including their performance under variations. Because these tasks are highly dependent on the nanomaterials used, we refer to them as nanocentric design. CAD evaluation enhances nanocentric design, allowing benchmark circuits to be tested on these architectures and trade-offs of architectural parameters to be explored. The end results are new nanoelectronic designs with satisfactory performance in terms of area, delay, and yield. These designs provide an early assessment of the feasibility of nanomaterial-based devices and circuits.

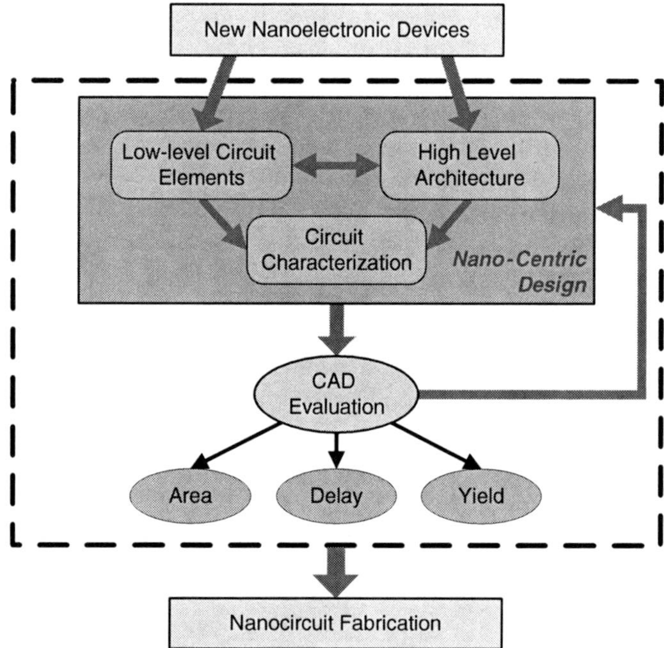

Fig. 1 Overall design flow

4 Nanoelectronic Devices

4.1 Carbon Nanotubes

The first carbon nanotubes were discovered in 1991 by Sumio Iijima at NEC laboratory in Japan [15]. They were created as a byproduct of fullerene synthesis using an electric arc discharge method. Since the discovery, carbon nanotubes have emerged as a promising material for use in future electronics due to their nanoscale dimensions, 1D structure, ballistic electron conduction, low electromigration, and high thermal conductivity.

Carbon nanotubes can be categorized into two groups: single-walled carbon nanotubes (SWCNTs) and multiwalled carbon nanotubes (MWCNTs) (Fig. 2). SWCNTs are molecular devices with a diameter of roughly 1 nm. They are composed of hexagonal carbon rings, which form a seamless cylinder. MWCNTs, on the other hand, are composed of a number of nested hollow cylinders inside one another like Russian dolls and typically have dimensions of several tens of nanometers. SWCNTs draw more attention in the research community because they exhibit unique electric properties and are considered a promising candidate to replace silicon as electronic device feature sizes scale to atomic dimensions.

A CNT can be thought of as a graphene sheet of a certain size that is wrapped to form a seamless cylinder. Due to cylinder symmetry, there are a discreet set of directions the sheet can be rolled (Figs. 2 and 3). To characterize each direction, two atoms in the graphene sheet are chosen, one of which serves the role as origin. The sheet is rolled until the two atoms coincide. The vector pointing from the first atom toward the other is called the chiral vector and its length is equal to the circumference of the nanotube.

Given a SWCNT, its electrical properties can be determined by its chiral vector (n, m) or, in other words, the direction in which the graphene sheet has been rolled. A SWCNT with a chiral vector (n, m) indicates that during rolling, the carbon atom at the origin is superimposed on the carbon atom at lattice location (n, m). Figure 3a illustrates different chiral vectors. Depending on the rolling method, three distinct types of SWCNT can be synthesized (Fig. 3b): the zigzag nanotube with $n = 0$ or $m = 0$, the armchair nanotube with $m = n$, and chiral nanotubes with $n \neq m \neq 0$. The conductivity of SWCNT can be determined by the vector. Metallic SWCNTs have $n - m = 3q$ where q is an integer, and other SWCNTs are semiconducting with a bandgap dependent on the carbon nanotube diameter [17, 18].

Metallic SWCNTs have high electron mobility and robustness and can carry a current density $\sim 1,000\times$ larger than Cu, making them attractive for use in nanoscale interconnect and nanoelectromechanical systems (NEMS). Semiconducting SWCNTs, on the other hand, have ideal characteristics for use in FETs [19]. In the past, it was difficult to mass produce CNT-based circuits because of the inability to accurately control nanotube growth and position. However, recent research has demonstrated the fabrication of dense, perfectly aligned arrays of linear SWCNTs [19] and wafer-scale CNT-based logic devices [20, 21].

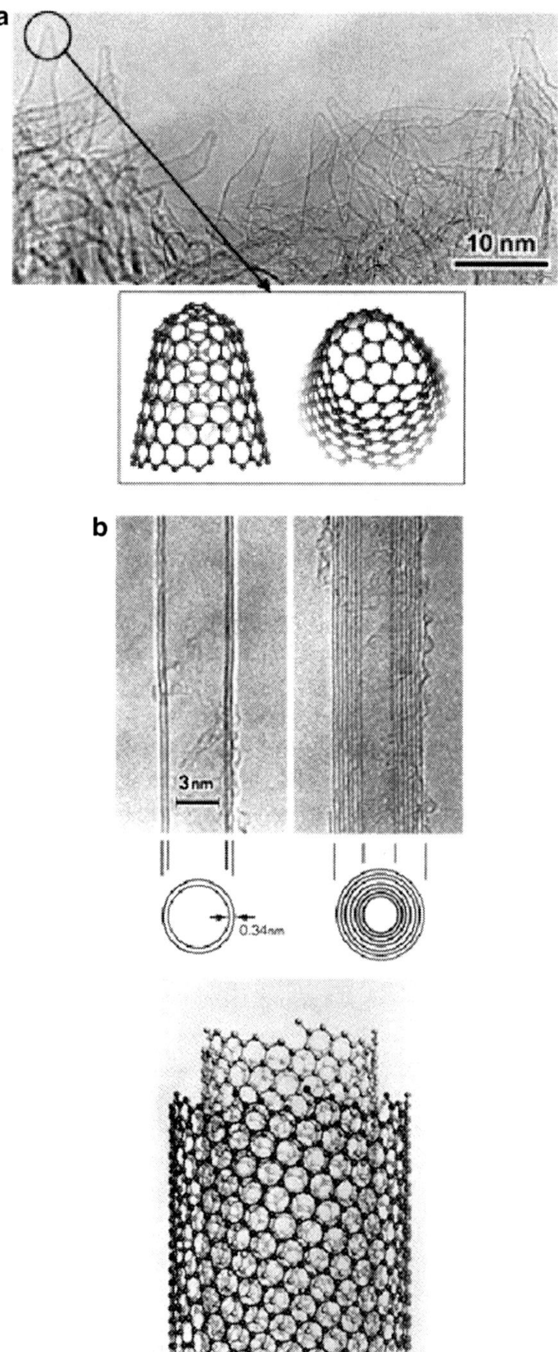

Fig. 2 (**a**) Single-walled carbon nanotube (SWCNT). (**b**) Multiwalled carbon nanotubes (MWCNTs)- [16]

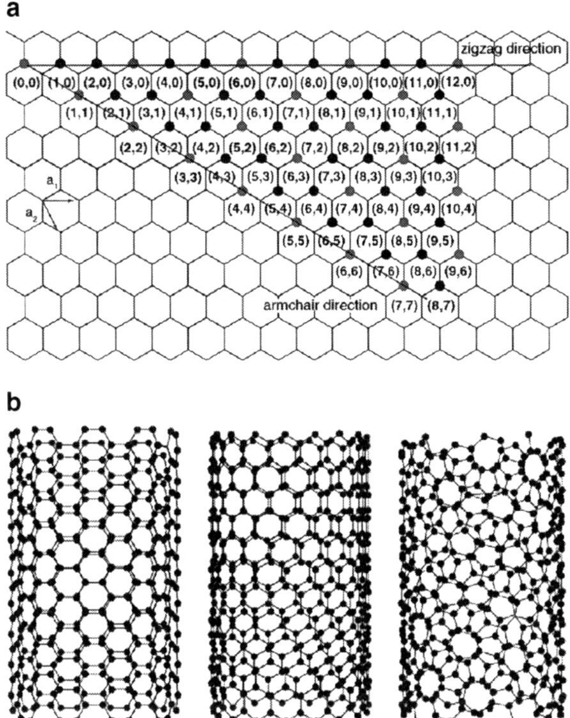

Fig. 3 (a) Chiral vectors for general CNTs -[17]. (b) Armchair, zigzag and chiral CNTs -[18]

4.2 CNT Field-Effect Transistors

The first reports of room-temperature operation of CNT field-effect transistors (CNFETs) were from Delft University of Technology [22] and IBM [23] in 1998. The structure of this CNFET is shown in Fig. 4a; a single nanotube behaves as the channel region and is connected to two gold electrodes. The silicon substrate is used as a back gate. By applying a gate voltage, the channel conductance can be varied by more than five orders of magnitude, similar to a silicon MOSFET. However, without special treatment, the obtained CNFETs are always of p-type.

In 2001, the group from Delft University of Technology enhanced its previous CNFET design by using aluminum local gates to control individual transistors (Fig. 4b, c) [24]. This enhancement allows the integration of multiple FETs on the same substrate. In this work, the thick SiO_2 gate insulator has been replaced by a thin Al_2O_3 layer to create a strong capacitive coupling between the gate and the nanotube. In the meantime, a group from IBM discovered that n-type CNFETs can be obtained by simply annealing p-type CNFET in a vacuum [25].

In 2002, a group from IBM [26] fabricated a CNFET in a structure similar to that of silicon MOSFET with a gate electrode over the conducting channel separated by

                                                                                               C. Dong et al.

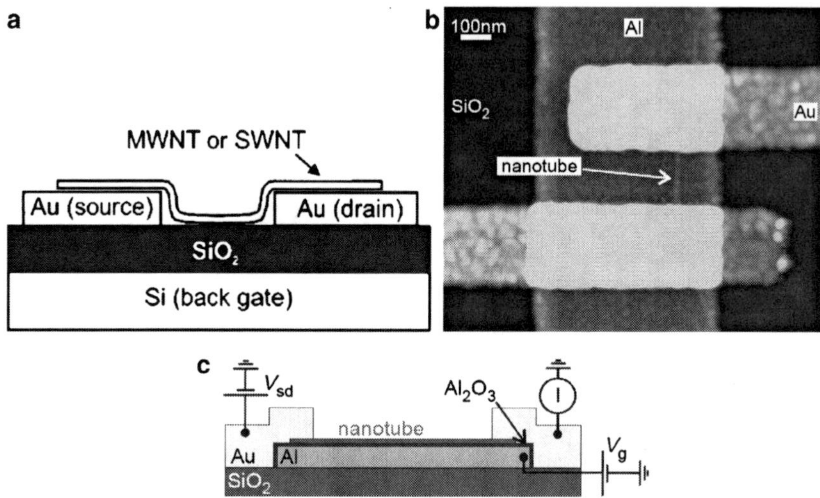

Fig. 4 (**a**) Cross-section of the back-gated CNFET. A single nanotube connects two gold electrodes [23]. (**b**) Atomic force microscope image of a single-nanotube transistor [24]. (**c**) CNFET with individual Al back gate [24]

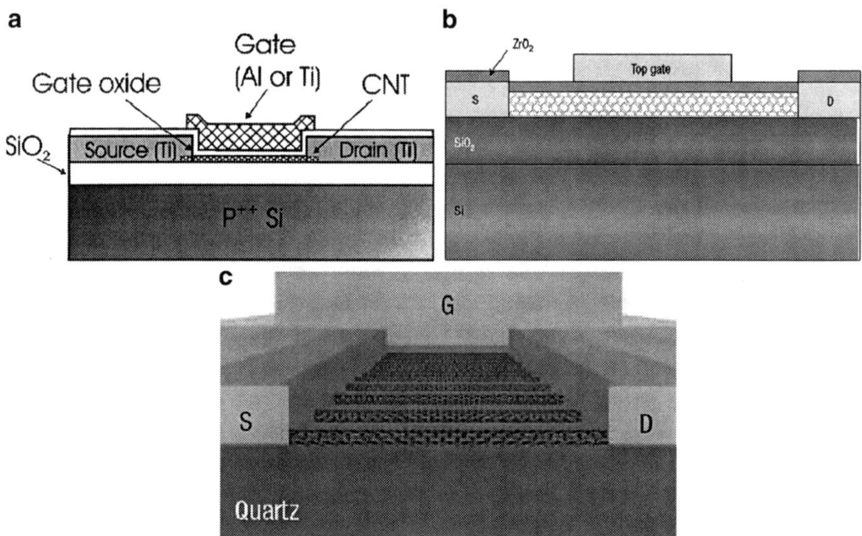

Fig. 5 (**a**) Cross-section of a top-gated CNFET [26]. (**b**) Cross-section of a high-k CNFET with ALD-ZrO₂ as the gate dielectric [27]. (**c**) Cross-section of a CNFET with multiple parallel tubes as channel region [28]

a thin layer of SiO_2 dielectric. The structure of this top-gated design is shown in Fig. 5a. Unlike previously developed back-gated devices, which normally require a high gate voltage to switch the device on, top-gated CNFETs are more compatible with standard CMOS technology in terms of architecture and operating voltage. Excellent electrical characteristics, including steep subthreshold slope, transconductance, and drive current per unit width of CNFETs, have been measured. Results show that CNFETs can outperform state-of-the-art planar silicon devices by several times and have significant potential for future applications. The top-gated structure also encapsulates the CNT channel in dielectrics, which increases reliability. Meanwhile, high-k dielectrics have been proven as an effective technique to boost MOSFET performance and are widely used in modern IC designs. Integration of thin films of ZrO_2 high-k dielectrics into CNFETs has been reported in [27] (Fig. 5b). By using high-k gate insulator, the transconductance and carrier mobility of CNFET was reported to reach 6,000 S m^{-1} and 3,000 cm^2 V^{-1} s^{-1}, respectively. High voltage gains of up to 60 were obtained for complementary CNFET inverters.

Another innovation in the development of CNFETs is the use of densely packed, perfectly aligned horizontal arrays of nonoverlapping SWCNTs as an effective channel, as shown in Fig. 5c [28]. This structure has several advantages. First of all, multiple parallel transport paths in these arrays provide a large drive current. Second, multiple CNTs statistically average out the performance differences among CNTs, which leads to smaller device-to-device variations and increased reliability. However, this implementation relies on the fabrication of perfectly aligned defect-free CNT arrays, which is difficult under current technology. Challenges involve the precise control of CNT size and location during fabrication, as well as the removal of metallic CNTs in the channel to enhance the device ON/OFF ratios. Recently, a newly proposed technique, VLSI-compatible metallic-CNT removal (VMR), combines layout design with CNFET processing and is able to efficiently remove metallic-CNTs in the circuit at the chip scale. As a result, VMR enables the first experimental demonstration of complex cascaded CNFET logic circuits and represents a promising fabrication technique for large-scale CNT-based circuits [29].

4.3 CNT Logic

As CNFETs have demonstrated promise as future electronic devices, the research community is making a significant effort to integrate simple CNFET devices into complex logic circuits. The first CNFET logic gates were demonstrated in [24]. A range of digital logic operations were demonstrated, including an inverter, a logic NOR, a static random-access memory cell, and an ac ring oscillator. This first work, however, was implemented using resistor–transistor logic, in which the CNFETs were connected to large off-chip resistors. The ring oscillator was implemented by connecting three inverters in series (Fig. 6a) and achieved a frequency of 5 Hz. This low frequency was determined by the gigaohm resistance and a 100 pF parasitic capacitance from the wires connecting to the off-chip bias resistors.

                                          C. Dong et al.

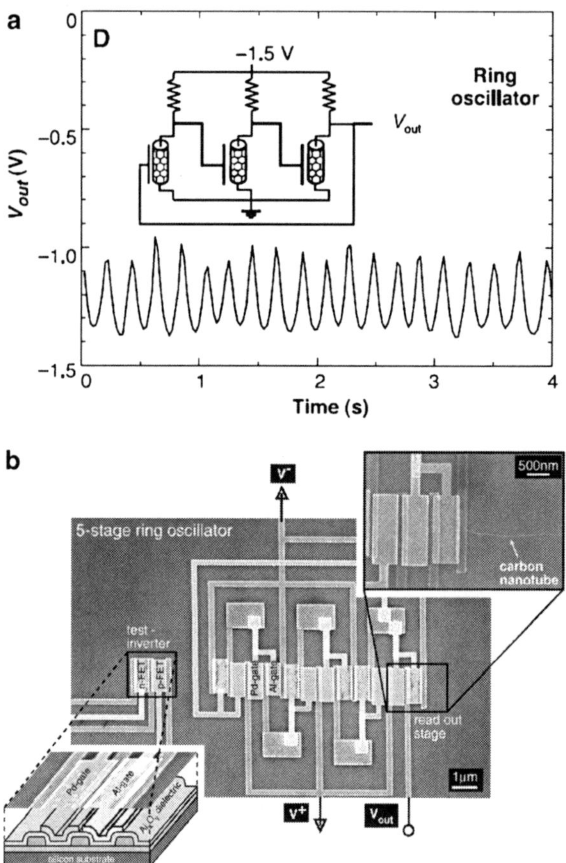

Fig. 6 (a) Three-stage ring oscillator consisting of p-type CNFETs and resistors [24]. (b) Scanning electron microscope image of an SWCNT ring oscillator [30]

The performance of the ring oscillator was enhanced in a following work [31]. SWCNT arrays were synthesized by chemical vapor deposition (CVD) on substrates prepatterned with a catalyst, and local gating was obtained using tungsten metal back gates. More importantly, local doping was applied to convert p-type CNFETs into n-type, allowing complementary CNFET logic to be created for the first time. The resulting three-stage ring oscillator had a measured frequency of 220 Hz, much higher than previously achieved with resistor–transistor logic. This speed can be further optimized by reducing interconnect parasitic capacitance and contact resistance.

Recently, a multistage top-gated complementary CNFET ring oscillator has been built on a single 18-mm-long SWCNT (Fig. 6b) [30]. This ring oscillator consists of 12 individual CNFETs, 6 p-type FETs (purple) with Pd metal gates, and 6 n-type FETs (blue) with Al gates. Five inverter stages were used for oscillation and another inverter was used for reading. A frequency response up to 52 MHz was measured. This measured frequency was still limited by the parasitics rather than by intrinsic nanotube speed.

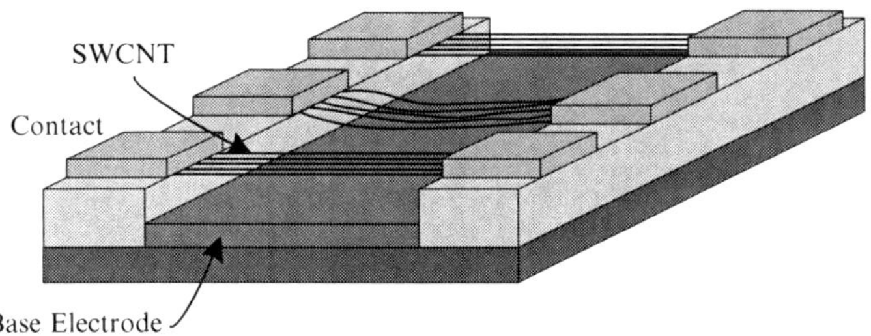

Fig. 7 Architecture of an NRAM memory cell

4.4 Nram

NRAM is a nonvolatile NEMS memory device formed by the suspension of metallic CNTs over a trench which contains a base electrode (Fig. 7). Bistable ON/OFF states at the crosspoints are related to the two minimum energy points observed on the total energy curve, which is given as [10]:

$$E_T = E_{\mathrm{vdw}} + E_{\mathrm{elas}} + E_{\mathrm{elec}},$$

where E_T is the total energy of the crossbar elements, E_{vdw} is the van der Waals energy (vdW), E_{elas} is the elastic energy, and E_{elec} is the electrostatic energy. When the nanotubes are freely suspended (a finite separation from the bottom electrode), the elastic energy is minimized, producing the first minimum total energy location. This represents the OFF state, when the junction resistance between separated nanotubes and electrode is very high. When the suspended nanotubes are deflected into contact with the lower base electrode, the attractive van der Waals force is maximized and a second minimum total energy location is created. This second location represents the ON state, where the junction resistance will be orders of magnitude lower. Since these interactions are purely molecular, no power is consumed when the memory is at rest. Programming is accomplished by applying either attractive or repulsive voltages at the CNT and base electrode. This creates an electro-mechanically switchable, bi-stable memory device with well-defined off and on states [10, 11].

4.5 CNT-Bundle Interconnect

As integrated circuit dimensions scale down, the resistivity of Cu interconnect increases due to electron surface scattering and grain-boundary scattering, leading to a communication bottleneck. Metallic CNTs are a promising replacement because they offer superior conductivity and current-carrying capabilities [32–34]. Since individual SWCNTs can have a large contact resistance, a rope or bundle of SWCNTs is used to transfer current in parallel.

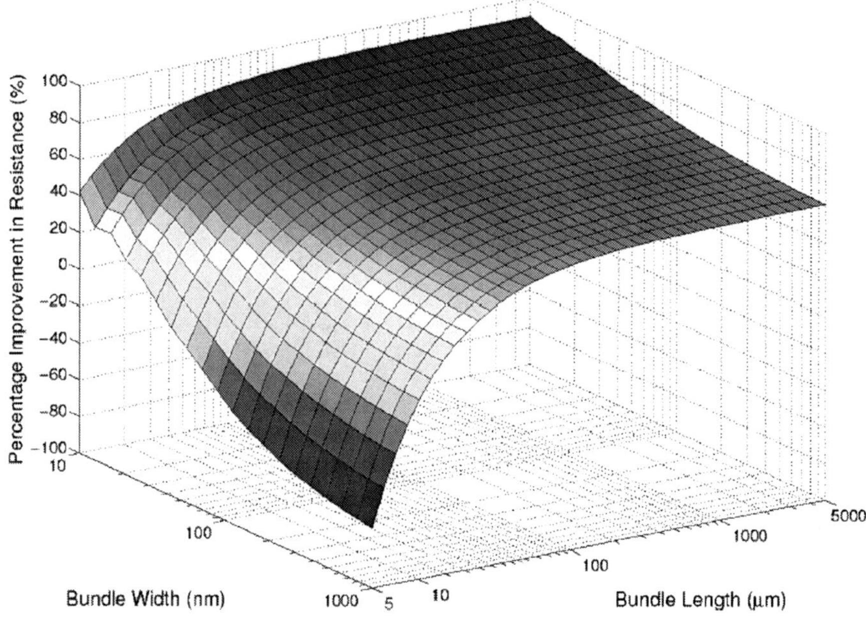

Fig. 8 CNT bundle interconnect resistance [32]

The performance improvement of the SWCNT-bundle interconnect over Cu interconnect is shown in Fig. 8, assuming the SWCNT bundle consists of densely packed SWCNTs with diameters of 1 nm. It has been concluded in [32] that:

- The best application for a SWCNT bundle is a long interconnect with small dimensions. This is because for a long SWCNT bundle, ohmic resistance is dominant and the contact resistance is insignificant. In the meantime, Cu suffers from increasing resistivity as it scales down. For a width of approximately 22 nm, the improvement in resistance is 82%.
- For long bundles with large widths, the contact resistance of the bundle is still insignificant, but Cu has resistivity close to its bulk value. The overall improvement of the SWCNT-bundle interconnect is, therefore, decreased to 61% relative to copper.
- For short bundle lengths, although a SWCNT bundle has large contact resistance, it can still outperform Cu because Cu has exponentially increasing resistivity due to scattering at narrow widths.
- SWCNT bundles are at a disadvantage for short interconnect lengths and large widths. The contact resistance is dominant compared to the ohmic resistance, and the resistivity of the Cu interconnect is low.

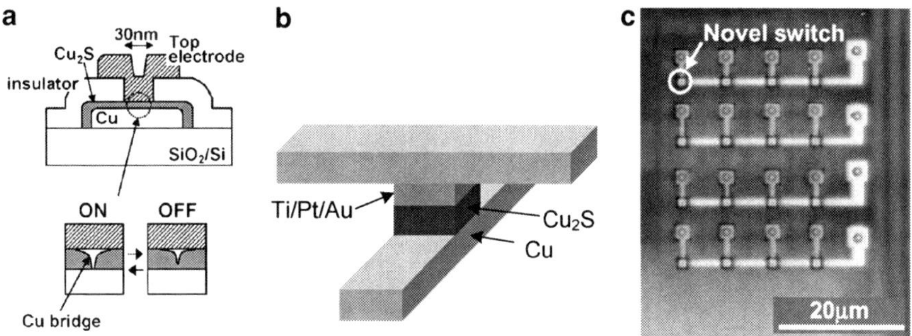

Fig. 9 (**a**) Solid-electrolyte switch. (**b**) Implementation in metal interconnect. (**c**) SEM image of a 4 × 4 crossbar switch array [35]

4.6 Solid-Electrolyte Nanoswitches

Solid-electrolyte switches are a nanoscale switch developed by [35]. A solid-electrolyte switch is created by sandwiching a layer of Cu_2S between two metals, a top electrode (Ti, Pt, or Au) and bottom layer of Cu (Fig. 9a).

When a negative voltage is applied at the top electrode, Cu ions in the Cu_2S are electrochemically neutralized by the electrons coming from that electrode and a conductive bridge between the two electrodes is created, turning the switch on. An on-state resistance of as low as 50 Ω can be achieved by continually applying negative voltage to make the nanobridge thicker. Similarly, the bridge can be ionized and dissolved by applying a positive voltage to the top electrode, turning the switch off.

Because this design does not depend on a substrate, the switches can be manufactured between the higher layers of metal interconnect that are used for routing, as shown in Fig. 9b. This figure shows how a Cu interconnect line can serve as the bottom layer in the electrolyte switch. In addition to individual devices, crossbars can be made from switch arrays. An SEM image of a prototype 4 × 4 crossbar is shown in Fig. 9c [35].

5 FPCNA Architecture

In this section, we describe the FPCNA architecture in detail. We begin with the introduction of the LUT design, which is based on CNT devices. Then we describe a basic logic element (BLE) that can be created using this LUT design. Finally, we discuss FPCNA's high-level architecture, including the design of local and global routing.

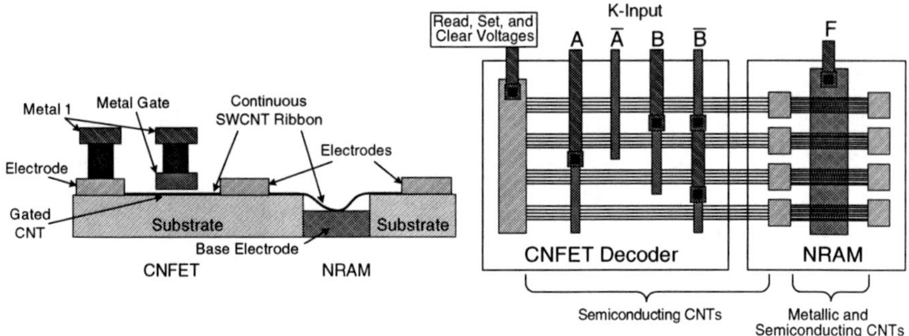

Fig. 10 CNT-based LUT

5.1 CNT-Based LUT

A K-input LUT (K-LUT) is the basic unit of programmable logic in modern FPGAs.
For FPCNA, a novel K-LUT design is used that is based entirely on CNT devices.
Profile and overhead views of this device are shown in Fig. 10. This design uses
parallel ribbons of SWCNTs held in place by metal electrodes and crossed by metal
gates. PMOS CNFET devices are formed at the crossing points of the CNT ribbons
and the metal gates, creating a CNFET decoder. At points where the CNT ribbons
pass over a trench in the substrate, NRAM devices are formed. This CNT memory
is used to store the truth table of the BLE's logic function. By applying K inputs to
the decoder, a reading voltage is sent to the corresponding memory bit whose output
can then be read from the base electrode.

One of the key innovations of this LUT design is that it builds the decoder and
memory on the same continuous CNT ribbons. This structure allows for high logic
density and simplifies the manufacturing process. For comparison, the work in [14]
uses an LUT memory based on individually crossed nanotubes that is addressed by
a CMOS multiplexor tree. In addition to being more costly in area, this design
suffers from fabrication issues because it requires the alignment and interfacing of
individual nanotubes in two dimensions.

Because of the use of CNT ribbons, each device contains multiple tubes. This
adds fault tolerance that addresses the high defect rates of nanotube fabrication and
increases the chance that a CNFET or NRAM device will contain functioning
nanotubes. Thus, the design is more reliable than the one in [14] where a device
fails if either of the two nanotubes is defective.

5.2 BLE Design

In Fig. 10, a 2-to-4 (2 input, 4 NRAM cell) LUT is shown for illustration purposes.
In modern FPGAs, each BLE typically contains a 4-to-16 LUT, as well as a flip-flop
and multiplexor (MUX) to allow registered output. When scaled to K inputs, the

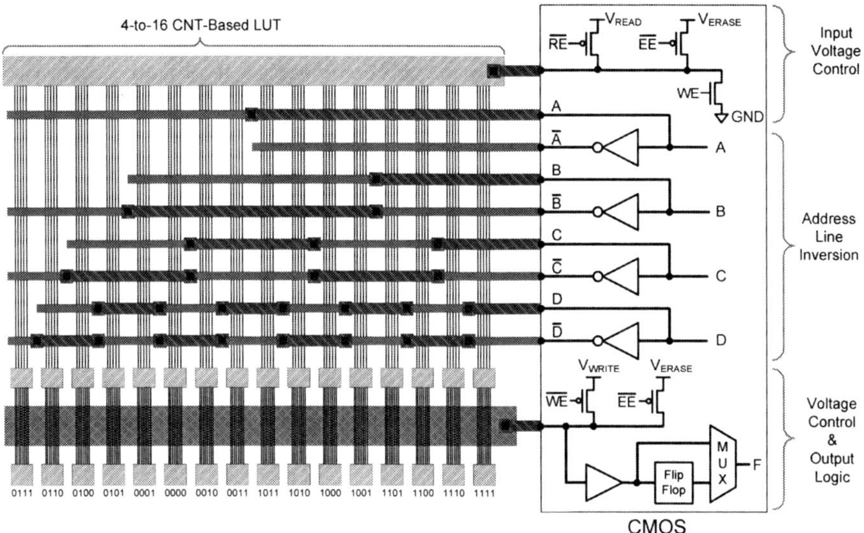

Fig. 11 FPCNA BLE with a 4-to-16 CNT-based LUT

Table 1 NRAM operating modes

Mode	RE	WE	EE	Ribbon voltage	Base electrode voltage
Reading	High	Low	Low	V_{READ} (1 V)	Output
Writing	Low	High	Low	Ground (0 V)	V_{WRITE} (1.6 V)
Erasing	Low	Low	High	V_{ERASE} (+2.5 V)	V_{ERASE} (+2.5 V)

LUT contains 2^K CNT ribbons. The BLE design used for FPCNA is shown in Fig. 11. In this figure, we expand the LUT to four inputs and add supporting CMOS logic for voltage control, address line inversion, and registered output.

In the decoder, Gray address decoding is used to minimize the number of gate-to-metal-1 transitions. Compared to binary decoding, this reduces the number of vias by 46% (from 48 to 26). Since the LUT depends on both normal and complemented inputs, inverters are added to each of the address line inputs. A buffer is used to restore the output signal before it passes to the flip-flop and MUX.

To read and program the NRAM, three different voltage configurations are needed. CMOS pass transistors are used to configure the three modes. Table 1 shows the pass transistor enable signals and voltages during each mode. Most often, the circuit is in the reading mode with the RE (read enable) signal set. This allows V_{READ} to pass through the decoder and select the appropriate NRAM bit. If the NRAM bit is set, the signal passes through the relatively low resistance of the nanotubes contacting the base electrode (logical 1). If the NRAM bit is not set, a multiple $G\Omega$ resistance prevents transmission (logical 0).

To program a value, RE is deactivated, and either WE (write enable) or EE (erase enable) is activated. When WE is set, the selected CNT ribbon is grounded, and V_{WRITE} is applied to the base electrode. The difference in potential creates an attractive force, which pulls the ribbon down into the trench. In [11], Nantero reported a threshold voltage measurement of 1.4 ± 0.2 V. Hence, we assume a V_{WRITE} of 1.6 V. When erasing, the same voltage (V_{ERASE}) is applied to both the ribbon and the base electrode. The like voltages repel each other, releasing the ribbon from the trench floor and allowing it to return to an unbent state. V_{ERASE} must be somewhat larger than V_{WRITE} [10, 11]; hence, we assume a value of $+2.5$ V. There is a risk that this voltage, when applied to the base electrode, could attract an unselected ribbon, causing an unintentional write. This can be avoided by erasing all of the NRAM bits. As each bit is erased, its ribbon is temporarily charged to $+2.5$ V, repelling it from the electrode during the erasure of the remaining bits. Then the individual bits that need to be set as logic 1 can be written to realize the new configuration. This erasing scheme shares some similarity with that used in flash-cell-based FPGAs.

Example 1 *Considering the 4-to-16 bit architecture of Fig. 11, describe how to perform a write for location 1101 and an erase for location 0101. Assume the LUT starts in the reading mode.*

Answer: For the write:

- Select memory device 1101 by setting A, B, and D high and C low.
- Perform the write by setting RE low and then WE high.
- Return to reading mode by setting WE low and then RE high.

For the erase:

- Read each memory location by cycling through A, B, C, and D, and remember which are high, ignoring location 0101.
- Enter erase mode by setting RE low and then EE high.
- Erase all memory addresses by cycling through A, B, C, and D.
- Return to reading mode by setting EE low and then RE high.
- Rewrite each location that was high using the procedure described above.

5.3 *Logic Block Design*

For FPCNA, a cluster-based configurable logic block (CLB) design is used. Each CLB contains N of the BLEs described in the previous section (where N is the cluster size), as well as the local routing used to connect the BLEs together. In conventional CMOS FPGA designs, the routing is often MUX-based. An example of this is shown in Fig. 12, a CLB schematic from [1]. While the same approach could be adopted, using CMOS for the MUXes and NRAM to store MUX configuration bits, a greater logic density can be achieved by using solid-electrolyte switch crossbars.

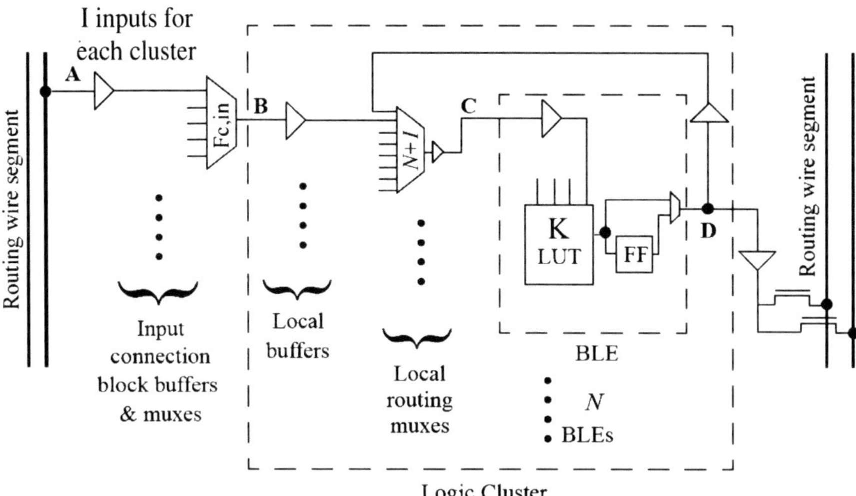

Fig. 12 Schematic of a CLB (logic cluster)

Figure 13 shows a simplified CLB design to illustrate this technique. The CLB in this figure contains four BLEs made from CNT-based LUTs. The local routing is created with solid-electrolyte switches created at the crosspoints of the vertical and horizontal routing wires. By programming the nanoswitch points, a BLE output can be routed to any BLE input. In Fig. 13, one of the input signals to BLE 1 is identified with a dashed line labeled "Input to BLE." The black dots at crosspoints indicate that solid-electrolyte switches at those locations are turned on. By using more switches, the same signal can be routed to multiple BLE inputs. Output from a BLE can connect to the inputs of other BLEs or be output from the CLB. Note that Fig. 13 shows the local routing positioned between BLEs for clarity. In an actual implementation, the local routing and routing switches can be made above the BLEs, and the area calculations reflect this.

Example 2 *A CLB contains 10 BLEs, and each BLE contains a 4-input LUT. Compute the number of local routing tracks and number of programmable switches in the CLB. Assume the CLB is fully connected, i.e. each BLE output can be programmed to connect to any other BLE input in the same CLB. The number of CLB input pins can be computed as $pin\_num = k(N + 1)/2$, where k is the LUT size and N is the cluster size.*

Answer: Based on the equation, a CLB with $N = 10$ and $k = 4$ has $p = k(N + 1)/2 = 4 \times (10 + 1)/2 = 22$ input pins.
Total CLB pins $=$ input pins $+$ output pins $= 22 + 10 = 32$.
Each LUT contains four inputs and one output. Hence, 10 LUTs have a total of 50 pins.

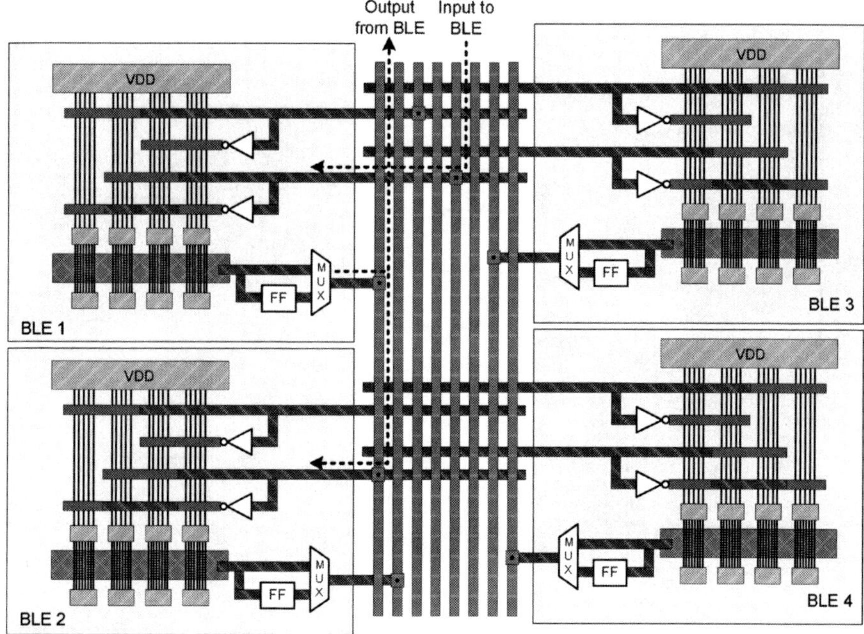

Fig. 13 CNT-based CLB with nanoswitch local routing

Therefore, the local routing crossbar has dimensions of 32 × 50, and there are 1,600 programmable switches.

5.4 High-Level Architecture and Global Routing

A conventional island-based FPGA architecture is adopted for the high-level organization of FPCNA. The basic structural unit is a tile, consisting of one programmable switch block (SB), two connection blocks (CB), and one CLB. This tile is replicated to create the FPGA fabric, as shown in Fig. 14.

The global routing structure consists of two-dimensional segmented interconnects connected through programmable SBs and CBs. The CLBs are given access to these channels through connections in the CBs. The parameter I represents the number of inputs to a CLB, and F_c defines the number of routing tracks a CLB input can connect to. CNT-bundle interconnects are used for global routing because they have been shown to be superior to Cu in terms of current density and delay [32].

In a traditional CMOS-based FPGA, the SBs and CBs take up the majority of the overall area [36]. For example, if the CLB size is 10 and the BLE size is 4 (popular parameters for commercial FPGA products), global routing takes 57.4% of the area,

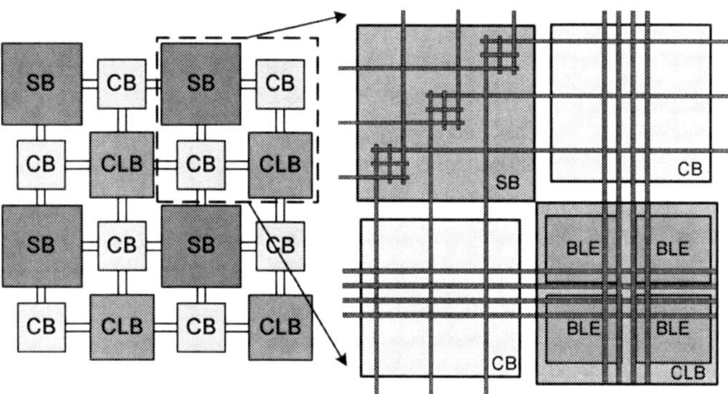

Fig. 14 High-level layout of FPCNA

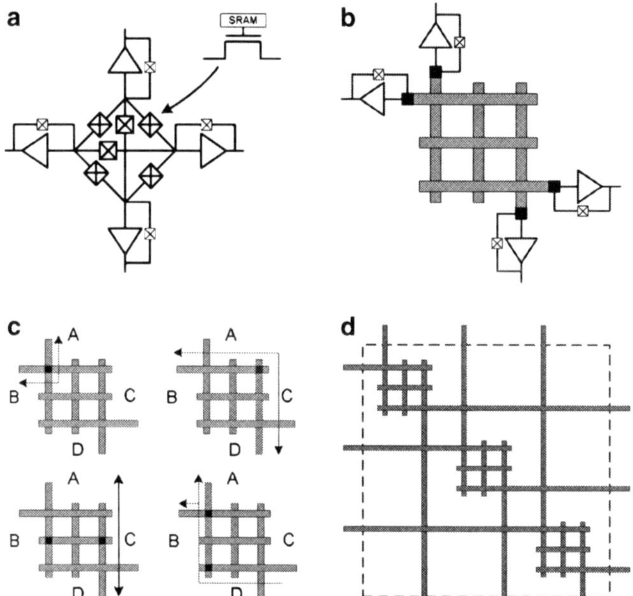

Fig. 15 (**a**) CMOS switch point. (**b**) Nanoswitch-based switch point with CMOS driving buffers. (**c**) Example switching scenarios. (**d**) 3 × 3 switch block (driving buffers not shown)

with the CLBs occupying the remaining 42.6% [36]. To reduce the size of global routing in FPCNA, the traditional CB is replaced with a solid-electrolyte switch crossbar, and a new nanoswitch-based SB design is used.

The new SB design is shown in Fig. 15. Instead of using six SRAM-controlled pass transistors for each switch point, as in conventional CMOS designs (Fig. 15a [1]), six perpendicular wire segments are used with solid-electrolyte nanoswitches at the crosspoints. In this design, the driving buffers and input control pass transistors are implemented in CMOS, as shown in Fig. 15b. By programming nanoswitches at the crosspoints of the wire segment array, a signal coming from one side of the block can be routed to any or all of the other three sides. To demonstrate how routing connections can be made, four switching scenarios are illustrated in Fig. 15c. In the figure, arrows represent signal directions and black dots indicate the activated switches. The upper-left scenario shows how signals A and B are connected using a single switch. A multipath connection is demonstrated in the lower-right scenario, where a signal from C is driving both A and B. By turning on the appropriate nanoswitches, any connection of signals can be made. Using these switch points, larger SBs can be constructed. For example, the 3×3 universal-style SB in Fig. 15d is made from three nanoswitch-based switch points. This design can be scaled to any routing channel width and significantly reduces the SB area. In a conventional CMOS switch point (Fig. 15a, center), six $10\times$ pass transistors are controlled by six SRAM cells, which normally requires an area of 88.2T (where T is the area of a minimum-size transistor). When using nanoswitch-based switch points, the same routing function can be achieved in approximately 9T area.

6 Nanotube LUT Fabrication

Recent progress in the fabrication of CNTs has enabled the use of CNT-based structures in FPCNA's LUT design. To demonstrate the feasibility of this design at the 32 nm technology node, some of the fabrication issues involved are addressed in this section.

The first step in manufacturing the LUT is to define the NRAM trench in the silicon wafer using a process similar to the one described in [11]. Then the nanotubes are grown on separate quartz wafers using CVD. Since the desired CNT ribbons are all aligned in the same direction, an array-based CNT growth process can be used. In [19], researchers report a technique for fabricating dense, perfectly aligned arrays of CNTs using photolithographically defined catalytic seeds, which achieves an alignment of up to 99.9%. The aligned nanotubes can then be transferred to a silicon wafer using a stamping process like the one developed in [37]. These techniques create nanotubes that are suitable for the transistors and NEMS devices used in the LUT. In addition, it is possible to improve nanotube density on the silicon wafer by performing multiple consecutive transfers. This analysis assumes a multiple transfer process is used that provides a CNT pitch of 4 nm.

After the nanotubes have been transferred to the substrate, parallel ribbons are then made from the continuous nanotube array by using an etching process similar to the one used in [20]. The distance between ribbons is set to 96 nm to allow spacing for contacts, and this resolution is assumed to be achievable in the target process technology. By using etching to define the ribbons, there is an added advantage of making the ribbons misalignment immune. This is because any nanotubes crossing the border of a ribbon will be removed during the etching

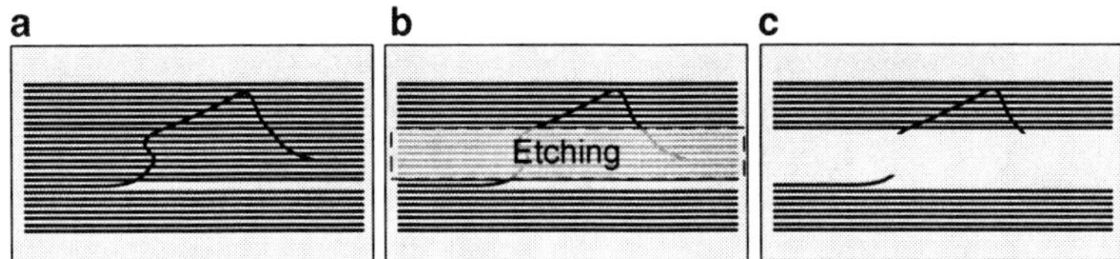

Fig. 16 CNT ribbon etching

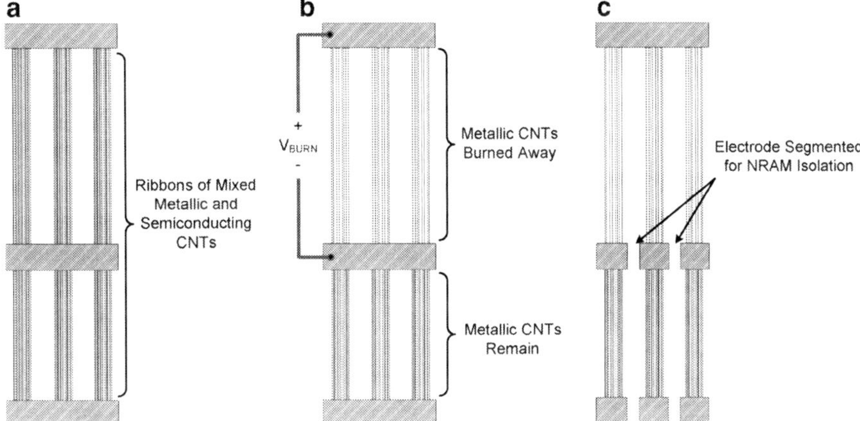

Fig. 17 Metallic CNT removal

process. Figure 16 demonstrates this concept, where (a) shows a misaligned tube, (b) shows the etched area, and (c) shows the resulting CNT ribbons.

The next major step in fabrication is to disable the metallic nanotubes inside the decoder region. Since metallic CNTs act as a short between source and drain, they need to be removed to create CNFET transistors with desirable on/off current ratios. Electrical burning [38] is an effective method to selectively disable the metallic CNTs. In this technique, a large voltage is applied across the array which heats the conducting metallic nanotubes to a breakdown temperature of $\sim$600°C and causes irreversible oxidization. Because this is done when the CNTs are still exposed to air, a minimum power dissipation of 0.05 mW is needed to achieve breakdown [38].

Since metallic nanotubes are used for NRAM operation, the burning must only be done in the decoder region. One way to remove the metallic CNTs from the decoder, but keep them for the NRAM devices, is shown in Fig. 17. In this figure, (a) shows vertical ribbons of mixed metallic and semiconducting CNTs held in place by horizontal metal electrodes. The middle and bottom electrodes are used to hold the ribbons in place during NRAM operation. In (b), a thermal breakdown voltage, V_{BURN}, is applied between the top and middle electrodes. This burns away the metallic tubes in the decoder region but leaves them in the NRAM region.

Because the NRAM memory devices need to be individually addressable, the electrode is segmented to provide electrical isolation (c).

After the CNT ribbons are defined and processed, the gate and source/drain formation is similar to a regular CMOS process. Based on these techniques and the existing CNT fabrication work [19–21], the proposed nanotube-based LUT design is believed to be implementable.

7 Circuit Characterization

7.1 CNFET and CNT-Based LUT Variation

As mentioned in Sect. 4, CNFETs have many properties that make them attractive for use in future electrical circuits. Ideally, the channel region of these CNFETs would consist of identical, well-aligned semiconducting CNTs with the same source/drain doping levels. However, it is difficult to synthesize nanotubes with exactly controlled chirality using known fabrication techniques. HiPco synthesis techniques yield around $50 \pm 10\%$ metallic CNTs [39]. This means the number of semiconducting CNTs per device is stochastic, causing drive current variations even after the metallic CNTs are burned away. Meanwhile, CNFETs are also susceptible to variations in diameter and source/drain region doping [40].

In a traditional MOSFET, Gaussian distributions are often assumed when modeling variation sources such as channel length and gate width. These models are then used in the delay or power characterization of the MOSFET. A similar approach can be used to characterize CNFETs. To quantify the effects of CNFET variations, a Monte Carlo simulation of CNFET devices with 2,000 runs is performed. The sources of variation that are considered are listed in Table 2, with two scenarios for the number of CNTs in a channel: 8 ± 3 and 6 ± 2, both normally distributed. The diameter range, doping level range, and CNFET model are suggested in [40].

The results of the simulation show that the delay distribution of a CNFET device under these variations fits the Gaussian distribution. Figure 18 illustrates this distribution for a CNFET with 8 ± 3 semiconducting nanotubes in its channel.

Using the CNFET model, the performance of the CNT-based LUT design can also be evaluated. The LUT decoder consists of multiple stages of p-type CNFETs, simulated under the variations mentioned in Table 2. The contact resistance between an electrode and a single nanotube is assumed to be 20 KΩ based on

Table 2 Sources of CNFET variation	Parameter	Mean	Variation (3σ)
	CNTs per channel case 1	8	± 3
	CNTs per channel case 2	6	± 2
	CNT diameter	1.5 nm	± 0.3 nm
	Doping level	0.6 eV	± 0.03 eV

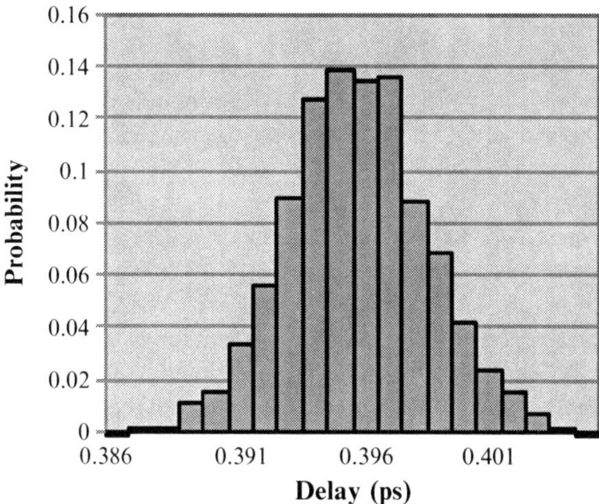

Fig. 18 Delay distribution of a CNFET under process variation

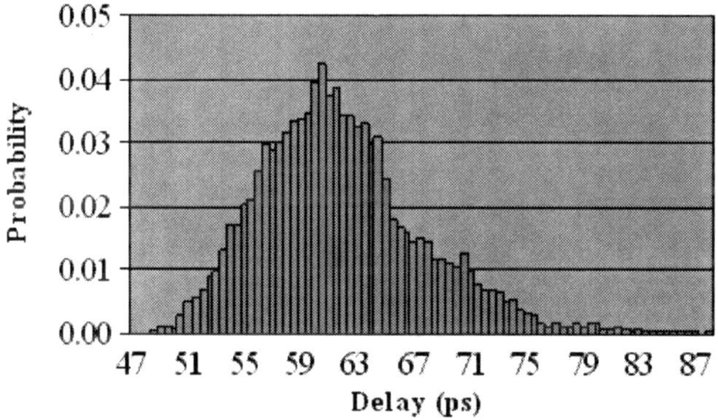

Fig. 19 CNT-based LUT delay considering variation

[32]. In a ribbon, multiple CNTs operate in parallel; hence, the ribbon contact resistance is considered to be inversely proportional to the number of semiconducting nanotubes. For NRAM devices, a contact resistance between a bending nanotube and the base electrode of 20 KΩ is assumed, based on the measurements in [12]. Since these CNTs also operate in parallel, the total ribbon NRAM contact resistance is treated as inversely proportional to the number of metallic nanotubes in the ribbon. The resulting LUT delay distribution generated by Monte Carlo simulation in HSPICE is shown in Fig. 19.

The average delay is 60.94 ps, which is 41% smaller than that of a traditional 32 nm CMOS LUT, which has a delay of 103.8 ps. Unlike the CMOS LUT, the delay of the nanotube-based LUT has a distribution similar to log-normal.

7.2   Crossbar Characterization

As described in Sect. 5, the routing in FPCNA is implemented using crossbars. We capture the delay and variation of these crossbars using HSPICE. The CNT-bundle interconnect is assumed to be 32 nm in width, with an aspect ratio of 2. We set the dielectric constant of the insulation material around the crossbar at 2.5 and derive a unit resistance of 10.742 $\Omega/\mu m$ and capacitance of 359.078 aF/μm for the CNT bundles. The interconnects are evaluated for 10% geometrical variation of wire width, wire thickness, and spacing according to [41]. The CNT-bundle interconnect variation also considers a 40–60% range on percentage of metallic nanotubes inside a bundle. The solid-electrolyte switches between interconnect layers are considered with a 100 Ω ON resistance [35] and 10% variation to capture via contact resistance.

7.3   Timing Block Evaluation

To support the evaluation CAD flow, various circuit models are needed to capture characteristics of the FPCNA architecture. In the architecture specification file of VPR, the delay values for certain combinational circuit paths are specified to enable accurate timing analysis. For example, in Fig. 12, there are paths A→B, B→C, and D→C, etc. In FPCNA, these paths contain buffers, metal wires, and solid-electrolyte switches, making them susceptible to process variations.

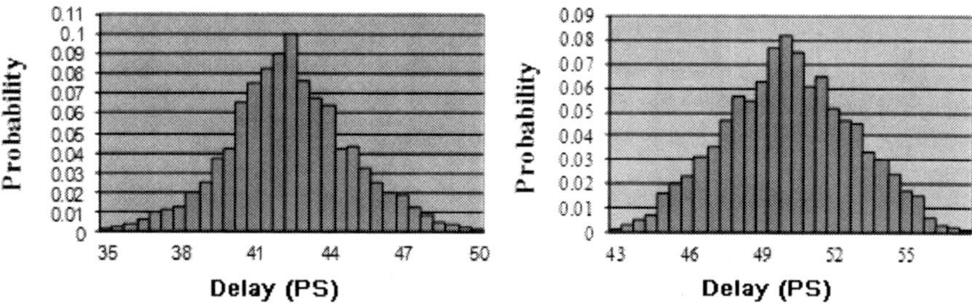

Fig. 20 Delay distribution of wire track to CLB inpin (*left*) and subblock opin to subblock inpin (*right*)

Table 3 Delay comparison between baseline CMOS and FPCNA

Paths	CMOS-Baseline		FPCNA	
	μ (ps)	σ (ps)	μ (ps)	σ (ps)
A→B	141.66	7.13	42.24	2.48
B→C	107.59	5.37	30.45	2.21
D→C	107.59	5.37	49.96	2.92
D→Out	28.48	1.22	29.91	2.28

The delay and variation of these paths in FPCNA are computed by performing a Monte Carlo simulation of 1,000 runs, varying the CNFET parameters and CNT contact resistance for each run. Figure 20 illustrates the resulting delay distributions of wire track to CLB inpin connections (A→B) and subblock opin to subblock inpin connections (D→C).

Based on these results, we observe that the timing blocks follow a normal-like distribution. Therefore, the mean (μ) and variation (σ) of each delay path can be calculated, as shown in Table 3. An equivalent design in CMOS is measured as a baseline for comparison, assuming 12% channel width variation, 8% gate dielectric thickness variation, and 10% doping variation (values from [42] for 32 nm CMOS). These values are also shown in the table.

8   CAD Flow

In the previous sections, we show that both deep submicron CMOS and nanoscale devices are susceptible to variation. In traditional SSTA, it is assumed that all circuit elements have deterministic delay. This approach cannot correctly capture the variability of the fabrication process. The worst-case analysis commonly used by industrial designs satisfies yield but is overly pessimistic. On the other hand, the nominal case produces low yield due to variation-based timing failures. To maximize yield without sacrificing performance, it is necessary for CAD tools to consider the statistical information of circuit elements during timing analysis.

In this work, a timing-driven, variation-aware CAD flow is used, as shown in Fig. 21. Each benchmark circuit goes through technology-independent logic optimization using SIS [43] and is technology-mapped to 4-LUTs using DAOmap [44]. The mapped netlist then feeds into T-VPACK and VPR [1], which perform timing-driven packing (i.e. clustering LUTs into the CLBs), placement, and routing. To take variation into consideration, the VPR tool [1] is enhanced to make it variation-aware.

Existing works have shown that statistical optimization techniques are useful during the physical design stage. Variation-aware placement is implemented in [45] and variation-aware routing is developed in [46]. Based on the ideas presented in these works, a complete variation-aware physical design flow is implemented. In this holistic solution, the placer calls the variation-aware router to generate delay estimates for its timing cost calculations.

From the Monte Carlo simulation results in Sect. 7, we observe that the CNT-based LUT delay follows a non-Gaussian distribution. Reference [32] also reports a non-Gaussian distribution for the CNT-bundle interconnect. However, all of the existing CAD work targeting CMOS assumes normally distributed random variables [45–47]. The Gaussian-based SSTA algorithms that these works use to evaluate CMOS are not suitable for modeling the nonnormal variables of molecule-based architectures. Therefore, a statistical timing analyzer is utilized that can

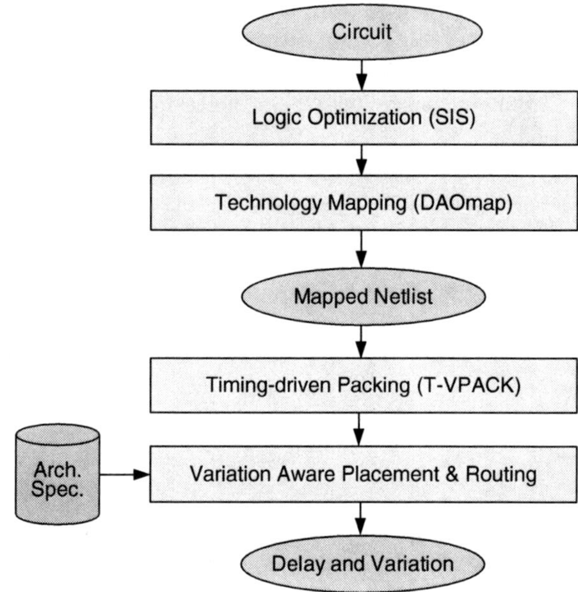

Fig. 21 FPCNA evaluation flow

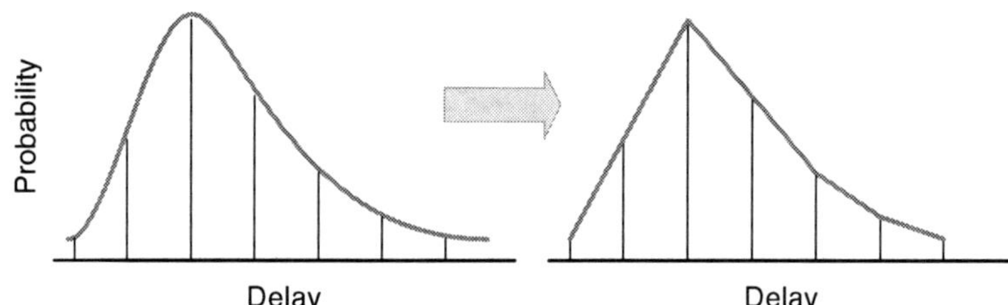

Fig. 22 Discretization process of a log-normal probability density function

handle any arbitrary distribution, based on discretization techniques adapted from [48, 49].

One such technique is the probabilistic event propagation [48], in which discretized random variables of cell delays are used for timing analysis. As illustrated in Fig. 22, a non-Gaussian probability density function can be represented as a set of delay-probability pairs, which contain the time t and the probability a signal will arrive at time t. In [49], ADD and MIN operations are developed for propagating multiple event groups. These operations are used, and a MAX operation is defined for use in the statistical timing analyzer. In Fig. 23, we illustrate how the discretized MAX operation is performed using an example point.

During the MAX operation, all possible timing points at the output are evaluated, computing their probability based on the input sets of delay-probability pairs.

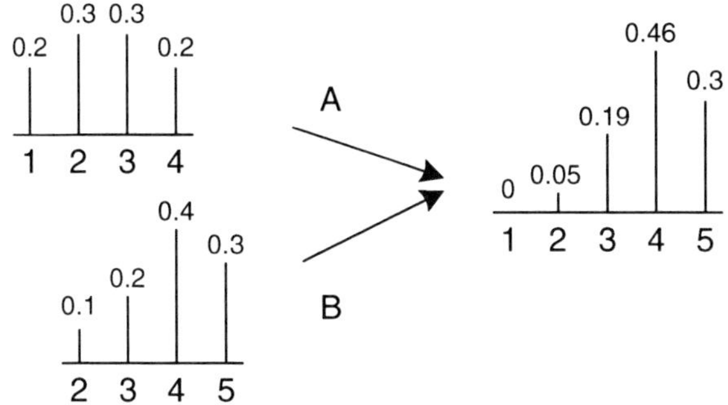

Fig. 23 The discretized MAX operation

For each timing point t, the probability that both inputs arrive is defined as $P(t)$. $P(t)$ can be derived using conditional probability as the sum of:

1. The probability that both A and B arrive at t
2. The probability that A arrives at t and B arrived before t
3. The probability that B arrives at t and A arrived before t

The accuracy of this technique is dependent on the number of points used for piecewise linear approximation. It is shown in [48] that seven points are sufficient to obtain an accuracy of less than 1% error compared to Monte Carlo. Therefore, a seven-point sampling is used throughout the discretized SSTA.

Example 3 *Two inputs A and B with probability distribution are given in Fig. 23. Derive discrete probability density of MAX(A, B).*

Answer: The previous definition of $P(t)$ *can be formulated as:*
$P_t(MAX) = P_t(A_t B_t) + P_t(A_t|B_{t'<t}) + P_t(B_t|A_{t'<t})$, *where* A_t *and* B_t *are the probabilities that input A and input B have arrived at time* t, *respectively.*

$$P_1(MAX) = P_1(A_1 B_1) + P_1(A_1|B_{t'<1}) + P_1(B_1|A_{t'<1})$$
$$= 0.2 \times 0 + 0.2 \times 0 + 0 \times 0 = 0,$$

$$P_2(MAX) = P_2(A_2 B_2) + P_2(A_2|B_{t'<2}) + P_2(B_2|A_{t'<2})$$
$$= 0.3 \times 0.1 + 0.3 \times 0 + 0.1 \times 0.2 = 0.05,$$

$$P_3(MAX) = P_3(A_3 B_3) + P_3(A_3|B_{t'<3}) + P_3(B_3|A_{t'<3})$$
$$= 0.3 \times 0.2 + 0.3 \times 0.1 + 0.2 \times (0.2 + 0.3) = 0.19,$$

$$P_4(MAX) = P_4(A_4 B_4) + P_4(A_4|B_{t'<4}) + P_4(B_4|A_{t'<4})$$
$$= 0.2 \times 0.4 + 0.2 \times (0.1 + 0.2) + 0.4 \times (0.2 + 0.3 + 0.3) = 0.46,$$

```
Set all nets i and sinks j, Crit(i,j) =1.0;
while (overused routing resources exist) do
      for (each net, i) do
            Rip-up routing tree of net i;
            for (each sink j of net i in decreasing Crit(i,j) order) do
                  Find the least cost route of sink j;
                  for (all nodes in the path from i to j) do
                  | Update congestion;
                  end
                  Update delay and standard deviation of the route;
                  (Update discretized delay for FPCNA;)
            end
      end
      Update historic congestion;
      Compute mean and standard deviation of delay for each net(i);
      (Compute discretized delay for each net(i) in FPCNA;)
      Update Crit(i,j);
end
```

Fig. 24 Pseudo-code of the modified VPR router

$$
\begin{aligned}
P_5(MAX) &= P_5(A_5B_5) + P_5(A_5|B_{t'<5}) + P_5(B_5|A_{t'<5}) \\
&= 0 \times 0.3 + 0 \times (0.1 + 0.2 + 0.4) + 0.3 \times (0.2 + 0.3 + 0.3 + 0.2) \\
&= 0.3.
\end{aligned}
$$

Figure 24 shows the pseudo-code of the variation-aware router. The routing is iterative. During the first iteration, the criticality of each pin in every net is set to 1 (highest criticality) to minimize the delay of each pin. For the CMOS architecture, the Gaussian delay mean (μ) and standard deviation (σ) of each path are computed during the routing of each net. For FPCNA, the discretized delay distribution of each path is computed. If congestion exists, more routing iterations are performed until all of the overused routing resources are resolved. At the end of each routing iteration, criticality and congestion information are updated before the next iteration starts.

To consider variation, new formulas to capture the criticality of sink j of net i are derived. For the CMOS architecture with a Gaussian distribution, we express the arrival time of pin j in net i as $\mathrm{arr}(i, j) = (t_a, \sigma_a)$ and the required time as $\mathrm{req}(i, j) = (t_r, \sigma_r)$. The mean and standard deviation of slack $\mathrm{slack}(i, j) = (t_s, \sigma_s)$ can be derived as:

$$
t_s = t_r - t_a \quad \sigma_s = \sqrt{\left| \sigma_a^2 + \sigma_r^2 \right|}.
$$

The criticality of pin j in net i can then be computed by taking both slack and slack variation into consideration:

$$
\mathrm{Crit}(i, j) = 1 - \frac{t_s - 3\sigma_s(i, j)}{t_{\mathrm{crit}} + 3\sigma_{\mathrm{crit}}}.
$$

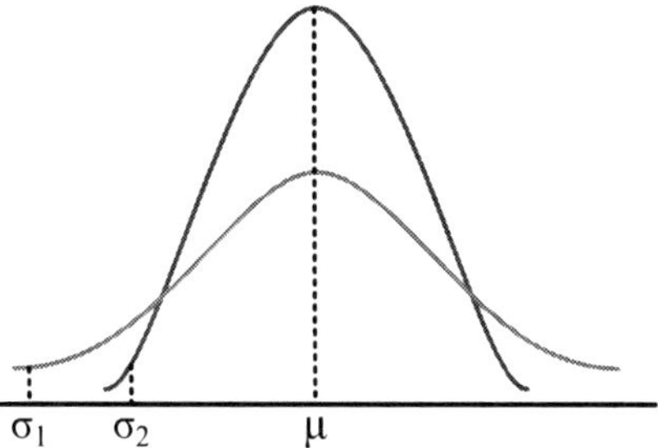

Fig. 25 Criticality estimation

The original VPR cost function is modified this way so that when two slacks have similar means but different variations, the $t_s - 3\sigma_s(i,j)$ term assigns a larger criticality to the path with the greater variation to weigh it more heavily in the next routing iteration. This is illustrated in Fig. 25, where the distribution with slack variation σ_1 will be assigned a higher criticality than the distribution with slack variation σ_2, even though they have the same mean. This cost function also considers the critical path variation with the $t_{crit} + 3\sigma_{crit}$ term.

In discretized routing, the expected values of the slack and critical path discretized points are computed and used in the following criticality function:

$$\text{Disc\_Crit}(i, j) = 1 - \frac{E[\text{disc\_slack}(i, j)]}{E[\text{disc\_}t_{crit}]}$$

Example 4 *Given a critical path delay with $\mu_d = 2.5$ ns and $\sigma_d = 0.25$ ns, compute the criticality of net j with* slack$_j(\mu = 0.6$ ns$, \sigma = 0.05$ ns$)$ *and net k with* slack$_k(\mu = 0.7$ ns$, \sigma = 0.15$ ns$)$. *Which net is statistically more critical? How does this compare to the deterministic solution? When should each type of evaluation be used?*

Based on the definition of criticality:

$$\text{Crit}_j = 1 - \frac{\mu - 3\sigma}{\mu_d + 3\sigma_d} = 1 - \frac{0.6 - 3 \times 0.05}{2.5 + 3 \times 0.25} = 0.86$$

$$\text{Crit}_k = 1 - \frac{\mu - 3\sigma}{\mu_d + 3\sigma_d} = 1 - \frac{0.7 - 3 \times 0.15}{2.5 + 3 \times 0.25} = 0.92$$

Deterministically, net j will have a smaller slack value which makes it more critical to the router than net k. However, when slacks are viewed statistically, net k becomes more critical due to its larger variation. Deterministic evaluation is faster than statistical evaluation and should be used when speed is important or when variation is small. Statistical analysis provides higher accuracy and becomes necessary when analyzing designs with large variation.

After each routing iteration, SSTA is executed by traversing the updated timing graph to calculate the new slack and critical path delay. The variation-aware placer also uses these criticality functions to calculate the timing cost of each move during simulated annealing. In the placement cost function, the criticality value is raised by the exponent β. The optimal value of β is determined to be 6 for this design. This differs from the original VPR method of incrementing β from 1 to 8 and from [45] where a β value of 0.3 was used. As in [45], the variation is calculated during the delta array creation, and these precalculated values are also stored in the delta arrays for use in placement. The main difference is that a variation-aware router is used to generate the delay and store sets of discretized delay-probability points for each delay value in addition to the mean and variation.

9 Experimental Results

9.1 Experimental Setup

Because this CAD flow is flexible, we can experiment with various architectural parameters and determine their impact. To evaluate FPCNA, we use a fixed LUT input size $K = 4$ and explore logic cluster sizes of $N = 4$, 10, and 20. We also explore the difference between using an average of 16 CNTs per ribbon and 12 CNTs per ribbon, to see the effect on area. The number of CLB inputs is set based on the cluster size so that it equals $2N + 2$. F_c is kept at 0.5, a typical value which connects the CLB input to half of the routing tracks in the channel.

It is shown in [1] that a mixture of different-length interconnects can provide improved performance. Two popular wire length mixtures are evaluated: an equal mixture of length-4 and length-8 wire segments (wires crossing either four CLBs or eight CLBs) and a mix of 30% length-1, 40% length-2, and 30% length-4 wire segments.

For each configuration of the above parameters, a binary search is performed to determine the routing channel width needed to successfully route the largest benchmark, and then that width is used to evaluate all of the benchmarks.

9.2 Area Reduction

Due to the high density of CNT-based logic and solid-electrolyte switch-based routing, the footprint of FPCNA is significantly smaller than the equivalent CMOS FPGA. To calculate the area, the architectural parameters defined above are used, and a transistor feature size of 32 nm (2λ) is assumed for both CNT and CMOS-based transistors. The area of the CNT-based LUT (Fig. 11) is determined by the size and spacing of the CNT-ribbons and addressing lines. Since an average CNT pitch of 4 nm is assumed, the CNT ribbons are 64 nm wide for the 16-tube per ribbon experiments and 48 nm wide for the 12-tube per ribbon experiments. To accommodate gate layer to metal-1 layer vias, the nanotube ribbons are spaced 96 nm (6λ) apart. LUT addressing gate metal is 32 nm (2λ), with a spacing of 80 nm (5λ) between adjacent lines. Gate to metal-1 via size is assumed to be 64 nm (4λ) square. The nanotube memory, NRAM, offers a much smaller area than an SRAM cell. The NRAM trench is assumed to be 180 nm in width and 18 nm in height. These dimensions are conservative estimates based on fabrication results in [37]. The trench-to-electrode spacing is set to 90 nm for each side. All of the LUT electrodes are assumed to be 64 nm wide. The area of the 32 nm CMOS BLE is calculated using a technique from [1] by counting minimum width transistor area. In the FPCNA design, each BLE also contains CMOS components, including four size-2 buffers, one MUX, and one flip-flop, as shown in Fig. 13. The total BLE logic area is the sum of both the CMOS logic area and the CNT-based LUT area. Figure 26 demonstrates design details for both CMOS and FPCNA LUT cells. For simplicity, only a 2-input LUT is shown.

Local CLB interconnect crossbars are assumed to have a line thickness of 64 nm (4λ) and spacing of 64 nm (4λ). The routing crossbars are created on the metal layers above the CLB logic, so they do not add to the overall CLB area (assuming the crossbar area is smaller than the logic area, which was true in all of the experiments). Since the routing path is controlled by nonvolatile solid-electrolyte switches, the SRAM cells used in the baseline CMOS FPGA can be eliminated in FPCNA. By replacing the MUX-based routing with crossbars and switching to CNT-based LUTs, a large overall area reduction is seen. For an architecture with a cluster size of 10 and wire segmentation of length 4 and 8, we estimate the footprint of a baseline CMOS FPGA tile to be 34,623T. Using a minimum-width transistor area of $T = 0.0451$ μm^2 for a 32-nm transistor gives us a tile area of 1,561.5 μm^2. When we calculate the area for an equivalent FPCNA tile under the area assumptions above, only 307.99 μm^2 is used. These calculations show that FPCNA can achieve an area reduction of roughly $5\times$ over CMOS.

The area breakdowns for a single LUT and an architecture tile are shown in Table 4. In this table, the CB area is the sum of both CB. Table 5 shows the area breakdowns of various FPCNA architectures. The first row describes global routing

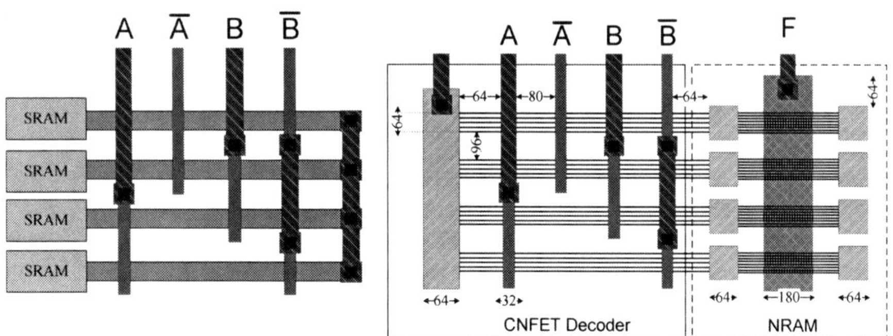

Fig. 26 Design details for the LUT cells

Table 4 Area reduction of FPCNA

	CMOS FPGA (μm^2)	FPCNA (μm^2)	Reduction
Single LUT area	10.88	2.15	5.06×
LUT addressing area	5.68	1.52	3.73×
LUT memory area	5.20	0.63	8.24×
Tile area	1,561.5	307.99	5.07×
CLB area	665.2	63.290	10.51×
CB area	337.7	82.5	4.1×
SB area	558.6	162.2	3.4×

with 30% length-1, 40% length-2, and 30% length-4 wire segments. The second row is for 50% length-4 and 50% length-8 interconnects. As seen in the table, routing occupies the majority of FPCNA's overall area. Due to the size of the SB area, wire segmentation has a significant impact on the overall area. Shorter wire segments have better flexibility during routing, but require a larger number of switch points in each SB, which greatly increases the tile size.

9.3 Performance Gain

In this section, we evaluate the experimental CAD flow presented in Sect. 8, quantifying the overall performance improvement of FPCNA from the baseline CMOS counterpart. When considering variation, performance evaluation becomes complicated. The critical path delay can no longer serve as the absolute measure of performance. Due to variations, near-critical paths may actually be statistically critical. This is illustrated by PO3 in Fig. 27. In addition, setting a clock period based only on the most statistically critical path is not appropriate. Consider the case in Fig. 27, where the target clock period is set to a 95% guard-band of PO3. This means that for 95% of chips made, PO3 will not generate a timing failure. However, at this clock period, the other POs may also fail due to variation, making the overall yield less than 95%. Because of this phenomenon, it is necessary to consider the statistical delay of every path in yield analysis. We express the performance yield as a delay-probability pair (t, p), so that by setting the clock period t, we can evaluate the system yield p. This allows us to compare the performance of the statistical information generated by our experiments.

We calculate performance yield for both Gaussian and non-Gaussian distributions using the flow in Fig. 28. After selecting a target clock period T_c, the yield of each of the POs is computed. For a Gaussian distribution, the yield is calculated by computing the inverse cumulative distribution function (CDF) of the delay random variable. In a non-Gaussian delay distribution, the delay is represented by a group of points, so the yield is computed by converting the piecewise linear PDF into a piecewise linear CDF (Fig. 29). The overall system yield is determined by multiplying all of the path yields. If the system yield is not satisfied, we increase

Table 5 Area of various FPCNA architectures

		Cluster 4		Cluster 10		Cluster 20	
		16 Tubes	12 Tubes	16 Tubes	12 Tubes	16 Tubes	12 Tubes
1–2–4 Wire segments	CLB area (μm^2)	25.316	23.784	63.29	59.46	126.58	118.921
	CB area (μm^2)	23.46	23.46	75.01	75.01	203.438	203.438
	SB area (μm^2)	205.06	205.06	444.49	444.49	829.437	829.437
	Total tile area (μm^2)	253.84	252.31	582.79	578.96	1,159.46	1,151.8
	Tile edge length (μm)	15.932	15.884	24.141	24.062	34.051	33.938
4–8 Wire segments	CLB area (μm^2)	25.316	23.784	63.29	59.46	126.58	118.921
	CB area (μm^2)	27.07	27.07	82.5	82.5	435.94	435.94
	SB area (μm^2)	83.94	83.94	162.2	162.2	1,087.86	1,087.86
	Total tile area (μm^2)	136.33	134.8	307.99	304.16	1,650.38	1,642.72
	Tile edge length (μm)	11.676	11.61	17.55	17.44	40.625	40.531

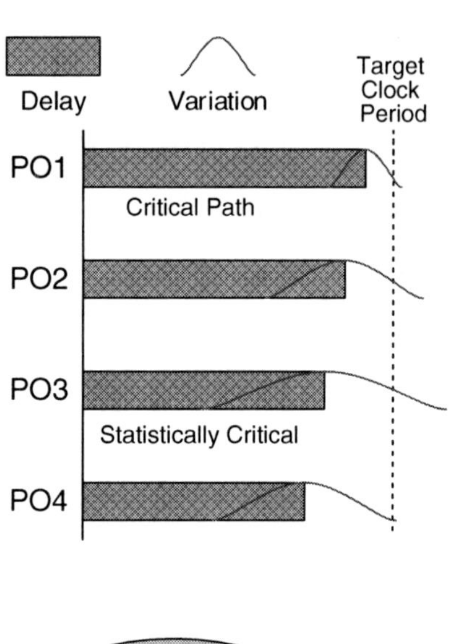

Fig. 27 The effect of variation on critical path and yield

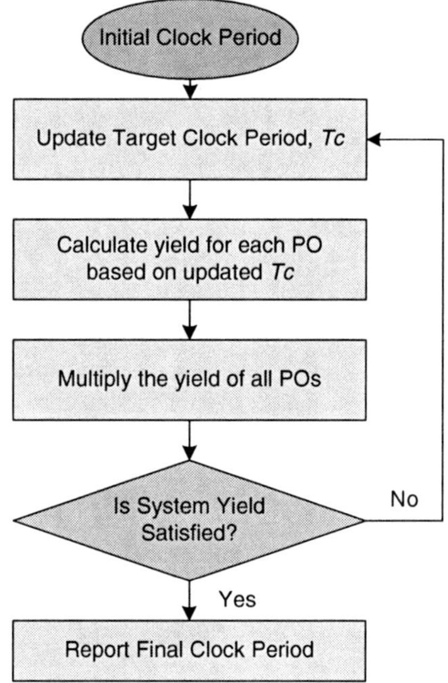

Fig. 28 Performance yield estimation

the T_c and repeat the process until the desired yield is obtained. We then report the final clock period, which guarantees the targeted yield.

Using the variation-aware CAD flow, we evaluate the achievable clock period of 20 MCNC benchmarks and report the results in Table 6. For a rough comparison to a deterministic solution, we evaluate the CMOS design using VPR [1], with a

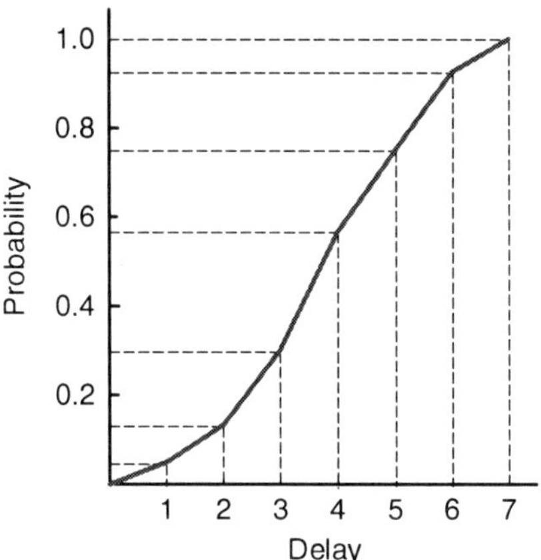

Fig. 29 Piecewise linear CDF in discretized timing analysis

worst case delay guard-band of 3σ added to each component in VPR's architecture file. This equates to a component yield of roughly 99%.

The table shows the deterministic CMOS results, variation-aware CMOS results, and two versions of variation-aware FPCNA results; one with 16 nanotubes per CNT ribbon and the other with 12 nanotubes per ribbon. For each variation-aware flow, we calculate the clock period for performance yield at both 95 and 99%. We also generate the performance gain of the FPCNA architectures over the baseline CMOS. In this table, both CMOS and FPCNA are configured with a cluster size of 10 and interconnect wire segmentation of 50% length-4 and 50% length-8. Average delays are calculated using the geometric mean. At a 95% performance yield, the FPCNA designs have an average gain of 2.75× and 2.65× over the CMOS counterpart, for 16 and 12 ribbons, respectively. This significant improvement in performance is achieved by the synergistic combination of CNT logic, CNT-bundle interconnects, and routing crossbar design in FPCNA.

As shown in Table 5, reducing the number of tubes inside the nanotube ribbon can reduce tile footprint, which will reduce the length of global interconnect and should therefore enhance performance. However, as we see in Table 6, the overall performance is actually degraded. This is because with fewer tubes, each CNFETs has less driving capability, which increases the LUT delay enough to overcome any global interconnect savings. To develop a better understanding of how the FPCNA architecture affects performance, we evaluated different architectural combinations of wire segmentation, cluster size, and nanotube ribbon size for the 20 benchmarks. The average results, again using the geometric mean, are plotted in Fig. 30.

As seen in Fig. 30a, for small and medium cluster sizes (4 and 10), long interconnects are preferable, because they can make connections to CLBs which are far away. For the larger cluster size of 20, shorter wire segments are preferred (Fig. 30b). Note that in Fig. 30a, the performance degrades rapidly at cluster size 20 because there are an increased number of connections between neighboring CLBs

Table 6 System clock period needed to achieve target performance yield

MCNC bench-marks	CMOS with determi-nistic CAD flow	CMOS with variation-aware CAD flow		FPCNA with variation-aware CAD flow (16 CNTs Per Ribbon)			FPCNA with variation-aware CAD flow (12 CNTs Per Ribbon)		
	99% Component yield (ns)	95% Performance yield (ns)	99% Performance yield (ns)	95% Performance yield (ns)	99% Performance yield (ns)	Perf. gain over CMOS at 95% yield	95% Performance yield (ns)	99% Performance yield (ns)	Perf. gain over CMOS at 95% yield
Alu4	9.262	7.338	7.469	2.559	2.698	2.87×	2.678	2.812	2.74×
apex2	10.51	8.444	8.587	3.235	3.313	2.61×	3.263	3.307	2.59×
apex4	9.796	7.602	7.726	3.666	3.706	2.07×	3.460	3.756	2.2×
Bigkey	4.580	4.336	4.416	1.480	1.502	2.93×	1.474	1.495	2.94×
Clma	20.55	18.98	19.18	5.666	5.720	3.35×	6.790	6.818	2.8×
Des	8.900	8.853	8.994	2.884	2.921	3.07×	3.027	3.058	2.92×
diffeq	7.241	6.351	6.448	2.736	2.978	2.32×	2.827	3.070	2.25×
dsip	4.790	4.856	4.954	1.643	1.647	2.96×	1.668	1.682	2.91×
elliptic	14.87	11.26	11.39	3.342	3.483	3.37×	3.810	3.967	2.96×
ex1010	16.39	12.99	13.15	5.215	5.363	2.49×	4.801	5.046	2.71×
Ex5p	9.885	8.693	8.847	3.760	3.812	2.31×	4.500	4.554	1.93×
frisc	16.11	14.99	15.15	3.908	4.367	3.84×	5.114	5.316	2.93×
misex3	8.284	6.543	6.649	3.092	3.284	2.12×	2.709	2.899	2.42×
pdc	17.25	16.13	16.32	4.637	4.863	3.48×	4.770	4.957	3.38×
s298	15.14	14.10	14.25	3.822	3.857	3.69×	4.029	4.134	3.5×
s38417	10.97	10.62	10.74	4.314	4.370	2.46×	3.463	3.590	3.07×
s38584.1	8.456	7.024	7.140	2.816	2.894	2.49×	2.884	3.019	2.44×
Seq	10.78	7.859	7.987	3.203	3.344	2.45×	3.634	3.757	2.16×
Spla	15.20	12.04	12.20	4.643	4.730	2.59×	4.826	4.864	2.49×
Tseng	8.851	6.700	6.804	2.692	2.785	2.49×	2.835	2.917	2.36×
Average	10.59	9.070	9.203	3.293	3.404	2.75×	3.417	3.536	2.65×

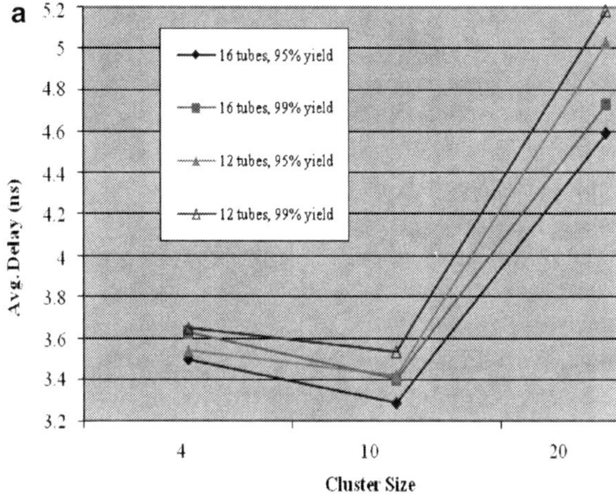

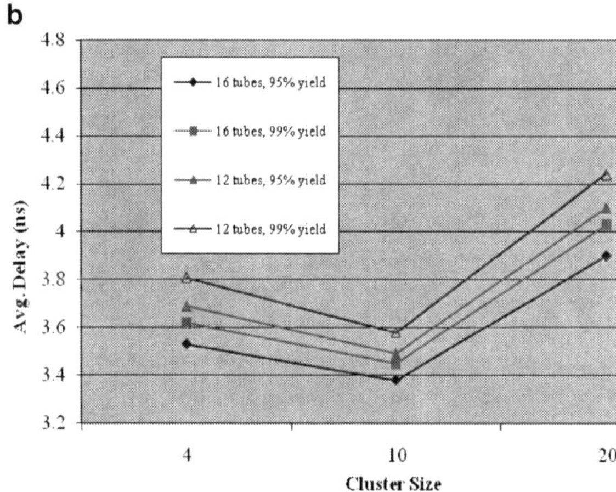

Fig. 30 Average delay for different architecture parameters at 95 and 99% yield

and a limited number of short wire segments. The experiments also show that medium-sized clusters with longer interconnects have the best performance for FPCNA. This is because a medium-sized cluster will take advantage of both CNT-bundle interconnect and local routing.

10 Conclusion and Future Work

In this chapter, we described a nanomaterial-based FPGA architecture called FPCNA. We introduced CNT-based and nanoswitch-based building blocks, such as the CNT-based LUT, and characterized them under nanospecific process variations such as CNT diameter and CNT doping level. We also covered some of the issues in CNT device fabrication. A variation-aware CAD flow was used which handles arbitrary delay distributions during variation-aware placement and routing. Experimental results show that FPCNA offers a 5.07× footprint reduction and a 2.75× performance gain (targeting a 95% yield) compared to a baseline CMOS FPGA at the same technology node. This clearly demonstrates the potential for using nanomaterials to build next-generation FPGAs.

While only random variation was studied in this work, a more accurate variation model could be obtained by considering correlated variation. Another consideration is that the universal memory market is still in its infancy. In the future, alternatives to solid-electrolyte nanoswitches and NRAM might become widely adopted and cost-effective, and then should also be considered.

Exercises

1. To compute the propagation delay of a routing crossbar, the first step is to extract the coupling capacitance correctly. A good approach is to use an electromagnetic field solver such as *Fast Field Solvers* from *http://www.fas-tfieldsolvers.com/*. In the 3 × 3 crossbar shown in Ex. Figure 1, each wire has a width of 50 nm, height of 50 nm and length of 250 nm. The spacing between two adjacent wires is 50 nm. The spacing between two layers is 50 nm as well. Find the Maxwell capacitance matrix using *Fast Field Solvers* and derive the charge-capacitance equations of wire 1 to wire 6.

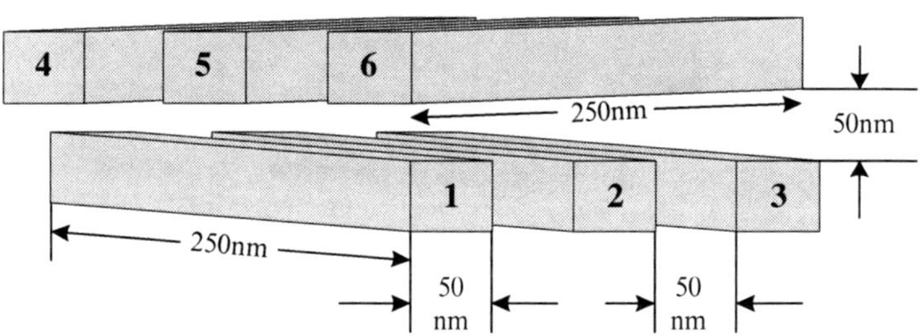

Ex. Figure 1 3 × 3 Crossbar

2. Using HSpice, measure propagation delay of AOI222 gate shown in Ex. Figure 2. Assuming falling delay has input vector [ABCDEF] switching from [101010] to [111010] and rising delay has input vector switching from [111010] to [101010].

   (a) CMOS. Using PTM 45 nm technology (*http://ptm.asu.edu/*), with all NMOS W = 130 nm and PMOS W = 260 nm.

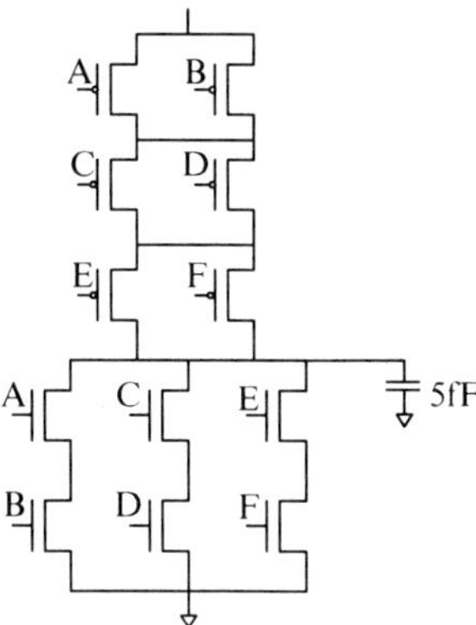

Ex. Figure 2 AOI222 Gate

(b) CNFET. Using the CNFET model from Stanford University (*http://nano. stanford.edu/models.php*), assuming 8 tubes per device.

3. Find the delay distribution of a 15-stage CNT ring oscillator using Monte Carlo simulation in HSpice. Process parameters are listed in Table 2.
4. Given delay distributions of five primary outputs as below:

	Mean (s)	3-Sigma (s)
PO1	7.25e-09	7.66e-10
PO2	7.17e-09	8.27e-10
PO3	6.75e-09	7.57e-10
PO4	6.34e-09	6.42e-10
PO5	5.91e-09	6.31e-10

Compute the clock period that satisfies 90%, 95%, and 99% yield, respectively.
5. Compute statistical MIN in Ex. Figure 3:

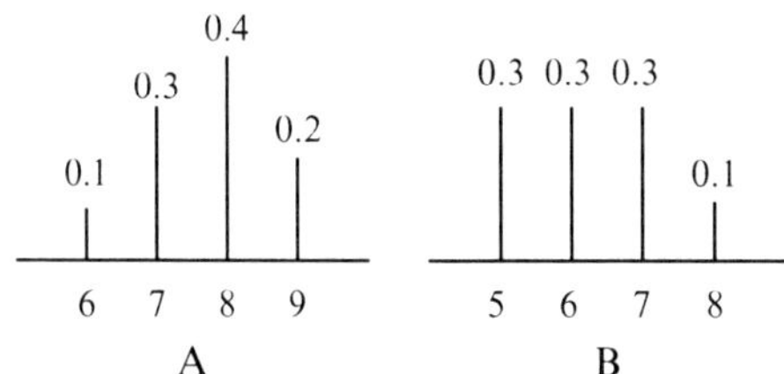

Ex. Figure 3

6. For CNT-based LUT design shown in Fig. 10, explain the reason why metallic CNTs need to be removed in the decoder but remain in the NRAM?

7. Why does a modern FPGA group several LUTs into one CLB? What are the advantages and disadvantages of large CLB size?

8. What are the benefits of using SSTA compared to static timing analysis? What are the differences between path-based SSTA and block-based SSTA?

9. Read the routing algorithm of VPR and answer the following questions:

   (a) How does VPR handle routing congestion?
   (b) Using criticality function $\mathrm{Crit}(i,j) = 1 - \frac{ts - 3\sigma_s(i,j)}{t_{crit} + 3\sigma_{crit}}$ during routing may have a potential issue when many nets have large variations. Can you explain what may happen?

10. How many levels does Stanford CNFET Spice model have? What are the differences among them?

Acknowledgments This work was partially supported by NSF Career Award CCF 07–46608, NSF grant CCF 07–02501, and a gift grant from Altera Corporation. We also appreciate the helpful discussions with Prof. John Rogers of the University of Illinois at Urbana Champaign and Prof. Subhasish Mitra of Stanford University.

References

1. V. Betz, J. Rose, and A. Marquardt, *Architecture and CAD for Deep-submicron FPGAs*, Kluwer Academic Publishers, Feb. 1999.
2. S. C. Goldstein and M. Budiu, "NanoFabric: Spatial Computing using molecular electronics," in *Proc. Int. Symp. on Computer Architecture*, 2001.
3. A. DeHon, "Nanowire-based programmable architectures," *ACM Journal of Emerging Technologies in Computing Systems*, vol. 1, no. 2, pp. 109–162, 2005.
4. G. Snider, P. Kuekes, and R. S. Williams, "CMOS-like logic in defective nanoscale crossbars," *Nanotechnology*, vol. 15, 2004.
5. A. Gayasen, N. Vijaykrishana, and M. J. Irwin, "Exploring technology alternatives for nanoscale FPGA interconnects," in *Proc. Design Automation Conference*, 2005.
6. R. M. P. Rad and M. Tehranipoor, "A new hybrid FPGA with nanoscale clusters and CMOS routing," in *Proc. Design Automation Conference*, 2006.
7. D. B. Strukov and K. K. Likharev, "CMOL FPGA: A reconfigurable architecture for hybrid digital circuits with two-terminal nanodevices," *Nanotechnology*, vol. 16, no. 888–900, 2005.
8. G. Snider and S. Williams, "Nano/CMOS architecture using a field-programmable nanowire interconnect," *Nanotechnology*, vol. 18, 2007.
9. C. Dong, D. Chen, S. Haruehanroengra, and W. Wang, "3-D nFPGA: A reconfigurable architecture for 3-D CMOS/nanomaterial hybrid digital circuits," *IEEE Transactions on Circuits and Systems I*, vol. 54, no. 11, pp. 2489–2501, Nov. 2007.
10. T. Rueckes, K. Kim, E. Joselevich, G. Y. Tseng, C. Cheung, and C. M. Lieber, "Carbon nanotube-based nonvolatile random access memory for molecular computing," *Science*, vol. 289. no. 5476, pp. 94–97, July 2000.
11. J. W. Ward, M. Meinhold, B. M. Segal, J. Berg, R. Sen, R. Sivarajan, D. K. Brock, and T. Rueckes, "A nonvolatile nanoelectromechanical memory element utilizing a fabric of carbon nanotubes," in *Proc. Non-Volatile Memory Technology Symposium*, pp. 34–38, Nov. 2004.

12. R. F. Smith. T. Rueckes, S. Konsek, J. W. Ward, D. K. Brock, and B. M. Segal, "Carbon nanotube based memory development and testing," in *Proc. Aerospace Conference*, pp. 1–5, Mar. 2007.

13. W. Zhang, N. K. Jha, and L. Shang, "NATURE: A hybrid nanotube/CMOS dynamically reconfigurable architecture," in *Proc. Design Automation Conference*, 2006.

14. Y. Zhou, S. Thekkel, and S. Bhunia, "Low power FPGA design using hybrid CMOS-NEMS approach," in *Proc. International Symposium on Low Power Electronics and Design*, Aug. 2007.

15. S. Iijima, "Helical microtubules of graphitic carbon," *Nature*, vol. 354, no. 6348, pp. 56–58, 7 Nov. 1991.

16. S. Iijima, "Carbon nanotubes: Past, present, and future," *Physica B: Condensed Matter*, vol. 323, no. 1–4, pp. 1–5, Oct. 2002.

17. H. Nejo, *Nanostructures – Fabrication and Analysis*, Springer, 2007.

18. C. Dupas, P. Houdy, and M. Lahmani, *Nanoscience*, Springer Berlin Heidelberg, 2007.

19. S. J. Kang et al., "High-performance electronics using dense, perfectly aligned arrays of single-walled carbon nanotubes," *Nature Nanotechnology*, vol. 2, no. 4, pp. 230–236, 2007.

20. N. Patil, A. Lin, E. Myers, H. S. -P. Wong, and S. Mitra, "Integrated wafer-scale growth and transfer of directional carbon nanotubes and misaligned-carbon-nanotube-immune logic structures," in *Proc. Symp. VLSI Technology*, 2008.

21. W. Zhou, C. Rutherglen, and P. Burke, "Wafer scale synthesis of dense aligned arrays of single-walled carbon nanotubes." *Nano Research*, vol. 1, pp. 158–165, Aug. 2008.

22. S. J. Tans, A. R. M. Verschueren, and C. Dekker, "Room-temperature transistor based on a single carbon nanotube." *Nature*, vol. 393, no. 6680, pp. 49–52, 7 May 1998.

23. R. Martel, T. Schmidt, H. R. Shea, T. Hertel, and Ph. Avouris, "Single- and multi-wall carbon nanotube field-effect transistors," *Applied Physics Letters*, vol. 73, no. 17, p. 2447, 26 Oct. 1998.

24. A. Bachtold, P. Hadley, T. Nakanishi, and C. Dekker, "Logic circuits with carbon nanotube transistors," *Science*, vol. 294, no. 5545, pp. 1317–1320, 9 Nov. 2001.

25. V. Derycke, R. Martel, J. Appenzeller, and Ph. Avouris, "Carbon nanotube inter- and intramolecular logic gates," *Nano Letters*, vol. 1, no. 9, pp. 453–456, 2001.

26. S. J. Wind, J. Appenzeller, R. Martel, V. Derycke, and Ph. Avouris, "Vertical scaling of carbon nanotube field-effect transistors using top gate electrodes," *Applied Physics Letters*, vol. 80, no. 20, pp. 3817–3819, May 2002.

27. A. Javey, H. Kim, M. Brink, Q. Wang, A. Ural, J. Guo, P. Mcintyre, P. Mceuen, M. Lundstrom, and H. Dai, "High-k dielectrics for advanced carbon nanotube transistors and logic gates," *Nature Materials*, vol. 1, no. 4, pp. 241–246, Dec. 2002.

28. S. J. Kang, C. Kocabas, T. Ozel, M. Shim, N. Pimparkar, M. A. Alam, S. V. Rotkin, and J. A. Rogers, "High-performance electronics using dense, perfectly aligned arrays of single-walled carbon nanotubes," *Nature Nanotechnology*, vol. 2, no. 4, pp. 230–236, Apr. 2007.

29. N. Patil, A. Lin, J. Zhang, H. Wei, K. Anderson, H. S. P. Wong, and S. Mitra, "VMR: VLSI-compatible metallic carbon nanotube removal for imperfection-immune cascaded multi-stage digital logic circuits using carbon nanotube FETs," in *Proc. IEEE Intl. Electron Devices Meeting* pp. 573–576, 2009.

30. Z. Chen, J. Appenzeller, Y. Lin, J. Sippel-Oakley, A. Rinzler, J. Tang, S. Wind, P. Solomon, and P. Avouris, "An integrated logic circuit assembled on a single carbon nanotube," *Science*, vol. 311, no. 5768, p. 1735, Mar. 2006.

31. A. Javey, Q. Wang, A. Ural, Y. Li, and H. Dai, "Carbon nanotube transistor arrays for multistage complementary logic and ring oscillators," *Nano Letters*, vol. 2, no. 9, pp. 929–932, Sept. 2002.

32. Y. Massoud and A. Nieuwoudt, "Modeling and design challenges and solutions for carbon nanotube-based interconnect in future high performance integrated circuits," *ACM Journal on Emerging Technologies in Computing Systems*, vol. 2, pp. 155–196, 2006.

33. B. Q. Wei, R. Vajtai, and P. M. Ajayan, "Reliability and current carrying capacity of carbon nanotubes," *Applied Physics Letter*, vol. 79, no. 8, pp. 1172–1174, 2001.
34. N. Srivastava and K. Banerjee, "Performance analysis of carbon nanotube interconnects for VLSI applications," in *Proc. International Conference on Computer-Aided Design*, pp. 383–390, 2005.
35. S. Kaeriyama et al., "A nonvolatile programmable solid-electrolyte nanometer switch," *IEEE Journal of Solid-State Circuits*, vol. 40, no. 1, pp. 168–176, Jan. 2005.
36. E. Ahmed and J. Rose, "The effect of LUT and cluster size on deep-submicron FPGA performance and density," *IEEE Transactions on VLSI*, vol. 12, no. 3, pp. 288–298, Mar. 2004.
37. S. J. Kang, C. Kocabas, H. S. Kim, Q. Cao, M. A. Meitl, D. Y. Khang, and J. A. Rogers, "Printed multilayer superstructures of aligned single-walled carbon nanotubes for electronic applications," *Nano Letters*, vol. 7, no. 11, pp. 3343–3348, Nov. 2007.
38. E. Pop, "The role of electrical and thermal contact resistance for Joule breakdown of single-wall carbon nanotube," *Nanotechnology*, vol. 19, 2008.
39. Y. Li et al., "Preferential growth of semiconducting single-walled carbon nanotubes by a plasma enhanced CVD method," *Nano Letters*, vol. 4, p. 317, 2004.
40. J. Deng et al., "Carbon nanotube transistor circuits: Circuit-level performance benchmarking and design Options for living with imperfections," in *Proc. International Solid-State Circuits Conference*, 2007.
41. D. Boning and S. Nassif, "Models of process variations in device and interconnect," *Design of High-Performance Microprocessor Circuits*, Wiley-IEEE Press, ISBN: 978–0–7803–6001–3, 2000.
42. International Technology Roadmap for Semiconductors, http://www.itrs.net/.
43. E. M. Sentovich et al. "SIS: A system for sequential circuit synthesis," Dept. of Electrical Engineering and Computer Science, University of California, Berkeley, CA 94720, 1992.
44. D. Chen and J. Cong, "DAOmap: A depth-optimal area optimization mapping algorithm for FPGA designs," in *Proc. International Conference on Computer-Aided Design*, Nov. 2004.
45. Y. Lin, M. Hutton, and L. He, "Placement and timing for FPGAs considering variations," in *Proc. Field Programmable Logic and Applications*, pp. 1–7, Aug. 2006.
46. S. Sivaswamy and K. Bazargan, "Variation-aware routing for FPGAs," in *Proc. Int. Symp. on Field Programmable Gate Arrays*, 2007.
47. C. Visweswariah et al., "First-order incremental block-based statistical timing analysis," in *Proc. Design Automation Conference*, pp. 331–336, 2004.
48. A. Devgan and C. Kashyap, "Block-based static timing analysis with uncertainty," in *Proc. International Conference on Computer-Aided Design*, pp. 607–614, 2003.
49. J. Liou, K. Cheng, S. Kundu, and A. Krstic, "Fast statistical timing analysis by probabilistic event propagation," in *Proc. Design Automation Conference*, pp. 661–666, 2001.

Graphene Transistors and Circuits

Kartik Mohanram and Xuebei Yang

1 Introduction

Graphene, which is a monolayer of carbon atoms packed into a two-dimensional (2D) honeycomb lattice, has demonstrated high mobility for ballistic transport, high carrier velocity for fast switching, monolayer thin body for optimum electrostatic scaling, and excellent thermal conductivity [1–5].

The potential to produce wafer-scale graphene films with full planar processing for devices promises high integration potential with conventional CMOS fabrication processes [6–8]. Various graphene-based devices have been proposed, studied, and fabricated, including micron-long, micron-wide graphene transistors, graphene nanoribbon (GNR) transistors, and bilayer graphene transistors. These devices exhibit extraordinary performance and have attracted strong interest as potential candidates for analog and digital applications.

We begin with a brief overview of graphene fabrication in Sect. 2. Since graphene is a zero band-gap material, which is not preferred in digital applications, we next summarize the techniques that have been proposed to open a band-gap and also introduce the structure of different graphene transistors. The chapter then covers the specifics of graphene transistors for analog applications in Sect. 3. Unlike digital applications, where a band-gap is essential for circuit operation, analog circuits focus on other parameters, such as gain and transconductance, and have been shown to work using zero band-gap graphene transistors. The chapter then covers the specifics of graphene transistors for digital applications in Sect. 4. Since digital applications require a sufficiently large I_{on}/I_{off} ratio for reliable and efficient operation, only transistors built using graphene with a band-gap are considered for digital applications. Different modeling methods that address different graphene transistors are covered in Sect. 5. Section 6 draws conclusions and describes directions for future research.

K. Mohanram (✉)
Electrical and Computer Engineering, Rice University, Houston, TX, USA
e-mail: kmram@rice.edu

N.K. Jha and D. Chen (eds.), *Nanoelectronic Circuit Design*,
DOI 10.1007/978-1-4419-7609-3_10, © Springer Science+Business Media, LLC 2011

2 Fabrication

Graphene is a single atomic plane of graphite, which is one of the best known allotropes of carbon. Although such monolayer atomic planes are constituents of bulk crystals, they have to be sufficiently isolated from the environment to be considered freestanding. However, since crystal growth requires high temperatures and the associated thermal fluctuations are detrimental to the stability of low-dimensional objects, the growth and study of one-atom-thick materials, such as graphene, remained unknown for a long time.

Graphene was first isolated using what is now commonly called the scotch-tape method [4] (see Fig. 1a). This technique, based on mechanical exfoliation, splits strongly layered materials such as graphite into individual atomic planes. This process produces 2D crystals of high structural and electronic quality, and techniques, including optical microscopy, atomic force microscopy (AFM), scanning electron microscopy (SEM), and micro-Raman spectroscopy, are used to locate and study the graphene samples. Further refinements, such as the use of ultrasonics, and chemical processes, such as intercalation, can loosen the atomic planes and provide pathways for graphene production on a large scale. However, although this method can produce high-quality graphene, the graphene sheet is usually small, typically below 100 μm^2 [9], and it is hard to control its position, geometry, and orientation.

The alternate strategy that shows great promise for large-scale integration, with implications for the future of electronics, is based on the epitaxial growth of graphitic layers on other crystals, such as SiC, illustrated in Fig. 1b. Using such chemical synthesis techniques, graphene sheets of the order of cm^2 have been produced using this method [8]. Such growth is actually 3D during which the layers remain bound to the substrate and the thermal fluctuations that lead to bond-breaking are effectively suppressed. When the epitaxial structure is eventually cooled following the growth process, the substrate can be removed using chemical etching techniques.

Based on this approach, wafers of continuous few-layer graphene have already been grown on polycrystalline Nickel (Ni) films and transferred onto plastic and Silicon (Si)

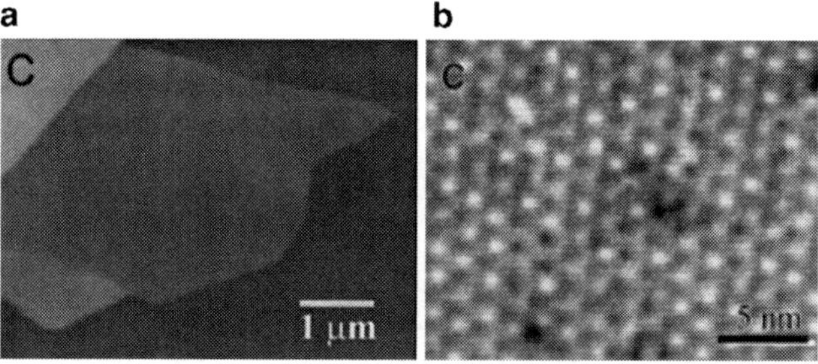

Fig. 1 (**a**) AFM image of graphene on SiO_2 [4] and (**b**) STM image of graphene grown on SiC [1]

wafers [10, 11]. With a carrier mobility of up to 4,000 cm^2 V^{-1} s^{-1}, the results are comparable to the high quality of graphene samples obtained using the exfoliation techniques. Further optimization of the epitaxial growth process, the substrate, and the transfer procedure will provide a pathway to wafer-scale processing and integration with traditional CMOS-based manufacturing.

The methods described in [7, 12] use a Tungsten (W) wafer on top of which a thin layer of Ni is epitaxially grown. This is followed by chemical vapor deposition (CVD) of a carbon monolayer, since the growth of graphene on Ni can be self-terminating with little lattice mismatch [7, 12]. In this manner, wafer-scale single crystals of graphene that are chemically bound to Ni have been grown in practice[7]. In order to remove the Ni, a polymer or another film is deposited on top, and Ni is etched away as a sacrificial layer leaving a graphene monolayer on an insulating substrate.

In summary, the production of graphene can be broadly classified into techniques based on mechanical exfoliation and chemical synthesis. Mechanical exfoliation techniques are relatively simple and yield good-quality graphene samples for small-scale use in laboratories. However, mechanical exfoliation cannot produce graphene for large-scale use, since it is labor-intensive to locate the individual graphene samples. Chemical synthesis, on the other hand, is less labor-intensive and can be used to produce graphene on a large scale. However, the resulting graphene is of relatively inferior quality and it is not easy to transfer the produced graphene from the materials that it is grown on to a Si wafer. Chemical synthesis techniques are still under development and refinement, and it is anticipated that the challenges of process controllability and quality will be surmounted in the near future.

2.1 Techniques to Open a Band-Gap

Two-dimensional graphene is a zero band-gap semi-metal, and the graphene flakes and sheets realized using these fabrication techniques usually have a large area and a band-gap that is close to zero. If transistors are built on these graphene sheets directly, the lack of a band-gap will result in a low I_{on}/I_{off} ratio. Whereas this is not a significant issue for analog applications, it is a serious problem for digital applications. Therefore, several techniques to open a band-gap in graphene sheets have been proposed and demonstrated in literature.

One class of methods advocates the use of bilayer graphene, where two sheets of monolayer graphene are stacked on top of each other. It has been shown theoretically and experimentally [13–15] that a band-gap of 100–300 meV can be obtained in bilayer graphene, and it is further demonstrated that the size of the band-gap is proportional to the potential difference between the two graphene planes [14]. The fabrication of bilayer graphene is very similar to that of single-layer graphene, and can be achieved either by mechanical exfoliation or chemical synthesis.

However, whereas bilayer graphene obtained from mechanical exfoliation usually exhibits crystalline ordering with a characteristic AB stacking, it is often misoriented when grown epitaxially on silicon carbide [16]. It has also been shown in [17] that the energy band-gap obtained by applying a potential difference between the graphene monolayers may decrease in misoriented bilayer graphene.

An alternative technique to open a band-gap is based on patterning the single-layer graphene sheet into graphene nanoribbons (GNRs). It has been both theoretically proven and experimentally demonstrated that the band-gap of a GNR is in general inversely proportional to its width [18, 19]. However, width confinement down to the sub-10 nm scale is essential to open a band-gap that is sufficient for room temperature transistor operation. Unlike carbon nanotubes (CNTs), which are mixtures of metallic and semi-conducting materials, recent samples of chemically derived sub-10 nm GNRs have exhibited all-semiconducting behavior [20] generating considerable excitement for digital circuits applications.

Two basic approaches to produce GNRs have been proposed in literature. The first method is based on lithography [18]. In this approach, e-beam resist and hydrogen silsesquioxane (HSQ) is first spun on the graphene sheet to form the mask that defines the nanoribbons. Oxygen plasma is then used to etch away the unprotected graphene, leaving the GNRs under the HSQ mask. However, the quality of the fabricated GNRs is limited by the lithographic resolution that is available with current technologies. The width of the best GNRs obtained using lithographic techniques is still above 10 nm, and strong edge roughness has also been reported in such GNRs. The other approach uses chemical techniques to produce GNRs. For example, in [20], the graphene sheet is first heated to 1,000°C in argon gas containing 3% hydrogen for 60 s and then dispersed in a 1,2-dichloroethane (DCE) solution of poly (*m*-phenylenevinylene-*co*-2,5-dioctoxy-*p*-phenylenevinylene) (PmPV) for 30 min. Centrifugation is then used to remove large pieces of materials from the supernatant, including large graphene pieces and incompletely exfoliated graphite flakes. AFM is then used to locate and characterize the produced GNRs. Sub-10 nm GNRs have been successfully fabricated and studied using this method, as illustrated by the images in Fig. 2. Recently, other approaches for fabricating GNRs have also been reported. For example, in [21, 22], GNRs are produced by unzipping CNTs using chemical techniques. In [23], a CVD approach has been proposed to fabricate GNRs.

Depending on the edge geometry of the GNR, there are two main types of GNRs: zigzag-edge GNRs (ZGNRs) and armchair-edge GNRs (AGNRs), as illustrated in Fig. 3a, b, respectively. ZGNRs are predicted to be metallic by a simple tight-binding model, but a band-gap exists in more advanced, spin-unrestricted simulations [25]. For digital applications, the focus has been on using AGNRs as the channel material. AGNRs have an electronic structure that is closely related to that of zigzag CNTs, and both materials exhibit a band-gap with a period-three modulation in the confined in-plane direction [19].

The band-gap in AGNRs originates from quantum confinement and edge effects play a critical role [25]. Edge-bond relaxation and nearest neighbor effects are

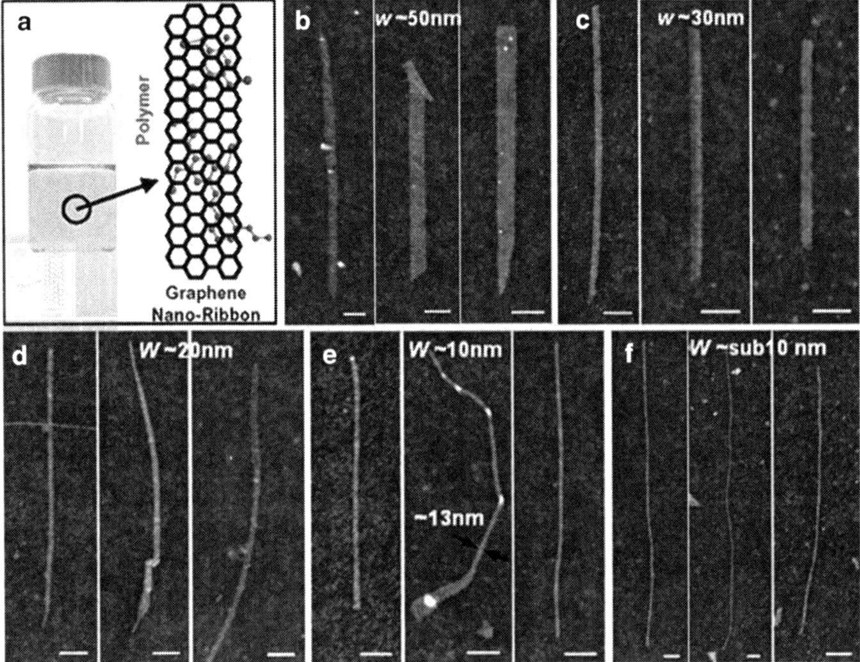

Fig. 2 (**a**) Illustration of GNRs derived using the chemical techniques described in [20]. (**b–e**) AFM images of GNRs ranging in width from 50 nm to sub-10 nm

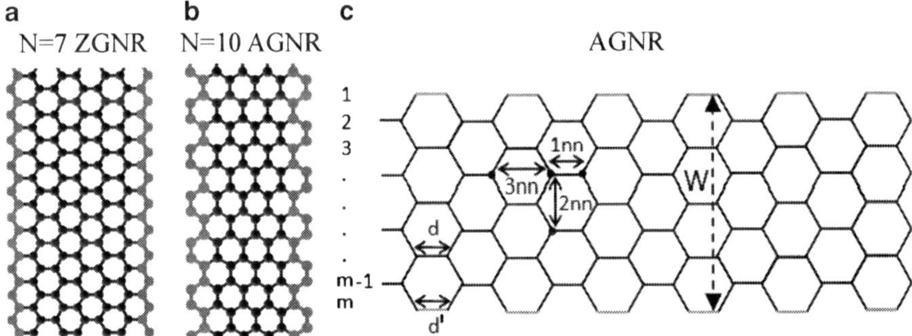

Fig. 3 The schematic sketch of (**a**) zigzag GNRs (ZGNRs) and (**b**) armchair-edge GNRs (AGNRs). (**c**) This figure illustrates an AGNR where the edge of the honeycomb lattice is hydrogen terminated. The edge bonds (*colored lines*) have a different bonding length and bonding parameter from those in the middle of the AGNR due to edge bond relaxation. The interaction between the first nearest neighbor (1 nn), the second nearest neighbor (2 nn), and the third nearest neighbor (3 nn) atoms are also shown [24]

illustrated for an AGNR in Fig. 3c. Note that the band-structure of monolayer GNRs may also be strongly influenced by the substrate, such as SiO_2, when the edges of the ribbon are oxygen-terminated and not passivated using hydrogen [26].

Finally, it is worth noting that realizing a large band-gap is an intense area of research in graphene electronics. Recent developments, such as the use of uniaxial strain [27] and the graphene nanomesh [28], showcase the unique opportunities and challenges in this direction. However, these methods have been proposed very recently and their prospects and scalability remain the focus of current research.

2.2 Graphene Transistors

Depending on whether the channel material is a micron-wide graphene sheet, bilayer graphene, or a GNR, the graphene transistors exhibit different electrical properties. However, the basic transistor structures and the fabrication processes are very similar.

The commonly fabricated graphene transistors today deploy a back-gated structure wherein graphene serves as the channel material. The advantage of the back-gated structure is that the transistor can be fabricated easily and the mobility of graphene can be maintained. For a back-gated device, the wafer is heavily doped to act as the back gate and subsequently covered by a layer of silicon dioxide, which functions as the gate dielectric of the transistor. In most back-gated transistors reported in literature, the thickness of the silicon dioxide is around 300 nm. This thickness is preferred because the graphene is easily visible due to light reflection. However, the thick silicon dioxide greatly reduces the gate capacitance and it has been demonstrated in [29] that a silicon dioxide thickness of 10 nm can significantly increase the I_{on} of GNR transistors. Figure 4 presents the schematic and AFM image for this back-gated transistor [29]. The graphene sheet/GNRs/bilayer graphene are either directly produced on the silicon wafer, or produced on other crystals and then transferred onto the silicon wafer. One drawback of the back-gated structure is that all the transistors on a wafer share the same gate voltage and, hence, cannot be operated independently.

In contrast to the back-gated structure, top-gated transistor structures have also been proposed and demonstrated. A commonly fabricated top-gated transistor first uses atomic layer deposition/physical vapor deposition/CVD to produce dielectric materials, such as polymethyl methacrylate (PMMA), Al_2O_3, HfO_2, and evaporated SiO_2, onto the graphene surface [30–33]. Finally, the gate electrode Cr/Au or Ni/Cu is made by electron beam evaporation. The top gate enables better control of the electronic properties of the graphene transistor and may also help achieve current saturation; however, it also reduces the mobility of graphene, mostly because of the scattering that is introduced between graphene and the gate dielectric. The schematic and SEM image for the top-gated graphene transistor described in [32] is presented in Fig. 5.

Most of the graphene transistors demonstrated so far use metals such as Pd, Ti, and/or Au to fabricate the drain and source contacts. A common fabrication process for the contacts uses e-beam lithography to define the source/drain contact regions, followed by e-beam evaporation to realize the metal electrodes. Transistors with

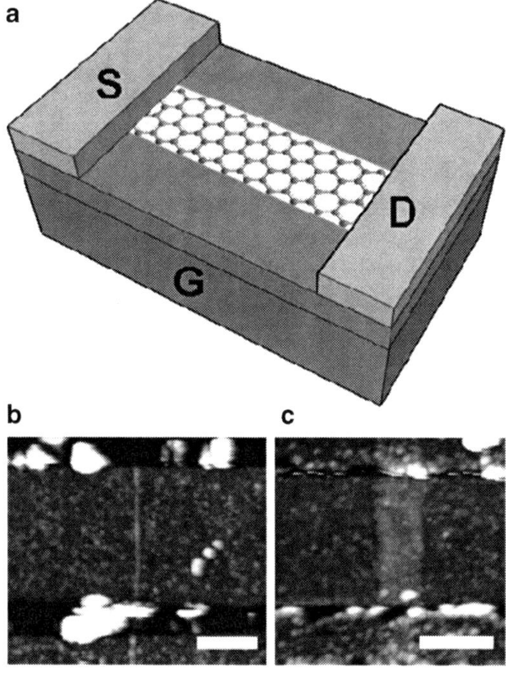

Fig. 4 (**a**) Schematic of back-gated GNRFET. (**b**) and (**c**) are AFM images of the GNRFET [29]

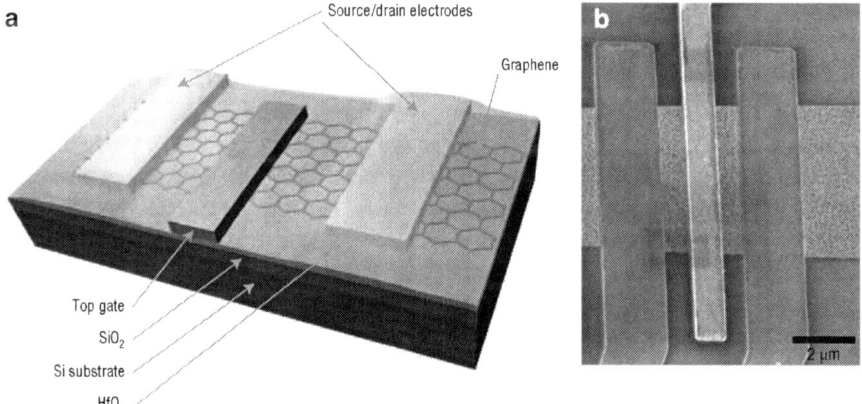

Fig. 5 (**a**) Schematic and (**b**) SEM image of the top-gated graphene transistor described in [32]

metal contacts and intrinsic graphene channel usually exhibit ambipolar behavior due to the formation of Schottky barriers (SBs) at the contacts, i.e., they conduct both electrons and holes, showing a superposition of p-type and n-type behavior. Ambipolar conduction may lower both I_{on} and the I_{on}/I_{off} ratio of the transistors.

In GNRFETs, the two SBs that are formed at the contacts reduce the probability that the carriers can enter the transistor channel. However, it has been reported in [29] that by choosing contact electrodes with a proper workfunction, ambipolar conduction can be reduced. For example, Pd can minimize the SB height for holes in p-type transistors as compared to Ti/Au, and the fabricated GNRFETs with channel lengths of the order of 200 nm delivered about 21% of the ballistic current at $V_{DS} = 1$ V and about 4.5% of the ballistic current at low $V_{DS} < 0.1$ V.

Other techniques to eliminate ambipolar conduction have also been proposed. For example, simulation studies of GNRFETs with doped reservoirs (MOSFET-type GNRFET) have been reported in literature [34–36]. In [34, 37], SB-type GNRFETs were compared to MOSFET-type GNRFETs. There is consensus that in ideal devices, MOSFET-type GNRFETs show better device characteristics over SB-type GNRFETs: larger maximum achievable I_{on}/I_{off}, larger I_{on}, larger transconductance, and better saturation behavior. It is further proposed that bilayer devices exhibit different but more favorable I–V characteristics in comparison to the monolayer devices.

However, doping GNRs remains one of the most important challenges to be addressed in graphene electronics. The small thickness of the graphene sheet immediately rules out the use of traditional doping technologies. New techniques, such as the use of chemical doping, have been proposed and demonstrated. For example, in [38], graphene is n-doped by high-power electrical annealing (e-annealing) in NH_3. However, no experimental data for MOSFET-type GNRFETs with doped contacts have been collected and reported in literature to date.

3 Analog Circuits

Graphene has the highest intrinsic carrier mobility at room temperature among known materials, making it a very promising material for analog circuit applications. Furthermore, the drain current of a graphene transistor can be easily increased by increasing the width of the graphene sheet, which is a significant advantage over other carbon-based materials, such as CNTs. Therefore, it is widely believed that graphene-based transistors will surpass silicon MOSFETs and provide a means to achieve much higher operating frequencies.

Most of the graphene transistors fabricated to study analog circuit performance utilize a wide graphene sheet that is usually 100–1,000 nm wide as the channel material, instead of a GNR or bilayer graphene. This is because analog applications typically place a higher emphasis on transconductance and current drive of the transistor over the I_{on}/I_{off} ratio. We begin this section with a survey of the "fastest" graphene transistors – as measured using the cutoff frequency f_T – reported in literature.

A top-gated graphene transistor with $f_T = 14.7$ GHz was reported in [39]. The transistor was fabricated using a mechanically exfoliated graphene sheet with a channel width of 2.5 μm and channel length of 0.5 μm. Thirty nanometer HfO_2 is deposited on the surface of the graphene channel as the top-gate insulator.

The authors also proposed a model based on charge-collection to describe the I–V behavior of the device, which will be introduced later in this chapter.

A top-gated graphene transistor with $f_T = 26$ GHz was reported in [40]. The graphene sheet was mechanically exfoliated and 12-nm-thick Al_2O_3 was used as the top-gate insulator. Transistors with various gate lengths ranging from 150 to 500 nm were fabricated and measured. It was found that f_T is inversely proportional to the square of the channel length, and the maximum f_T was obtained at a channel length of 150 nm. The authors also project that THz operation can be achieved for a channel length of 50 nm using these transistors.

A top-gated graphene transistor with $f_T = 4.4$ GHz was reported in [41]. The graphene sheet was synthesized using graphitization of semi-insulating Si-face 6H-SiC (0001) substrates. The fabricated transistor had a channel length of 2 μm and used 20-nm-thick Al_2O_3 as the gate insulator.

A top-gated graphene transistor with $f_T = 9$ GHz was reported in [42]. The graphene sheet was mechanically exfoliated and the channel length of the transistor was 1.1 μm. NFC 1400-3CP was used as the buffer layer between the gate insulator, which is 10 nm thick HfO_2, and the graphene surface. The authors concluded that the use of a buffer layer can, to a large extent, mitigate the mobility degradation caused by the gate insulator. They also projected that the f_T of the transistors fabricated using this approach will exceed what was reported in [40] at comparable channel lengths.

The fastest graphene transistor reported to date was described in [43]. In this paper, graphene was epitaxially grown on the Si face of a semi-insulating, high-purity SiC wafer, and a 10-nm thick HfO_2 was used as the gate insulator. For a channel length of 240 nm, the authors reported a f_T as high as 100 GHz.

While many groups are trying to increase the frequency of graphene transistors to the THz domain, design approaches based on these graphene transistors for analog applications focused on its ambipolar conduction property have also been reported in literature. Ambipolar conduction is illustrated using the I_d–$V_{gs\text{-top}}$ curve for the transistor from [32] in Fig. 6. In this figure, it is clear that both hole and

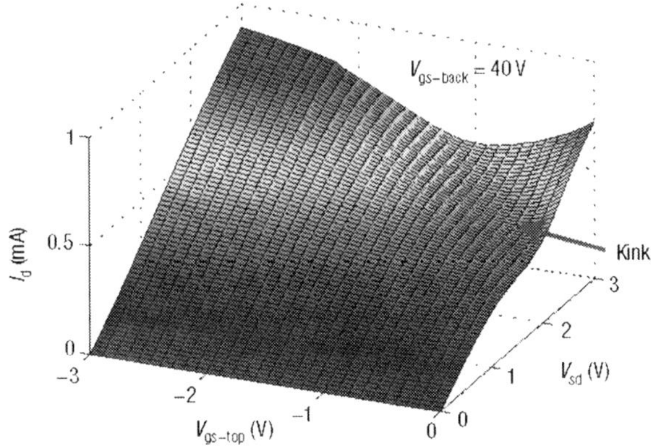

Fig. 6 Ambipolar I–V characteristics for top-gated graphene transistor described in [32]

electron conduction is feasible, depending on the applied bias [2]. By properly adjusting the gate–source and drain–source voltages, the transistor can be switched from n-type to p-type, with electron and hole conduction dominating the current, respectively. At the minimum conduction point V_{min}, the electron current is equal to the hole current, and the total current is at its minimum. When the gate–source voltage is less than V_{min}, the current is mainly based on hole conduction, and the device behaves as a p-type transistor; at higher voltages, on the other hand, electron conduction dominates and the device behaves as an n-type transistor. Since conventional analog circuits are based on unipolar devices, this novel ambipolar behavior was initially considered undesirable. However, recent work has shown that the ability to control device polarity (p-type or n-type) in the field can present new design opportunities.

In [44], the first single-transistor frequency multiplier and full-wave rectifier that does not require any filters was implemented. The authors used a fabricated back-gate device, with the channel length of 1.3 µm and mean width of 2.7 µm. The circuit was configured as a traditional common-source amplifier that requires one transistor and one resistor. When the transistor is biased at the minimum conduction point of the ambipolar curve and a small signal AC input is applied at the gate, the positive part of the signal sees a negative gain while the negative part sees a positive gain, thus realizing frequency multiplication and full-wave rectification. Compared to traditional approaches for frequency multiplication, the single-transistor graphene amplifier not only has a much simpler structure, but also promises high spectral purity (94%), signal gain, and potential for THz operation.

Indeed, the concept and the same circuit structure can be further generalized, depending on the choice of bias points as follows (the results are also validated for the transistor described in [32] and presented in Fig. 7). When $V_{bias} > V_{min}$, the device is biased at the right branch. The current is mainly due to electron

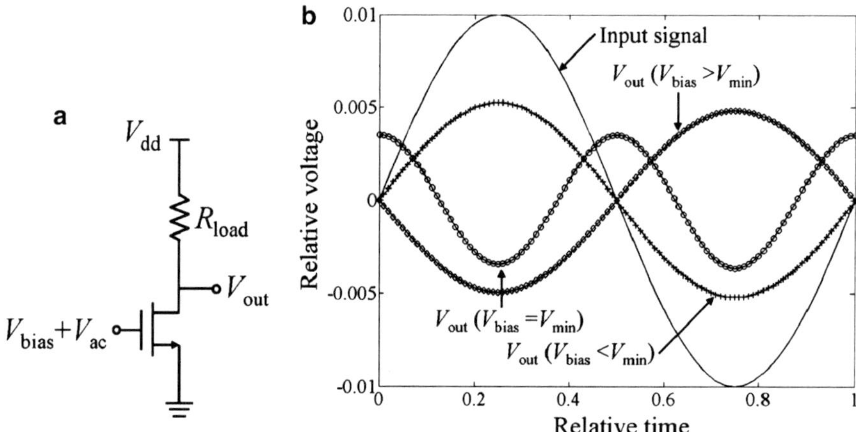

Fig. 7 (a) The circuit structure of the graphene-based amplifier. (b) The AC coupled signals at the amplifier output for $V_{bias} > V_{min}$, $V_{bias} = V_{min}$, and $V_{bias} < V_{min}$

conduction, so the transistor can be regarded as n-type. Under this situation, the circuit is configured as a common-source amplifier, the gain of which is $A = -|g_m|$ R_{load}, so the small signal gain is negative. When $V_{\text{bias}} < V_{\text{min}}$, the device is biased at the left branch. The current is mainly due to hole conduction, so the transistor can be regarded as p-type. In this situation, the circuit is configured as a common-drain amplifier, and its gain can be roughly expressed as $A = |g_m|R_{\text{load}}/(|g_m|R_{\text{load}} + 1)$, so the small signal gain is positive.

In summary, depending on the choice of the bias points, the single-transistor graphene amplifier can realize positive gain, negative gain, or frequency multiplication. Since this provides a unique opportunity to control the amplifier configuration in the field, the single transistor can benefit several applications when a large square wave signal is applied as the bias and a sinusoidal signal is applied as the small-signal AC input:

1. If the bias voltage is chosen so that the circuit is biased in either the positive gain or the negative gain mode, the small-signal AC output will exhibit a phase shift of 180°, realizing the signal modulation necessary for phase shift keying (PSK).
2. If the bias voltage is chosen so that the circuits is biased in either the frequency multiplication mode or one of the common-source/common-drain mode, the frequency of the small-signal AC output will either double or remain the same, realizing the signal modulation necessary for frequency shift keying (FSK).
3. If the square wave signal and the small signal AC input have the same frequency but a phase difference ϕ, by choosing the bias voltage so that the circuit is biased in either the positive gain or the negative gain mode, the DC component of the output can be made proportional to $\cos(\phi)$ realizing phase detection.

Compared to the traditional approaches to implement PSK, FSK, or phase detection, where an analog multiplier is usually used, the graphene circuits have the following advantages. First, the graphene amplifier is a single-transistor design, while conventional multipliers require multiple transistors and/or filters. Second, it is difficult to achieve a bandwidth over 10 GHz with Si CMOS and Si BJT multipliers because of the inherent low f_T of the devices. However, the potential of THz graphene [40] promises a much larger bandwidth. Finally, the simple structure of these single-transistor circuits will potentially consume less power. When the circuits were simulated with the model proposed in [32], the power consumption was 2–4 mW, which is lower than the DC power consumption of 150 mW to 2.5 W of high electron-mobility transistor (HEMT) and heterojunction bipolar transistor (HBT) multipliers for which a bandwidth of 20–40 GHz has been achieved. This triple-mode amplifier circuit has been experimentally demonstrated in [80].

4 Digital Circuits

Transistors for digital applications usually need to provide a high $I_{\text{on}}/I_{\text{off}}$ ratio. Therefore, only transistors built on graphene with a band-gap are considered for digital applications. Recently, GNRFETs with an $I_{\text{on}}/I_{\text{off}}$ of the order of 10^6 at room temperature have been reported [29, 38, 44]. However, in order to achieve a sufficient band-gap, GNRs have to be patterned to widths less than 10 nm. Although it has been experimentally reported in [29] that all sub-10 nm GNRs are semiconducting, which is a huge advantage over CNTs, where roughly a third are metallic, this still faces significant manufacturing challenges. Moreover, such GNRFETs also suffer from the problem of low I_{on} because of the nanometer-wide channel. In order to increase I_{on}, it is commonly agreed that a GNRFET will have to be fabricated with a parallel array of multiple GNRs in the channel, as shown using the images of the devices from [45] and [18] in Fig. 8.

In contrast, bilayer graphene FETs can usually be more easily fabricated because the width requirements are not as stringent and because bilayer graphene FETs potentially deliver larger I_{on}. However, several simulation studies [46, 47] have shown that the band-gap of bilayer graphene is not large enough to suppress band-to-band tunneling. As a result, bilayer graphene FETs have a high off current and a low I_{on}/I_{off} ratio. To date, the best state-of-the-art bilayer graphene transistors have only exhibited an I_{on}/I_{off} ratio of 100 at room temperature [48]. There have been attempts to study the potential of multilayer GNRs, e.g., in [49], the authors simulated multilayer graphene FETs with one to ten layers of 2.7-nm-wide GNRs. The simulation results indicated that ten-layer GNRFETs achieve both high I_{on}/I_{off} and large I_{on}. However, this technique also poses significant fabrication challenges, and more experimental data are necessary for validation.

Because no complex digital circuit containing tens of graphene transistors has been fabricated to date, we discuss the performance of GNRFETs in digital applications based on simulation data. The results and data can thus be regarded as establishing the bounds on the performance that can be achieved using GNRFETs. In addition, we also present some novel applications that exploit the novel ambipolar conduction property of SB-type FETs. Although proposed in the context of carbon nanotube FETs (CNTFETs), the concepts and designs can be readily applied in the context of ambipolar graphene transistors.

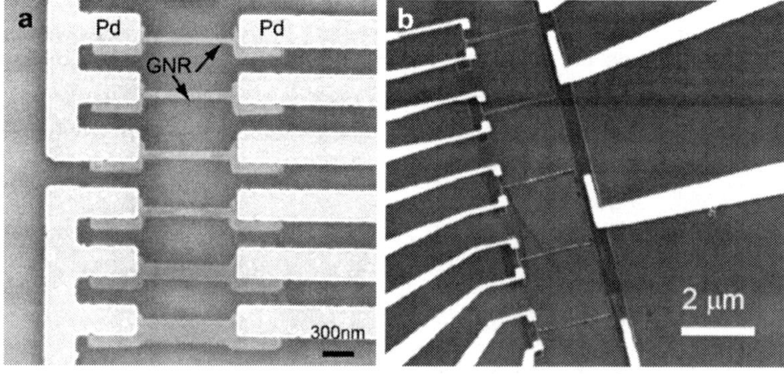

Fig. 8 GNRFET with multiple GNRs fabricated in a parallel array in the channel from (**a**) [45] and (**b**) [18]

4.1 GNRFET Digital Circuits

In [50], the authors use simulation data to study the performance of GNRFETs in digital circuits. The simulations consider a 15-nm-long AGNR as the channel material, and the GNR index is varied from $N = 9$ to $N = 18$, corresponding to widths from 1.1 to 2.2 nm. Double-gate geometry is implemented through a 1.5-nm-thick SiO_2 gate insulator ($\varepsilon_r = 3.9$). The source and drain contacts are metals, and the SB height is equal to half the band-gap of the channel GNR ($\phi_{Bn} = \phi_{Bp} = E_g/2$). The threshold of the transistor is tuned by varying the work function of the gate. The structure of the device and range of values for the extrinsic parameters are shown in Fig. 9. Note that it is assumed that each GNRFET has an array of four GNRs in the transistor channel to increase the current.

The authors first simulate a 15-stage ring oscillator based on $N = 12$ GNRFETs. By tuning the supply voltage V_{DD} and the threshold voltage V_t, the authors determine the best operation point, denoted by energy-delay-product (EDP) and static noise margin (SNM), to be $V_{DD} = 0.4$ V and $V_t = 0.13$ V. In order to compare with traditional CMOS technology, the authors further simulate the same circuit using $V_{DD} = 0.8, 0.6,$ and 0.4 V for 45-, 32-, and 22-nm CMOS technology. It is reported that whereas CMOS circuits slightly outperform GNRFET circuits in SNM, the optimum EDP for CMOS circuits is 40–168 times higher than that for the GNRFET circuits. This demonstrates that GNRFETs potentially offer significant advantages over CMOS in digital applications.

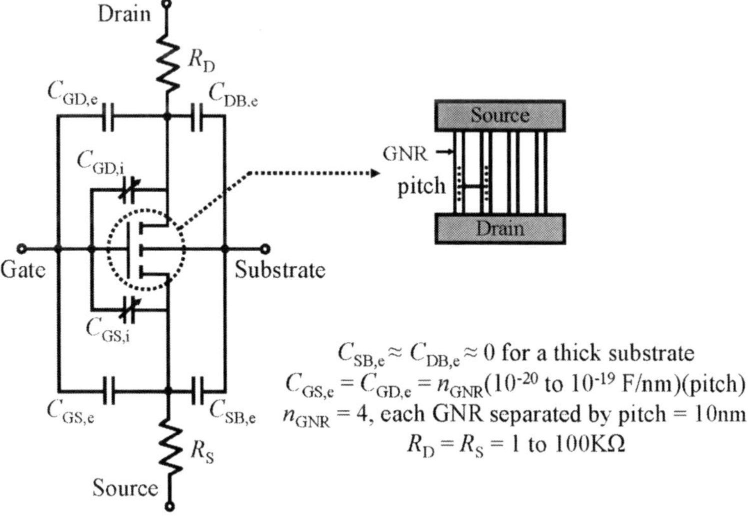

Fig. 9 Model used for simulation of GNRFET circuits. The intrinsic GNRs are simulated using quantum simulators based on the NEGF approach. These data are incorporated into a lookup table simulator that allows the modeling of extrinsics, such as parasitic capacitance and contact resistance [50]

Considering the small dimensions of a GNR, defects and variability are bound to have a large impact on circuit performance. Two major sources of variability and defects, GNR width and charge impurity, were determined by performing simulation on a wide variety of variability and defect mechanisms. The effect of GNR width on circuits was analyzed by varying GNR width from $N = 9$ to $N = 18$. It was shown that in the worst case, the static and dynamic power consumption of the circuit may increase by 313–643% and 27–217%, respectively, the delay may increase by 77%, and the SNM may decrease by 80%. The effect of charge impurities was also studied by placing a point charge of $-2q$, $-q$, $+q$, and $+2q$ near the source contact. It was reported that in the worst case, circuit delay increases by 90%, static and dynamic power consumption increases by 11 and 37%, and SNM decreases by 40%.

These results indicate that GNRFET circuits are vulnerable to process variability and defects. Therefore, although GNRFET circuits promise exciting advantages over CMOS circuits, the effects of process variability and manufacturing defects must be carefully considered in the performance assessment and design optimization for future graphene-based electronics technology.

4.2 Ambipolar Logic Circuits

Recent work has shown that it is possible to exploit the unique in-field controllability of the device polarity of ambipolar SB-type carbon CNTFETs to design a technology library with higher expressive power than conventional CMOS libraries. The ambipolar SB-type CNTFETs can be controlled by an additional terminal, called the polarity gate, which sets the p-type or n-type device polarity, while the actual gate terminal controls the current flow through the transistor [51]. If the polarity gate voltage is high, the device exhibits n-type behavior, and the actual gate voltage needs to be high as well to turn on the device, whereas if the polarity gate voltage is low, the device exhibits p-type behavior, and the actual gate voltage needs to be low to turn on the device.

This novel feature of ambipolar SB-type CNTFETs was investigated in [52], where a compact and in-field reconfigurable universal eight-function logic gate was described. Furthermore, the use of ambipolar SB-type CNTFETs to implement in-field reconfigurable generalized NOR (GNOR) gates, combining NOR and XOR operations, was described in [53]. It was shown that GNOR gates can be utilized in a two-level programmable logic array (PLA) to reduce circuit area, to map logic functions into compact and fast Whirlpool PLAs [54], and to realize AND–XOR planes that efficiently map n-bit adders [55].

In [56, 57], the authors described a family of full-swing static logic gates based upon SB-type CNTFETs in a transmission gate configuration. Based on generalized NOR–NAND–AOI–OAI primitives, the proposed library of static ambipolar CNTFET gates efficiently implements XOR functions, provides full-swing outputs, and is extensible to alternate forms with area-delay tradeoffs. Since the design of the gates can be regularized, the ability to functionalize them in the field opens up opportunities for novel regular fabrics based on ambipolar FETs.

It is worth noting that to date, many designs that exploit ambipolar conduction have been proposed. The next step would be to fabricate such ambipolar logic circuits and to validate the circuits experimentally. Moreover, though these designs are all based on CNTs, graphene also exhibits ambipolar conduction and it is reasonable to anticipate that ambipolar SB-type graphene transistors can be utilized as the substrate to implement these novel designs.

4.3 Tunneling FETs

Besides being used as the channel material for transistors in conventional structures, graphene has also attracted strong interest for use in tunneling FETs (TFETs). A tunneling transistor has p-doped drain, intrinsic channel, and n-doped source, and its conduction is dominated by band-to-band tunneling. The p–i–n structure can greatly decrease the subthreshold swing, hence improving the I_{on}/I_{off} ratio of the transistor. In [58], a 52.8 mV/dec subthreshold swing has been experimentally demonstrated in a Si-based TFET. However, Si-based TFETs suffer from low I_{on}, and there is interest in using narrow band-gap materials with smaller effective masses to enhance I_{on}.

Recently, both GNR TFETs [59–61] and bilayer graphene TFETs [62] have been studied theoretically, and larger I_{on} as well as lower subthreshold swing have been reported. For graphene bilayer TFETs, the simulation results indicate that the subthreshold swing could be lower than 20 mV/dec, while a minimum subthreshold swing of 0.19 mV/dec can potentially be achieved in GNR TFETs. Another recent study combines atomistic quantum transport modeling with circuit simulations to explore GNR TFET circuits for low-power applications [81]. In summary, these theoretical results indicate that graphene-based TFETS offer advantages over Si-based TFETs, and merit further experimental demonstration and validation.

5 Modeling and Simulation of Graphene Transistors

Modeling of transistors using micron-wide and micron-long graphene as channel material usually takes the traditional charge-collection approach, as presented in [32]. To model GNRFETs and graphene bilayer transistors with a channel length of the order of 10 nm, however, the traditional approach is not suitable. This is mainly because (1) the traditional approach is based on the assumption that there is significant scattering inside the channel and that the mean-free-path (MFP) for carriers is much smaller than the channel length [63], an assumption that is losing validity as devices shrink, and (2) the energy band structure of bilayer graphene can be tuned by the gate bias, which cannot be addressed in the traditional modeling approach.

Modeling approaches for graphene transistors take the form of: (1) computationally intensive, quantum-theory-based nonequilibrium Green's function (NEGF) approaches [64, 65] or (2) simpler semi-classical approaches [24, 66]. NEGF-based approaches are highly accurate but extremely time-consuming. Further, they provide limited intuition necessary for circuit design and optimization with multiple transistors. In comparison to NEGF-based approaches, the simpler semi-classical approaches are computationally very efficient. Further, they are physics-based and parameterizable, providing good intuition to designers. It has been shown that for MOSFET-type transistors, a semi-classical description is valid for a channel length down to about 10 nm [67].

In Sect. 5.1, we will first introduce a charge-collection-based approach for micron-wide graphene FETs presented in [32]. In Sect. 5.2, the quantum theory based approach will be introduced. In Sect. 5.3, a basic analytical theory of GNRFETs will be presented based on the top-of-the-barrier approach. This approach is suitable for simulating MOSFET-type GNRFETs, but is insufficient for handling SB-type GNRFETs because it cannot describe transport through the SB. A modeling approach that can handle SB-type GNRFETs is described in Sect. 5.4.

5.1 Charge-Collection Model

In [32], a charge-collection model was proposed for a double-gate micron-wide graphene FET. In the charge-collection model, the drain current is evaluated using the following formula:

$$I_{\mathrm{d}} = \frac{W}{L} \int_0^L en(x) v_{\mathrm{drift}}(x) dx,$$

where W and L are the channel width and length, $n(x)$ is the carrier density inside the channel, and $v_{\mathrm{drift}}(x)$ is the velocity of the carriers. $v_{\mathrm{drift}}(x)$ is evaluated using a velocity saturation model as follows:

$$v_{\mathrm{drift}}(x) = \frac{\mu E}{1 + \frac{\mu E}{v_{\mathrm{sat}}}},$$

where μ is the mobility of graphene, E is the electrical field inside the channel, and v_{sat} is the saturation velocity. The reader is referred to the supplementary material provided with [32] for details on how to estimate the saturation velocity v_{sat}. The carrier density $n(x)$ is calculated using a field-effect model:

$$n(x) = \sqrt{n_0^2 + \left(\frac{C_{\text{top}}(V_{\text{gs-top}} - V(x) - V_0)}{e}\right)^2},$$

$$V_0 = V_{\text{gs-top}}^0 + \left(\frac{C_{\text{back}}}{C_{\text{top}}}\right)\left(V_{\text{gs-back}}^0 - V_{\text{gs-back}}\right),$$

where n_0 is the minimum sheet carrier concentration as determined by disorder and thermal excitation, C_{back} and C_{top} are the back-gate and top-gate capacitance, $V_{\text{gs-top}}^0$ and $V_{\text{gs-back}}^0$ are top-gate-to-source and back-gate-to-source voltage at the Dirac point, respectively, and $V(x)$ is the potential inside the channel. By replacing $n(x)$ with its expression in the integral and changing the integral from dx to dV, the final expression for drain current can be written as:

$$I_{\text{d}} = \frac{\frac{W}{L} e \mu \int_0^{V_{\text{ds}}} \sqrt{n_0^2 + \left(\frac{C_{\text{top}}(V_{\text{gs-top}} - V(x) - V_0)}{e}\right)^2} \, dV}{1 + \frac{\mu V_{\text{ds}}}{L v_{\text{sat}}}}.$$

In summary, the simple charge-collection model makes it easy to perform device simulation and validate circuit designs, especially for analog circuit applications. Figure 10 presents results that show excellent agreement between the basic charge-collection model and experimental data. To simulate the device, the following parameters are required: W, L, μ, n_0, C_{top}, C_{back}, $V_{\text{gs-top}}^0$, $V_{\text{gs-back}}^0$, and v_{sat}. In [32], these specific values are given for a fabricated top-gated device and the implementation of the model is left as an exercise at the end of this chapter.

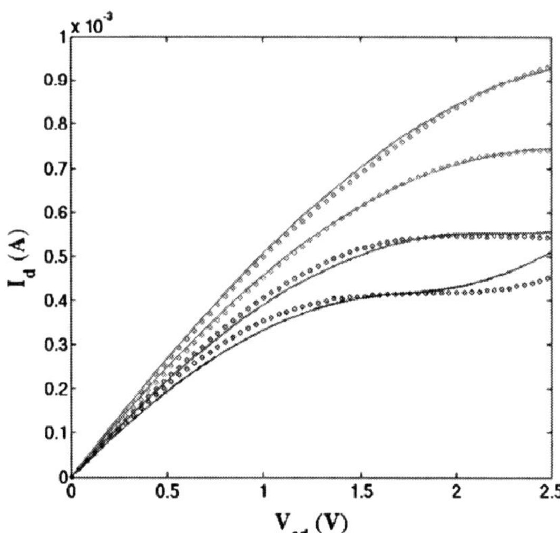

Fig. 10 Results for charge collection model (*solid lines*) compared with experimental data (*dotted lines*) for the graphene transistor described in [39]

5.2 *Quantum Simulation Techniques*

Quantum simulation is by far the most computationally demanding approach for the simulation of GNRs, and it is also the only approach for bilayer graphene transistors. Traditionally, such studies have used the quantum transport simulation framework based on the NEGF formalism, which provides an ideal approach for bottom-up device modeling and simulation [68] for the following reasons: (1) atomistic descriptions of devices can be readily implemented, (2) open boundaries can be rigorously treated, and (3) multiphenomena (e.g., inelastic scattering and light emission) can be modeled.

Figure 11 summarizes the procedure to apply the NEGF approach to a generic transistor. The transistor channel, which can be a piece of silicon, a GNR, a nanowire, or a single molecule, is connected to the source and drain contacts. The gate modulates the conductance of the channel. One first identifies a suitable basis set and derives the Hamiltonian matrix H for the isolated channel. Then, the self-energy matrices Σ_1, Σ_2, and Σ_S are computed. The self-energy matrices describe how the channel couples to the source contact, the drain contact, and the dissipative processes. Next, the retarded Green's function, including the self-consistent electrostatic potential U, is computed as:

$$G^r(E) = [(E + i0^+)I - H - U - \Sigma_1 - \Sigma_2 - \Sigma_S]^{-1}.$$

Finally, the physical quantities of interest, such as the charge density and current, are computed from the Green's function.

Most works in literature use such an approach to simulate the DC characteristics of ballistic GNRFETs (MOSFET-type FETs, tunneling FETs, as well as SB-type FETs) by solving open-boundary Schrodinger equation in an atomistic p_z orbital basis set using the NEGF formalism. A nearest-neighbor TB parameter of $t_0 = -2.7\,\mathrm{eV}$ is used, and edge bond relaxation is implemented by setting $t_0 = C_{\mathrm{edge}}t_0$ for the edge bonds, where $C_{\mathrm{edge}} = 1.12$, as parameterized to the *ab initio* band-structure simulations. The atomistic transport equation is self-consistently solved with a 3D Poisson equation using the finite-element method, which is efficient at treating a device with multiple gates because it can easily handle an arbitrary grid for complex geometries.

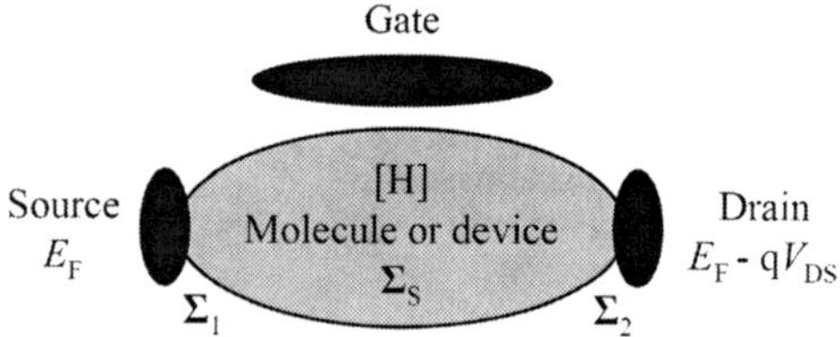

Fig. 11 Illustration of the application of the NEGF formalism to a generic transistor [50]

One such publicly available simulator, capable of simulating both GNRFETs and CNTFETs, is NANOTCAD ViDES [65]. Exercise 7 at the end of this chapter focuses on the simulation of ballistic GNRFETs with SB contacts using the decks that are provided through the nanoHUB [69].

5.3 Semi-classical Top-of-the-Barrier Modeling

The top-of-the-barrier approach is a semi-classical model suitable to treat MOSFET-type GNRFETs with Ohmic contacts at the channel ends instead of SB contacts. Figure 12 illustrates the basic principles of the top-of-the-barrier approach as applied to a generic transistor. Since the bottom of the conduction band is at zero at equilibrium, the application of a bias will shift the bottom of the conduction band to U_{scf}, which is the "top of the barrier." In the top-of-the-barrier approach, the current is evaluated using the Landauer formula as follows:

$$I = \frac{2q}{h} \int_{U_{scf}}^{\infty} \left(f(E - E_{f1}) - f(E - E_{f2}) \right) T(E) dE,$$

where h is the Planck's constant, I is the drain current, $f()$ is the Fermi function, E_{f1} and E_{f2} are the Fermi levels at the two contacts, and T is transmission coefficient, which is the probability that a carrier from one contact can reach the other contact.

In the ideal case where a ballistic channel is assumed, the transmission coefficient remains 1 above the barrier and 0 below the barrier. However, in order to include the effects of scattering in an actual device, the reader is referred to [24] where edge scattering and optical phonon scattering are modeled and incorporated into the top-of-the-barrier approach to study GNRFETs.

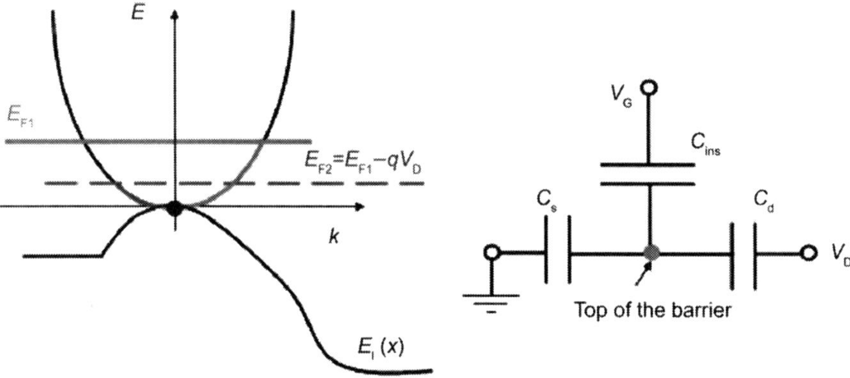

Fig. 12 Illustration of the top-of-the-barrier modeling approach [24]

When scattering effects are included, the transmission coefficient is scaled down by a factor determined by the channel length. The top-of-the-barrier U_{scf} is determined by two factors: (1) the applied terminal bias and (2) the change in carrier concentration in the channel. Therefore, the self-consistent potential can be written as $U_{\text{scf}} = U_{\text{L}} + U_{\text{P}}$, where U_{L} is called the Laplace potential due to the applied terminal biases and U_{P} is the potential due to the change in carrier concentration.

The Laplace potential U_{L} is easier to calculate because the applied terminal biases are known:

$$U_{\text{L}} = -q\frac{C_G V_G + C_D V_D + C_S V_S}{C_{sum}},$$

where C_G, C_D, and C_S are the gate, drain, and source capacitances and C_{sum} is given by the sum $C_G + C_D + C_S$. Since C_D and C_S are usually negligible in comparison to C_G, the effects of V_D and V_S on U_{L} can usually be ignored.

U_{P}, however, is more difficult to calculate because we cannot easily obtain the change in carrier concentration. When the terminal biases are zero, the electron density in the channel is:

$$N_0 = \int D(E)f(E - E_f)\mathrm{dE},$$

where $D(E)$ is the density of states at the bottom of the conduction band and $f(E - E_f)$ is the Fermi function. The density of states $D(E)$ of graphene is modeled in [24] and both edge bond relaxation and third nearest neighbor interactions are also considered.

When terminal biases are not zero, the device is not at equilibrium and the states at the bottom of the conduction band are filled by two different Fermi levels. States with positive and negative velocity are filled according to source Fermi level E_{f1} and drain Fermi level E_{f2}, respectively:

$$N^{+(-)} = \frac{1}{2}\int D(E - U_{scf})f(E - E_{f1(f2)})\mathrm{dE},$$

$$U_p = \frac{q^2}{C_{sum}}\Delta N = \frac{q^2}{C_{sum}\,(N^+ + N^- - N_0)}.$$

Therefore, U_{P} and N are self-consistently solved to obtain U_{scf}. In summary, the top-of-the-barrier model is a commonly used approach in modeling GNRFETs. In order to implement the top-of-the-barrier model for MOSFET-type GNRFETs, the reader needs to obtain the gate, source and drain capacitance, and the density of states of the GNR as described in [24].

5.4 Semi-classical Model with Tunneling

Whereas the top-of-the-barrier approach is sufficient to model MOSFET-type GNRFETs, it lacks the ability to handle the tunneling effects at SBs. In order to model SB-type GNRFETs, where tunneling current dominates the conduction, the basic framework is similar to the top-of-the-barrier approach, but the transmission coefficients in the Landauer formulation need to be evaluated according to the tunneling probability through the SBs.

As in the case of the top-of-the-barrier approach, the current through the SB-type FET is given by the following expressions:

$$I_{\text{Electron}} = \frac{2q}{h} \int_{-\infty}^{\infty} (f(E - E_{\text{FS}}) - f(E - E_{\text{FD}}))T_{\text{e}}(E)\text{dE}, I_{electron}$$

$$= \frac{2q}{h} \int_{-\infty}^{\infty} (f(E - E_{\text{FS}}) - f(E - E_{\text{FD}}))T_{\text{e}}(E)\text{dE},$$

$$I = E_{\text{Electron}} + E_{\text{Hole}}.$$

Note that in SB-type FETs, both electron and hole currents are important, and T_{e} (T_{h}) is the transmission coefficient for electrons (holes). At each contact, the transmission coefficients for electron current and hole current need to be calculated, for a total of four coefficients: T_{Se}, T_{Sh}, T_{De}, and T_{Dh}. The final transmission coefficients for electrons T_{e} and holes T_{h} are obtained by combining T_{Se}, T_{De} and T_{Sh}, T_{Dh}. The current is divided into two components: thermionic current, which flows above the SBs and tunneling current, which flows through the SB. Whereas the transmission coefficient is 1 for the thermionic current, it is given by the tunneling probability for the tunneling current. The tunneling probability is calculated using the simple but accurate Wentzel–Kramers–Brillouin (WKB) approach that has been widely used in literature [70]. Based on the WKB approach, the tunneling probability T is given by

$$T = \exp\left(-2\int_{z_{\text{init}}}^{z_{\text{final}}} k_z(z)dz\right),$$

where z_{init} and z_{final} are the classical turning points and k_z is the parallel momentum related to the E–k relationship of GNRs. Consider electron tunneling, for example. k_z is given by the expression:

$$k_z(z) = \frac{\hbar}{v\left(n\frac{E_g}{2} + |E_{\text{C}}(z) - E|\right)},$$

where n denotes the nth subband, E_g is the energy band gap, $E_{\text{C}}(z)$ is the bottom of the conduction band in the channel direction, and v is the Fermi velocity of graphene [66]. For the double-gate geometry, the conduction band E_C near the two contacts is modeled in [66]. Neglecting phase coherence, the overall transmission coefficient is given by [68]:

$$T_{e(h)} = \frac{T_{Se(Sh)}T_{De(Dh)}}{T_{Se(Sh)} + T_{De(Dh)} - T_{Se(Sh)}T_{De(Dh)}}.$$

Because of the existence of SBs, the states with positive velocity and negative velocity N^+ and N^- are also influenced. They are not filled solely by carriers from source and drain, respectively, but are filled by a mixture of carriers from both contacts. This can be interpreted as a carrier from the source with a positive velocity that is scattered back by the SB at the drain side, thus having a negative velocity. Therefore, N^+ and N^- are filled according to f^+ and f^-, respectively, where f^+ and f^- are:

$$f^+ = \frac{T_S f_S + T_D f_D - T_S T_D f_D}{T_S + T_D - T_S T_D},$$

$$f^- = \frac{T_S f_S + T_D f_D - T_S T_D f_S}{T_S + T_D - T_S T_D}.$$

The simulation results for a SB-type double-gated GNRFET with a $N = 12$ AGNR channel and insulator thickness of 2 nm is shown in Fig. 13. In the figure, the I–V curves obtained using this model is compared to quantum simulations based on

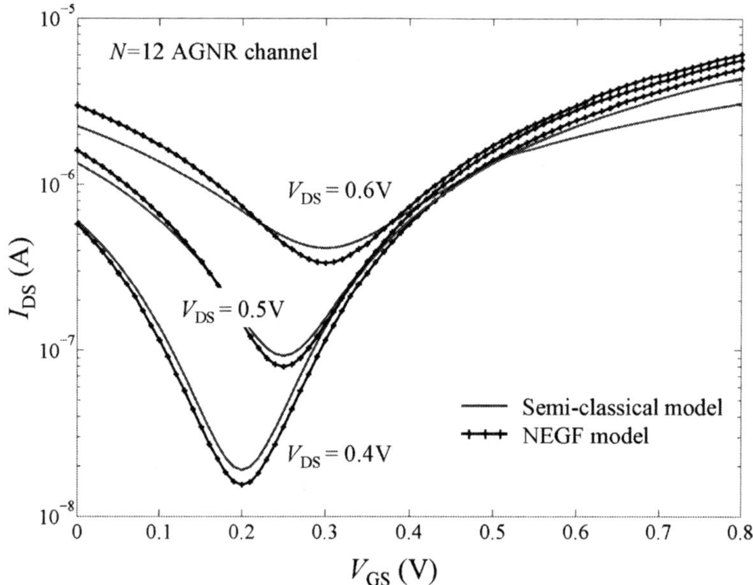

Fig. 13 Simulation results for a SB-type GNRFET comparing the semi-classical model to the NEGF approach [71]

the NEGF approach and there is an excellent agreement between the curves. In summary, by incorporating tunneling effects, the top-of-the-barrier modeling approach can be extended to treat SB-type GNRFETs in a parameterizable and computationally efficient and accurate manner.

6 Conclusions and Prospects

Graphene has evolved to where experimental and theoretical advances indicate its high potential as a candidate material for nanoelectronics applications. Several challenges spanning materials and fabrication, device modeling and simulation, and circuit design and optimization remain to be addressed and are the focus of intensive concurrent interdisciplinary research.

On the materials and fabrication front, it is necessary to develop techniques to manufacture high-quality graphene on a large scale, with high yield follow-through steps to realize devices based on monolayer, bilayer, and nanoribbon channels. Furthermore, concerns arising from edge quality, width variations, substrate impurities, source/drain contact resistance, and contact-limited effects will play an important role in determining the optimum device configuration.

On the modeling and simulation front, it is necessary to calibrate the various approaches described in this chapter to ensure accuracy and integration of the various phenomena that influence quantum transport in such ultimately scaled devices. Furthermore, it is also necessary to develop computationally efficient models so that it is easy to identify, model, and explore the effect of different variability and defect mechanisms in complex circuits comprising more than just one transistor. This will provide expedient means to systematically understand and predict the effects of variability and defects on the performance and reliability of practical graphene-based circuits.

The challenge of moving beyond ideal device structures to real-world devices – in a bottom-up manner, with a better understanding of practical limits – is critical to the success of graphene-based electronics. Finally, circuit design and optimization will need to explore and leverage new transistor characteristics arising from contact-limited transport and quantum effects. There is already significant progress by way of unique single-transistor analog circuits, ambipolar logic circuits, and ultra-steep subthreshold logic circuits. Further research and concurrent optimization of technology and design has the potential to result in the advances necessary to harness the early science of graphene-based electronics into practical design technologies.

Exercises

1. Implement the simple tight binding (TB) model to plot the AGNR band-gap against AGNR width, as described in [61]. Reproduce the curves for the $3p$, $3p + 1$, and $3p + 2$ AGNRs, as shown in Fig. 3 from this paper.

2. Extend the simple TB model to incorporate edge-bond relaxation and third nearest-neighbor effects, as described in [61]. Reproduce Fig. 3 from this paper, which illustrates how these effects influence AGNR band-gap in comparison to the original simple TB model.

3. Implement the charge-collection model for graphene transistors described in [32]. Additional parameters and details necessary to implement the model are available as part of the supplementary material that is provided with the paper.

4. Use the model that you implement in Excercise 3 to demonstrate frequency multiplication, as proposed in [44].

5. Implement the I–V model for GNRFETs proposed in [66]. This model assumes that the device operates in the quantum capacitance limit, hence eliminating the need for self-consistent solution.

6. Implement the top-of-the-barrier model for GNRFETs proposed in [61].

7. Register and open an account on the nanoHUB to get access to the NanoTCAD VIDES package. Simulate a basic SB-type GNRFET using the deck provided in the package. Vary GNR width and oxide thickness to simulate the effect of parameters that influence GNRFET operation.

8. Simulate a basic inverter using the I–V data that you obtain from Excercise 7. Set GNR $N = 12$ and $V_{DD} = 1$ V. Look at the transfer curve of the inverter. Does the inverter function properly?

9. Implement the Boolean function $f = a'bc' + ab'c + c'de' + c'd'e$ using a single ambipolar logic gate using the transmission gate design approach described in [56]. Compare the design against the number of transistors and levels of logic required for a traditional optimized implementation in CMOS technology.

10. Using the logic style proposed in Fig. 2 of [53], implement the function $f = a'b' + abcd$. If *only* the voltage of the polarity gate of each of the transistors is configurable (the regular inputs are fixed per the structure that you derive), how many functions of four inputs can the same circuit implement?

References

1. C. Berger, Z. Song, X. Li, X. Wu, N. Brown, C. Naud, D. Mayou, T. Li, J. Hass, A. N. Marchenkov, E.H. Conrad, P.N. First, and W.A. de Heer, "Electronic confinement and coherence in patterned epitaxial graphene," *Science*, 312, 1191–1196, 2006.

2. A.K. Geim and K.S. Novoselov, "The rise of graphene," *Nature Mater.*, 6, 183–191, 2007.

3. R.V. Noorden, "Moving towards a graphene world," *Nature*, 442, 228–229, 2006.

4. K.S. Novoselov, A.K. Geim, S.V. Morozov, D. Jiang, Y. Zhang, S.V. Dubonos, I.V. Grigorieva, and A.A. Firsov, "Electric field effect in atomically thin carbon films," *Science*, 306, 666–669, 2004.

5. Y. Zhang, Y.-W. Tan, H.L. Stormer, and P. Kim, "Experimental observation of the quantum Hall effect and Berry's phase in graphene," *Nature*, 438, 201–204, 2005.

6. W.A. de Heer, C. Berger, E. Conrad, P. First, R. Murali, and J. Meindl, "Pionics: The emerging science and technology of graphene-based nanoelectronics," in *Proc. Intl. Electron Devices Meeting*, pp. 199–202, 2007.

7. A. Gruneis and D.V. Vyalikh, "Tunable hybridization between electronic states of graphene and a metal surface," *Phys. Rev. B.*, 77, 193407, 2008.

8. X. Li, W. Cai, J. An, S. Kim, J. Nah, D. Yang, R. Piner, A. Velamakanni, I. Jung, E. Tutuc, S.K. Banerjee, L. Colombo, and R.S. Ruoff, "Large-area synthesis of high-quality and uniform graphene films on copper foils," *Science*, 324, 5932, 1312–1314, 2009.

9. S. Unarunotai, Y. Murata, C.E. Chialvo, H.-S. Kim, S. MacLaren, N. Mason, I. Petrov, and J.A. Rogers, "Transfer of graphene layers grown on SiC wafers to other substrates and their integration into field effect transistors," *Appl. Phys. Lett.*, 95, 202101, 2009.

10. J.C. Meyer, A.K. Geim, M.I. Katsnelson, K.S. Novoselov, T.J. Booth, and S. Roth, "The structure of suspended graphene sheets," *Nature*, 446, 60–63, 2007.

11. A. Reina, X. Jia, J. Ho, D. Nezich, H. Son, V. Bulovic, M.S. Dresselhaus, and J. Kong, "Large area, few-layer graphene films on arbitrary substrates by chemical vapor deposition," *Nano Lett.*, 9, 1, 30–35, 2009.

12. C. Oshima and A. Nagashima, "Ultra-thin epitaxial films of graphite and hexagonal boron nitride on solid surfaces," *J. Phys. Condens. Matter*, 9, 1, 1–20, 1997.

13. E.V. Castro, K.S. Novoselov, S.V. Morozov, N.M.R. Peres, J.M.B. Lopes dos Santos, J. Nilsson, F. Guinea, A.K. Geim, and A.H. Castro Neto, "Biased bilayer graphene: Semiconductor with a gap tunable by the electric field effect," *Phys. Rev. Lett.*, 99, 21, 216802, 2007.

14. J. Nilsson, C.A.H. Neto, F. Guinea, and N.M.R. Peres, "Electronic properties of bilayer and multilayer graphene," *Phys. Rev. B, Condens. Matter*, 78, 4, 045405-1–045405-34, 2008.

15. T. Ohta, A. Bostwick, T. Seyller, K. Horn, and E. Rotenberg, "Controlling the electronic structure of bilayer graphene," *Science*, 313, 5789, 951–954, 2006.

16. H. Schmidt, T. Lüdtke, P. Barthold, E. McCann, V.I. Fal'ko, and R.J. Haug, "Tunable graphene system with two decoupled monolayers," *Appl. Phys. Lett.*, 93, 172108, 2008.

17. J.M.B. Lopes dos Santos, N.M.R. Peres, and A.H. Castro Neto, "Graphene bilayer with a twist: Electronic structure," *Phys. Rev. Lett.*, 99, 256802, 2007.

18. M.Y. Han, B. Özyilmaz, Y. Zhang, and P. Kim, "Energy band-gap engineering of graphene nanoribbons," *Phys. Rev. Lett.*, 98, 206805, 2007.

19. C.T. White, J. Li, D. Gunlycke, and J.W. Mintmire, "Hidden one-electron interactions in carbon nanotubes revealed in graphene nanostrips," *Nano Lett.*, 7, 825–830, 2007.

20. X. Wang, L. Zhang, S. Lee, and H. Dai, "Chemically derived, ultrasmooth graphene nanoribbon semiconductors," *Science*, 319, 1229–1232, 2008.

21. L. Jiao, L. Zhang, X. Wang, G. Diankov, and H. Dai, "Narrow graphene nanoribbons from carbon nanotubes," *Nature*, 458, 877–880, 2009.

22. K.V. Kosynkin, A.L. Higginbotham, A. Sinitskii, J.R. Lomeda, A. Dimiev, B.K. Price, and J.M. Tour, "Longitudinal unzipping of carbon nanotubes to form graphene nanoribbons," *Nature*, 458, 872–876, 2009.

23. J. Delgado, J.M.R. Herrera, X. Jia, D.A. Cullen, H. Muramatsu, Y.A. Kim, T. Hayashi, Z. Ren, D.J. Smith, Y. Okuno, T. Ohba, H. Kanoh, K. Kaneko, M. Endo, H. Terrones, M.S. Dresselhaus, and M. Terrone, "Bulk production of a new form of sp^2 carbon: Crystalline graphene nanoribbons," *Nano Lett.*, 8, 9, 2773–2778, 2008.

24. P. Zhao, M. Choudhury, K. Mohanram, and J. Guo, "Computational model of edge effects in graphene nanoribbon transistors," *Nano Res.*, 1, 5, 395–402, 2008.

25. Y.-W. Son, M.L. Cohen, and S.G. Louie, "Energy gaps in graphene nanoribbons," *Phys. Rev. Lett.*, 97, 216803, 2006.

26. P. Shemella and S.K. Nayak, "Electronic structure and band-gap modulation of graphene via substrate surface chemistry," *Appl. Phys. Lett.*, 94, 032101, 2009.

27. Z. Ni, T. Yu, Y.H. Lu, Y.Y. Wang, Y.P. Feng, and Z.X. Shen, "Uniaxial strain on graphene: Raman spectroscopy study and band-gap opening," *ACS Nano*, 2, 11, 2301–2305, 2008.

28. J. Bai, X. Zhong, S. Jiang, Y. Huang, and X. Duan, "Graphene nanomesh," *Nat. Nanotechnol.*, 5, 190–194, 2010.

29. X. Wang, Y. Ouyang, X. Li, H. Wang, J. Guo, and H. Dai, "Room temperature all semiconducting sub-10 nm graphene nanoribbon FETs," *Phys. Rev. Lett.*, 100, 206803, 2008.

30. B. Huard, N. Stander, J.A. Sulpizio, and D. Goldhaber-Gordon, "Evidence of the role of contacts on the observed electron-hole asymmetry in graphene," *Phys. Rev. B*, 78, 121402, 2008.
31. M.C. Lemme, T.J. Echtermeyer, M. Baus, and H. Kurz, "A graphene field-effect device," *Electron Device Lett.*, 28, 4, 282–284, 2009.
32. I. Meric, M.Y. Han, A.F. Young, B. Ozyilmaz, P. Kim, and K.L. Shepard, "Current saturation in zero-bandgap, top-gated graphene field effect transistors," *Nat. Nanotechnol.*, 3, 1–6, 2008.
33. J.R. Williams, L. DiCarlo, C.M. Marcus, "Quantum hall effect in a gate-controlled p-n junction of graphene," *Science*, 317, 5838, 638–641, 2007.
34. G. Fiori and G. Iannaccone, "Simulation of graphene nanoribbon field effect transistors," *IEEE Electron. Device Lett.*, 28, 760–762, 2007.
35. G. Liang, N. Neophytou, M. Lundstrom, and D.E. Nikonov, "Ballistic graphene nanoribbon metal oxide semiconductor field-effect transistors: A full real-space quantum transport simulation," *J. Appl. Phys.*, 102, 054307, 2007.
36. G. Liang, N. Neophytou, D.E. Nikonov, and M. Lundstrom, "Performance projections for ballistic graphene nanoribbon field-effect transistors," *IEEE Trans. Electron. Device*, 54, 677–682, 2007.
37. Y. Yoon, G. Fiori, S. Hong, G. Iannaccone, and J. Guo. "Performance comparison of graphene nanoribbon FETs with Schottky contacts and doped reservoirs," *IEEE Trans. Electron Device*, 55, 2314–2323, 2008.
38. X. Wang, X. Li, L. Zhang, Y. Yoon, P.K. Weber, H. Wang, J. Guo, and H. Dai, "*N*-doping of graphene through electrothermal reactions with ammonia," *Science*, 324, 5928, 768–771, 2009.
39. I. Meric, N. Baklitskaya, P. Kim, and K.L. Shepard, "RF performance of top-gated, zero-bandgap graphene field effect transistors," in *Proc. Intl. Electron Devices Meeting*, pp. 1–4, 2008.
40. Y. Lin, K.A. Jenkins, A. Valdes-Garcia, J.P. Small, D.B. Farmer, and Ph. Avouris, "Operation of graphene transistors at gigahertz frequencies," *Nano Lett.*, 9, 1, 422–426, 2009.
41. J.S. Moon, D. Curtis, M. Hu, D. Wong, C. McGuire, P.M. Campbell, G. Jernigan, J.L. Tedesco, B. Vanmil, R. Myers-Ward, C. Eddy, and D.K. Gaskill, "Epitaxial-graphene RF field-effect transistors on Si-face 6H-SiC substrates," *IEEE Electron Device Lett.*, 30, 6, 650–652, 2009.
42. D.B. Farmer, H.-Y. Chiu, Y.-M. Lin, K.A. Jenkins, F. Xia, and Ph. Avouris, "Utilization of a buffered dielectric to achieve high field-effect carrier mobility in graphene transistors," *Nano Lett.*, 9, 12, 4474–4478, 2009.
43. Y. Lin, C. Dimitrakopoulos, K.A. Jenkins, D.B. Farmer, H.-Y. Chiu, A. Grill, and Ph. Avouris, "100-GHz transistors from wafer-scale epitaxial graphene," *Science*, 327, 5966, 662, 2010.
44. H. Wang, D. Nezich, J. Kong, and T. Palacios, "Graphene frequency multipliers," *IEEE Electron Device Lett.*, 30, 5, 547–549, 2009.
45. Z. Chen, Y.-M. Lin, M.J. Rooks, and Ph. Avouris, "Graphene nano-ribbon electronics," *Physica E*, 40, 228, 2007.
46. G. Fiori and G. Iannaccone, "On the possibility of tunable-gap bilayer graphene FET," *IEEE Electron Device Lett.*, 30, 261–264, 2009.
47. Y. Ouyang, P. Campbell, and J. Guo, "Analysis of ballistic monolayer and bilayer graphene field-effect transistors," *Appl. Phys. Lett.*, 92, 063120, 2008.
48. F. Xia, D.B. Farmer, Y.-M. Lin, and Ph. Avouris, "Graphene field-effect transistors with high on/off current ratio and large transport band gap at room temperature," *Nano Lett.*, 10, 715–718, 2010.
49. Y. Ouyang, H. Dai, and J. Guo, "Projected performance advantage of multilayer graphene nanoribbons as a transistor channel material," *Nano Res.*, 3, 8–15, 2010.
50. M. Choudhury, Y. Yoon, J. Guo, and K. Mohanram, "Technology exploration for graphene nanoribbon FETs," in *Proc. Design Automation Conference*, pp. 272–277, 2008.

51. Y.-M. Lin, J. Appenzeller, J. Knoch, and Ph. Avouris, "High-performance carbon nanotube field-effect transistor with tunable polarities," *IEEE Trans. Nanotechnol.*, 4, 481–489, 2005.

52. I. O'Connor, J. Liu, F. Gaffiot, F. Pregaldiny, C. Lallement, C. Maneux, J. Goguet, F. Fregonese, T. Zimmer, L. Anghel, T.-T. Dang, and R. Leveugle, "CNTFET modeling and reconfigurable logic-circuit design," *IEEE Trans. Circuits Syst. I*, 54, 11, 2365–2379, 2007.

53. M.H. Ben-Jamaa, D. Atienza, Y. Leblebici, and G. De Micheli, "Programmable logic circuits based on ambipolar CNFET," in *Proc. Design Automation Conference*, pp. 339–340, 2008.

54. F. Mo and R.K. Brayton, "Whirlpool PLAs: A regular logic structure and their synthesis," in *Proc. Intl. Conference Computer-Aided Design*, pp. 543–550, 2002.

55. T. Sasao, *Switching Theory for Logic Synthesis*, Kluwer Academic Publishers, Netherlands, 1999.

56. M.H. Ben-Jamaa, K. Mohanram, and G. De Micheli, "Novel library of logic gates with ambipolar CNTFETs: Opportunities for multi-level logic synthesis," in *Proc. Design Automation and Test in Europe Conference*, pp. 622–627, 2009.

57. M.H. Ben-Jamaa, K. Mohanram, and G. De Micheli, "Power consumption of logic circuits in ambipolar carbon nanotube technology," in *Proc. Design Automation and Test in Europe Conference*, pp. 622–627, 2010.

58. W.Y. Choi, B.-G. Park, J.D. Lee, and T.-J.K. Liu, "Tunneling field-effect transistors (TFETs) with subthreshold swing (SS) less than 60 mV/dec," *IEEE Electron Device Lett.*, 28, 8, 743–745, 2007.

59. M. Luisier and G. Klimeck, "Performance analysis of statistical samples of graphene nanoribbon tunneling transistors with line edge roughness," *Appl. Phys. Lett.*, 94, 223505, 2009.

60. Q. Zhang, T. Fang, H. Xing, A. Seabaugh, and D. Jena, "Graphene nanoribbon tunnel transistors," *IEEE Electron Device Lett.*, 29, 1344–1346, 2008.

61. P. Zhao, J. Chauhan, and J. Guo, "Computational study of tunneling transistor based on graphene nanoribbon," *Nano Lett.*, 9, 684–688, 2009.

62. G. Fiori and G. Iannaccone, "Ultralow-voltage bilayer graphene tunnel FET," *IEEE Trans. Electron Devices*, 55, 10, 1096–1098, 2008.

63. M.P. Anantram, M.S. Lundstrom, and D.E. Nikonov, "Modeling of nanoscale devices," *Proc. IEEE*, 96, 1511–1550, 2008.

64. Y. Ouyang, Y. Yoon, and J. Guo "Scaling behaviors of graphene nanoribbon FETs: A 3D quantum simulation," *IEEE Trans. Electron Devices*, 54, 2223, 2007.

65. G. Fiori and G. Iannaccone, "NanoTCAD ViDES," 2008. DOI: 10254/nanohubr5116.3.

66. D. Jimenez, "A current-voltage model for Schottky-barrier graphene-based transistors," *Nanotechnology*, 19, 345204, 2008.

67. A. Rahman, J. Guo, S. Datta, and M.S. Lundstrom, "Theory of ballistic nanotransistors," *IEEE Trans. Electron Devices*, 50, 1853–1864, 2003.

68. S. Datta, *Electronic Tansport in Mesoscopic Systems*, New York: Cambridge University Press, 1995.

69. http://www.nanohub.org

70. N.H. Frank and L.A. Young, "Transmission of electrons through potential barriers," *Phys. Rev.*, 38, 80–86, 1931.

71. X. Yang, G. Fiori, G. Iannaccone, and K. Mohanram, "Physics-based semi-analytical model for Schottky-barrier carbon nanotube and graphene nanoribbon transistors," in *Proc. Great Lakes Symposium on VLSI*, 2010.

72. D. Gunlycke and C.T. White, "Tight-binding energy dispersions of armchair-edge graphene nanostrips," *Phys. Rev. B*, 77, 115116, 2008.

73. S. Heinze, J. Tersoff, R. Martel, V. Derycke, J. Appenzeller, and Ph. Avouris, "Carbon nanotubes as Schottky barrier transistors," *Phys. Rev. Lett.*, 89, 106801, 2002.

74. S. Koswatta and M. Lundstrom, "Influence of phonon scattering on the performance of p-i-n band-to-band tunneling transistors," *Appl. Phys. Lett.*, 92, 043125, 2008.

75. S. Koswatta, M.S. Lundstrom, and D.E. Nikonov, "Performance comparison between p-i-n tunneling transistors and conventional MOSFETs," *IEEE Trans. Electron Devices*, 56, 456–465, 2009.
76. M. Lundstrom, J. Guo, *Nanoscale Transistors: Device Physics, Modeling and Simulation*, New York: Springer, 2006.
77. K. Nakada and M. Fujita, "Edge state in graphene ribbons: Nanometer size effect and edge shape dependence," *Phys. Rev. B*, 54, 17954–17961, 1996.
78. K. Natori, "Ballistic metal oxide semiconductor field effect transistor," *J. Appl. Phys.*, 76, 4879–4890, 1994.
79. X. Yang, X. Dou, A. Rouhanipour, L. Zhi, H.J. Räder, and K. Müllen "Two-dimensional graphene nanoribbons," *J. Am. Chem. Soc.*, 130, 4216–4217, 2008.
80. X. Yang, G. Liu, A.A. Balandin, and K. Mohanram, "Triple-mode single-transistor graphene amplifier and its applications," ACS Nano, 4, 10, 5532–5538, 2010.
81. Yang-glsvlsi-10, "Graphene tunneling FET and its applications in low-power circuit design," *Proc. Great Lakes Symposium on VLSI*, 263–268, 2010.

Study of Performances of Low-k Cu, CNTs, and Optical Interconnects

Kyung-Hoae Koo and Krishna C. Saraswat

Abstract Carbon Nanotubes (CNTs), Graphene Nanoribbons (GNRs), and Optical Interconnects and carbon nanotubes present promising options for replacing the existing Cu-based global/semiglobal and local wires. In this chapter, the performance of these novel interconnects is quantified and compared with Cu/low-κ wires for future high-performance integrated circuits. For a local wire, a CNT bundle exhibits a smaller latency than Cu for a given geometry. In addition, by leveraging the superior electromigration properties of CNT and optimizing its geometry, the latency advantage can be further amplified. GNRs outperforms Cu wire below the width of 7nm. For semiglobal and global wires, both optical and CNT options are compared with Cu in terms of latency, energy efficiency/power dissipation, and bandwidth density. The above trends are studied with technology node. In addition, for a future technology node, we compare the relationship between bandwidth density, power density, and latency, thus alluding to the latency and power penalty to achieve a given bandwidth density.

Keywords Carbon nanotube · Graphene nanoribbons · Optical interconnects · Latency · Power · Bandwidth density · Cu · Global interconnect · Local interconnect

1 Introduction

The ever-degrading performance of on-chip copper (Cu) wires threatens to greatly impede the continued IC improvement along Moore's law. All wire metrics, including latency, power dissipation, bandwidth density, and reliability, for local and global wires deteriorate with scaling. Specifically, electron scattering from interfaces and grain-boundaries dramatically increases Cu resistivity as dimension

K.C. Saraswat (✉)
Department of Electrical Engineering, Stanford University, Stanford, CA 94305, USA
e-mail: saraswat@stanford.edu

N.K. Jha and D. Chen (eds.), *Nanoelectronic Circuit Design*,
DOI 10.1007/978-1-4419-7609-3_11, © Springer Science+Business Media, LLC 2011

scales down [1]. Reduced interconnect cross-section also severely degrades electromigration reliability [2]. In addition, on the demand side, the advent of multicore architectures has placed a premium on high-bandwidth density (bandwidth per unit distance), Φ_{BW}, and low-latency links between cores. Thus, now more than ever, it is imperative to examine novel interconnect technologies.

Optical and *carbon nanotube* (CNT)-based wires constitute two compelling, novel, alternatives. Optical interconnects differ fundamentally from the electrical schemes (CNT and Cu) in that (1) a large part of the latency and the entire power dissipation is in the end-devices instead of the medium, and (2) the nature of power dissipation is mostly static instead of dynamic. These differences, when coupled with favorable wire architectures, present new opportunities for optical wires. Although, promising for most wire metrics, optics does suffer from the drawback of a relatively large transmission medium (waveguide) pitch ($\sim$0.6 μm). The resulting Φ_{BW} limitation can be surmounted using the unique *wavelength division multiplexing* (WDM) option available for optical wires. However, given the wire size scales and the complexity of WDM, it is only suitable for global/ semiglobal wiring. In contrast, a *single-walled CNT* (SWCNT) bundle can be used for all wires, including local, semiglobal and/or global, as it can be implemented in the same size scale as Cu wires ($\sim$50 nm). The CNT advantage is that it bears a much lower resistivity (large mean free path, MFP) [3], and a much higher electromigration tolerance and thermal conductivity than Cu. A 1-GHz CNT-integrated oscillator has already been demonstrated, expediting the advent of high-performance CNT-based interconnect fabric [4].

In this chapter, we seek to investigate the fundamental materials' or systems' properties for future interconnect candidates and compare their performance metrics. While comparing them, this chapter attempts a comprehensive quantification of advantages associated with novel technologies using metrics such as latency, energy per bit/power dissipation, and bandwidth density, Φ_{BW} – which is becoming more and more important as the performance of multicore architectures rely on the maximum bandwidth that can be extracted from the limited periphery of a core [5].

2   Circuit Parameter Modeling

2.1   Modeling Parameters for Copper

As wire cross-section dimensions and grain size become comparable to the bulk MFP of electrons in Cu, two major physical phenomena occur to the current-carrying electrons in Cu in addition to bulk phonon, as illustrated in Fig. 1a. One is interface scattering, (1) [6], and the other is grain boundary scattering, (2) [7]. These increase Cu resistivity more than the ideal bulk resistivity ($\rho_o = 1.9$ μΩ cm) [8]. The Fuchas–Sondheimer model and the theory of Mayadas and Shatzkes

Fig. 1 (**a**) Schematic illustration of the surface and grain boundary scatterings, and the barrier effect. (**b**) Impact of scaling on the barrier effect. Cu can be scaled while the barrier cannot

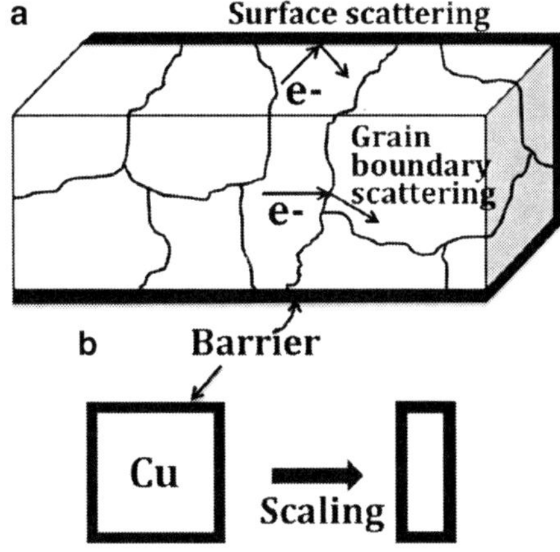

quantify these two effects. Equations (1) and (2) describe Fuchas–Sondheimer (ρ_{FS}) model and Mayadas–Shatzkes (ρ_{MS}) effect, respectively.

$$\frac{\rho_{FS}}{\rho_o} = 1 + \frac{3}{4}(1-p)\frac{l_o}{w} \tag{1}$$

$$\frac{\rho_o}{\rho_{MS}} = 3\left[\frac{1}{3} - \frac{\alpha}{2} + \alpha^2 - \alpha^3 \ln\left(1 + \frac{1}{\alpha}\right)\right] \tag{2}$$

$$\alpha = \frac{l_o}{d}\frac{R}{1-R}$$

Here, ρ_o is the bulk Cu resistivity and p is the electron fraction scattered specularly at the interface (assumed to be 0.6 [1]). The value of p ranges from 0 to 1 and approaches 1 as electrons have more interface scattering. The value w is the wire width (from ITRS [9]), and l_o is the bulk MFP of electrons. For (2), R is the reflectivity coefficient representing the electron fraction not scattered at grain boundaries (assumed to be 0.5 [1]). This coefficient also ranges from 0 to 1, approaching 0 as electrons have more grain boundary scatterings. The value d is the average grain size ($d-w$).

Based on the above models, at the 22-nm node, Cu resistivity for minimum width wire increases to 5.8 $\mu\Omega$ cm ($\sim3'$ that of bulk). On the other hand, Cu interconnects typically need a diffusion barrier, which could come in the form of Ta-, Ru-, and Mg-based materials. Because the resistivity of these materials is much higher than that of Cu, in effect, the barrier reduces the useful interconnect cross-section. One way to capture this effect is to define the problem in terms of an increased effective resistivity, which would be applicable to the original cross-sectional area. Figure 1b shows that scaling exacerbates the barrier problem as barrier thickness does not scale proportionately to the aggressive interconnect cross-sectional scaling. This, in turn, results in an increase in the effective

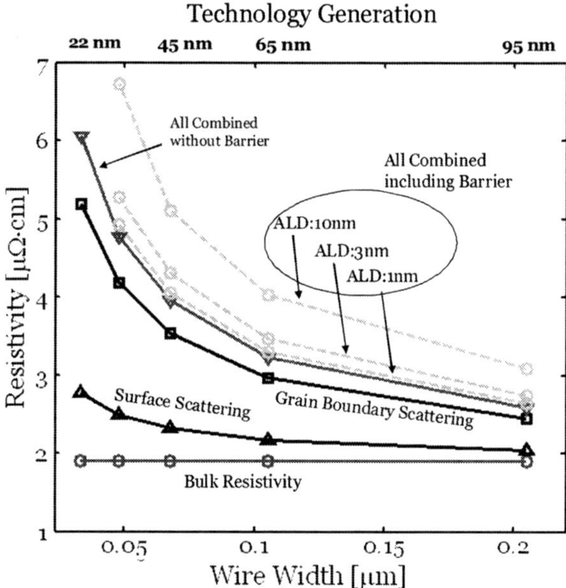

Fig. 2 Cu resistivity in terms of wire width, taking into account the surface and grain boundary scattering and barrier effect. The barrier layer is assumed to be uniformly deposited, e.g., using atomic layer deposition (ALD)

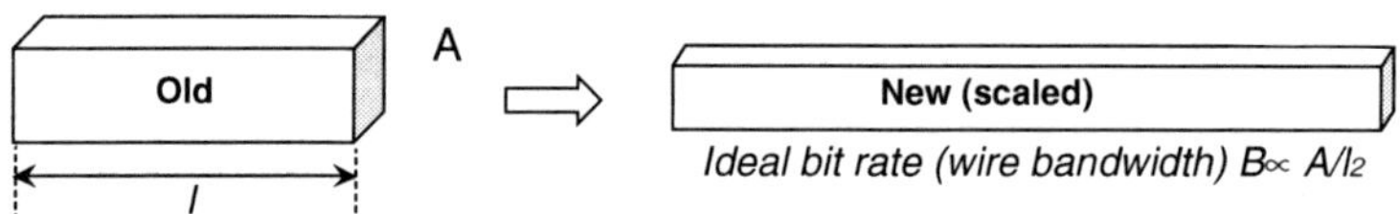

Fig. 3 The impact of interconnect scaling. Scaled wire with lower A and longer l has higher resistance, resulting in higher delay, increased power, and reduced bandwidth

resistivity with the technology node. Three dashed curves in Fig. 2 quantify this effect. The effective resistivity here also captures both the surface and grain boundary scatterings. It is clear that with technology scaling, effective resistivity increases dramatically. Further, lowering the barrier thickness (ALD: 3 vs. 1 nm) has a big impact on effective resistivity [8].

Figure 3 shows the scaling trend of the electrical wire dimensions and its impact on the bandwidth and power budget of the wire. The increase in wire length (l) in addition to the reduction in cross-section area (A) further exacerbates wire resistance, subsequently limiting the signal rise time and the bandwidth. This can be well understood from the simple relationship between the ideal bit rate (B), and the cross-sectional area and the wire length in Fig. 3. Typically, the wire buffering with multiple repeaters mitigates the bandwidth shortfall. However, it swallows a significant portion of the power budget (this will be explained in detail in "Sect. 2.1.1"). Thus, for the electrical interconnect, it becomes more difficult to meet the bandwidth requirement and power budget simultaneously [10].

Example 1. Figure 3 shows that the ideal bit rate is inversely proportional to l^2. How does the dynamic energy consumption (0.5 CV^2) scale as a function of l, d, and A? Assume that the wire is surrounded by two ground lines on both sides, and that the top and bottom ground lines are positioned at an infinite distance. The spacing between lines is equal to the wire width and has the same scaling factor as the wire dimension. Assume the wire cross-section is square-shaped. Neglect any fringing capacitance.

Answer:
The capacitance of the wire has only intermetal capacitance because the top and bottom layers are located at infinite distance. A general expression for the parallel plate capacitance is

$$C = \varepsilon \frac{\text{Area}}{\text{distance}}$$

The wire height or width is $\sqrt{A}$ from the description. The spacing between lines is also $\sqrt{A}$.

Thus, the capacitance of the wire is

$$C_{\text{wire}} = 2 \times \varepsilon \frac{\sqrt{A} \cdot l}{\sqrt{A}} \propto l$$

2.1.1 Conventional Interconnect Circuit and Its Performance Limit

In general, the interconnect performance has been optimized in terms of delay such that repeater distance (l) and size (w) are leveraged for equal delay from the repeater and wire [11–12]. However, this approach suffers from associated problems. While this methodology can significantly reduce delay, it results in inordinate power consumption. This excessive power expenditure has a close relationship with increasing number of repeaters, which is the consequence of meeting the small delay requirement [13].

Figure 4 shows this phenomenon, assuming a minimum power dissipation budget defined by ITRS [9]. Furthermore, a massive number of repeaters leads to the

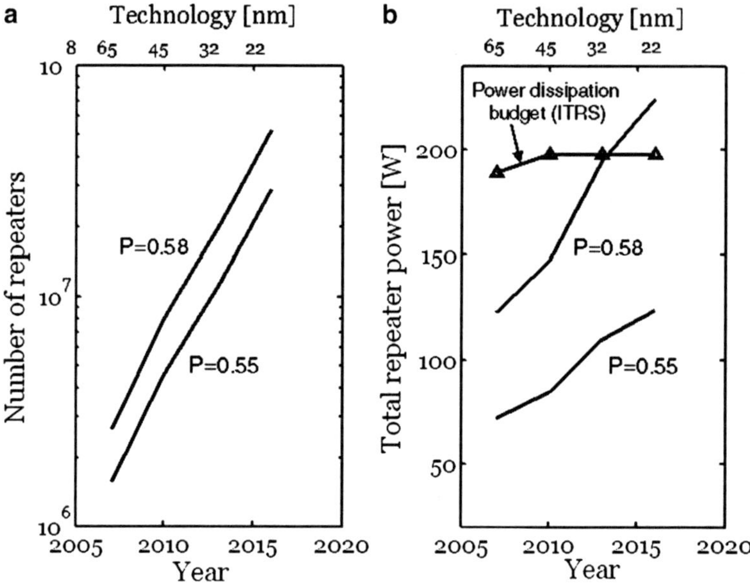

Fig. 4 (**a**) Number of global wire repeaters versus wire dimensions. (**b**) Total energy per bit (pJ) in global wire versus wire dimension

via blockage problem, which severely reduces the wiring efficiency [14–15]. On the other hand, if the global wires do not scale at all, staying at a constant pitch to avoid these problems, it will cause a rapid increase in the number of metal layers [15].

2.1.2 Interconnect Application of Carbon Nanotubes

CNTs have attracted great attention because of their interesting physical and electrical properties. Their near one-dimensional shape supports ballistic transport, making them potentially useful in many applications, such as transistors, sensors, and interconnects. In addition, CNTs offer great mechanical strength due to their strong sigma (σ) bonds between neighboring carbon atoms. These excellent physical properties have been proven theoretically and experimentally by intensive research for more than a decade. CNTs can be categorized as semiconducting or metallic depending on their chiral configurations.

Figure 5 shows the bandstructures of armchair and zigzag CNTs corresponding to metallic and semiconducting nanotubes, respectively. For the interconnect application, we only consider metallic carbon nanotubes. Depending on the shape, CNTs can be categorized as SWCNT or multiwalled carbon nanotube (MWCNT). An SWCNT is constructed by wrapping a graphene layer into a cylinderical shape. Its diameter ranges from 0.4 to 4 nm. Its typical diameter is around 1 nm. An MWCNT is formed of multiple layers of SWCNTs with different diameters. Its diameter ranges from 10 to 100 nm. Figure 6 shows the difference between SWCNT and MWCNT.

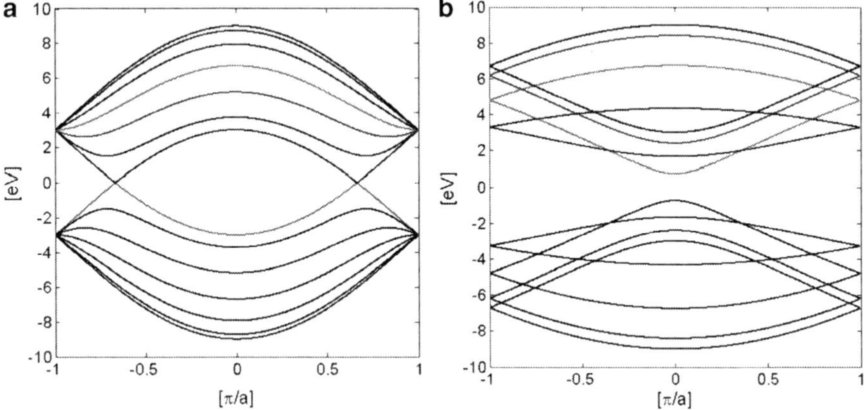

Fig. 5 (a) Armchair (6, 6) SWCNT, (b) Zigzag (7, 0) SWCNT

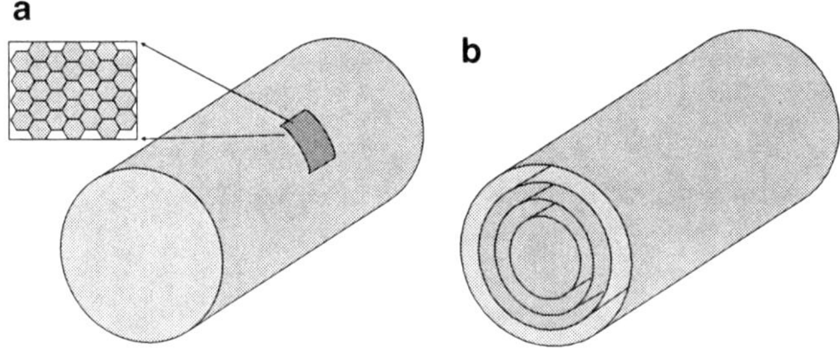

Fig. 6 (a) SWCNT, (b) MWCNT

2.1.3 Modeling Parameters for SWCNT Bundles

For an SWCNT, in the absence of any type of scattering, the maximum quantum
conductance is limited as shown in (3)

$$G_{\text{SWCNT}} = \frac{2q^2}{\pi\hbar} \tag{3}$$

where $\hbar$ is the Planck constant and q is a charge of one electron. The multiple two
accounts for the two channels due to electron spin and another two channels due to
sub-lattice degeneracy. Thus, the quantum resistance of an SWCNT is 6.45 kΩ.
This resistance is fairly large to allow use in interconnect applications.

One way to get around this problem is to use a bundle of SWCNTs, as illustrated
in Fig. 7. The resistance of a CNT bundle depends on the total wire cross-section
area and the fractional packing density (PD) of metallic CNTs within it (Fig. 7).
Henceforth, PD is assumed to be 33% unless specified otherwise. This is because
given an equal probability of the wrapping vectors, about one-third and two-thirds

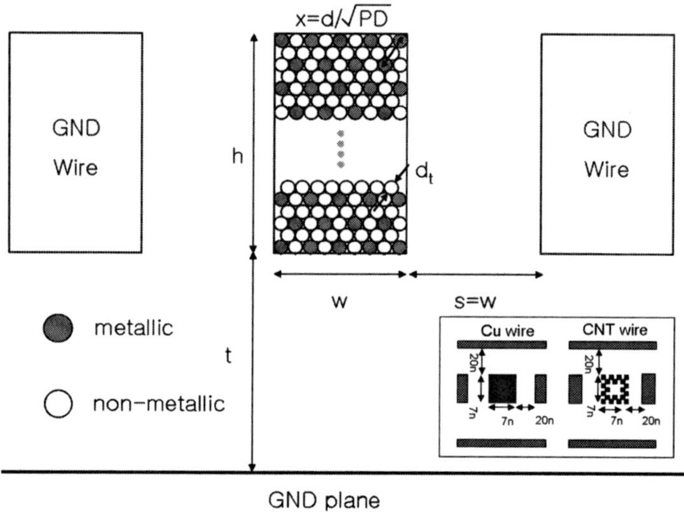

Fig. 7 Schematic of the interconnect geometry with CNT bundles, taking packing density (PD) into account. x is the distance between the closest metallic SWCNTs, d_t is the diameter of an SWCNT. (*Inset*) CNT and Cu interconnect cross-section geometry for Fastcap simulation

of SWCNTs turn out to be metallic and semiconducting, respectively [16]. The number of SWCNTs in a bundle is given by

$$n_w = \frac{w - d_t}{x}, n_h = \frac{h - d_t}{(\sqrt{3}/2)x} + 1 \quad \text{is even} = n_w n_h - \frac{n_h - 1}{2} \text{ if } n_h \text{ is odd} \quad (4)$$
$$n_{CNT} = n_w n_h - n_h/2 \text{ if} \{n_h$$

where n_w is the number of "columns," n_h is the number of "rows," and n_{CNT} is the number of SWCNTs in a bundle.

The quantum resistance of an SWCNT (R_Q) is 6.45 KΩ, as explained earlier. However, in practice, electrons do get scattered by either defects or phonons, hence, possess a finite MFP. Therefore, a linear dependence of resistance on length, based on recent results on high quality (long MFP) CNTs, can be defined as [17]:

$$R_{SWCNT} = R_Q \left(1 + \frac{l}{l_o}\right) \quad \text{and} \quad R_w = \frac{R_Q}{n_{CNT}} \left(1 + \frac{l}{l_o}\right) \quad (5)$$

where R_{SWCNT} and R_w are the resistances of one SWCNT and a bundle, respectively; l is wire length and l_o is the electron MFP; l_o is proportional to the SWCNT diameter with theoretically and experimentally derived proportionality constants of 2.8 μm/nm (henceforth, ideal model) [18] and 0.9 μm/nm (henceforth, practical model) [17], respectively. Thus, for a 1-nm-diam SWCNT, ideal and practical models yield an l_o of 2.8 and 0.9 μm, respectively.

These values are valid at a small voltage drop across the CNT interconnect (<0.16 V) [19]. Operation at higher voltage drops can degrade performance. An additional resistance component in the form of contact resistance has recently been shown to be reduced down to a few kilohms per tube [19, 20]. The total resistance of a CNT bundle is simply all the resistance components of a single tube divided by the number of metallic tubes in the bundle (dictated by PD), as shown in (5).

Example 2. Compare the resistance of the 1-μm-long SWCNT and SWCNT bundle. The wire width is 22 nm, aspect ratio is 2.0, and CNT diameter is 1 nm. Assume the fractional packing density is one-third and MFP of an SWCNT is 1 μm.

Answer:

R_Q is the quantum resistance and expressed as $\pi\hbar/2q^2$–6.45 kΩ. From (5), resistance of one SWCNT with 1 μm length can be calculated as

$$R_{SWCNT} = R_Q\left(1 + \frac{l}{l_o}\right) = 6.45 \times 10^3 \times \left(1 + \frac{1 \times 10^{-6}}{1 \times 10^{-6}}\right) = 12.9\,k\Omega$$

In order to calculate the resistance of a bundle, we need to calculate the number of CNTs in a bundle. From the relationship of (4):

$$n_w = \frac{w - d_t}{x} = \frac{22 - 1\,nm}{1\,nm/\sqrt{1/3}} = 12$$

$$n_h = \frac{h - d_t}{(\sqrt{3}/2)x} = \frac{22 \times 2 - 1\,nm}{(\sqrt{3}/2) \times (1\,nm/\sqrt{1/3})} = 29$$

$$n_{CNT} = n_w n_h - \frac{n_h - 1}{2} = 334$$

where n_{CNT} includes only the metallic portion among all SWCNTs inside the bundle. Thus, R_w of the bundle is

$$R_w = \frac{R_Q}{n_{CNT}}\left(1 + \frac{l}{l_o}\right) = \frac{6.45 \times 10^3}{334}(1 + 1) = 38.6\,\Omega$$

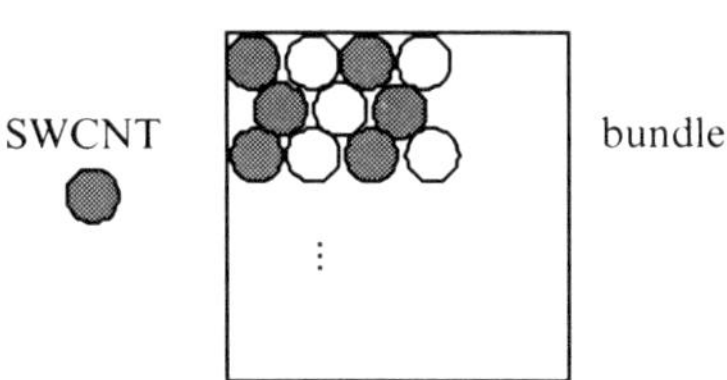

An SWCNT has two capacitance components: electrostatic (C_E), and the additional quantum capacitance (C_Q) [21] due to a reduced 2-D density of states for electrons. To add an electron in an SWCNT, one must add it at an available quantum state above the Fermi energy (E_F) due to Pauli's exclusion principle. C_E and C_Q can be described by (6) and (7), respectively.

$$C_E = \frac{2\pi\varepsilon}{\cosh^{-1}(2t/d_t)} = \frac{2\pi\varepsilon}{\ln(t/d_t)} \tag{6}$$

$$C_Q = \frac{e^2}{\pi\hbar v_f} \tag{7}$$

where ε is the permittivity of the dielectric between a wire and a ground plane, t is the distance between the ground and nanotube, d_t is the diameter of a single nanotube, $\hbar$ is the Planck constant, and v_f is the Fermi velocity of an SWCNT. For $d_t = 1$ nm, $t = 4h = 340$ nm, and ε_r (relative permittivity for the 22-nm technology node) $= 2.0$, the electrostatic capacitance is about 190 fF/mm. C_Q is around 100 fF/mm under the same condition as above, which is of the same order of magnitude as its electrostatic counterpart.

For a single tube, C_Q is comparable to C_E. However, in a bundle, the quantum components are added in series, making the bundle's total C_Q negligible compared to its C_E [22]. We simulated two geometries for capacitance, corresponding to CNT and Cu using FastCap (3-D field solver) [23]. Figure 7 (inset) shows the simulated geometries, with CNT exhibiting more roughness due to its bundled nature. The capacitance of the two geometries was within 4% agreement with the one presented in [24]. Thus, the total CNT bundle capacitance is approximately the same as the electrostatic capacitance of the Cu wire.

The importance of inductance (RC vs. RLC model) depends on the relative magnitude of the inductive reactance and the wire resistance. An SWCNT has an extra kinetic inductance (L_{kin}) in addition to magnetic inductance (L_{mag}). L_{kin} results from a higher electron kinetic energy because of a lower density of states in CNTs [22]. Per tube, it is about four orders of magnitude higher than L_{mag}. However, in a bundle, the total bundle L_{kin} reduces dramatically because CNTs are in parallel, as shown in Fig. 8a [22, 25] The total magnetic inductance (L_{mag}), which is the sum of self (L_{self}) and the mutual inductance (L_{mut}) between the tubes, is relatively constant with wire width. This is because at interconnect dimensions of interest, the L_{self} component of L_{mag} becomes negligible (L_{self} is added in parallel and is proportionately reduced). While L_{mut} is relatively constant beyond a certain width, as the impact of mutual inductance from SWCNT, further than a certain distance, gets weaker, this value converges to L_{self}, when solved using a matrix approach described in [26]. Thus, the total inductance of a CNT bundle is given by

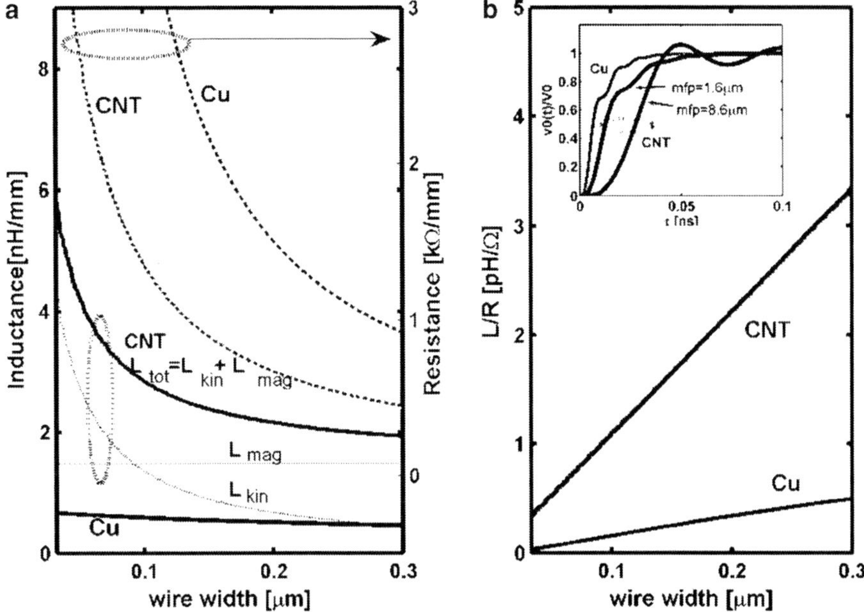

Fig. 8 (**a**) Inductance and resistance of Cu and CNT versus wire width. Wire width is in the global interconnect regime. PD is assumed to be 33%. l_0 is 1.6 μm. (**b**) Inductance-to-resistance ratio versus wire width. (**b**) (*Inset*) step response of Cu and CNT wire with optimized spacing (*h*)

$$L_{tot} = \frac{L_{kin}}{n_{CNT}} + L_{mag},$$

$$\text{where,} \tag{8}$$

$$L_{mag} = \left(L_{mut} + \frac{L_{self}}{n_{CNT}} \right) = L_{self}$$

Here, L_{mag} obtained using the model in [26] is $\sim$1.5 nH/mm. Figure 8a plots L_{tot} of a CNT bundle along with its components. Smaller widths render a larger L_{tot} because of an increase in L_{kin}. Cu L_{tot} is lower than CNT L_{tot} for all widths. In addition, Fig. 8 shows that the CNT bundle resistance is also lower than that of Cu due to a longer MFP. The above results exhibit approximately a 6× larger inductance-to-resistance ratio (*L*/*R*) for the CNT bundle compared to Cu wires, insinuating a more pronounced impact of inductance in the case of the CNT bundle. A simulated step response of a repeated CNT wire indeed shows a larger overshoot than Cu (Fig. 8b, inset), corroborating the importance of inductance in CNTs. Figure 8b also shows that the *L*/*R* ratio is higher for larger widths. Thus, a full RLC model is necessary for CNT bundles, especially for larger width, global wires. Inductance can be ignored for smaller width, local wires.

2.1.4 Secondary Effect of SWCNT Resistance Contacts

Contact resistance arises mainly due to scatterings at the contacts when interfaces are poorly connected. Some reports indicate that it can be of the order of hundreds of kilohms. This effect becomes more prominent when the diameter of nanotubes is smaller than 1 nm because it may form a poor bonding between CNTs and metal. In addition, metallic CNTs tend to have a small bandgap opening as the diameter is smaller than 1 nm. These may give rise to hundreds of kilohms of contact resistance. However, prior works have reported that contact resistances of nanotubes with a diameter larger than 1 nm are on the order of only few kilohms or even hundred ohms. This indicates that the contact resistance can be lowered to such a level that it can be neglected compared to the basic quantum resistance ($\sim$6.45 kΩ).

High bias voltages. CNTs are 1-D quantum wires and, in general, are considered to provide ballistic transport. However, electrons in these quantum wires tend to get backscattered as the length becomes longer. At small bias voltages, defects and acoustic phonons only act as scatterers. In this case, the electron MFP can be as large as 1.6 μm in high-quality nanotubes. However, at higher bias voltages, optical and zone boundary phonons, which have energies around $\hbar\Omega = 0.16$ eV, contribute to electron backscattering [19]. Once an electron with energy E finds an available state with energy $E-\hbar\Omega$, it emits a phonon with the energy of $\hbar\Omega$. An electron should be accelerated by electric field E to the length of $l_\Omega = \hbar\Omega/qE$ in order to gain this energy. Once it achieves this energy, it travels on average $l_o = 30$ nm before it scatters and emits an optical or zone-boundary phonon. The effective MFP, l_{eff}, can be described by

$$\frac{1}{l_{\text{eff}}} = \frac{1}{l_e} + \frac{1}{l_\Omega + l_o} \tag{9}$$

where l_e is the low-bias MFP. The resistance of an SWCNT is given by

$$R = R_c + R_Q\left(1 + \frac{L}{l_{\text{eff}}}\right) \tag{10}$$

where R_c is the contact resistance, R_Q is the fundamental quantum resistance and L is the tube length. The electric field across a tube will vary depending on the applied bias and should be written in an integral form as [24]:

$$R = R_c + R_Q + \left(R_Q\frac{l}{l_e} + \int_{x=0}^{x=l} \frac{dx}{(I_0/E(x)) + (l_o/R_Q)}\right) \tag{11}$$

where $E(x)$ is the position-dependent electric field, and $I_0 = 25$ μA.

2.1.5 Graphene Nanoribbon Interconnects

A graphene sheet is an ideal 2-D carbon honeycomb structure. Graphene nanoribbons, abbreviated as GNRs, are edge-terminated graphene sheet. They are just equivalent to unrolled SWCNTs. GNRs share most of the physical and electrical characteristics with CNTs such that they become semiconducting or metallic depending on the chirality of GNRs' edge [27]. GNRs can be fabricated in a more controllable process, such as optical lithography, while CNTs' growth results in a random chiral distribution. Thus, it is necessary to look at GNRs' performance as an interconnect application. To briefly look at the conductance model of a GNR interconnect, the MFP of a GNR should be analyzed. l_n is the distance that electrons in the nth mode move forward until they hit one of the GNR's side edges. This can be described as

$$l_n = \frac{k_\parallel}{k_\perp} W = W \sqrt{\left(\frac{E_F/\Delta E}{n + \beta}\right)^2 - 1} \tag{12}$$

where W is the width of GNR, and $k_\parallel$ and $k_\perp$ are the wave vectors in transport and non-transport direction, respectively. E_F is the Fermi level. ΔE is the energy gap between subbands in GNR. β is zero in a metallic GNR. Then the conductance of GNRs can be described by the total number of transmission modes and Mattiessen's rule, as shown in (9). The number of transmission modes can be determined by counting all subbands below the Fermi energy level:

$$G = \frac{q^2}{\pi \hbar} \sum_n \frac{1}{1 + L(1/l_D + 1/l_n)} \tag{13}$$

where l_D is the MFP caused by the acoustic phonon and defects, as in CNTs. If we use more mathematical approximations, (13) can be simplified further as (14):

$$G = \frac{q^2 l_D}{\pi \hbar L} + 1.5 \left(\frac{q^2 W^{2.5}}{\pi \hbar L}\right)\left(\frac{E_F}{\pi \hbar v_F}\right)^{1.5} \tag{14}$$

In the graph in Fig. 9, metallic GNR cannot outperform the Cu interconnect until the width is lower than 7 nm. These results explain that diffusive edge scattering significantly limits the performance of the GNR interconnect. In addition, the resistance of GNR is higher than that of the monolayer SWCNT for all ranges of wire width. This is an obvious result because physical properties of a GNR are similar to those of a CNT and only one GNR layer should be able to compete with a monolayer bundle of CNTs. However, if we can use a multilayered GNR interconnect, it would improve the performance.

Example 3. In Fig. 9, metallic GNR shows better resistance than Cu when its width is below 7 nm. Considering this factor, describe possible technological issues that arise.

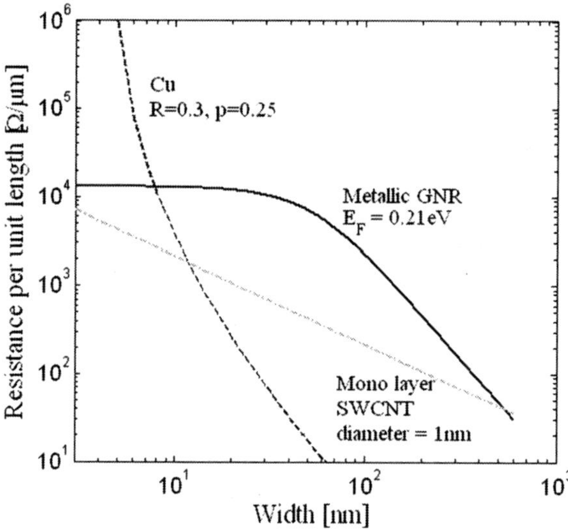

Fig. 9 Resistance comparison between GNR, monolayer SWCNT, and Cu. The Fermi-level is assumed to be 0.21 eV, as reported in an experimental result [27]

Answer:

Unlike CNTs, a GNR can be patterned with optical lithography giving a desired chiral configuration. However, it would be extremely hard to define the perfect chiral configuration when it comes to a small resolution feature, such as 7 nm or smaller than that. At very small widths (in the nanometer scale), a few atom vacancy on the edge can result in distortion of quantum-mechanical properties in GNRs, e.g., a metallic GNR can become a semiconducting GNR due to opening of the bandgap.

2.1.6 Optical Interconnects

Optics has many potential benefits to offer. The very high carrier frequency of optics, under a low-loss medium (waveguide or free space), gives a velocity near the speed of light with negligible power consumption in the waveguide [10]. In addition, optics can achieve a high bandwidth using the WDM scheme. On the other hand, highly efficient and manufacturable end devices (optical modulator and detector) are essential for high-speed, small-loss, and low-power electrical-to-optical (EO) or optical-to-electrical (OE) conversions [10, 28–30].

For EO conversion, the quantum confinement stark effect (QCSE) of *III–V* quantum well modulator can provide fast and efficient modulation [31–32]. However, this suffers from a lack of silicon compatibility. Recently, a Si/SiGe quantum well modulator has been demonstrated, providing a high potential for silicon-compatible manufacturability [33]. In addition to the quantum well-

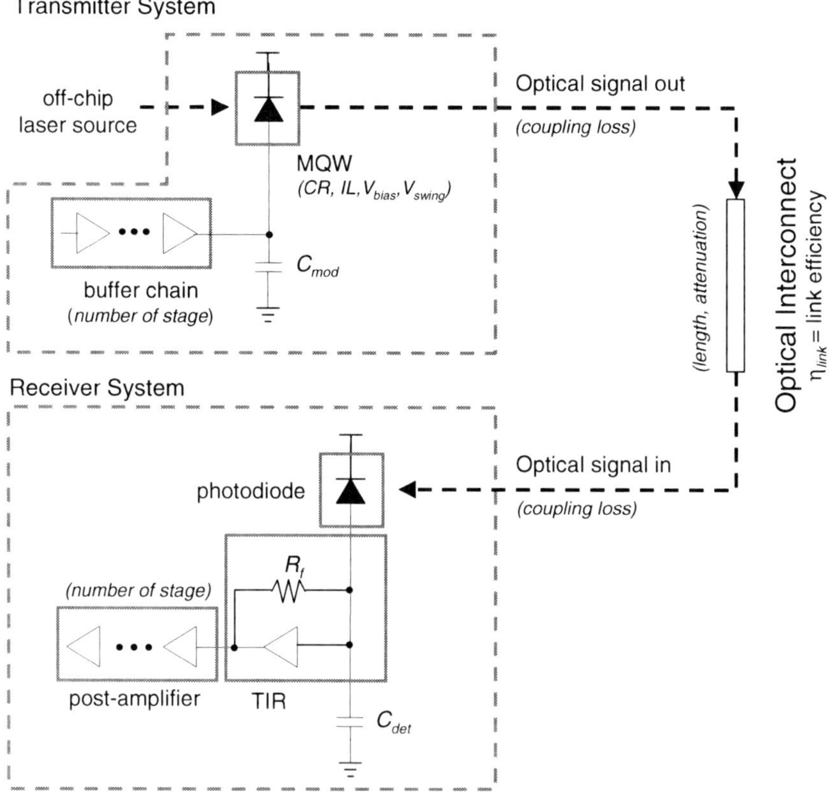

Fig. 10 The schematic of quantum-well modulator-based optical interconnect. The modulator parameters assumed for this optical interconnect were taken from ref. [38]

structured modulator, a Si ring resonant-type modulator also offers promising performance; however, its vulnerability of high Q (quality factor) to temperature and process variations needs to be resolved [34–35] For OE operation, a photodetector with Ge on Si substrate can provide high speed with low parasitic capacitance and coupling loss [36–37]. Figure 10 illustrates the schematic view of the optical interconnect system.

Transmitter. The transmitter system of an optical interconnect consists of an optical modulator and buffer chain. An off-chip laser source injects a continuous wave (CW) at the input of the transmitter system via coupling devices. An optical modulator determines the on and off states of the optical signal, depending on the applied field across the modulator. This field is driven by multiple buffer stages. Here, we choose a multi-quantum-well structure for the modulator because it is more conducive to low transmitter power dissipation and Si-compatible monolithic integration with the Si/Ge fabrication technique. It is critical to minimize the power dissipation of the transmitter system without losing too much speed. The size of modulator capacitance, contrast ratio (CR), and modulator's insertion loss (IL) are key parameters to determine the optimum power dissipation. Modulator capacitance is the parasitic outcome of modulator's structure. CR is the power ratio of the on and off stated. IL is absorbed power during the on state.

Waveguide medium. The waveguide medium guides the mode of optical waves. The medium can be either free space or guiding structures with core and cladding. The performance of the waveguide can be characterized by coupling loss and attenuation, which determines optical power transfer efficiency (η_{link}). Here, we choose the Si/SiO_2 (core/cladding) waveguide for the interconnect medium.

Receiver. The optical receiver consists of a photodetector, transimpedence amplifier (TIA), and post-amplifying stages. The photodetector converts photons from incoming optical signals to electrons. TIA converts the electrical current generated at the detector to provide voltage with enough gain. Then the post-amplifier recovers the full swing. The important design parameters in the receiver are the size of input stage of TIA, the feedback resistance of the front-end, and the number of gain stages at a given input optical power and detector capacitance. This is also constrained by bandwidth, bit error rate (BER) through receiver noise, and the output swing level.

3 Circuit Modeling

Figure 11 illustrates the schematic representation of a buffered interconnect. If we consider an interconnect of total length L, it is buffered at length l. In this section, we present the methodology for optimally buffering the interconnect in order to minimize delay. Figure 11b shows a distributed RC network, indicating one segment of the repeated interconnect with length l, where V_{tr} is the voltage source at the input stage, R_{tr} is the driver resistance that has dependence on the transistor

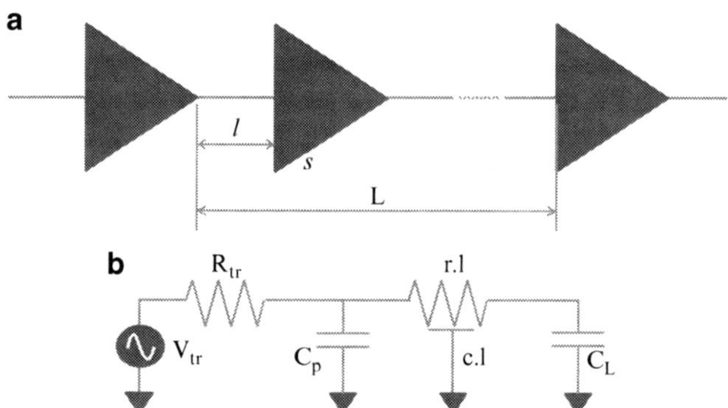

Fig. 11 (a) Schematic of optimally buffered interconnect. The total length is L, and l is the optimal distance between repeaters to minimize delay. S refers to the optimal size of the input transistor. Each repeater has a fanout of one (FO1). (b) The equivalent circuit of one segment with l and s

size, C_p is the output parasitic capacitance of the driver, C_L is the load capacitance of the receiving end, and r and c are the interconnect resistance and capacitance per unit length, respectively. The segment delay of this interconnect, τ_0, is given by [11, 39]

$$\tau_0 = bR_{tr}(C_L + C_P) + b(cR_{tr} + rC_L)l + arcl^2 \tag{15}$$

where a and b are the switching-dependent parameters. For instance, if the delay is measured at 50% of total voltage swing, where $a = 0.4$, $b = 0.7$, and r_0 is the resistance of a minimum-sized driver. Henceforth, R_{tr} can be defined as r_0/s. c_0 and c_p are the input and parasitic output capacitances of a minimum-sized driver, respectively. Thus, C_p and C_L are proportional to the size of a driver as $C_p = sc_p$, and $C_L = sc_0$. If the total interconnect length L is divided into n segments of length $l = L/n$, the delay with overall length of L, τ_d, is given by

$$\tau_d = \frac{L}{l}b(x)r_0(c_0 + c_P) + b(x)\left(c\frac{r_0}{s} + src_0\right)L + a(x)rclL \tag{16}$$

Then, the optimum values l_{opt} and s_{opt}, and the delay of the optimally buffered interconnect with total length L, τ_d, can be obtained as

$$l_{opt} = 1.32\sqrt{\frac{r_0(c_0 + c_p)}{rc}} \quad \text{and} \quad s_{opt} = \sqrt{\frac{r_0 c}{rc_0}}, \quad \text{and} \quad \tau_d = 2.049L\sqrt{rct_{FO1}} \tag{17}$$

where $t_{FO1} = 2r_0c_0$ (generally $c_0 = c_p$), representing the delay of an inverter with a fanout load of one (FO1).

Example 4. Derive l_{opt}, s_{opt}, and τ_d in (17) from (16).

Answer:
We can take partial derivatives of (16) with respect to s and l and set them equal to zero

$$\frac{\partial \tau_d}{\partial l} = -Lbr_o(c_o + c_P)l^{-2} + arcL = 0$$

So

$$l_{opt} = \sqrt{\frac{br_0(c_0 + c_p)}{arc}} = 1.32\sqrt{\frac{r_0(c_0 + c_p)}{rc}} \tag{18}$$

$$\frac{\partial \tau_d}{\partial s} = -cr_0 s^{-2} + rc_0 = 0$$

So

$$s_{opt} = \sqrt{\frac{r_0 c}{rc_0}} \tag{19}$$

Plugging (18) and (19) into (16) in order to get the optimized wire delay, we obtain

$$\tau_d = 2.049 L \sqrt{rct_{FO1}}$$

The wire capacitance is modeled as follows [40]:

$$C_w = \varepsilon \left[\begin{array}{c} 1.15\dfrac{w}{t} + 2.80\left(\dfrac{h}{t}\right)^{0.222} \\[2ex] + \left(0.66\dfrac{w}{t} + 1.66\dfrac{h}{t} - 0.14\left(\dfrac{h}{t}\right)^{0.222}\right) \cdot \left(\dfrac{t}{s}\right)^{1.34} \end{array} \right] \tag{20}$$

where ε is the dielectric permittivity, s is the interwire spacing (assuming $s = w$), h is wire height ($h = w \times$ aspect ratio), and t is the intermetal layer spacing (assuming $t = 4h$). For global wires, we included inductance, given by (19) [41]. At the 22-nm technology node, this value is about 0.5 nH/mm.

$$L_w = 2 \times 10^{-7} l \left(\ln \frac{2l}{w+h} + 0.5 + \frac{w+h}{3l} \right) \tag{19}$$

The dynamic power dissipation of repeaters and wire is given by

$$P = \alpha(s(c_p + c_0) + l \cdot c)V_{DD}^2 f_{clk} \tag{20}$$

where α is the switching activity (0–1) and f_{clk} is the clock frequency.

Example 5. Suppose you have an interconnect with length L. If you divide this into five segments with ideal repeaters, which do not have any parasitic components, what would be the total delay? Assume r and c are the resistance and capacitance per unit length, respectively.

Answer:

Without repeaters, the total resistance and capacitance would be linearly proportional to the wire length. Thus, total delay is $t = RC = rcL^2$. However, if the wire is repeated, then each segment has resistance and capacitance linearly dependent on the segment length $L/5$. In this case, the total delay is just obtained by summing up all segment delays. As a result, the total delay of the repeated wire is $t_{rep} = 5R_{seg}C_{seg} = rcL^2/5$

4   Local Interconnects: CNT Bundles Versus Cu

Figure 12 shows that for minimum width and both short and long wires, the CNT bundle exhibits a lower latency than Cu at all aspect ratios (AR). In addition, the difference in latency between the two technologies is smaller for shorter wires and higher ARs because these conditions render the wire resistance (R_w) lower than the driver resistance (R_{tr}). A dominant R_{tr} obviates the CNT R_w advantage. Figure 12 also shows that both Cu and CNT wires exhibit a minimum in latency with a changing AR. The initial reduction in AR lowers the delay because at large ARs, a low R_w results in the delay being dominated by the wire capacitance (C_w) and R_{tr} product; further, C_w reduces with AR. However, below a certain AR, R_w becomes dominant over R_{tr}, and the delay is now dictated by the R_w and C_w product. R_w rises with a smaller AR and C_w cannot reduce as fast as the increase in R_w because its reduction is limited by a constant interlevel dielectric (ILD) and the fringe component. Figure 12b plots wire resistance (R) and capacitance (C) ratio between CNT and Cu as a function of AR.

Although both technologies exhibit an optimum AR for minimizing delay, in practice, only CNT can be operated at the optimum because of EM considerations. Figure 12 (right y-axis) shows the increase in the wire current density with smaller AR. A maximum allowed current density of 14.7×10^6 A/cm^2 in Cu limits its minimum AR to about 1.5 for the ITRS-dictated minimum widths. On the other hand, a single CNT tube can support a current density of up to 10^9 A/cm^2. Thus, CNTs can be operated at their optimum AR, pointing to a different geometry than Cu.

Figure 13a explicitly shows these geometric differences. The Cu optimum aspect ratio (AR$_{opt}$) is lower than what is allowed by the Cu EM limit, and is higher than CNT AR$_{opt}$. In addition, AR$_{opt}$ increases with wire length for both Cu and CNT. This is because AR$_{opt}$ represents a point where R_{tr} and R_w are comparable, and

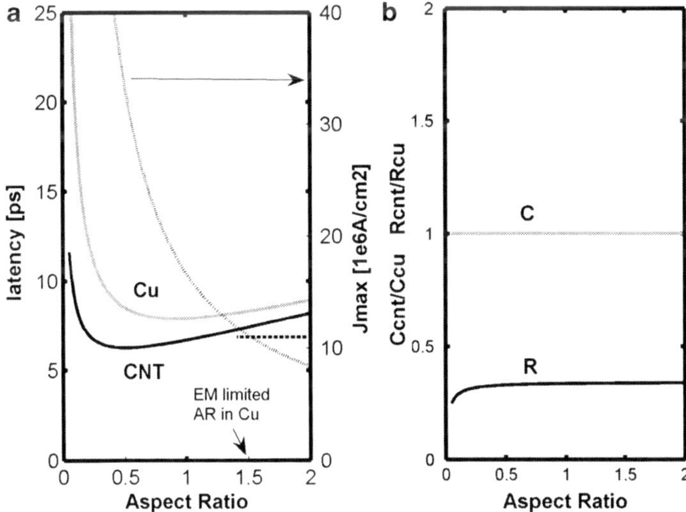

Fig. 12 (a) Local interconnect delay versus aspect ratio (height-to-width ratio) for 10 μm wire length (left y-axis). Current density versus aspect ratio (right y-axis). Wire width corresponds to the minimum specified at 22 nm in ITRS [9]. The simulations assume a driver resistance of 3 kΩ, the CNT MFP of 0.9 mm (conservative), and a contact resistance and packing density of 10 kΩ and 33%, respectively. (b) Wire resistance (R) and capacitance (C) ratio between CNT and Cu

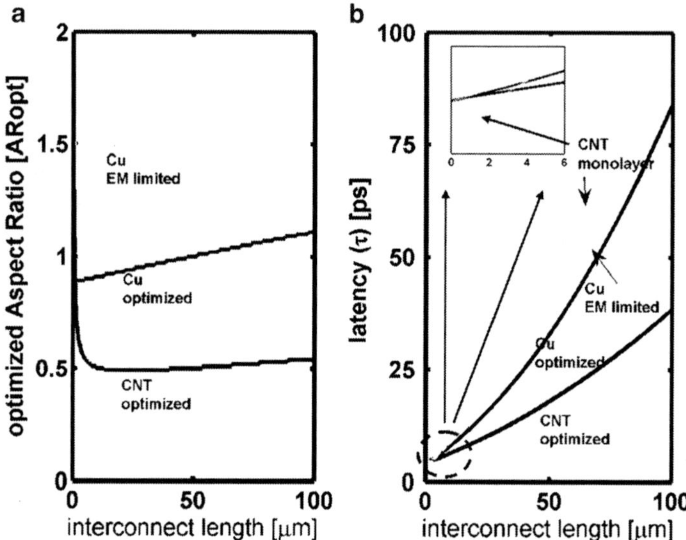

Fig. 13 (**a**) The plot of optimum aspect ratio (AR_{opt}) for minimum latency as a function of length. The dotted line depicts the EM-limited AR for Cu. (**b**) The plot of latency as a function of length. Four different curves are shown: two for CNT (τ with AR_{opt} and τ with monolayer) and two for Cu (τ with EM limited AR and τ with AR_{opt}). Inset: Zoomed-in view from 0 to 6 μm of local interconnect length

longer wires (because of their larger resistance) reach comparable values to R_{tr} at larger cross-section area (higher AR). Figure 13b) explicitly plots the latency versus length for four cases. This includes the lowest possible latency using AR_{opt} for both the CNT bundle and Cu, as well as the latency of a monolayer SWCNT and EM-limited Cu. The optimized CNT-bundle shows better latency than optimized Cu for all local length scales. This advantage increases to about 2× when we compare the optimized CNT (leveraging its EM robustness) with the EM-limited Cu. Finally, we observe (inset of Fig. 13b) that a monolayer CNT gives the smallest delay for very short local interconnects as it is dominated by the time of flight rather than RC time constants. In addition to latency, a smaller AR_{opt} for the CNT bundle would also bring a dramatic power reduction compared to Cu because of a lower capacitance.

5 Global and Semiglobal Interconnects

5.1 *Cu/CNT and Optical Global Wire Circuit Models*

We should use RLC models for global/semiglobal Cu and CNT wires and include repeater insertion for delay reduction (Fig. 14). The values assumed for R, L, and C are as discussed in "Sect. 2." Unlike the RC wire model, an RLC model does not

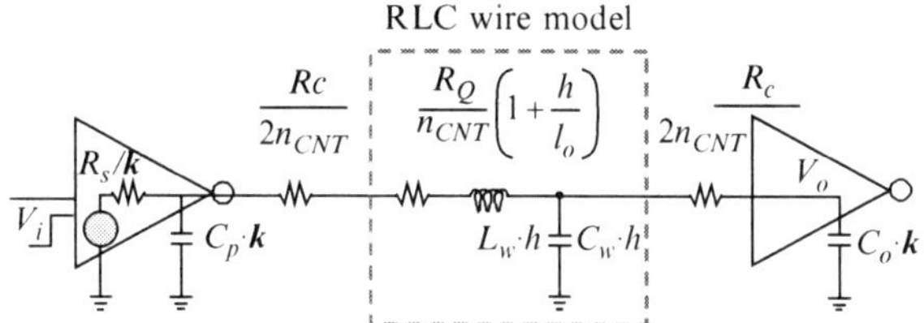

Fig. 14 Equivalent circuit model of a repeater segment for CNTs, where s is the optimized repeater area, l is the optimized wire length per repeater, R_s is the inverter output resistance, R_c and R_Q are the contact resistance and quantum resistance of the SWCNT, respectively, L_w and C_w are the inductance and the capacitance of the CNT bundle, respectively, and C_p and C_0 are the input capacitance and the output parasitic capacitance of the inverter, respectively

have a closed-form solution for optimizing the repeater size (s) and its spacing (l) for minimum delay. Thus, these parameters can be optimized using fourth-order Páde expansion and the Newton–Raphson numerical method [42].

The optical link, on the other hand, consists of an off-chip laser, a quantum-well modulator at the transmitter (converts CMOS gate output to optical signal), a waveguide consisting of a silicon core (refractive index $\sim$3.5), SiO_2 cladding as the transmission medium, and a transimpedance amplifier (TIA) followed by gain stages at the receiver (Fig. 10). The total delay of an optical wire is the sum of the transmitter, waveguide, and the receiver delays.

The transmitter delay arises from the CMOS gate driving the capacitive modulator load. It is minimized using a buffer chain and is dependent on the fan-out-of-four (FO4) inverter delay of a particular technology node. The waveguide delay is dictated by the speed of light in a dielectric waveguide ($\sim$11.3 ps/mm). Finally, the receiver delay is calculated based on circuit considerations. It assumes the input pole (the node at the input of TIA) to be dominant and is optimized to meet the bandwidth and BER (10^{-15}) criteria. The total power dissipation for the optical interconnect is calculated by optimizing the sum of the receiver and transmitter power dissipation, as outlined in ref. [43], in the context of off-chip interconnects.

5.2 Latency and Energy per Bit as a Function of Scaling

Figure 15 compares the interconnect latency of CNT, Cu, and optics-based links as a function of the technology node for semiglobal ($\sim$1 mm) and global ($\sim$10 mm) wires. For MFPs (l_o) of 0.9 and 2.8 μm, CNT wires show 1.6× and 3× latency improvement, respectively, over Cu at all technology nodes. The optical wire shows an advantage over both CNT and Cu for longer lengths ($\sim$10 mm) because a large fraction of the delay occurs in end-devices. For 1-mm long wires, optics becomes advantageous over CNT only at smaller technology nodes. This is because with scaling, CNT and Cu latency increases, whereas, optical delay reduces due to an improvement in transistor performance (transmitter and receiver).

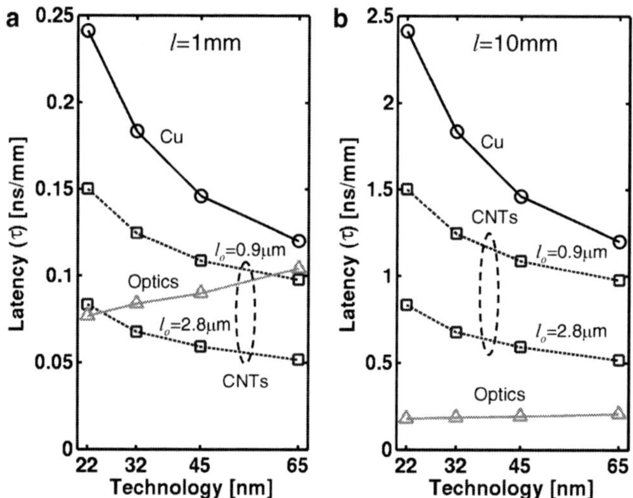

Fig. 15 Latency in terms of technology node for two different interconnect lengths. l_o is the MFP and PD is the packing density of metallic SWCNTs in a bundle. The SWCNT diameter (d_t) is 1 nm. For optics, the capacitance of monolithically integrated modulator/detector (C_{det}) is 10 fF [44,45]

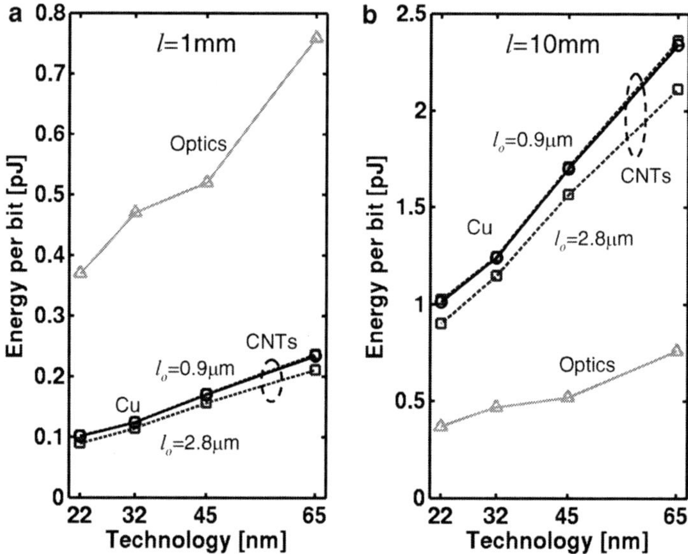

Fig. 16 Energy per bit versus technology node for two different interconnect lengths corresponding to global and semiglobal wire length scales. For CNTs, PD is 33% and the wire diameter, d_t, is 1 nm. For optics, the capacitance of monolithically-integrated modulator/detector capacitance (C_{det}) is 10 fF

Figure 16 compares the energy per bit requirements of the three technologies. For Cu/CNT wires, the dominant energy is the switching energy: CV^2 (C is total capacitance that includes the wire and repeater components, V is the voltage), while for optics it arises from the static power dissipation in the end-device amplifiers. For 10 mm global wires, optics is most energy-efficient, while for 1 mm semiglobal wires, both Cu and CNT present a better efficiency than optics. At all length scales and at the 22 nm node, CNTs with $l_o = 2.8$ μm are 20% more energy-efficient compared to Cu. This is because CNT operates in the RLC region, where a smaller resistance results in a smaller optimum repeater size, hence a smaller total repeater capacitance [46]. This is in contrast with an RC wire, where the total optimum repeater capacitance is a constant fraction of the wire capacitance, irrespective of the resistance. The 0.9 μm l_0 CNT exhibits similar energy per bit as Cu because even though l_0 is larger than that for Cu, the sparse (33%) PD results in a resistance similar to that of Cu.

In Fig. 17(a), the dependence of latency on the wire length between Cu/CNTs and optics is explored. The delay in all three technologies linearly increases because Cu/CNTs are buffered with repeaters in a delay-optimized fashion and for optics, the latency of the medium is linearly dependent on the length. Optics clearly shows the lowest latency for the reasons stated above. CNTs with $l_o = 0.9$ μm and $l_o = 2.8$ μm give 1.6′ and 3′ latency improvement over Cu, respectively, as shown in Fig. 17a.

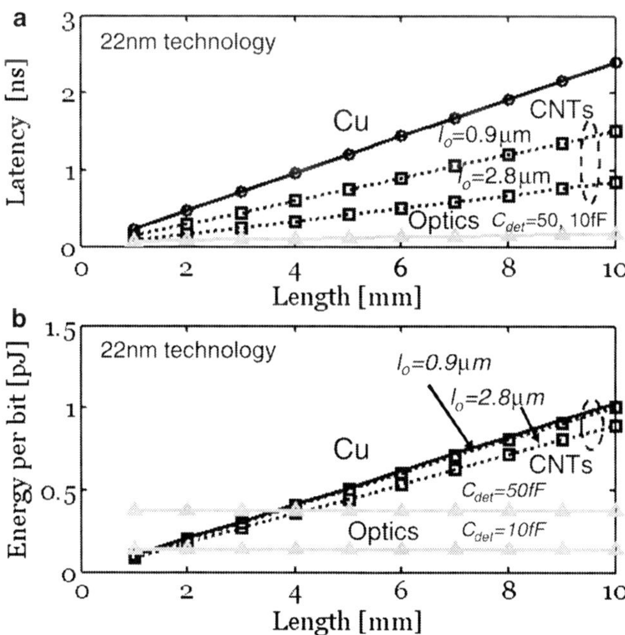

Fig. 17 Latency and energy per bit in terms of wire length for the 22-nm technology node. l_o is the MFP. For optics, the detector capacitance (C_{det}) is 50 and 10 fF

5.3 Latency, Power Density and Bandwidth Density

Modern multicore processors can be limited by the intercore communication bandwidth. This coupled with a limited processor perimeter, puts a premium on ϕ_{BW}. An ideal wire technology should provide a large ϕ_{BW}, at the lowest possible latency and power density. We now take a system's top-down view, and compare the power density and the latency penalty incurred to achieve an architecture-dictated global-wire ϕ_{BW} for all three technologies (Fig. 18).

For Cu and CNT wires, a larger ϕ_{BW} is achieved by implicitly reducing the wire pitch, while keeping bandwidth per wire constant (assumed to be dictated by the global clock: 10 Gbps) using wave pipelining between repeater stages. Thus, the maximum achievable ϕ_{BW} is limited by the minimum wire width (ITRS [9]) and is also shown in Fig. 18. The minimum wire pitch (maximum ϕ_{BW}) can be limited to a larger (smaller) value by practical considerations, such as excessive area of the repeaters. For optical interconnects, a larger ϕ_{BW} is achieved using a different mechanism because the waveguide size cannot be reduced beyond a certain point. Instead, optical wires deploy denser WDM and the maximum ϕ_{BW} is limited by the degree (number of wavelengths) of WDM.

In Fig. 18, optical links exhibit the lowest latency, followed by CNT bundle and Cu at all ϕ_{BW}. Further, the optical latency is independent of ϕ_{BW}, since a higher ϕ_{BW} is achieved just by adding more wavelengths, which does not impact the

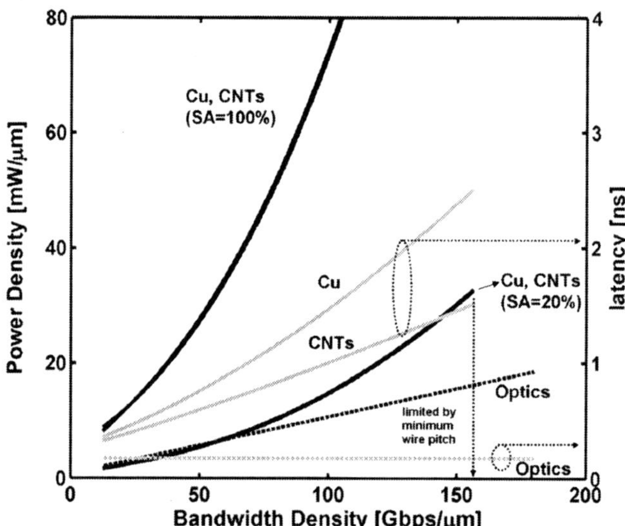

Fig. 18 Power density and latency versus bandwidth density (Φ_{BW}) comparisons among Cu, CNTs, and optical interconnect for different switching activities at 10 Gb/s global clock frequency (f_{clk}). The bandwidth density is implicitly changed by wire pitch for Cu and CNTs and number of channels of WDM for the optical interconnect. For CNTs, l_0 is 0.9 μm and PD is 33%. For optics, the detector capacitance (C_{det}) is 10 fF (wire length = 10 mm, 22 nm transistor technology node)

latency of each wavelength. In contrast, both CNT and Cu latency increases with ϕ_{BW}, as a smaller wire pitch increases resistance and the resulting number of repeater stages. Thus, the optical link latency advantage becomes more pronounced to about 13′ and 8′ better than Cu and CNT, respectively, at a very high intercore ϕ_{BW} requirement.

In contrast to latency, the power density rises linearly with ϕ_{BW} for an optical wire (Fig. 18) as the additional bandwidth arises from equally power-consuming wavelength channels. However, the CNT and Cu power density increases nonlinearly with ϕ_{BW}, following the dependence of total wire capacitance (repeater and wire capacitance) on wire pitch: at high ϕBW (small wire pitch), the inter-metal dielectric (IMD) capacitance dominates, whereas at smaller ϕ_{BW} (large pitch), the ILD capacitance is larger. Figure 18 also shows that CNT ($l_o = 0.9$ μm) and Cu have equal power density as a result of a similar wire and repeater capacitance. However, a higher CNT l_o leads to a lower power density than Cu (not shown), as a smaller resistance in the RLC regime results in a smaller optimum repeater size. The coupling capacitance plays an important role in determining both power and delay. However, for power calculation, on average, the nominal capacitance is a good approximation.

Finally, the power dissipation in the electrical schemes has a strong switching activity (SA) dependence. The optical link, to the first order, is independent of SA, as most of its power is static in nature. The optical links exhibit a lower power density than both Cu and CNTs for SA = 100%. However, at SA = 20%, there is a critical ϕ_{BW} beyond which the optical links are more power-efficient. Thus, optical interconnects are more power-efficient at larger required ϕ_{BW}, larger SA, and larger lengths.

Figure 19 explicitly plots the power density as a function of *SA* for each of the three technologies, showing the critical SA beyond which optical wire power density is superior. These trends in SA point to the importance of architectural choices to maximally exploit the potential of optical wires.

5.4 *Impact of CNT and Optics Technology Improvement*

The performance of CNT is a strong function of l_0 and PD, whereas optical wire performance critically depends on capacitances associated with the detector and modulator (C_{det}, C_{mod}). In this section, we quantify the impact of these device and materials parameters on the comparisons. Such an analysis is useful for technologists to get an idea of the required parameters from a system's standpoint, to make their technology competitive.

Figure 20 (SA = 20%) illustrates that an improvement in CNT l_0 from 0.9 μm (practical) to 2.8 μm (ideal) and in PD from 0.3 to 1 results in some reduction in power density for all ϕ_{BW}. Moreover, the improvement in l_0 shows a stronger impact than an improvement in PD. This is because a PD increase results in a smaller increase in the inductance-to-resistance ratio (L/R) as L_{kin} also decreases

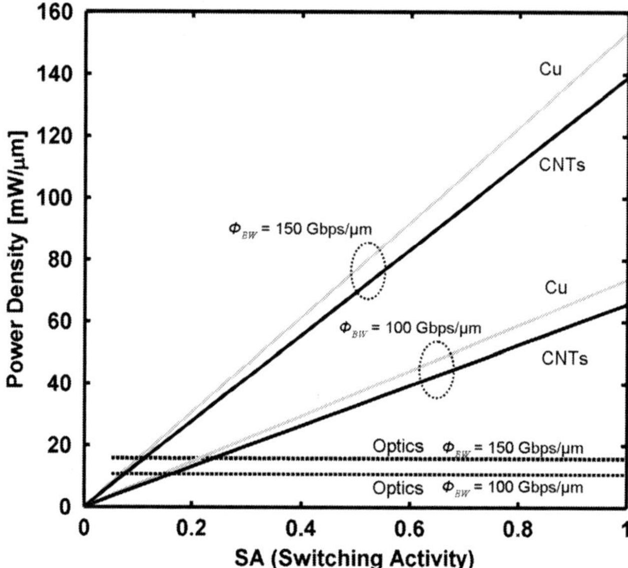

Fig. 19 Power density versus SA (switching activity) with different bandwidth density (Φ_{BW}) at 10 Gb/s global clock frequency. For CNT, PD is 33% and l_0 is 2.8 mm. For optics, the detector capacitance (C_{det}) is 10 fF (wire length = 10 mm, 22-nm transistor technology node)

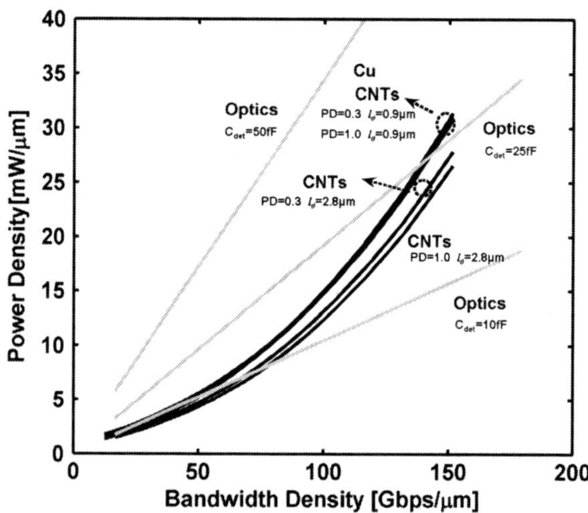

Fig. 20 The impact of CNT and optics technology improvements on power density versus bandwidth density curves (SA = 20%). C_{det} reduction for optics results in a large improvement in power density (wire length = 10 mm, 22 nm transistor technology node, f_{clk} = 10Gb/s)

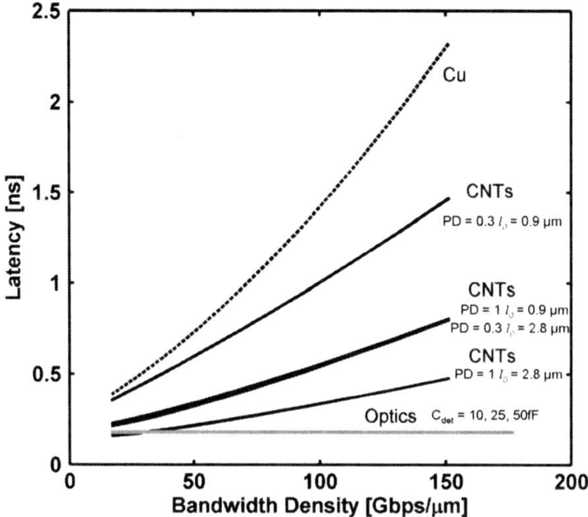

Fig. 21 The impact of CNT and optics technology improvement on latency versus bandwidth density. CNT parameter improvement results in a very large improvement in latency over Cu. CNT with ideal parameters has comparable latency to that of optical wires at very low Φ_{BW} (wire length = 10 mm, 22 nm transistor technology node, f_{clk}=10Gb/s)

along with R. For this SA and for C_{det} = 25 fF, CNT, and Cu outperform optics. However, if the C_{det} and C_{mod} can be brought down to 10 fF using a monolithic detector (as opposed to hybrid bonded *III–V* detector), optical wires outperform other technologies even at smaller SA.

Similarly, Fig. 21 captures the impact of technology parameters on the latency. For CNTs, the improvement in both l_0 and PD results in a substantial decrease in latency compared to Cu. With an ideal l_0 and PD, and at lower ϕ_{BW}, the latency performance is comparable to that of optical wires.

Exercises

1. In Fig. 3, the bitrate of an electrical wire (*B*) is shown to be proportional to A/l^2. Derive this relationship.

2. In Exercises A1, the bitrate is calculated under very simple assumptions. However, in practice, the wire resistance is far more exacerbated with the size effect, degrading the bitrate even more. Plot the resistivity of Cu wire as a function of wire width (20–100 nm) considering the surface and grain boundary scatterings and compare it with bulk resistivity. Assume AR is 2 for all widths.

3. Draw the equivalent ac circuit model of a SWCNT interconnect including all parameters: quantum conductance (two spins and two degenerate channels), electrostatic capacitance, quantum capacitance, magnetic inductance, and kinetic inductance.

4. At high bias voltages, CNTs suffer from optical and zone boundary phonons which have energies around $\hbar\Omega = 0.16$ eV in addition to scatterings from acoustic phonon and defects. This effect degrades the circuit performance of CNT interconnect even more. Equations (9)–(11) describe this phenomenon.

   1. Draw the equivalent circuit of the CNT interconnect reflecting (9).
   2. From (11), derive the relationship between CNT interconnect resistance (R) and applied bias (V), taking into account the fact that l_o is only 30 nm.

5. Derive the analytical formula of the optimal segment length (l_{opt}) and optimal repeater width (w_{opt}) for a SWCNT interconnect bundle. Use (5) and (16). Assume that the electrostatic capacitance of the SWCNT bundle is equal to that of the Cu interconnect.

6. Figure 12 shows that a CNT bundle allows more aspect ratio reduction to a local interconnect for capacitance reduction because it is more tolerant to electromigration than Cu. We want to investigate the optimal AR point for the CNT local interconnect via this problem. Calculate the minimum AR of the local CNT interconnect. Assume that the driver resistance (R_{tr}) is 3 kΩ and parasitic capacitance ($C_L + C_p$) is 2 fF at fixed driver width, contact resistance between CNTs and metal is zero, wire width is 25 nm, total wire length is 10 μm, dielectric permittivity (ε) is 2 $\times$ ε_0 (low-k material), interwire spacing (s) is equal to the wire width (w), and intermetal layer spacing (t) is four times the wire width (w). CNT MFP is 1.6 μm. Neglect the barrier scattering effect and fringing capacitance.

7. You are a circuit designer. You have to layout global interconnects that run across 10 mm. You are only given 1.2 μm periphery to layout wires in parallel and want to maximize the bandwidth density (φ_{BW}). You want to use either an optical interconnect or electrical interconnect for this global communication.

Optical interconnect	Cu interconnect
Wave guide width = 600 nm	Maximum wire bit rate = 4 Gbps (double edge)
Modulation bit = 12 Gbps	Wire width = 40 nm
	Wire spacing distance = 40 nm
	AR = 2.5
Maximum number of wavelengths in WDM = 6	Wire capacitance per length = 0.2 fF/μm
	Driver resistance/minimum size = 100 kΩ
Time of flight in waveguide = 11.3 ps/mm	$c_o = c_p = 0.02$ fF/minimum size
	p (in (1)) = 0.6
Laser power = 1.3 mW	R (in (2)) = 0.5
	Switching activity (α) = 0.2 (100%)
	$V_{DD} = 0.8$ V
Transceiver power = 0.66 mW	(Ignore leakage and short-circuit power for
Transmitter delay = 40 ps	simplicity)
Receiver delay = 40 ps	

   1. What are the bandwidth densities (φ_{BW}) for optical and electrical interconnects, respectively?

2. Calculate the delay and power dissipation of each interconnect scheme. Which one would you prefer to use for global on-chip communication?

8. The bandwidth of a single wire with length L can be defined as $1/\tau$ where τ is the delay of a wire.

   1. How will you define the bandwidth of delay-optimal repeated global interconnect? Derive the analytical relationship from (15) or (16). Neglect any intersymbol interference (ISI) effect.
   2. Let us assume that global interconnects on the top metal layer run in parallel from the one side of the chip to the other with equal distance. Derive the total bandwidth of the top-layer global interconnects from (1) and given parameters as follows. The chip is square-shaped and has an area of A; d is the width of the wire. Assume that the distance between interconnects is equal to the width of the wire.

References

1. W. Steinhogl, G. Schindler, and M. Engelhardt, "Size-dependent resistivity of metallic wires in the mesoscopic range," *Phys. Rev. B*, vol. 66, p. 075414, 2002.
2. C. Ryu, K.-W. Kwon, A.L.S. Loke, H. Lee, T. Nogami, V.M. Dubin, R.A. Kavari, G.W. Ray, and S.S. Wong, "Microstructure and reliability of copper interconnects," *IEEE Trans. Electron Dev.*, vol. 46, pp. 1113–1120, June 1999.
3. C.T. White and T.N. Todorov, "Carbon nanotube as long ballistic conductors," *Nature*, vol. 393, pp. 240–242, 1998.
4. G.F. Close, S. Yasuda, B. Paul, S. Fujita, H.-S.P. Wong, "1-GHz integrated circuit with carbon nanotube interconnects and silicon transistors," *Nano Lett.*, vol. 8, vol. 2, pp. 706–709, 2008.
5. A. Naeemi, R. Venkatesan, and J.D. Meindl, "Optimal global interconnects for GSI," *IEEE Trans. Electron Dev.*, vol. 50, no. 4, 2003.
6. E.H. Sondheimer, "The mean free path of electrons in metals," *Adv. Phys.*, vol. 1, no. 1, pp. 1–2, 1952.
7. A.F. Mayadas and M. Shatzkes, "Electrical-resistivity model for polycrystalline films: The case of arbitrary reflection at external surfaces," *Phys. Rev. B*, vol. 1, pp. 1382–1389, 1970.
8. P. Kapur, J.P. McVittie, and K.C. Saraswat, "Technology and reliability constrained future copper interconnects. Part I: Resistance modeling," *IEEE Trans. Electron Dev.*, vol. 49, no. 4, pp. 590–597, 2002.
9. International technology roadmap of semiconductor, 2005 version, http://www.itrs.net.
10. D.A.B. Miller, "Physical reasons for optical interconnection," *Int. J. Optoelectron.*, vol. 11, pp. 155–168, 1997.
11. H.B. Bakoglu, *Circuits, Interconnections, and Packaging for VLSI*, New York: Addison-Wesley, 1990.
12. P. Kapur, J.P. McVittie, and K.C. Saraswat, "Technology and reliability constrained future copper interconnects. Part II: Performance implications," *IEEE Trans. Electron Dev.*, vol. 49, no. 4, pp. 598–604, 2002.
13. P. Kapur, "Scaling induced performance challenges/limitations of on-chip metal interconnects and comparisons with optical interconnects," Stanford University, PhD thesis, 2002.
14. Q. Chen, J.A. Davis, P. Zarkesh-Ha, and J.D. Meindl, "A novel via blockage model and its implications," in *Proc. Int. Interconnect Technology Conference*, pp.15–18, June 2000.

15. J.A. Davis, R. Venkatesa, A. Kaloyeros, M. Beylansky, S.J. Souri, K. Banerjee, K.C. Saraswat, A. Rahman, R. Reif, and J.D. Meindl, "Interconnect limits on gigascale integration (GSI) in the 21st century," *Proc. IEEE*, vol. 89, no. 3, pp. 305–324. 2001.
16. M.S. Dresselhaus, G. Dresselhaus, and P. Avouris, eds., *Topics in Applied Physics, Carbon Nanotubes: Synthesis, Structure, Properties and Applications*, New York: Springer, 2000.
17. J.Y. Park, S. Rosenbelt, Y. Yaish, V. Sazonova, H. Ustunel, S. Braig, T.A. Arias, and P.L. McEuen, "Electron–phonon scattering in metallic single-wall carbon nanotubes," *Nano Lett.*, vol. 4, pp. 517–520, 2004.
18. A. Nieuwoudt and Y. Massoud, "Evaluating the impact of resistance in carbon nanotube bundles for VLSI interconnect using diameter-dependent modeling techniques," *IEEE Trans. Electron Dev.*, vol. 53, no. 10, pp. 2460–2466, 2006.
19. Z. Yao, C.L. Kane, and C. Dekker, "High-field electrical transport in single-wall carbon nanotubes," *Phys. Rev. Lett.*, vol. 84, pp. 2941–2944, 2000.
20. O. Hjortstam, P. Isberg, S. Soderholm, and H. Dai, "Can we achieve ultra low resistivity in carbon nanotube-based metal composites?" *Appl. Phys. A*, vol. 78, pp. 1175–1179, 2004.
21. P.J. Burke, "Luttinger liquid theory as a model of the gigahertz electrical properties of carbon nanotubes," *IEEE Trans. Nanotechnol.*, vol. 1, no. 3, pp. 129–144, 2002.
22. A. Naeemi, R Sarvari, and J.D. Meindl, "Performance comparison between carbon nanotube and copper interconnect for gigascale integration (GSI)," *Electron Device Lett.*, vol. 26, no. 2, pp. 84–86, 2005.
23. FastCap: available on-line at http://www.rle.mit.edu/cpg/research_codes.htm
24. A. Naeemi and J.D. Meindl, "Design and performance modeling for single-wall carbon nanotubes as local, semi-global and global interconnects in gigascale integrated systems," *IEEE Trans. Electron Dev.*, vol. 54, pp. 26–37, 2007.
25. N. Srivastava and K. Banerjee, "Performance analysis of carbon nanotube interconnects for VLSI application," in *Proc. IEEE/ACM Int. Conf. on Computer-Aided Design*, pp. 383–390, 2005.
26. A. Nieuwoudt and Y. Massoud, "Understanding the impact of inductance of carbon nanotube bundles for VLSI interconnect using scalable modeling techniques," *IEEE Trans. Electron Dev.*, vol. 5, no. 6, 2006.
27. A. Naeemi and J.D. Meindl, "Conductance modeling of graphene nanoribbons (GNR) interconnects," *IEEE Electron Device Lett.*, vol. 28, no. 5, pp. 428–431, 2007.
28. P. Kapur and K.C. Saraswat, "Comparisons between electrical and optical interconnects for on-chip signaling," in *Proc. Int. Interconnect Technology Conference*, pp. 89–91, 2002.
29. H. Cho, K.H. Koo, P. Kapur, and K.C. Saraswat, "The delay, energy, and bandwidth comparisons between copper, carbon nanotube, and optical interconnects for local and global wiring application," in *Proc. Int. Interconnect Technology Conference*, p. 135, 2007.
30. K.H. Koo, H. Cho, P. Kapur, and K.C. Saraswat, "Performance comparisons between carbon nanotubes, optical, and Cu for future high-performance on-chip interconnect applications," *IEEE Trans. Electron Dev.*, vol. 54, no. 12, p. 3206, 2007.
31. D.A.B. Miller "Band-edge electroabsorption in quantum-well structures: The quantum-confined Stark effect," *Phys. Rev. Lett.*, vol. 53, pp. 2173–2176, 1984.
32. D.A.B. Miller "Rationale and challenges for optical interconnects to electronic chips," *Proc. IEEE*, vol. 88, pp. 728–749, 2000.
33. J.E. Roth, O. Fidaner, E.H. Edwards, R.K. Schaevitz, Y.H. Kuo, N.C. Helman, T.I. Kamins, J. S. Harris, and D.A.B. Miller, "C-band side-entry Ge quantum-well electroabsorption modulator on SOI operating at 1 V swing," *Electron. Lett.*, vol. 44, no. 1, pp. 49–50, Jan. 2008.
34. Q. Xu, B. Schmidt, S. Pradhan, and M. Lipson, "Micrometer-scale silicon electro-optic modulator," *Nature*, vol. 435, pp. 325–327, 2005.
35. Q. Xu, D. Fattal, and R.G. Beausoleil, "Silicon micro-ring resonators with 1.5-μm radius," *Opt. Exp.*, vol. 16, pp. 4309–4315, 2008.
36. O.I. Dosunmu, D.D. Cannon, M.K. Emsley, L.C. Kimerling, and M.S. Unlu. "High-speed resonant cavity enhanced Ge photodetectors on reflecting Si substrates for 1550 nm operation," *Photonics Technol. Lett.*, vol. 17, no. 1, pp. 175–177, 2005.

37. A.K. Okyay, A.M. Nayfeh, A. Marshall, T. Yonehara, P.C. McIntyre, and K.C. Saraswat, "Ge on Si by novel heteroepitaxy for high efficiency near infrared photodetection," in *Proc. Conf. on Lasers and Electro-Optics*, 2006.http://www.opticsinfobase.org/abstract.cfm?uri=CLEO-2006-CTuU5

38. O. Kibar, D.A.V. Blerkom, C. Fan, and S.C. Esener, "Power minimization and technology comparisons for digital free-space optoelectronic interconnects," *IEEE J. Lightw. Technol.*, vol. 17, no. 4, pp. 546–554, 1999.

39. K. Banerjee, S.J. Souri, P. Kapur, and K.C. Saraswat, "3-D ICs: A novel chip design for improving deep submicron interconnect performance and systems-on-chip integration," *Proc. IEEE*, vol. 89, pp. 602–633, May 2001.

40. T. Sakurai and T. Tamuru, "Simple formulas for two- and three-dimensional capacitances," *IEEE Trans.Electron Dev.*, vol. 30, pp. 183–185, 1983.

41. C.P. Yue and S.S. Wong, "Physical modeling of spiral inductors on silicon," *IEEE Trans. Electron Dev.*, vol. 47, no. 3, pp. 560–568, 2000.

42. K. Banerjee and A. Mehrotra, "Analysis of on-chip inductance effects for distributed RLC interconnects," *IEEE Tran. CAD Integr. Circuits. Syst.*, vol. 21, no. 8, pp. 904–914, 2002.

43. P. Kapur and K.C. Saraswat, "Comparisons between electrical and optical interconnects for on-chip signaling," in *Proc. Int. Interconnect Technology Conference*, pp. 89–91, 2002.

44. Y.H. Kuo, Y.K. Lee, Y. Ge, S. Ren, J.E. Roth, T.I. Kamins, D.A.B. Miller, and J.S. Harris, "Strong quantum-confined Stark effect in germanium quantum-well structures on silicon," *Nature*, vol. 437, pp. 1334–1336, 2005.

45. O.I. Dosunmu et al. "High-speed resonant cavity enhanced Ge photodetectors on reflecting Si substrates for 1550 nm operation," *Photonics Technol. Lett.*, vol. 17, no. 1, pp. 175–177, 2005.

46. Y.I. Ismail and E.G. Friedman, "Effects of inductance on the propagation delay and repeater insertion in VLSI circuits: A summary," *IEEE Circuits Syst. Mag.*, pp. 24–28, 2003.

Circuit Design with Resonant Tunneling Diodes

Pallav Gupta

1 Introduction

Advances in crystal growth and fabrication technologies are enabling researchers to explore novel device concepts that are based on the quantum mechanical features of electrons. These structures offer the possibility of circuit and device miniaturization well below those offered by conventional complementary metal-oxide semiconductor (CMOS) lithography techniques. The resonant tunneling diode (RTD) concept, which utilizes the electron-wave resonance in multi-barrier heterostructures, emerged as a pioneering device in the mid-1970s. It was predicted that such devices would offer picosecond device switching speeds and reduction in device counts per circuit function.

While the properties of RTDs were predicted over 30 years ago, only in the last 15–20 years have such devices been actually fabricated. This was mainly due to advancements in molecular-beam epitaxy – a technique for deposition and growth of crystals. In order to exhibit the resonant tunneling phenomenon, it is necessary to scale devices along a single dimension only. This can be achieved by nonlithographic processes in order to form the critical dimensions required for the tunneling effect. This implies that it is possible to integrate RTD-based devices with conventional (albeit, high-performance III-V) transistors.

The co-integration of RTDs with gallium-arsenide (GaAs) or silicon-germanium (SiGe) transistors has enabled some novel, high-performance digital logic families. Table 1 compares conventional CMOS logic with such families using similar feature sizes and device dimensions [1]. The entries for power per gate in Table 1 are for circuits operating at the maximum speed corresponding to the entry for delay per gate for that circuit. As can be seen, RTD-based circuits offer lower switching speeds and lower device count. However, they consume significantly higher power. The higher power consumption is due to the fact that silicon possesses a much higher hole mobility than GaAs. Furthermore, processing techniques for III-V

P. Gupta (✉)
Core CAD Technologies, Intel Corporation, Folsom, CA 95630, USA
e-mail: pallav.gupta@intel.com

N.K. Jha and D. Chen (eds.), *Nanoelectronic Circuit Design*,
DOI 10.1007/978-1-4419-7609-3_12, © Springer Science+Business Media, LLC 2011

Table 1 Comparison of process technologies

Parameter(per gate)	CMOS	GaAs-CHFET	RTD–HBT	RTD–MODFET
Power (mW)	0.2	0.1	0.5	0.3
Delay (ps)	500	250	40	200
Device count	Large	Large	Small	Small

Source: Ref. 1
CHFET complementary heterojunction field-effect transistor (FET), *MODFET* modulation-doped FET

materials (such as GaAs) are incompatible with those for silicon, and are expensive. These are some of the main reasons why GaAs logic circuits have been unable to compete with silicon logic circuits. In addition, SiGe technology is mainly used for heterojunction bipolar transistors (HBTs) which limits integration density.

Given the technological and economical barriers, it is difficult to envision RTDs grown on III-V materials as a candidate for post-CMOS technology. However, such RTDs will find applications in niche areas of high-speed, mixed-signal ICs or radio-frequency devices. Furthermore, if the resonant tunneling concept can be demonstrated in devices that are compatible with silicon technology, then that opens up an exciting area of research that could help sustain Moore's law further. Nevertheless, it is still in the reader's interest to understand the basic concept behind RTDs and get a brief overview of some of the research that has been done in this area.

The remainder of the chapter provides the reader with an introduction to RTDs. We first study RTD device characteristics. Next, we explore some circuit examples using RTDs. Since RTDs have the capability to implement threshold logic, we briefly discuss algorithms for threshold logic synthesis and testing.

2 RTD Fundamentals

As illustrated in Fig. 1a, an RTD is composed of alternate layers of heterogeneous semiconductors. In this figure, the structure is shown using the Si/SiGe material system, although other material systems are also possible. An RTD enables the tunneling effect that results in the folded device I-V characteristics shown in Fig. 1b. As the voltage across the RTD terminals is increased from zero, the current increases due to tunneling until the peak RTD voltage V_p is realized. The current corresponding to V_p is called the peak current I_p. When the voltage increases beyond V_p, the current through the RTD drops due to reduction in tunneling, until the valley voltage V_v is realized. The current corresponding to V_v is called the valley current I_v. Beyond V_v, the current increases once again following the characteristics of a conventional diode.

For a current that lies between the valley and peak currents (i.e., $[I_v, I_p]$), there are two possible stable voltages: $V_1 < V_p$ or $V_3 > V_v$. The voltage across the RTD, when it is operating in the first positive differential-resistance region and the current

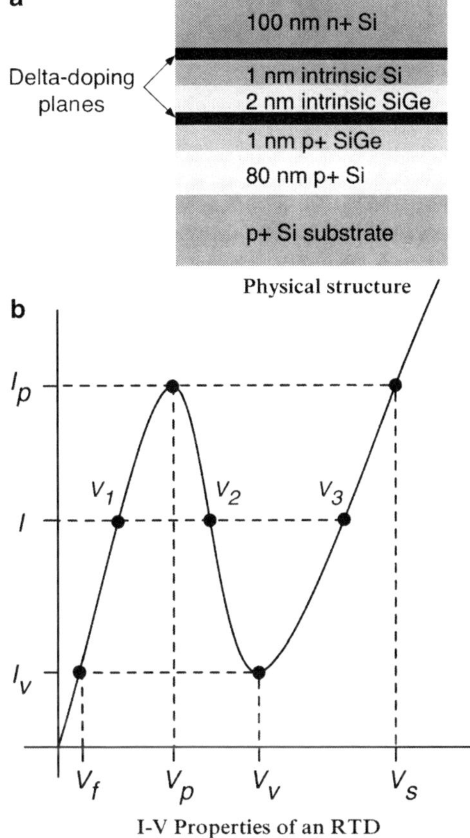

Fig. 1 Properties of an RTD

I-V Properties of an RTD

through it equals I_v, is called the first voltage V_f. Similarly, the voltage across the RTD, when it is operating in the second positive differential-resistance region and the current through it equals I_p, is called the second voltage V_s.

Based on these definitions, we can derive some important RTD parameters using elementary circuit theory. They are as follows:

$$R_{p1} = \frac{V_p}{I_p}, \tag{1}$$

$$R_{p2} = \frac{V_s - V_v}{I_p - I_v}, \tag{2}$$

$$|R_n| = \frac{V_v - V_p}{I_p - I_v}, \tag{3}$$

$$\text{PVVR} = \frac{V_{\text{p}}}{V_{\text{v}}}, \tag{4}$$

$$\text{PVCR} = \frac{I_{\text{p}}}{I_{\text{v}}}. \tag{5}$$

In (1)–(5), R_{p1} and R_{p2} are the first and second positive differential-resistance, respectively, R_{n} is the negative differential-resistance, and PVVR and PVCR are the RTD peak-to-valley voltage and current ratio, respectively [1].

In order for RTDs to be useful in high-performance circuits, it is necessary for them to exhibit a steep and narrow valley region. This allows for picosecond switching and sharp transitions. Furthermore, the folded characteristic shown in Fig. 1b enables the RTD to be used as a load device. Given proper biasing, it draws very small currents for both low and high outputs. It is worth mentioning that tunneling diodes have existed for a long time. In the past, they exhibited a wide valley region, which resulted in low switching speeds and, hence, were not conducive to high-performance circuit design.

2.1 *Bistable Logic Using RTDs*

To understand the operating principles of a new bistable element, using RTDs in conjunction with HBTs, let us consider the simplified circuit shown in Fig. 2a. There are n input transistors ($x_1, x_2, \ldots, x_n$), a clock transistor driving a single RTD load, and power supply V_{DD}. The input transistor can be either in the on state with collector current I_h or off state with no collector current. In addition, the clock transistor can be in two states: high with collector current I_{clkh} or low (quiescent) with collector current I_{clkq}.

Initially, all collector currents are zero in the global reset state. When the clock current is I_{clkq} (i.e., CLK = Q), the circuit has two possible stable operating points for every possible input combination. This can be seen in the load lines of Fig. 2b. However, when the clock current is I_{clkh} (i.e., CLK = H) there is only one stable operating point for the circuit when m or more inputs are logic high. Using Kirchhoff's current law, the sum of the collector currents is $mI_n + I_{\text{clkh}}$. This operating point corresponds to a logic-zero output voltage. It should now be evident that this circuit can be operated to implement any non-weighted threshold logic function $f'(x_1, x_2, \ldots, x_n; m)$, where $f(x_1, x_2, \ldots, x_n)$ is 1 if and only if $\sum_{i=1}^{n} x_i < m(\forall x_i \varepsilon \{0, 1\})$. This is known as the n-input $i = 1$ inverting majority or minority gate [1].

The operating sequence for an RTD bistable gate begins with resetting all the collector currents to zero. This causes the output voltage V_{OUT} to go logic high. Next, the reset signal is removed and the clock is made logic high so that the clock current is I_{clkh}. At this point, the inputs have their normal stable values. If m or more inputs are logic high, the output goes logic low as this corresponds to the second

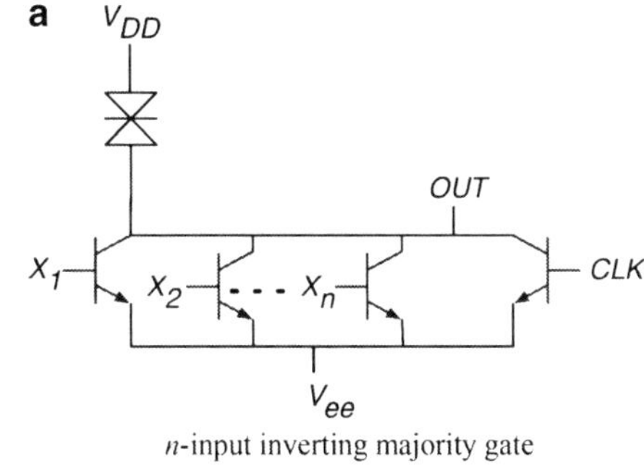

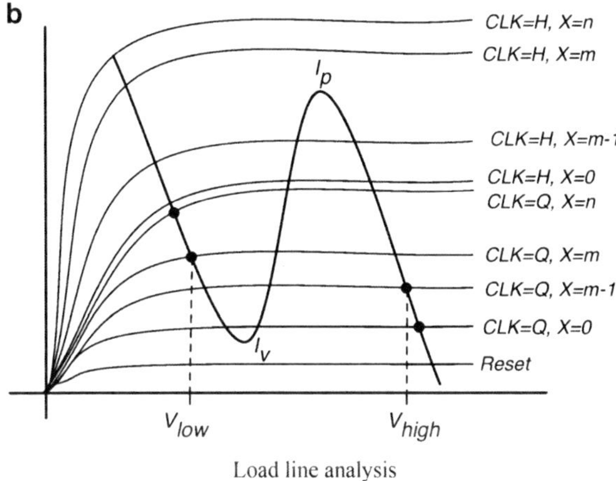

Fig. 2 Operating principle of a bistable RTD–HBT threshold logic gate

positive differential region (see Fig. 1b) corresponding to $V_{RTD} > V_v$, where V_{RTD} is the voltage across the RTD. However, if fewer than m inputs are logic high, the output remains at logic high as this corresponds to the first positive differential region where $V_{RTD} < V_p$. The final step is to bring the clock back to its quiescent state so that the clock current is I_{clkq}. This allows V_{OUT} to reach a stable voltage value corresponding to whether the RTD was in the first (logic high) or second positive (logic low) differential-resistance region [1].

As long as the clock remains in its quiescent state and the reset signal is not applied, the output does not change with the inputs. Once the inputs have stabilized, the correct output can be achieved using the evaluation procedure described above. Thus, it can be seen that the circuit functions as a clocked, self-latching threshold element. If multiple such elements are cascaded together, a very fine-grained pipelining mechanism called nanopipelining can be achieved. This basically converts each level in a multi-level logic circuit into a pipeline stage. Should RTDs become practical in the future, many traditional architectures will need to be reworked to take advantage of nanopipelining. Such a scheme will have tremendous benefits in high-throughput, data intensive applications.

2.2 Noise Margins of RTD–HBT Threshold Logic Gates

Let us now derive the noise margins of the RTD–HBT inverter shown in Fig. 3a. Recall that the high noise margin NM_H is defined to be the difference between the minimum high output voltage of the driving gate and the minimum high input voltage recognized by the receiving gate. Similarly, the low noise margin NM_L is defined to be the difference between the maximum low input voltage recognized by the receiving gate and the maximum low output voltage produced by the driving gate.

To derive an expression for the noise margins of an RTD–HBT inverter, we can use a simple, yet accurate, method of fitting a maximum area rectangle between the gate's normal and mirrored DC transfer characteristics [1]. However, we need to make some modifications because our gates employ a clock signal. When evaluating the inputs to a gate, the clock voltage increases to V_{clkh} and after evaluation, this level decreases to a quiescent value of V_{clkq}. If we consider two RTD–HBT inverters connected in series, the clocks of the two gates have to be out of phase for correction operation. That is, when the second gate is being evaluated by a high clock pulse, the clock to the previous gate has to be quiescent. In other words, the transfer characteristic should show that the second inverter sees the output of the first inverter with the clock at V_{clkq} and not at V_{clkh}.

The normal transfer curve of the RTD–HBT inverter is shown in Fig. 3a. The folded I–V characteristic produces a sharp transition (region 2), which can be assumed to be a perfect vertical drop. If we look at Fig. 3b, the largest rectangle within the enclosure formed between the normal and mirrored curves will be A′BCD. If we try to use the rectangle PQRS, we realize that such a rectangle will almost always be smaller than A′BCD, since regions 1 and 3 are virtually horizontal. Thus, we can consider A′BCD to be a valid approximation.

If the coordinates of points B and E are (x, y) and (x, y'), respectively, then the coordinates of point D is (y', x) since its lies on the mirrored curve. Thus, the noise margins can be expressed as

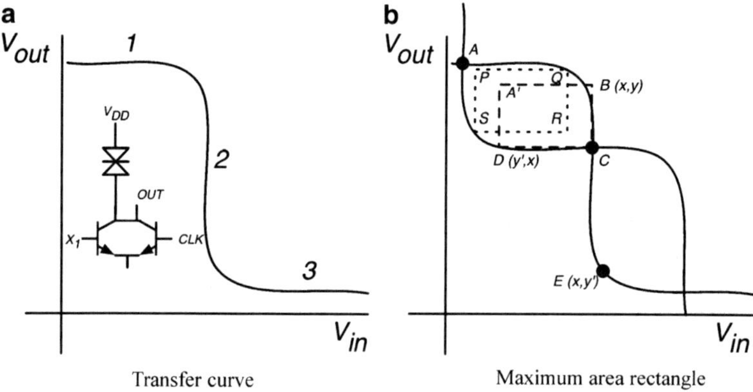

Fig. 3 Determining the noise margins of the RTD–HBT inverter

$$NM_H = y - x, \tag{6}$$

$$NM_L = x - y\prime > . \tag{7}$$

The sum of the noise margins can be rewritten as

$$NM_H + NM_L = y - y\prime > . \tag{8}$$

If we refer back to Fig. 1b, we can see that

$$y = V_{DD} - V_p + (I_{clkh} - I_{clkq})R_{p1}, \tag{9}$$

$$y\prime = V_{DD} - V_s - (I_{clkh} - I_{clkq})R_{p2} > . \tag{10}$$

If we substitute (9) and (10) into (8), we have

$$NM_H + NM_L = (V_s - V_p) + (I_{clkh} - I_{clkq})(R_{p1} - R_{p2}). \tag{11}$$

We can rewrite (11) as

$$NM_H + NM_L = (R_{p2} + |R_n|)(I_p - I_v) + (I_{clkh} - I_{clkq})(R_{p1} - R_{p2}), \tag{12}$$

by employing (11)–(2). For proper gate operation, it is required that $I_{clkh} < I_p$ and $I_{clkq} > I_v$. This implies that $I_{clkh} - I_{clkq} < I_p - I_v$. Now if $R_{p1} \geq R_{p2}$,

$$NM_H + NM_L < (R_{p1} + |R_n|)(I_p - I_v)$$

or

$$NM_H + NM_L < V_v\left(1 - \frac{PVVR}{PVCR}\right). \tag{13}$$

On the other hand, if $R_{p2} \geq R_{p1}$,

$$NM_H + NM_L < (R_{p2} + |R_n|)(I_p - I_v)$$

or

$$NM_H + NM_L < V_s - V_p. \tag{14}$$

Equations (13) and (14) provide us with important bounds on the sum of the two noise margins. Note that they are equal when $R_{p1} = R_{p2}$. These equations provide a simple first-order model to estimate the maximum noise margin that can be attained from a given RTD. They also provide the engineer with guidelines when designing RTDs for digital logic applications.

So far, we have only determined the upper bounds on the sum of the noise margins. To determine the actual expressions for NM_H and NM_L, we use the traditional Ebers–Moll model for an HBT describing the collector current I_c as a function of its base-emitter (V_{be}) and base-collector (V_{bc}) voltages:

$$I_c = I_s(e^{(V_{be}/V_T)} - e^{(V_{bc}/V_T)}), \tag{15}$$

where I_s is the saturation current and $V_T = kT/q$. Using this equation, we can write

$$x = V_T \ln\left(\frac{I_p - I_{clkh}}{I_s(1 - e^{-((V_{DD}-V_p)/V_T)})}\right). \tag{16}$$

Finally, the individual noise margins can be written as

$$NM_H = (V_{DD} - V_p) + (I_{clkh} - I_{clkq})R_{p1} - V_T \ln\left(\frac{I_p - I_{clkh}}{I_s(1 - e^{-((V_{DD}-V_p)/V_T)})}\right), \tag{17}$$

$$NM_L = V_T \ln\left(\frac{I_p - I_{clkh}}{I_s(1 - e^{-((V_{DD}-V_p)/V_T)})}\right) - (V_{DD} - V_s) - (I_{clkh} - I_{clkq})R_{p2}. \tag{18}$$

Figure 4a shows variations of an RTD–HBT inverter's noise margins as a function of I_{clkq}. The upper bound is given by (13). It can be seen that this upper bound is never reached since, as mentioned earlier, $I_{clkh} < I_p$ and $I_{clkq} > I_v$ for proper gate operation. We can see that for $I_{clkh} = 2.5$ mA ($I_p = 3.08$ mA), as I_{clkq} is increased from I_v to I_{clkh}, NM_H decreases while NM_L increases in accordance with (17). It should be mentioned that the derivation of (17) requires that $V_{DD} - V_s$ be positive. If $V_{DD} < V_s$, load lines drawn will show that the operating point after switching (from logic high to logic low) has to be such that the HBT will be in saturation mode. From this observation, it should be clear that (17) cannot be used if $V_{DD} < V_s$. Figure 4b shows the dependence of noise margins on I_{clkh}. While NM_H increases with I_{clkh}, NM_L decreases with I_{clkh}, for a given value of I_{clkq}.

It can be seen from above that the upper bound on the sum of the noise margins that can be achieved for a given RTD, irrespective of the types of transistors being used, can never exceed V_v. For reliable operation, it is necessary to have small values for R_{p1} and R_{p2}. This, combined with the need to have high PVCR and low PVVR [see (17)], suggests that the ideal RTD characteristic would have a very sharp "N" shape.

As a closing note, the reader should keep in mind that the equations derived above are first-order models. In practice, noise margins should be calculated using CAD tools such as SPICE which allows for a more detailed and precise modeling of an RTD.

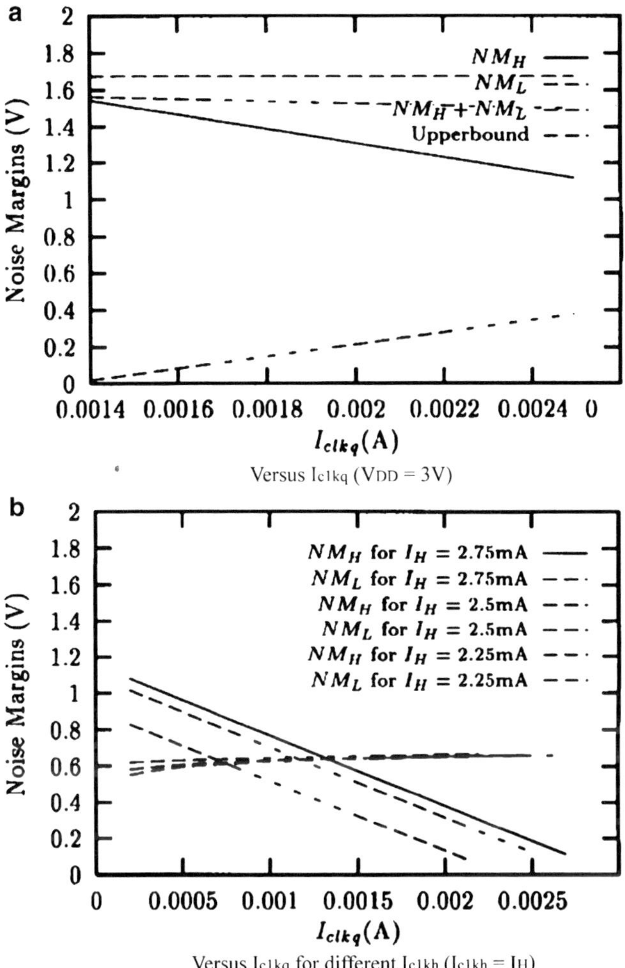

Fig. 4 Computer simulations of noise margins (*Source:* Ref. 1)

2.3 Monostable–Bistable Logic Elements

The main problem with RTD–HBT circuits is that they employ HBTs which seriously limits the device count. In order to achieve highly compact RTD-based circuits, it is necessary to have the ability to integrate RTDs and FET devices. That is, we should be able to connect an RTD and FET in parallel. So far, RTD–MOD-FET devices have been demonstrated in a III-V material system. A logic circuit comprising an RTD in series with the aforementioned RTD–MODFET structure is known as a monostable–bistable logic element (MOBILE) [2].

The operating principle of a MOBILE is demonstrated in Fig. 5. The gate is controlled by an oscillating bias voltage V_{bias}. As can been seen in Fig. 5a, there is

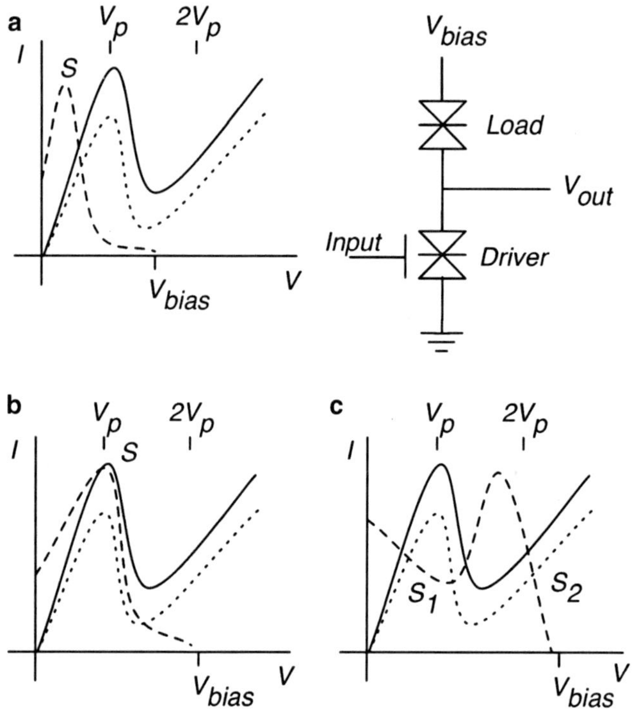

Fig. 5 Operating principle of a MOBILE. The load line diagrams are for (**a**) $V_{\text{bias}} < 2V_{\text{p}}$, (**b**) $V_{\text{bias}} = 2V_{\text{p}}$, and (**c**) $V_{\text{bias}} > 2V_{\text{p}}$

one stable point (monostable) when $V_{\text{bias}} < 2V_{\text{p}}$. This stable point evolves into two stable points (bistable: corresponding to logic high/low) when $V_{\text{bias}} > 2V_{\text{p}}$, as shown in Fig. 5c. The state that the circuit settles into after the monostable-to-bistable transition is determined by the difference in the peak currents between the driver and load devices. A small peak current in the driver device results in the stable point S_2 while a large peak current in the driver results in the stable point S_1. In order to use MOBILEs for logic operations, it is necessary to develop a gate-controlling scheme in which the gate voltage (input) associated with each RTD–MODFET pair can control the magnitude of the peak current and, therefore, determine the circuit state after the monostable–bistable transition.

Figure 6 illustrates the cross-sectional view of the monolithic integration of an RTD and MODFET. The RTD and MODFET are connected in parallel. Consequently, the total source-to-drain current I_{DS} is the sum of the currents passing through the RTD (I_{RTD}) and MODFET (I_{FET}). I_{RTD} remains unchanged for different gate voltages. However, I_{DS} is modulated as a result of changing I_{FET}. Thus, this structure can control the resonant tunneling peak current as the gate voltage is changed.

To demonstrate how logic gates can be constructed using MOBILEs, consider the two-input OR gate shown in Fig. 7a. Using threshold logic, we know that such a

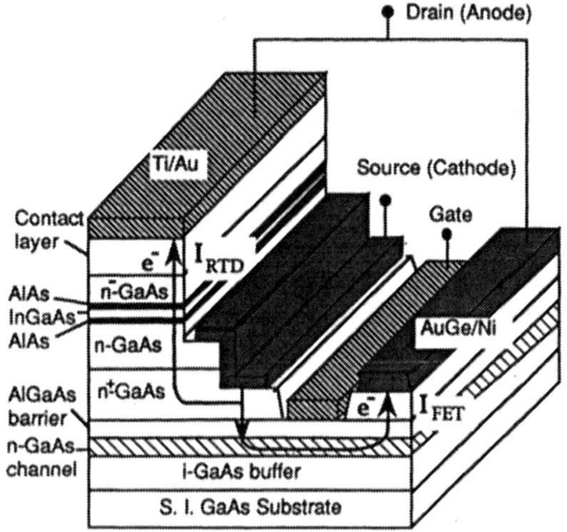

Fig. 6 Schematic cross-section of the RTD–MODFET device. The cathode (anode) of the RTD and source (drain) of the MODFET are connected. The parallel electron transport paths are also shown (*Source*: Ref. 2)

gate can be implemented with the weight-threshold vector ($w_1 = 1$, $w_2 = 1$; $T = 1$), where weight w_1 (w_2) corresponds to input x_1 (x_2) and the output is logic high if the weighted sum of the inputs is greater than or equal to threshold T, and logic low otherwise. To implement a threshold of unity, the driver RTD in Fig. 7a is a minimum-sized RTD. To implement the weights for inputs x_1 and x_2, two additional RTD–MODFET pairs are added in parallel to the load RTD. Figure 7b shows HSPICE simulation results for this gate. The first subplot shows the oscillating bias voltage. The next two subplots show the various input combinations. The reader can confirm that the output of the OR gate is computed correctly as shown in the final subplot.

A linear threshold function can be realized by a MOBILE, such as the one shown in Fig. 8. The input's weights are implemented by linearly scaling the area A_{RTD} of a minimum-sized RTD (i.e., an RTD with area λA_{RTD} implies $w = \lambda$), while the threshold is determined by the net difference in area between the driver and load RTDs relative to A_{RTD}. The modulation current ΔI applied at the output node determines what digital state the device transitions to. Using Kirchhoff's current law, we can express the current as:

$$\Delta I = \sum_{i=1}^{N_p} w_i I(V_{GS}) - \sum_{i=1}^{N_n} w_i I(V_{GS}), \tag{19}$$

where N_p and N_n are the number of positive and negative weighted inputs, respectively, and $I(V_{GS})$ is the peak current of a minimum-sized RTD [i.e., $I(V_{GS}) = I_p$]. If the gate-to-source voltage of input x_i exceeds the threshold voltage of the MODFET, the current $w_i I(V_{GS})$ is supplied either to the load or driver RTD current. The net RTD current for the load and driver is $I_t = T I(V_{GS})$.

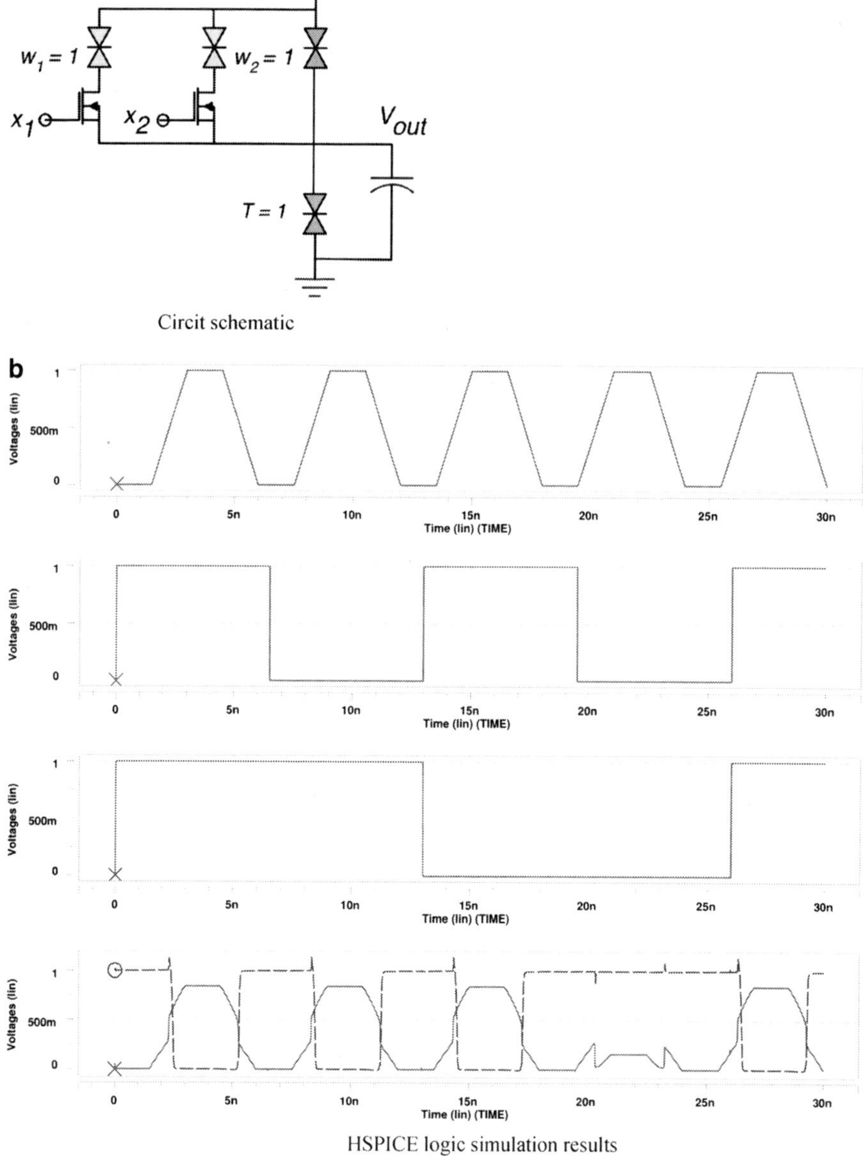

Fig. 7 Circuit (**a**) schematic and (**b**) simulation results of a MOBILE two-input OR gate (*Source:* Ref. 7)

Consequently, V_{OUT} is logic high if ΔI I_t is positive and logic low otherwise. Note that it was mentioned earlier that I_{RTD} does not change for different gate voltages. However, it is proportional to the size of the RTD (thus, can be used to implement the input weight).

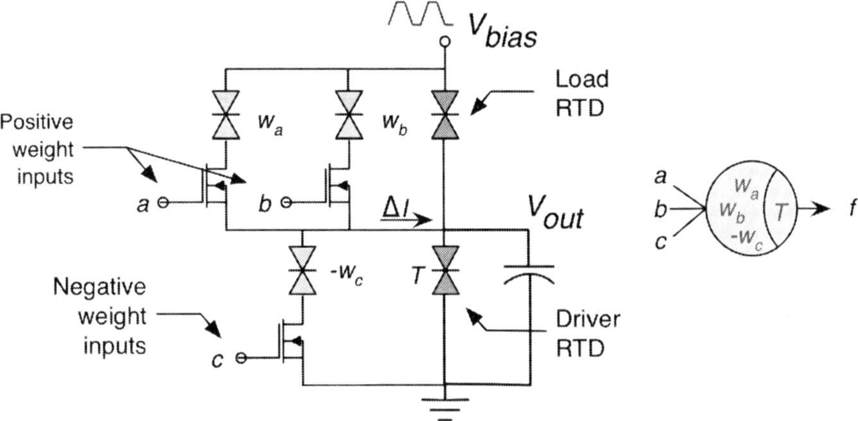

Fig. 8 Implementing a linear threshold function using a MOBILE

2.4 *Circuit Examples*

In this subsection, we review a few RTD-based circuit designs that have been
proposed in the literature. The reader should keep in mind that these are experimental circuits only. So far, with the exception of MOBILEs, there is no systematic
methodology for circuit design with RTDs. Thus, the reader should not despair for
the lack of consistency, but rather should appreciate the design possibilities.

Figure 9 shows the circuit of a MOBILE D flip-flop. It consists of two MOBILEs
whose outputs drive a set/reset flip-flop (SR-FF). The SR-FF configuration is similar
to that of MOBILEs discussed earlier. The only difference is that the RTD areas are
relatively small so that the MODFETs can switch the circuit state without requiring
a bias voltage. At the positive clock edge, depending upon the data input, the
MOBILE with a gate input attached to the lower RTD generates a reset pulse, or
the MOBILE with a gate input attached to the upper RTD generates a set pulse.
These pulses then switch the SR-FF, and the data are stored at the flip-flip output.

Another design of the D flip-flop is shown in Fig. 10a. In this case, the circuit
uses a current mode logic (CML) type MOBILE. CML employs a differential
circuit that uses two complementary signals (and the voltage swing between the
signals) to distinguish between logic states. The circuit comprises a CML-type
MOBILE core (Q_1, Q_2, RTD$_1$–RTD$_4$) generating the complementary return-to-zero
(RZ) mode outputs, and a CML-type set/reset latch (Q_3, Q_4, RTD$_5$–RTD$_6$). When
the clock transitions to logic high, the set and reset signals are generated by the
input data in the CML-type MOBILE core. They maintain their state as long as
the clock is logic high. When the clock is logic low, both signals are at logic low.
The CML-type latch senses and stores the set and reset signals. When the set
voltage is higher than the reset voltage, $I_{SET} = I_{EE2}$ and $I_{RESET} = 0$ due to the
emitter-coupled CML configuration. This results in OUT/OUT being logic high/
low. When both set and reset are logic low, the latch maintains it state with

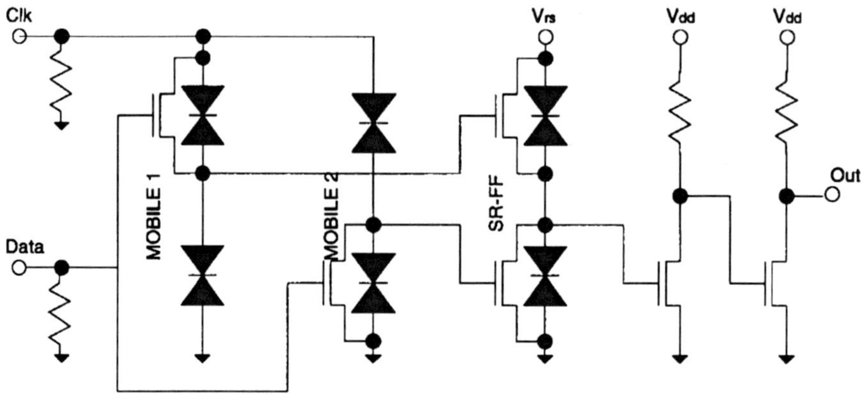

Fig. 9 A MOBILE D flip-flop (*Source:* Ref. 1)

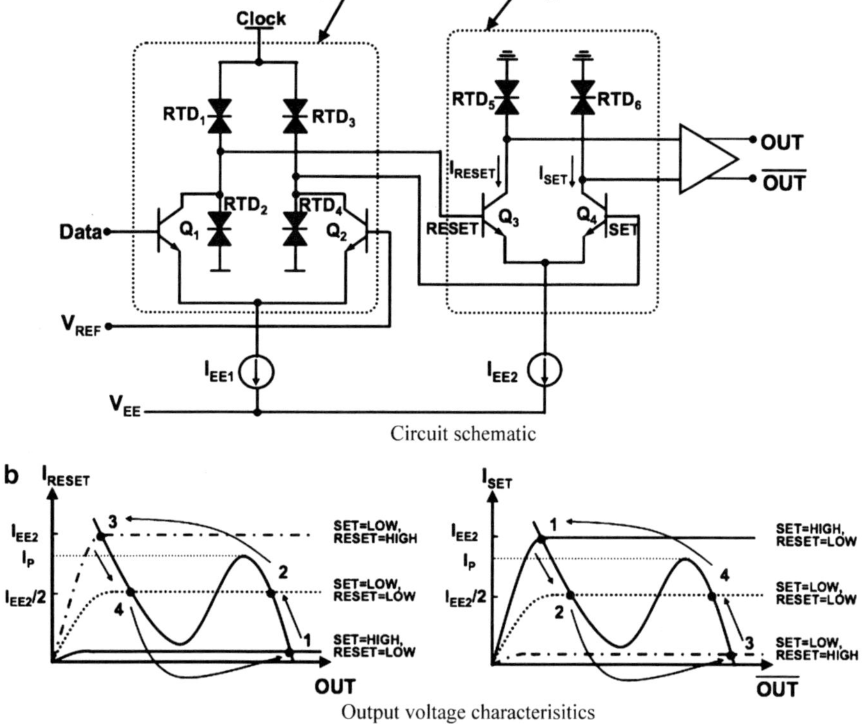

Fig. 10 A D flip-flop using CML-type MOBILEs (*Source:* Ref. 3.)

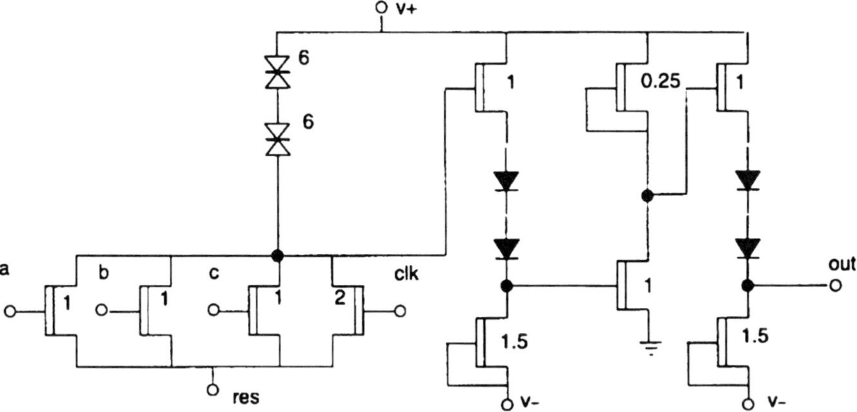

Fig. 11 A three-input majority gate (*Source:* Ref. 1)

$I_{SET} = I_{RESET} = I_{EE2/2}$. Figure 10b shows the output voltage characteristics with respect to the set and reset signals [3].

The final example of Fig. 11 shows a self-latching, three-input majority gate. Such a gate has a weight-threshold vector of (1, 1, 1; 2). The circuit comprises two stages. The first stage implements a three-input minority gate. The second stage is an inverter that inverts the result to get the majority function. In this case, the inverter has an MODFET load instead of an RTD load. When the input to the inverter is low, the bottom transistor (driver) is off and the output is logic high. When the inverter input is logic high, the output is pulled to logic low. The minority gate operates as discussed earlier. The numbers shown in the figure indicate area factors (transistor and RTD sizing).

3   Threshold Logic Synthesis for RTD-Based Devices

Since RTD-based devices implement threshold logic, it is necessary to develop CAD tools and methodologies to effectively utilize this capability. It is well known that threshold logic circuits are more compact than their Boolean counterparts. Furthermore, threshold logic circuits typically have two to six inputs while CMOS gates have a fanin limited to four. In the next few subsections, we will discuss threshold logic synthesis and testing methodologies that are directly applicable to RTD-based devices. These methodologies have been developed in a tool named ThrEshold Logic Synthezier (TELS) [7] which is built on top of an existing Boolean logic synthesis named SIS.

The goal of threshold logic synthesis is to synthesize efficient threshold logic circuits for arbitrary multi-output Boolean functions. The input is a set of n-input, m-output Boolean functions and the output is an optimized circuit consisting of threshold gates that implements the given Boolean function.

Definitions

In this subsection, we briefly review a few basic concepts that will be helpful in understanding the threshold logic synthesis methodology that will be presented later.

A linear threshold gate (LTG) has l two-valued inputs, $x_i \in \{0, 1\}$, and a single two-valued output f. Its internal parameters are a threshold T and weights w_i, $i \in \{1, 2, \ldots, l\}$, where w_i is associated with a particular input variable x_i. The input–output relation of an LTG is based on the fact that output f assumes the value 1 when the weighted sum of the inputs equals or exceeds the value of the threshold, T, and assumes the value 0 otherwise. That is,

$$f(x_1, x_2, \ldots, x_l) = \begin{cases} 1 & \text{if } \sum_{i=1}^{l} w_i x_i \geq T, \\ 1 & \text{if } \sum_{i=1}^{l} w_i x_i < T. \end{cases} \qquad (20)$$

A short-hand notation for the weights and threshold of the LTG is given by the weight-threshold vector $(w_1, w_2, \ldots, w_l; T)$. If we want to increase the robustness of the LTG, we can incorporate defect tolerance into the definition of an LTG, given by (20), as follows:

$$f(x_1, x_2, \ldots, x_l) = \begin{cases} 1 & \text{if } \sum_{i=1}^{l} w_i x_i \geq T + \delta_{\text{on}}, \\ 1 & \text{if } \sum_{i=1}^{l} w_i x_i \leq T - \delta_{\text{off}}, \end{cases} \qquad (21)$$

where parameters δ_{on} and δ_{off} represent defect tolerances that capture variations in the weights due to manufacturing defects or temperature deviations. Such variations can lead to a malfunction of an LTG. Generally, δ_{on} and δ_{off} take on positive values. We assume that $\delta_{\text{on}} = 0$ and $\delta_{\text{off}} = 1$ for the examples given later. However, the threshold logic synthesis methodology presented later can take into account any user-specified values for δ_{on} and δ_{off}.

A Boolean logic function that can be realized by a single LTG is called a threshold function. An l-input NAND and NOR gate can both be realized by a single LTG. Table 2 shows the basic Boolean logic gates and their equivalent LTGs. Because any arbitrary Boolean logic function can be realized solely by a collection of NAND or NOR gates, such gates are functionally complete. Consequently, LTGs are also functionally complete. However, not all Boolean logic functions can be realized by a single LTG. A network of LTGs is called a threshold network.

A Boolean logic function, $f(x_1, x_2, \ldots, x_l)$, is said to be positive (negative) in variable x_i if there exists a disjunctive or conjunctive expression of f in which x_i appears in uncomplemented (complemented) form only. If f is either positive or

Table 2 Boolean gates and their equivalent LTG

Gate type	Boolean function	Weight-threshold vector
BUF	$F = x_i$	$(1; 1)$
NOT	$F = x_i'$	$(-1; 0)$
n-Input AND	$F = x_1 x_2 \cdots x_n$	$(1, 1, \ldots, 1; n)$
n-Input OR	$F = x_1 + x_2 + \cdots + x_n$	$(1, 1, \ldots, 1; 1)$
n-Input NAND	$F = x'_1 x'_2 \cdots x'_n$	$(-1, -1, \ldots, -1; 1 - n)$
n-Input NOR	$F = (x_1 + x_2 + \cdots + x_n)'$	$(-1, -1, \ldots, -1; 0)$

negative in x_i, it is said to be unate in x_i. Otherwise, it is binate in x_i. For example, consider $f_1 = x_1 x_2 + x'_2 x_3$, $f_2 = x_1 x_2 + x'_3$, and $f_3 = x_1 x_2 + x_3 x_4$. Since variable x_2 appears in both positive and negative form in f_1, then f_1 is a binate function. However, all variables in f_2 and f_3 appear either in positive or negative phase, but not both. Consequently, f_2 and f_3 are unate functions. Unateness is a very important property in threshold logic synthesis because every threshold function must be unate. However, not all unate functions are threshold functions. For example, f_1 is not a threshold function because it is binate. f_2 is a threshold function with weight-threshold vector $(1, 1, -2; 0)$. However, f_3 is not a threshold function even though it is unate.

A Boolean logic function, $f(x_1, x_2, \ldots, x_l)$, is commonly expressed in sum-of-products (SOP) form, $\sum_{i=1}^{m} C_i$, where C_i is a cube which is a conjunction of literals (a literal is a variable or its complement). Function f is said to be algebraic if no cube, C_i, is contained within another cube. That is, $\forall i, j, i \neq j, C_i \not\subset C_j$. If f is not algebraic, it is Boolean. A factored form F is said to be algebraically factored if the SOP expression obtained by multiplying F out directly is algebraic, assuming identities $xx' = 0$ and $xx = x$ are not used and no cube contains another. Otherwise, F is Boolean-factored. For example, $f_1 = x_1 + x_2 x_3$ and $f_2 = (x_1 + x_4)(x_2 + x_3)$ are in algebraically factored form while $f_3 = (x_1 + x_2 + x_3)(x'_1 + x'_2 + x'_3)$ and $f_4 = (x_1 x_4 + x_2 + x_3)(x_1 x_5 + x_6 + x_7)$ are in Boolean-factored form.

In threshold logic synthesis, an algebraically factored network is preferred over a Boolean-factored network as a starting point simply because the nodes are more likely to be unate and, hence, possibly threshold functions. In addition, an algebraically factored network has interesting properties that have a significant impact on its testability.

3.1 Theorems for Threshold Logic

We present two useful theorems that will be utilized in the threshold logic synthesis methodology [7]. Along with the proof of each theorem, we demonstrate its application with an example.

Proposition 1 *Given a positive unate threshold function, $f(x_1, x_2, \ldots, x_l)$, with weight-threshold vector $(w_1, w_2, \ldots, w_l; T)$, if x_i is replaced by x'_i to obtain $g(x_1,$

$x_2, \ldots, x_{i-1}, x'_i, x_{i+1}, \ldots, x_l)$, *then the weight of* x_i *in g and threshold of g is* $-w_i$ *and* T w_i, *respectively.*

We negate the original weight of each variable that appears in negative phase in g to get its new weight. To obtain the new threshold, we subtract the sum of the negated weights from the original threshold.

Theorem 1 *Given a unate Boolean function,* $f(x_1, x_2, \ldots, x_l)$, *replace literal* x_i *by literal* x'_j, $i, j \in \{1, 2, \ldots, l\}$ *and* $i \neq j$, *resulting in* $g(x_1, x_2, \ldots, x_k)$, $k \in \{1, 2, \ldots, l\}$ *and* $k \neq i$. *If g is not a threshold function, then f is not a threshold function.*

Proof: We prove the contrapositive of the claim. That is, if f is a threshold function, then g is a threshold function. Assuming f is a threshold function with weight-threshold vector $(w_1, w_2, \ldots, w_l; T)$, we have

$$\sum_{k=1}^{l} w_k x_k \geq T + \delta_{\text{on}} \Rightarrow f = 1, \tag{22}$$

$$\sum_{k=1}^{l} w_k x_k \leq T - \delta_{\text{off}} \Rightarrow f = 0. \tag{23}$$

Note that (22) and (23) represent 2^l *inequalities for all value combinations of variables* $x_1, x_2, \ldots, x_l$. *By replacing* x_i *with* x'_j, *we obtain the following* 2^{l-1} inequalities:

$$\sum_{k=1, k \neq i}^{l} w_k x_k + w_i x'_j \geq T + \delta_{\text{on}} \Rightarrow g = 1, \tag{24}$$

$$\sum_{k=1, k \neq i}^{l} w_k x_k + w_i x'_j \leq T - \delta_{\text{off}} \Rightarrow g = 0. \tag{25}$$

Since $x'_j = 1 - x_j$, we obtain

$$\sum_{k=1, k \neq i, j}^{l} w_k x_k + (w_j - w_i) x_j \geq (T - w_i) + \delta_{\text{on}} \Rightarrow g = 1, \tag{26}$$

$$\sum_{k=1, k \neq i, j}^{l} w_k x_k + (w_j - w_i) x_j \geq (T - w_i) - \delta_{\text{off}} \Rightarrow g = 0. \tag{27}$$

If assignments for w_i and T exist such that the inequalities in (22) and (23) are satisfied, then the inequalities in (26) and (27) can also be satisfied with the weight-threshold vector $(w_1, w_2,\ldots, w_{i-1}, w_{i+1},\ldots, w_{j-1}, w_j \; w_i, w_{j+1},\ldots, w_l; T - w_i)$. $(x_1, x_2,\ldots, x_{i-1}, x_{i+1},\ldots, x_{j-1}, x_j, x_{j+1},\ldots, x_l)$ is the variable sequence corresponding to the weights. Thus, g is also a threshold function.

As an application of Theorem 1, consider $f = x_1 x_2 + x_3 x_4$. To determine if f is threshold or not, we replace x_3 by x'_1. This results in $g = x_1 x_2 + x'_1 x_4$. Since g is binate in x_1, it is not a threshold function and, therefore, f is not a threshold function.

Theorem 2 *If a Boolean function $f(x_1, x_2,\ldots, x_l)$ is a threshold function, then $h(x_1, x_2, \ldots, x_{l+k}) = f(x_1, x_2, \ldots, x_l) + x_{l+1} + x_{l+2} + \cdots + x_{l+k}$ is also a threshold function.*

Proof: A threshold function can always be represented in positive unate form by substituting negative variables with positive variables. For simplicity, we assume that f is already expressed in this form. There exists a weight-threshold vector, $(w_1, w_2,\ldots, w_l; T)$, for f since it is a threshold function. If any of the $x_{l+j}, j \in \{1, 2,\ldots, k\}$, equals 1, h equals 1. Otherwise, h is equal to f. If we set the weight w_{l+j} of x_{l+j} to a value no less than $T + \delta_{on}$, for example, $w_{l+j} = T + \delta_{on}$, then the output is 1 when x_{l+j} is 1. When x_{l+j} and some other $x_i, i \in \{1, 2,\ldots, l\}$, equal 1, the output is also 1. This is because we have represented the function in positive unate form, thereby guaranteeing that all the weights and threshold of f are positive. When all x_{l+j} equal 0, h equals f. Thus, a weight-threshold vector for h always exists. Therefore, h is a threshold function.

To illustrate Theorem 2, let $f_1(x_1, x_2, x_3) = x_1 x_2 + x_1 x_3$. Because f_1 is a threshold function with weight-threshold vector $(2,1,1;3)$, $h_1(x_1,x_2,x_3,x_4,x_5)=x_1 x_2+x_1 x_3+x_4+x_5$ is also a threshold function with weight-threshold vector $(2, 1, 1, 3, 3; 3)$. Taking another example, let $f_2(x_1, x_2) = x_1 x'_2$. First, we represent f_2 in positive unate form as $g_2(x_1, y_2) = x_1 y_2$, where $y_2 = x'_2$. Since g_2 is a threshold function with weight-threshold vector $(1, 1; 2)$, $h_2(x_1, y_2, x_3) = g_2(x_1, y_2) + x_3 = x_1 y_2 + x_3$ is also a threshold function with weight-threshold vector $(1, 1, 2; 2)$. Since $y_2 = 1 \; x_2$, $x_1 x'_2 + x_3$ is also a threshold function with weight-threshold vector $(1, -1, 2; 1)$.

3.2   Synthesis Methodology

We discuss a threshold logic synthesis methodology in detail next [7]. Figures 12 and 13 give a high-level overview of the main steps. The input is an algebraically factored multi-output combinational Boolean network G, and its output is a functionally equivalent threshold network G_t. This is a good starting point because powerful Boolean synthesis techniques and tools exist that can be applied to synthesize an optimized Boolean network. We can then perform threshold logic synthesis on such a network. The designer can specify the maximum fanin (number of inputs) restriction and defect tolerances $(\delta_{on}/\delta_{off})$ for the threshold gates during synthesis. Finally, the optimization criterion is minimizing the number of gates in the synthesized network.

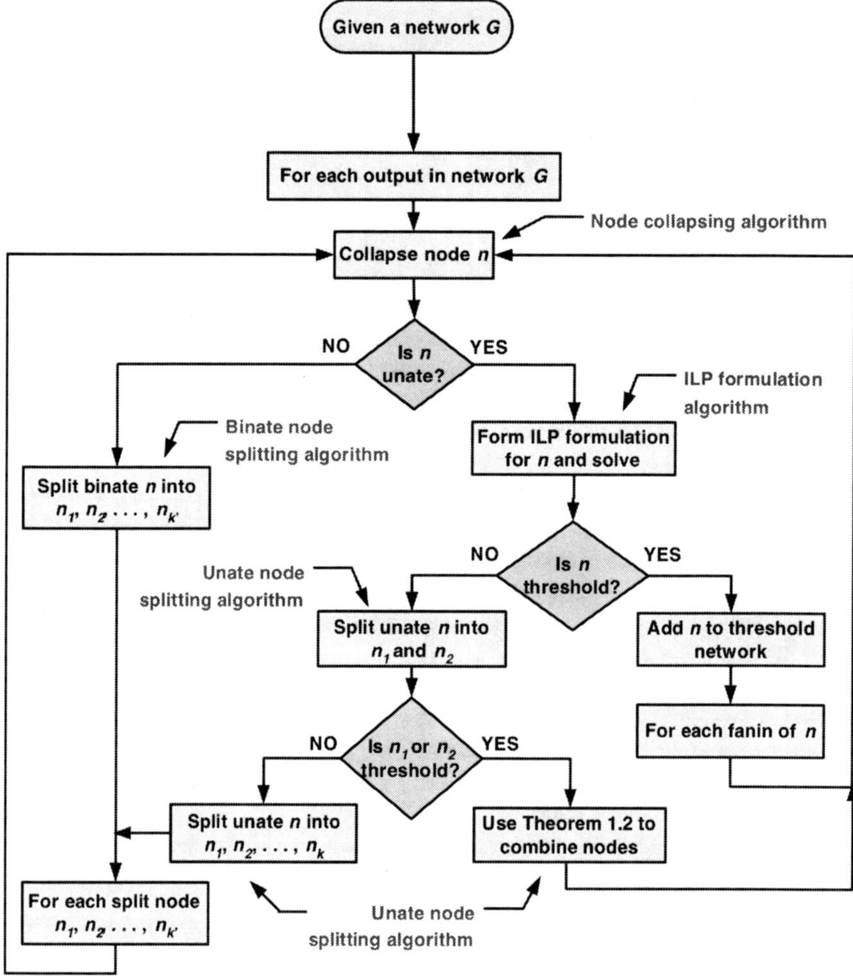

Fig. 12 The main method for threshold network synthesis

Two methods are used for synthesis as shown in Figs. 12 and 13. In the first method, shown in Fig. 12, the synthesis algorithm begins by processing each primary output of Boolean network G. First, the node representing a primary output is collapsed. If the node represents a binate function, it must be split into multiple nodes which are then processed recursively. This is because a binate function cannot be a threshold function. On the other hand, if the node is a unate function and that function is threshold, it is saved in the threshold network and the fanins of the node are processed recursively. Otherwise, the unate node is first split into two nodes. If either of the split nodes is a threshold function, Theorem 2 is applied as a simplification step. If neither of the split nodes is a threshold function, the original node is split into multiple nodes which are then processed recursively. The synthesis

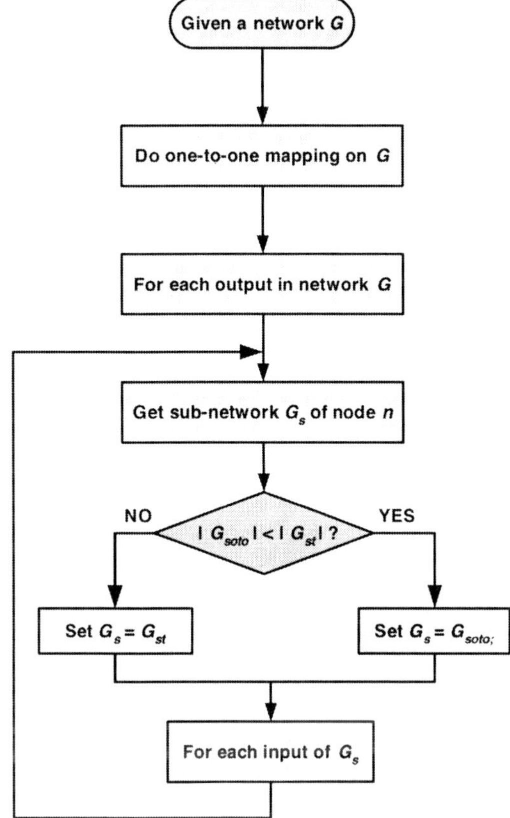

Fig. 13 The second method incorporates the synthesis method in Fig. 12

algorithm terminates when all the nodes in network G have been mapped into threshold nodes.

In the second method, shown in Fig. 13, a one-to-one mapping is first performed on network G. In such a mapping, we replace each Boolean gate with its threshold equivalent gate from Table 2. After that, starting from each primary output, we obtain the sub-networks. Threshold synthesis is then performed on each sub-network using the first synthesis algorithm. If the resulting threshold sub-network has fewer threshold gates than the sub-network obtained by the one-to-one mapping method, we choose the former. Otherwise, we choose the one-to-one mapped sub-network. Synthesis terminates when all the sub-networks have been processed. Finally, the two threshold networks synthesized by the two methods are compared and the network which contains fewer gates is selected as the final optimized network.

In the threshold logic synthesis methodology, there are four key operations that are required. As shown in Fig. 12, these are collapsing a node, determining if a node is a threshold function or not, splitting a unate node, and splitting a binate node. We consider each of these operations next.

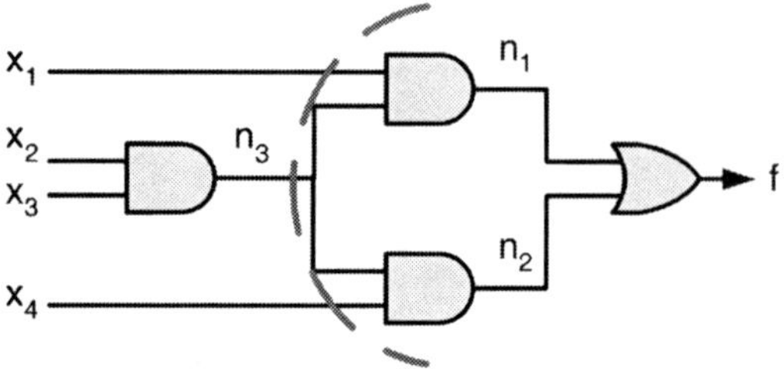

Fig. 14 An example network to demonstrate node collapsing on output f

Given a node n, if its fanin exceeds the fanin restriction, we split n by using the one-to-one mapping method. Otherwise, we keep collapsing it until either all of the fanins of n are primary inputs and/or fanout nodes or the fanin of n exceeds the fanin restriction. To demonstrate node collapsing, consider the network with output node f in Fig. 14. Assume that the fanin restriction is four. Since the inputs to f are not primary inputs, we first collapse n_1 to get $f = x_1 n_3 + n_2$. Since the fanin count of this expression is still less than four, we continue by collapsing n_2 to get $f = x_1 n_3 + n_3 x_4$. Now, we cannot collapse x_1 and x_4 since they are primary inputs. Furthermore, observing that n_3 is a fanout node, we do not collapse it either. Thus, the final result after collapsing is $f = x_1 n_3 + n_3 x_4$.

Note that node sharing is preserved during the collapsing process. This implicitly helps to maintain some of the original network structure and provides guidance for better network decomposition. The benefit is profound when the network contains many fanout nodes.

Once a node has been collapsed into a unate function, it is necessary to determine whether it is threshold or not. This can be done by formulating the problem as an integer linear programming (ILP) problem and then solving it using solvers like CPLEX. There are at most 2^l distinct cubes for a logic function of l variables and this leads to 2^l inequalities that represent the constraints (note that l cannot exceed the fanin restriction, and hence is not a large number). However, many of these constraints are redundant. A simple method has been devised to eliminate redundant constraints, which makes the ILP formulation smaller and possibly faster to solve. Even though in the worst case, an ILP problem may take exponential time to solve, in practice it is efficiently solved because of small values of l (typical values range from three to eight) for each threshold gate in the network.

The algorithm is best demonstrated by an example. Given a unate function, $f(x_1, x_2, \ldots, x_l)$, if it contains variables in negative phase, we first transform these variables into other variables in positive phase using variable substitution. Consider $f = x_1 x'_2 + x_1 x'_3$, where x_2 and x_3 are in negative phase. By replacing x'_2 with y_2 and x'_3 with y_3, we get the positive unate function $g = x_1 y_2 + x_1 y_3$. The ILP formulation for g is as follows:

Minimize:

$$w_1 + w_2 + w_3 + T, \tag{28}$$

subject to:

$$w_1 + w_2 \geq T + \delta_{\text{on}}, \tag{29}$$

$$w_1 + w_3 \geq T + \delta_{\text{on}}, \tag{30}$$

$$w_2 + w_3 \leq T - \delta_{\text{off}}, \tag{31}$$

$$w_1 \leq T - \delta_{\text{off}}, \tag{32}$$

$$w_i \geq 0, \text{integer}, i = 1, 2, 3. \tag{33}$$

The objective function for this ILP problem is defined as the summation of the weights and threshold since in some nanotechnologies, as we will see later, the area of the circuit is proportional to this sum. Since g has been transformed into a positive unate form, only 1 and don't care $(-)$ will appear in its ON-set cubes. The ON-set cubes are values for which the function equals 1. The ON-set cubes of g are $(1\ 1\ -)$ and $(1\ -\ 1)$, where x_1, y_2, and y_3 is the variable sequence. To transform the ON-set cubes into inequalities, the ith 1 value corresponds to weight w_i. We need not consider don't cares in the inequalities, because they represent redundancies in g. For example, the ON-set cube $(1\ 1\ -)$ corresponds to two inequalities, namely, $w_1 + w_2 \geq T + \delta_{\text{on}}$ and $w_1 + w_2 + w_3 \geq T + \delta_{\text{on}}$. The second inequality is redundant because once the first inequality is satisfied, the second inequality is automatically satisfied as well. Similarly, $w_1 + w_3 \geq T + \delta_{\text{on}}$ represents the constraint imposed by cube $(1\ -\ 1)$.

To compute the OFF-set cubes of g, we simply invert g to get g'. The OFF-set cubes are values for which the function equals 0. The ON-set cubes of g' correspond to the OFF-set cubes of g. Because g' is always in negative unate form, only 0 and $-$ appear in its ON-set cubes. Continuing with the earlier example, we invert g to get $g' = x'_1 + y'_2y'_3$ after simplification. The ON-set cubes for g' are $(0\ -\ -)$ and $(-\ 0\ 0)$. For the OFF-set inequalities, the ith don't care corresponds to weight w_i. Therefore, the ON-set inequalities for g' are $w_2 + w_3 \leq T\ \delta_{\text{off}}$ and $w_1 \leq T\ \delta_{\text{off}}$. Thus, the OFF-set inequalities for g are $w_2 + w_3 \leq T\ \delta_{\text{off}}$ and $w_1 \leq T\ \delta_{\text{off}}$, as given in (31) and (32).

By requiring the variables to be integer-valued [i.e., (33)], this ILP problem has an optimal solution. The weight-threshold vector for g is $(2, 1, 1; 3)$. Applying Proposition 1, the final weight-threshold vector for f is $(2, -1, -1; 1)$.

If the ILP problem for node n does not have a solution, the node must be split into multiple nodes to increase the likelihood of the split nodes being threshold functions. If all the variables appear only once, we simply split the node into two, with each node containing roughly equal number of cubes. When a variable appears in all the cubes, we split the node into two by factoring this variable out of the node. If the above two conditions are not met, we split the node using the most frequently appearing variable. For example, given $n = x_1x_2 + x_1x_3 + x_4x_5$, we split on x_1 to get $n_1 = x_1x_2 + x_1x_3$ and $n_2 = x_4x_5$, with $n = n_1 + n_2$.

Once a node has been split, we choose the larger node (i.e., the one with more cubes) and check that node to see if it is a threshold function. If it is, we apply Theorem 2. Looking at the last example, since $n_1 = x_1x_2 + x_1x_3$ is a threshold function with weight-threshold vector (2, 1, 1; 3), function $n = x_1x_2 + x_1x_3 + n_2$ is also a threshold function with weight-threshold vector (2, 1, 1, 3; 3). Now, n_2 is processed by further collapsing, threshold checking, or splitting. If neither of the split nodes is a threshold function, the original node is split into k smaller nodes, where k is the smaller of fanin restriction and the number of cubes in n. After splitting, $n = n_1 + n_2 + \cdots + n_k$, which is a threshold function with weight-threshold vector (1, 1,..., 1; 1). The split nodes, n_i, $I \in \{1, 2,..., k\}$, are then processed recursively.

If a node n is binate, we split it into at most k nodes where k is the smaller of fanin restriction and the number of cubes in n. The splitting stops when a unate node is generated. We first split the binate node on the most frequently appearing binate variable. If the split nodes are binate, we repeat the process. Otherwise, we stop splitting.

To demonstrate binate splitting, consider $n = x'_1x_4 + x_2x_3 + x'_2x_4x_5$, where the fanin restriction is four and the number of cubes is three. This node will be split into at most three nodes. First, it is split on the binate variable, x_2, to get $n_1 = x'_1x_4 + x_2x_3$ and $n_2 = x'_2x_4x_5$. Because n_1 and n_2 are unate nodes, we stop splitting. Thus, n is represented as $n = n_1 + n_2$, which is a threshold function with weight-threshold vector (1, 1; 1). Threshold network synthesis proceeds recursively by processing each of the split nodes.

3.3 An Example

To illustrate the synthesis methodology, we consider the Boolean network shown in Fig. 15a. Assume the fanin restriction for threshold logic synthesis is set to four. We first collapse $f = n_1 + n_2$ to get $n_3x_5 + x_6x_7$. Next, it must be determined if f is a threshold function or not. We formulate this problem as an ILP one and solve it. It turns out that the ILP formulation does not have a solution and, therefore, f is not threshold. Since f is a unate node, the unate node splitting algorithm is applied to split f into smaller nodes. Based on the heuristics described earlier, f is split as $f = n_1 + x_6x_7$ where $n_1 = n_3x_5$. Next, we proceed to synthesize n_1. Collapsing n_1, we get $n_1 = n_4x_5 + n_5x_5$. Again, we check to see if n_1 is a threshold function and in this case, it is indeed. Consequently, we apply Theorem 2 to $f = n_1 + x_6x_7$ to create a threshold gate as well. At this point, the only nodes that remain are n_4 and n_5. These two nodes are synthesized next and the final threshold network that is produced is shown in Fig. 15b. This network contains four gates and three levels. Had we performed one-to-one mapping by converting each Boolean gate in Fig. 15a into a threshold gate, the network would have contained seven gates and five levels (including the inverter).

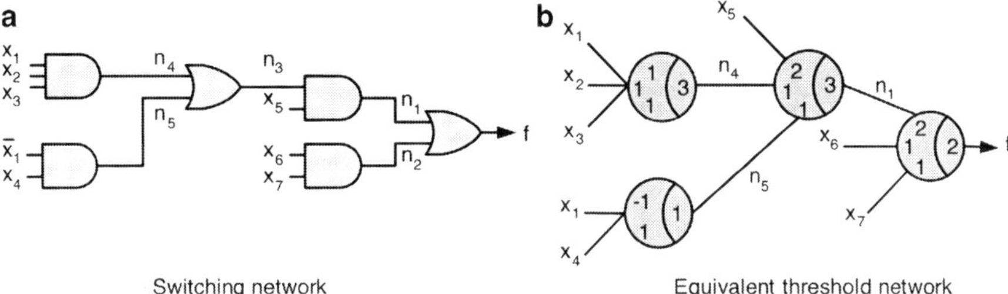

Fig. 15 An example to illustrate threshold logic synthesis methodology

4 Threshold Logic Testing for RTD-Based Devices

Now that we have a threshold logic synthesis methodology, we also want to develop a test generation framework that can be used to generate test vectors to detect manufacturing defects in RTD-based circuits. In this section, we discuss the theory that is needed to develop an automatic test pattern generation (ATPG) framework for combinational threshold networks [8]. We focus our attention mainly on the testing of MOBILE circuits.

4.1 Fault Modeling

In order to develop a fault model, it is important to identify the defects that are most likely to occur in MOBILEs. Given the RTD integration with MODFETs, cuts and shorts are a certainty. Furthermore, RTD area variation will cause peak current fluctuations which will cause logic faults. Figure 16a shows the cuts and shorts in a MOBILE that can be modeled as single stuck-at faults (SSFs) at the logic level. A SSF on a wire renders the wire permanently logic high, called a stuck-at 1 (SA1) fault, or logic low, called a stuck-at 0 (SA0) fault. A cut (sites 1, 2, and 3) on an MODFET or on a line connecting the RTD and MODFET will render it permanently nonconducting and is modeled as a SA0 fault. Similarly, a short across an RTD (site 4) or the driver RTD (site 8) is also modeled as an SA0 fault because in the former, the input weight will become zero, while in the latter, there will be a direct connection between the output and ground. A cut at site 6 represents either an SA0 or SA1 fault depending upon the threshold of the gate. If the threshold is less than zero, then the cut is modeled as an SA1 fault. Otherwise, it is modeled as an SA0 fault. On the other hand, faults at sites 5 and 7 are modeled as SA1 faults. A short across the MODFET will make it conduct permanently while a direct connection between the output and bias voltage will exist in the presence of a short across the load RTD, making the fault appear as an SA1 when the MOBILE is active.

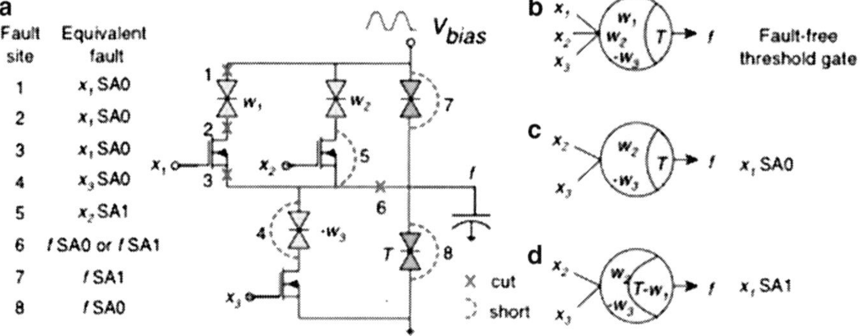

Fig. 16 (**a**) Fault modeling and equivalent LTG with (**b**) no faults, (**c**) an SA0 fault, and (**d**) an SA1 fault

To validate this fault model, HSPICE simulation results for defects present in the circuit of Fig. 17 are considered. Figures 17a–d show the behavior of the LTG in the presence of a single fault. For example, a fault at site 5 would lead to x_2 SA1. Consequently, the output will always be SA1 since this is an OR gate. Figure 17c confirms this assertion. Note that the SA1 behavior is manifested only when the clock is high.

4.2 Irredundant Threshold Networks and Redundancy Removal

The next step in test generation is redundancy identification and removal. In Boolean testing, irredundant networks are intricately linked to circuit testability. There are similar relationships that hold for threshold logic circuits. If no test vector exists for detecting a fault s in a threshold network G, then s is redundant. In such a case, the corresponding node or edge in the network can be removed without affecting the functionality of G. The results for removing a redundant fault in a threshold network are as follows:

1. If an SA0 fault on an edge is redundant, the edge in the network can be removed as shown in Fig. 16c.
2. If an SA1 fault on an edge is redundant, the edge in the network can be removed as shown in Fig. 16d. Furthermore, the threshold of the nodes in the edge's fanout must be lowered by the weight of the removed edge.

Finally, all nodes and edges in the sub-network that do not fan out and are in the transitive fanin of the removed edge can be removed from the network in both cases.

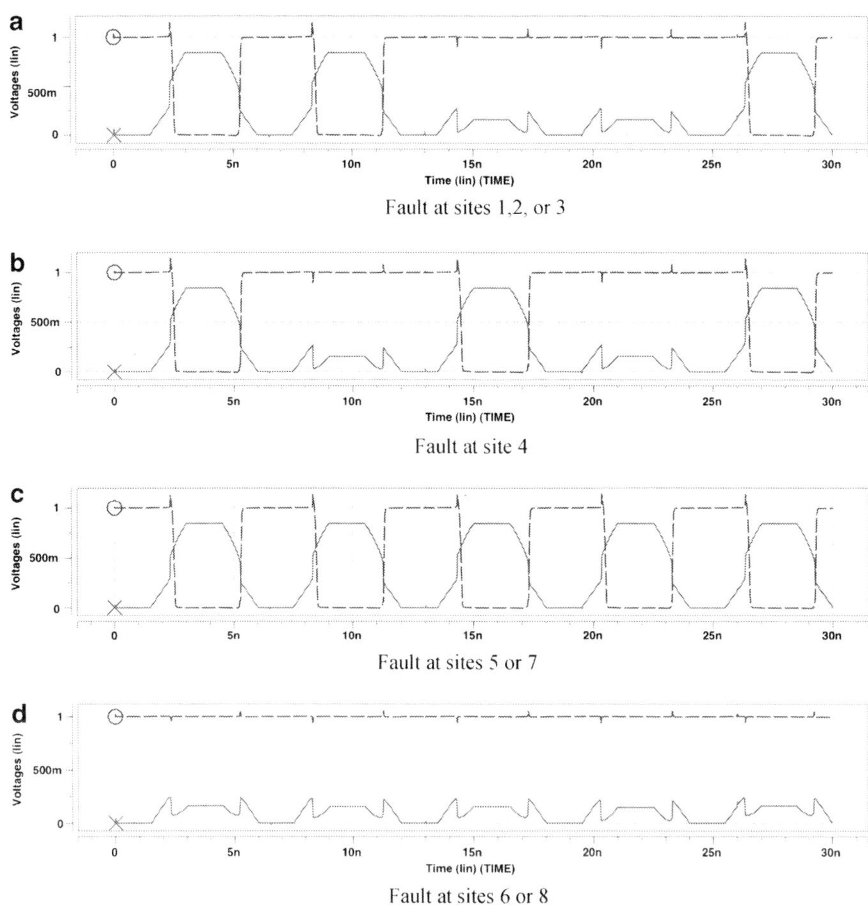

Fig. 17 HSPICE simulation of faulty LTG implementing a MOBILE two-input OR gate

4.3 Test Generation

The final step is the actual test generation step. In order to find a test vector for a fault at input x_i of a threshold gate, it is necessary that the term $w_i x_i$ in (20) be the dictating factor in determining the output value of the function. The following theorem gives the conditions the test vector has to satisfy.

Theorem 3 *To find test vectors for x_i SA0 and x_i SA1 in a threshold gate implementing the threshold function $f(x_1, x_2, \ldots, x_n)$, we must find an assignment on the remaining input variables such that one of the following inequalities is satisfied:*

$$T - w_i \leq \sum_{j=1,j\neq i}^{n} w_i x_i < T \tag{34}$$

or

$$T \leq \sum_{j=1,j\neq i}^{n} w_i x_i < T - w_i. \tag{35}$$

If an assignment exists, then $(x_1, x_2,\ldots, x_i = 1,\ldots, x_n)$ and $(x_1, x_2,\ldots, x_i = 0,\ldots, x_n)$ are test vectors for x_i *SA0* and x_i *SA1*, respectively. If no assignment exists, then both faults are untestable, and therefore, redundant.

Theorem 4 *In a threshold gate implementing the threshold function $f(x_1, x_2,\ldots, x_n)$, if there exist two (or more) inputs x_j and x_k such that $w_j = w_k$, then test vectors to detect x_k SA0 and x_k SA1 can be obtained simply by interchanging the bit positions of x_j and x_k in the SA0 and SA1 test vectors for x_j, respectively, assuming they exist.*

To understand these theorems, consider a gate that realizes the threshold function, $f(x_1, x_2, x_3) = x_1 x_2 + x_1 x_3$ with weight-threshold vector (2, 1, 1; 3). To test for x_1 SA0, the inequalities to be satisfied are $1 \leq (w_2 x_2 + w_3 x_3) < 3$ or $3 \leq (w_2 x_2 + w_3 x_3) < 1$. Obviously, the second inequality cannot be satisfied. However, the first inequality can be satisfied with three test vectors, namely 101, 110, and 111. The test vectors for x_1 SA1 can be easily obtained by replacing $x_1 = 1$ with $x_1 = 0$ in the original test vectors. Thus, vectors 001, 010, and 011 detect x_1 SA1. Finally, given that vector 110 is a test for x_2 SA0 and because $w_2 = w_3$, a test vector that detects x_3 SA0 is obtained by interchanging the bit positions of x_2 and x_3 in 110 to get 101.

A popular algorithm for Boolean testing is the D-algorithm. It employs the concepts of primitive D-cube of a fault (PDCF), propagation D-cubes, and singular covers. A PDCF in a threshold gate can be computed by obtaining a test vector for the fault that results in a D (1 in the fault-free case and 0 in the faulty case) or D' (0 in the fault-free case and 1 in the faulty case) at its output, using the theorems given above.

Propagation D-cubes are used to sensitize a path from the fault site to one (or more) primary outputs. Knowing the threshold function that is implemented by a threshold gate, we can use algebraic substitution to determine the propagation D-cubes using the D-notation. For example, to determine the propagation D-cubes from x_1 in $f(x_1, x_2, x_3) = x_1 x_2 + x_1 x_3$, substituting D for x_1 in f, we get $Dx_2 + Dx_3$. For the fault to propagate, it is required that only the cubes containing D (or D') get "activated" in f. In this case, because both cubes contain D, activating either or both cubes will result in a propagation D-cube. Thus, the propagation D-cubes for x_1 are {D10D, D01D, D11D} (the last D corresponds to output f). Of course, {D'10D', D'01D', D'11D'} are also propagation D-cubes.

Singular covers are used in test generation to justify the assignments made to the output of a threshold gate. They are easily obtained from the function of the

Fig. 18 Testing for x_1 SA1

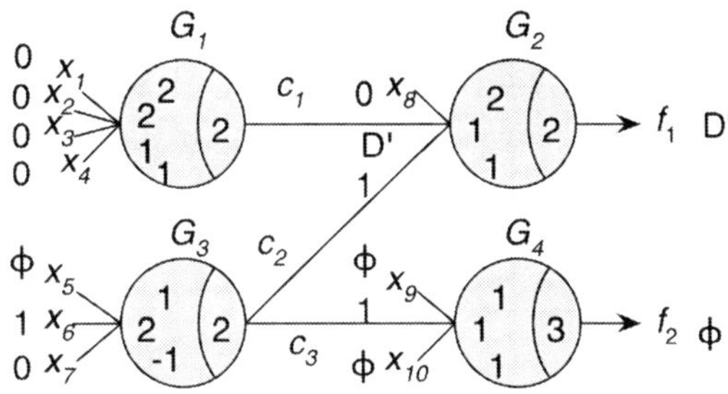

threshold gate. Consider the threshold network in Fig. 18. Suppose we want to derive a test vector for x_1 SA1. The PDCF for this fault in gate G_1 is 0000D'. Using the propagation D-cube of gate G_2 as shown, the fault effect can be propagated to circuit output f_1. This requires 1 to be justified on line c_2 through the application of the relevant singular cube to gate G_3 as shown. Thus, a test vector for the above fault is $(0, 0, 0, 0, \varphi, 1, 0, 0, \varphi, \varphi)$. Note that φ is a don't care as the fault is visible on output f_1.

To reduce the run-time, it is necessary to reduce the number of faults for which test generation needs to be done. This is known as fault collapsing and can be done by exploiting fault dominance relationships.

Fault 1 is said to dominate fault 2 if all test vectors for fault 2 are also test vectors for fault 1. Thus, if fault 2 has been targeted in test generation, fault 1 need not be targeted. The following fault dominance relationship hold in a threshold gate that implements the threshold function $f(x_1, x_2, \ldots, x_n)$.

Theorem 5 *An output f SA0 (SA1) fault dominates an x_i SA0 (SA1) fault if (34) is satisfied. An output f SA1 (SA0) fault dominates an x_i SA0 (SA1) fault if (35) is satisfied.*

The net effect of this theorem is that output faults do not need to be tested since a test vector for an input fault will also detect an output fault. Note that this assumes that the output does not fan out to other parts of the circuit. If the output branches, then it is still necessary to test for faults on the branches. To demonstrate this theorem, consider the threshold function $f(x_1, x_2, x_3) = x_1x_2 + x_1x_3$ again. If we apply the theorem, we see that f SA0 (SA1) dominates x_1 SA0 (SA1), x_2 SA0 (SA1), and x_3 SA0 (SA1). Hence, f SA0 (SA1) can be discarded from the fault list. Exploiting this theorem for maximum benefit leads to the following theorem on test generation for irredundant combinational threshold networks.

Theorem 6 *In an irredundant combinational threshold network G, any test set V that detects all SSFs on the primary inputs and fanout branches detects all SSFs in G.*

The reader will note that this is the equivalent of the famous checkpoint theorem in Boolean testing.

Exercises

1 Determine if the following functions are unate or binate:

(a) $f(x_1, x_2, x_3) = x_1x_2 + x_1x_3$
(b) $g(x_1, x_2, x_3, x_4) = x_1x_2 + x'_3x_2 + x_3x'_4$
(c) $h(x_1, x_2, x_3) = x_1x'_2 + x_3$

2. Based on examining linear inequalities, determine which of the following functions is a threshold function. For each one that is, find the corresponding weight-threshold vector.

(a) $f(x_1, x_2, x_3) = \Sigma(1, 2, 3, 5, 6, 7)$
(b) $g(x_1, x_2, x_3) = \Sigma(1, 3, 6, 7)$
(c) $h(x_1, x_2, x_3) = \Sigma(3, 5, 6, 7)$

3. Determine the truth table of the circuit shown in Fig. 19.

4. For each of the functions of Exercise 2 that is realizable by a single threshold gate, find a realization for $f'(x'_1, x_2, x_3)$.

5. A three-input minority gate (output is one if at-least two of the three inputs are zero) can be implemented using a threshold gate with weight-threshold vector $<1, -1, -1; -1>$. Implement the function $f = x'_1x'_2x_3 + x_1x'_3 + x_2x'_3 + x_1x_2$ with at most four minority gates.

6. For the network shown in Fig. 20, obtain all test vectors that detect x_2 SA0.

7. Prove that threshold logic is functionally complete.

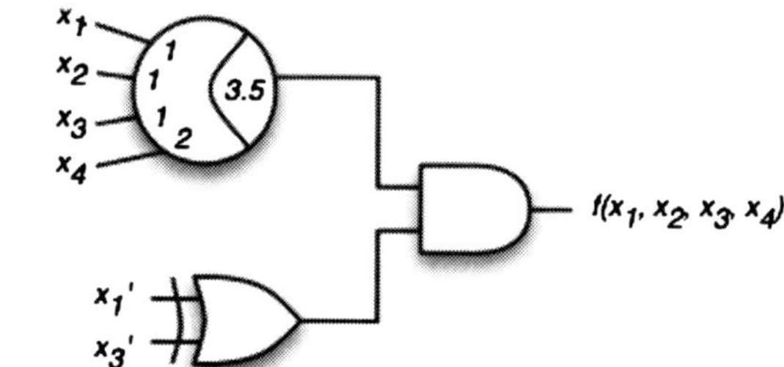

Fig. 19 Circuit for Exercise 3

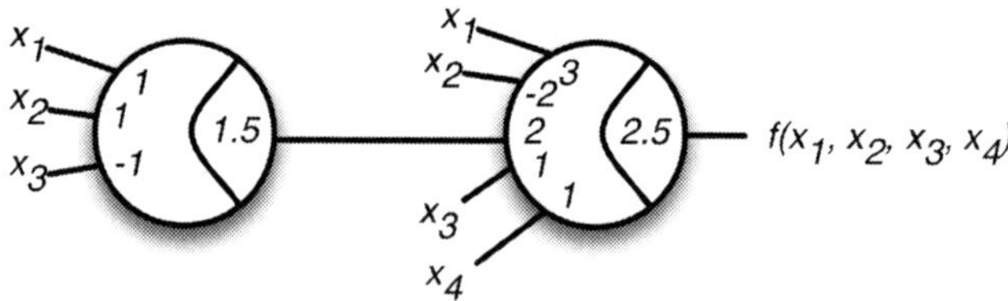

Fig. 20 Circuit for Exercise 6

8. Given a threshold gate implementing function $f(x_1, x_2, \ldots, x_n)$, show that

   (a) Output f SA0 (SA1) dominates x_i SA0 (SA1) if (34) is satisfied.
   (b) Output f SA1 (SA0) dominates x_i SA0 (SA1) if (35) is satisfied.

9. Find three recent scholarly articles on any aspect of RTD design. Prepare a 15-min presentation for your class.

References

1. P. Mazumder, S. Kulkarni, M. Bhattacharya, J.P. Sun, and G.I. Haddad, "Digital circuit applications of resonant tunneling devices," *Proc. IEEE*, vol. 86, no. 4, pp. 664–686, Apr. 1998.
2. K.J. Chen, T. Akeyoshi, and K. Maezawa, "Monostable–bistable transition logic elements (MOBILEs) based on monolithic integration of resonant tunneling diodes and FETs," *J. Appl. Phys.*, vol. 34, no. 2B, pp. 1199–1203, Feb. 1995.
3. T. Kim, Y. Jeong, and K. Yang, "A 45-mW RTD/HBT MOBILE D-flip flop IC operating up to 32Gb/s," in *Proc. Int. Conf. Indium Phosphide & Related Materials*, Princeton, NJ, pp. 348–351, May 2006.
4. J.P. Sun, G.I. Haddad, P. Mazumder, and J.N. Schulman, "Resonant tunneling diodes: models and properties," *Proc. IEEE*, vol. 86, no. 4, pp. 641–661, Apr 1998.
5. W. Williamson, S.B. Enquist, D.H. Chow, H.L. Dunlap, S. Subramaniam, P. Lei, G.H. Bernstein, and B.K. Gilbert, "12-GHz clocked operation of ultralow power interband resonant tunneling diode pipelined logic gates," *IEEE J. Solid-State Circuits*, vol. 32, no. 2, pp. 222–230, Feb. 1997.
6. M.T. Bjork, B.J. Ohlsoon, C. Thelander, A.I. Persson, K. Deppert, L.R. Wallenberg, and L. Samuelson, "Nanowire resonant tunneling diodes," *Appl. Phys. Lett.*, vol. 82, no. 23, pp. 4458–4460, Dec. 2002.
7. R. Zhang, P. Gupta, L. Zhong, and N.K. Jha, "Threshold network synthesis and optimization and its application to nanotechnologies," *IEEE Trans. Computer Aided Des.*, vol. 24, no. 1, pp. 107–118, Jan. 2005.
8. P. Gupta, R. Zhang, and N.K. Jha, "Automatic test pattern generation for combinational threshold logic networks," *IEEE Trans. VLSI Syst.*, vol. 16, no. 8, pp. 1035–1045, Aug. 2008.

Online Educational Links

Please visit the following links to learn more about RTDs and CAD tools for threshold logic synthesis and testing:

TELS: http://nanohub.org/resources/3353
RTD simulator: http://nanohub.org/tools/rtd
RTD modeling lectures: http://nanohub.org

Circuit Design with Quantum Cellular Automata

Pallav Gupta

Abstract The quantum cellular automata (QCA) concept uses a rectangular cell with a bistable charge configuration to represent information. We first study QCA and its fundamental building blocks in detail. We then proceed to understand logic design using QCA gates. Next, we explore testing of QCA gates. We briefly discuss current computer-aided design (CAD) tools available for QCA design. Finally, we discuss the current state of QCA fabrication technology and briefly comment on future research directions.

Keywords Quantum cellular automata · Design with QCA · Majority/minority synthesis · Testing of QCA

1 Introduction

Digital electronics is the by-product of two key ideas: (a) using binary numbers to represent information and (b) physically representing the binary values "0" and "1" as "off" and "on" states, respectively, of a current switch. *Complementary metal-oxide semiconductor* (CMOS) technology implements this functionality by cleverly using transistors as paired switch networks so that current flows only when the state of the pair is altered. Representing binary information in this manner has enabled technological advances that have revolutionized modern society.

The current-switch paradigm faces serious limitations as transistor dimensions are reduced. With smaller feature sizes, the transistor drive current decreases, thereby requiring a longer time to charge or discharge an output node or interconnect. Furthermore, noise margins decrease, which reduces its ability to switch "cleanly" between the two states. As gate-oxide thickness scales to maintain

P. Gupta (✉)
Core CAD Technologies, Intel Corporation, Folsom, CA 95630, USA
e-mail: pallav.gupta@intel.com

N.K. Jha and D. Chen (eds.), *Nanoelectronic Circuit Design*,
DOI 10.1007/978-1-4419-7609-3_13, © Springer Science+Business Media, LLC 2011

channel conductivity, sub-threshold leakage current increases exponentially due to the quantum-mechanical phenomenon of electron tunneling. Process variations affect transistor dimensions, which affect its characteristics – making the transistor behavior less certain, more statistical. Finally, since electrons move from the power supply (V_{DD}) to ground, energy dissipation occurs.

For a single transistor, the aforementioned problems are insignificant. However, with modern integrated circuit (IC) densities crossing a billion transistors, these barriers are becoming limiting factors in IC design. In the short term, innovative design and manufacturing techniques will continue to provide solutions to these challenges and sustain Moore's law. However, fundamental considerations and fabrication costs will ultimately limit device densities in the longer term. The need for developing alternative approaches that enable miniaturization of devices down to the ultimate limits of molecular dimensions is evident.

The *quantum cellular automata* (QCA) concept uses a rectangular cell with a bistable charge configuration to represent information. The cell contains two mobile electrons and four quantum dots. One configuration of the charge represents a binary "0," the other a "1." However, unlike the current switch model, no current flows into or out of the cell. The electric field due to the charge configuration of one cell changes the charge configuration of the neighboring cell. Given a clocking scheme that can modulate barriers between cells, this basic cell–cell interaction provides a general-purpose computing mechanism with very little power dissipation.

The rest of this chapter provides the reader with an introduction to QCA circuit design. We first study QCA and its fundamental building blocks in detail. We then proceed to understand logic design using QCA gates. Next, we explore testing of QCA gates. We briefly discuss current *computer-aided design* (CAD) tools available for QCA design. Finally, we discuss the current state of QCA fabrication technology and briefly comment on future research directions.

2 QCA Fundamentals

The main idea in QCA technology is that logic states are encoded as the configuration of individual electrons rather than voltage levels. A QCA cell can be conceptualized as a set of four quantum dots positioned at the corners of a square. The essential feature of a quantum dot is that it localizes charge. It is a region confined by potential barriers that are high enough so that we can approximate the charge contained within to be quantized to a multiple of the elementary charge e. The cell also contains two extra mobile electrons that can quantum mechanically tunnel from one dot to another, but not between cells.

In the ground state and in the absence of external stimuli, the electrons occupy the opposite corner positions to maximize their separation due to Coulombic repulsion. Figure 1a shows how the two possible charge configurations can be used to represent logic states. The cell polarization P measures the amount of charge that is aligned along one of the diagonal axes and is defined as

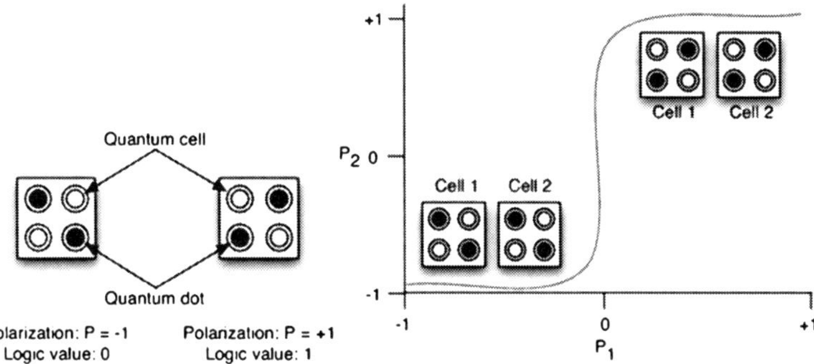

Fig. 1 QCA cells encode information using (**a**) charge configuration and have a (**b**) highly non-linear response

$$P = ((r_1 + r_3) - (r_2 + r_4))/(r_1 + r_2 + r_3 + r_4) \tag{1}$$

where r is the charge density on dot i. The range of (1) is $[-1, +1]$. $P = -1, +1$ represents a binary "0" and "1," respectively. For an isolated cell, both states have the same energy levels (i.e. they are energetically degenerate). However, in the presence of neighboring cells, one state becomes the cell ground state [1].

Figure 1b shows the cell-to-cell response function of a two-cell system. The polarization of cell 2 (P_2) is induced by the polarization of cell 1 (P_1). When $P_1 = -1$ (logic 0), P_2 is aligned with its neighbor to have the same polarization. When $P_1 = 1$ (logic 1), P_2 changes its polarization to match P_1. However, P_2's transition from -1 to $+1$ is abrupt, indicating a highly non-linear response. A slightly polarized input cell (e.g. $P_1 = 0.1$) will induce an almost fully polarized output cell. From a designer's perspective, the abrupt transition is a desirable feature as it leads to low drive strength, fast switching times, and signal restoration. However, the downside is that the cells are prone to transient errors from glitches and techniques need to be developed to properly modulate the potential barriers that control cell-to-cell interactions and information transfer.

2.1 Basic Logic Gates and Interconnect

The fundamental logic gates in QCA are the three-input majority gate and inverter (INV), shown in Fig. 2a, b. The majority gate can be realized using five QCA cells and is logic 1 if two or more of its inputs are logic 1. That is,

$$M(x_1, x_2, x_3) = x_1 x_2 + x_2 x_3 + x_1 x_3 \tag{2}$$

The complement of a majority gate is a minority gate. Note that if one of the inputs is set permanently to logic 1 or logic 0, (2) reduces to a two-input OR or

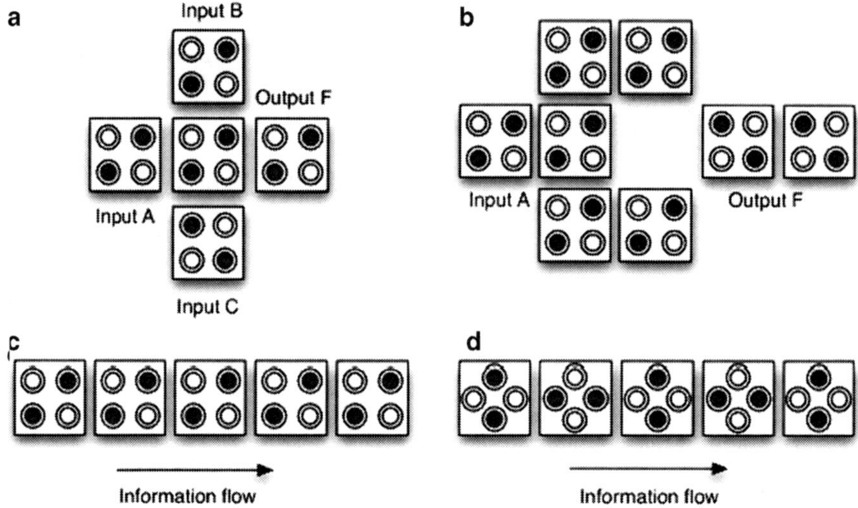

Fig. 2 A QCA (**a**) majority logic gate, (**b**) inverter, (**c**) binary wire, and (**d**) inverter chain

AND function, respectively. The 45° displacement in the two lines of merging cells in INV produce the input's complement. Unlike in CMOS, a QCA inverter consumes substantial real estate. Although, there are several other ways of obtaining the inverting function, the circuit in Fig. 2b is preferred as it is very robust.

It is known that AND/OR/INV form a functionally complete set capable of implementing any arbitrary Boolean function. Since QCA can implement these functions, it is also functionally complete by direct implication.

The binary wire and inverter chain, shown in Fig. 2c, d, are used as interconnect to propagate information. In a binary wire, a signal propagates from the input to the output. In an inverter chain, rotated (45°) cells cause the signal to alternate between the input value and its complement as it traverses the chain. In addition, both the original input and its complement can be realized using non-rotated (90°) cells in the middle of two rotated cells.

In CMOS, wires run in alternating directions (horizontal and vertical) in alternating layers to form the connections between transistors. Since QCA technology is still in the research phase and fabrication is confined to laboratory experiments, there is no multi-level layer stack for routing wires. It is not clear what the interconnect scheme will eventually look like. At this point, QCA circuits are purely two-dimensional and reside on a single plane. Consequently, the interconnects must be made to run horizontally and vertically without overlapping with each other. In QCA, crossing of wires in a plane is done by placing a binary wire between two inverter chains, as shown in Fig. 3. This allows two signals to cross without interfering with each other.

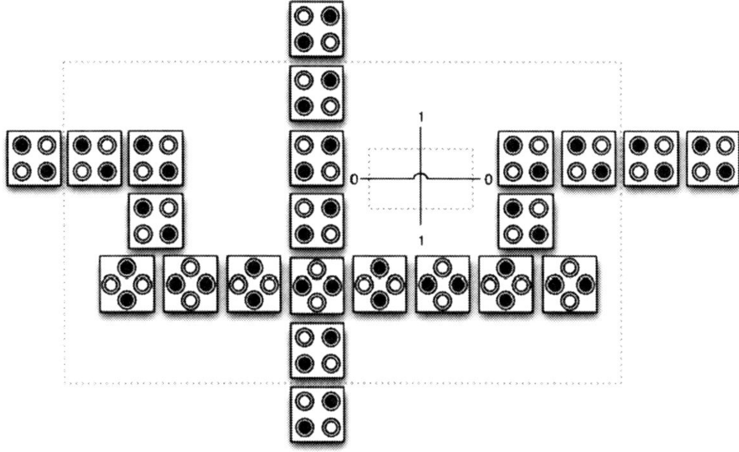

Fig. 3 Co-planar interconnect crossing of two QCA wires

2.2 Clocking Scheme

In CMOS, a reference signal (i.e. clock) is used to synchronize data and control of information flow in sequential circuits. In QCA, clocking is needed for both combinational and sequential circuits [5]. Clocking in QCA also provides the necessary benefit of restoring energy that is lost during computation. The clock signal, generated through an electric field, is applied to the cells to raise or lower the tunneling barrier between dots within a cell. When the barrier is low, the cells remain unpolarized; when the barrier is high, the cells cannot change state. The cells are allowed to change state only when the barrier is being raised.

Clocked QCA circuits use the six-dot tri-state cell shown in Fig. 4. The cell operates in the same manner as that in Fig. 1a with one notable difference. When the electrons are in the middle dots, the cell is in the *null* state. The clock signal can be used to either pull the electrons into the middle dots or push them into the corner dots.

The preferred method to clock QCA circuits is via *adiabatic switching*. The clocking scheme, shown in Fig. 5, is divided into four phases with each phase incurring a 90° delay. The phases are called *switch*, *hold*, *release*, and *relax*. This allows an array of cells to perform a certain computation, have its state frozen by raising of the barrier, and have the result of the computation serve as the input to the next array of cells. During the computation, the successor cell array is kept unpolarized so that it does not influence the computation.

For a clocked QCA circuit to function properly, it must be partitioned into clocking *zones* so that all the cells in a zone are controlled by the same clock. In the *relax* phase, the electrons are pulled to the middle dots so that the cell is in the *null* state. In the *switch* phase, the inter-dot potential barrier is slowly raised so that the cell can become polarized according to the state of their neighboring (input)

Fig. 4 Six-dot cell for
clocked QCA

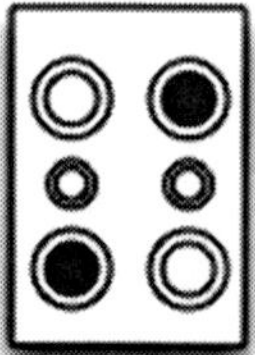

Logic value: 1

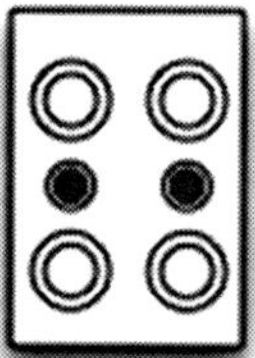

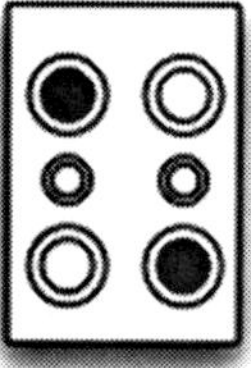

Logic value: 0

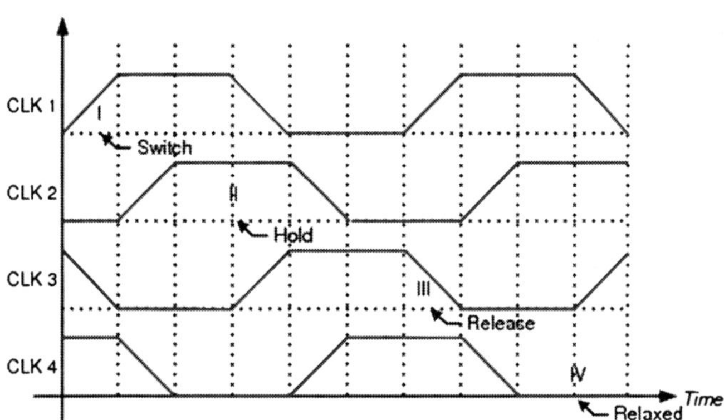

Fig. 5 A four-phase clocking scheme for QCA

cells. In the *hold* phase, the barrier remains high to allow the cell to maintain its
polarity and act as an input to the successor neighboring cells so that they can begin
their computation. In the *release* phase, the barrier is lowered so that the cell can
return to the *null* state.

It should be clear that in a clocked QCA circuit, information is transferred and processed via fine-grained pipelining. Clocking zones are arranged in a periodic fashion such that zones in the hold phase are followed by zones in release, relax, and switch phases. A signal gets latched when a clocking zone transitions into the hold phase.

Consider the binary wire in Fig. 2c using the clocking scheme of Fig. 5. Let the cells be named C_i ($i \in [0\ldots4]$) from left to right. Assume C_0 has a fixed polarization of 1 (logic 1). Initially, C_1 switches according to the fixed input and becomes 1. C_2 shows no polarization since it is in the relax phase. Then, C_1 enters the hold phase (i.e. signal is latched) and C_2 enters the switch phase and starts switching to 1. Since C_3 is in the relax phase, it does not affect the state of C_2. Next, C_1 enters the release phase; C_2 enters the hold phase; and C_3 starts switching to 1 as C_2 serves as its input. Then, C_3 enters the hold phase and C_4 can start switching to 1. At this point, C_0 enters the relax phase and is ready to perform the next computation. In this way, it can be seen how information propagates along the binary wire from left to right [5].

3 Logic Design with QCA

Most of the techniques developed for QCA logic design utilize a gate-based methodology. First, the logic function of a circuit is determined and then logic synthesis is performed to generate a netlist. A QCA library is needed to map the netlist onto a set of QCA gates. The library contains the majority gate and an inverter. Recall, that an AND/OR gate can be realized by permanently fixing the polarization of one of the inputs to -1 (logic 0)/1 (logic 1). The final phase is to map the cells onto a QCA layout and assign a clocking zone to each cell.

While tools and methods have been developed that can automatically synthesize a QCA logic netlist, there is currently no tool that can generate QCA layouts. Thus, most of the QCA circuits presented in the literature have been manually designed. However, this is to be expected given that QCA is still in its infancy. As our understanding of QCA grows, researchers will develop more powerful tools to automate QCA logic design.

3.1 Hand-Crafted Designs

In this subsection, we survey a few hand-crafted designs to understand the fundamentals of QCA gate-based design. The first example is the design of a one-bit full-adder. The Boolean logic function of the full-adder is,

$$C_{\mathrm{out}} = AB + BC_{\mathrm{in}} + AC_{\mathrm{in}} \tag{3}$$

$$S = ABC_{\text{in}} + A'B'C_{\text{in}} + A'BC_{\text{in}}' + AB'C_{\text{in}}'. \tag{4}$$

It is left an exercise for the reader to show that C_{out} and S can be implemented using majority and INV gates as follows:

$$C_{\text{out}} = M(A, B, C_{\text{in}}) \tag{5}$$

$$S = M\left(C_{\text{out}}', C_{\text{in}}, M(A, B, C_{\text{in}}')\right). \tag{6}$$

It can be seen that the QCA full-adder requires three majority gates and two INVs. One possible QCA layout is shown in Fig. 6a. The four colors, green, purple, cyan, and white, denote different clocking phases in the figure. The blue and yellow cells represent inputs and outputs, respectively. Note that there is an INV present for the inversion of C_{out}. However, an inverter chain is used to get C_{in}. Note that S is available one clock phase after C_{out}. Also, observe that all the inputs to the majority gate arrive in the same clock phase.

Next, consider the two-input XOR gate ($f = AB' + A'B$) shown in Fig. 6b. Note that one input of the majority gates is connected to logic 0/1 ($P = -1/P = +1$) to form AND/OR gates, respectively. Two INVs are present to complement A and B. Finally, what is different about this design from the full-adder is that it contains "crossover" (interconnect) cells that run on a different layer. They are connected to the main substrate layer using "vertical" cells (i.e. a QCA "via"). Researchers have proposed the crossover layer as a way to increase QCA cell densities and minimize dead space/interconnect. However, such a structure has not been fabricated yet and it is unclear if it will be feasible.

A 4:1 QCA multiplexer ($f = S_1'S_0'D_0 + S_1'S_0D_1 + S_1S_0'D_2 + S_1S_0D_3$) is shown in Fig. 6c. The circuit is composed of three 2:1 multiplexers. The final multiplexer selects between the outputs of the first two multiplexers which select between D_0, D_1 and D_2, D_3, depending upon the select signals S_0, S_1. Note that all the majority gates receive their inputs in the same clock phase. If the reader traces through the circuit, he or she will observe that the output is available two clock cycles (or eight clock phases) after S_0, D_0–D_3 have been applied. Also, note that for proper operation, S_1 should be delayed by one clock cycle. Although this is not shown explicitly in the figure, buffer cells will need to be added to S_1 to ensure this.

Figure 6d shows the circuit of a QCA D-latch ($Q^+ = DE_n + QE_n'$). When $E_n = 1, D$ is latched onto the output. When $E_n = 0$, the circuit maintains its previous state. Note that unlike in CMOS circuits, there is no clock signal to synchronize the elements. The clocking mechanism in QCA cells is provided by the clocking zones to which the cells are assigned. The feedback in the latch is achieved through the array of green cells shown at the bottom of the figure.

A 4:1 QCA decoder is shown in Fig. 7. Depending upon the select signals S_1, S_0, one of the outputs A_0–A_3 goes high. Four majority gates are used and function as AND gates with one input fixed at logic 0. What is interesting here is that inverter chains are used for the select signals. Depending upon where (odd or even location)

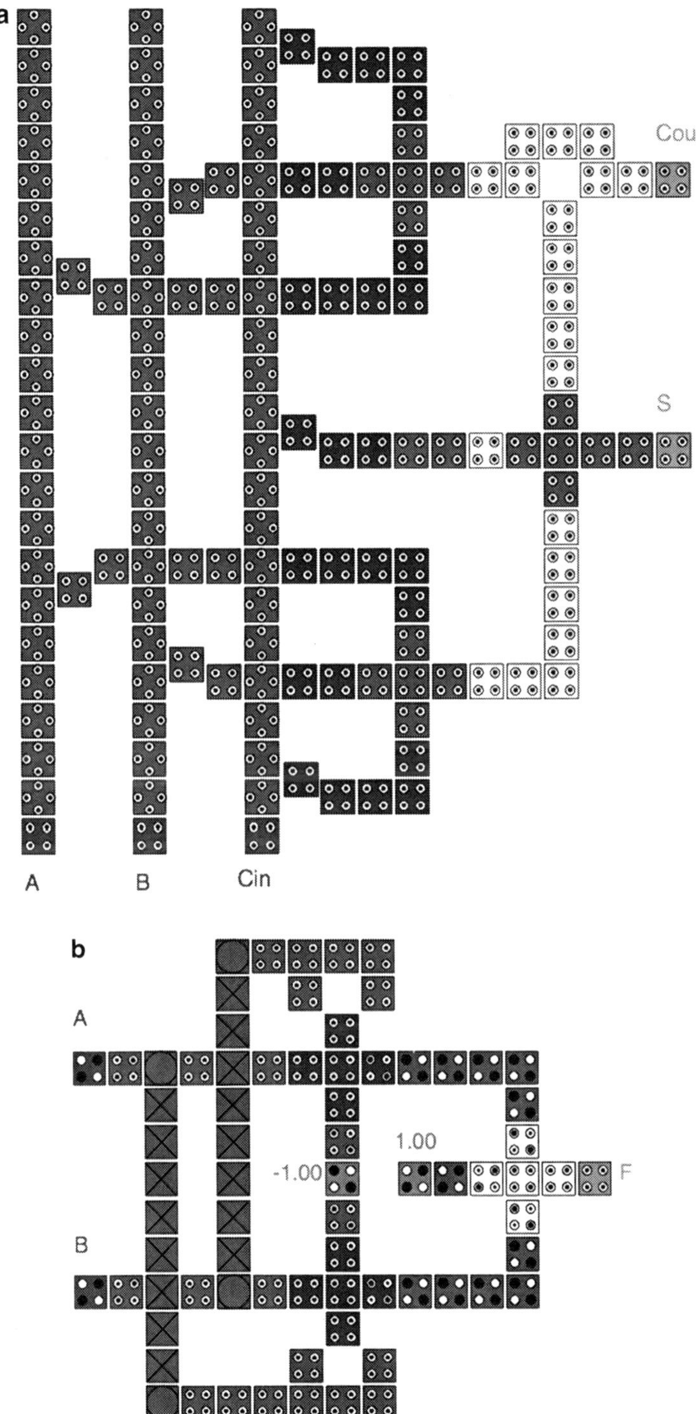

Fig. 6 Design of various QCA circuits: (**a**) one-bit full-adder, (**b**) XOR gate, (**c**) 4:1 multiplexer, and (**d**) D-latch

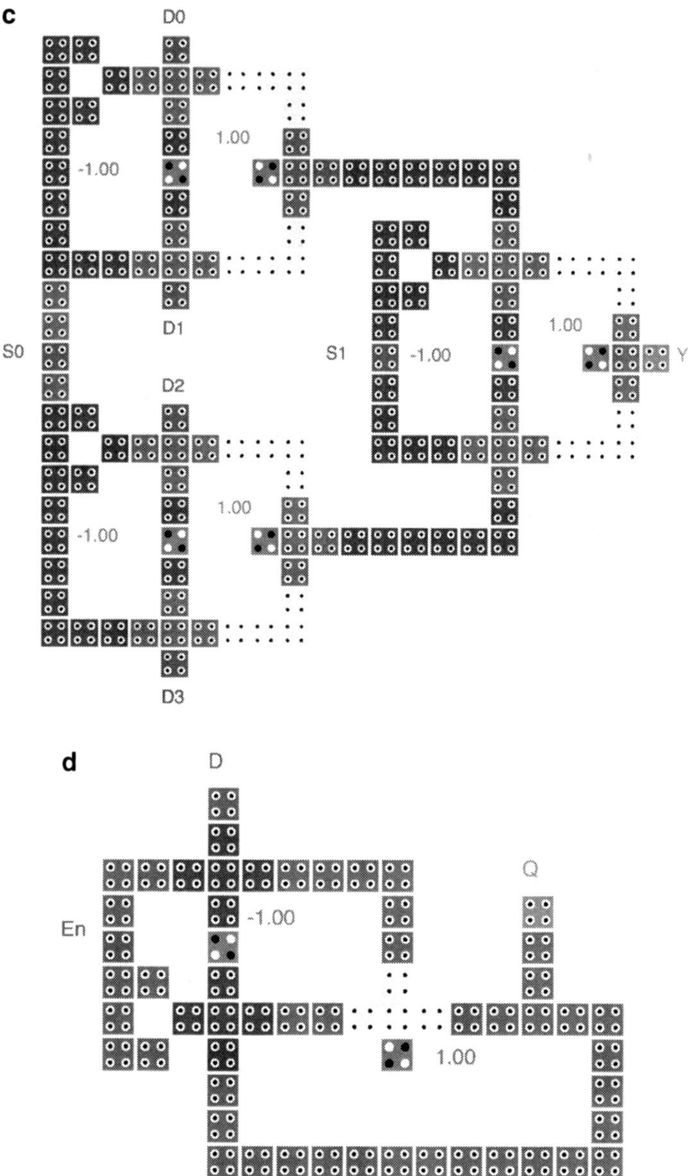

Fig. 6 (continued)

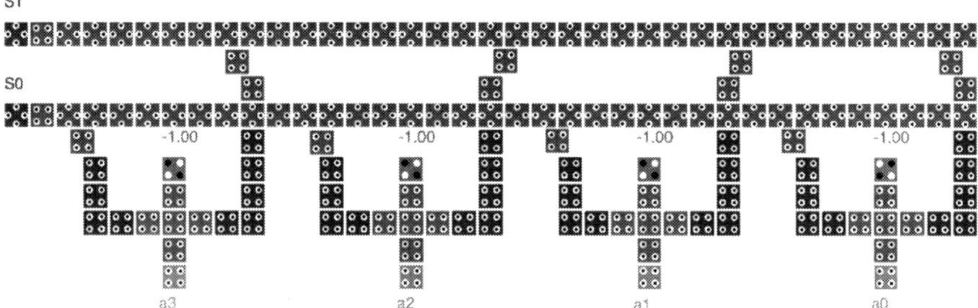

Fig. 7 A 4:1 QCA decoder

a QCA cell is placed at a 45° offset with respect to the inverter chain, it is possible to "rip" the value of the signal or its complement. For example, for S_1, we can see that "ripper" cells for A_3 and A_2 are placed in an odd-numbered (at 9 and 19, respectively, counting from left) position. This gives the same value of the signal. However, for A_1 and A_0, the cells are placed in an even-numbered location (at 28 and 36, respectively, counting from left). This results in the complement value of the signal. The "ripper" cells for S_0 alternate between odd- and even-numbered positions.

As a final example, consider the QCA memory cell shown in Fig. 8a. The operation of this circuit can be better understood if we construct the equivalent Boolean circuit shown in Fig. 8b. If $RS = 0$, the output is 0 and the entire circuit is cut off from the external environment. In order to read from or write to the cell, RS must equal 1. When $RS = 1$, if $WRB = 1$ (a write operation), then $G_1 = 1$ and $G_3 = 0$. This allows IN to propagate from G_2 to G_4 and, hence, be written to the cell. The output is not updated because $G_5 = 0$. If $WRB = 0$ (a read operation), then $G_1 = 0$ and $G_2 = 0$. Thus, the top input to G_4 is 0 while the bottom input to G_3 is 1. This allows Q to cycle through the feedback formed by G_3, G_4, and the cell maintains its state. Furthermore, the bottom input to G_5 is 1, which allows Q to be propagated to the output. It is worthwhile to note that a QCA memory cell is significantly larger than the conventional six-transistor static random access memory (SRAM) cell used in CMOS [6].

3.2 QCA Logic Synthesis

In the previous subsection, we saw designs of some rudimentary QCA circuits. They were obtained manually as they are small and/or regularly structured circuits and, hence, easy to implement. However, it is very cumbersome to manually design large circuits or those that comprise random logic. Boolean logic synthesis tools have been developed to solve this problem. They take in a description of a circuit and produce a gate-level netlist. Commercial tools have become sophisticated enough such that entire designs can be synthesized.

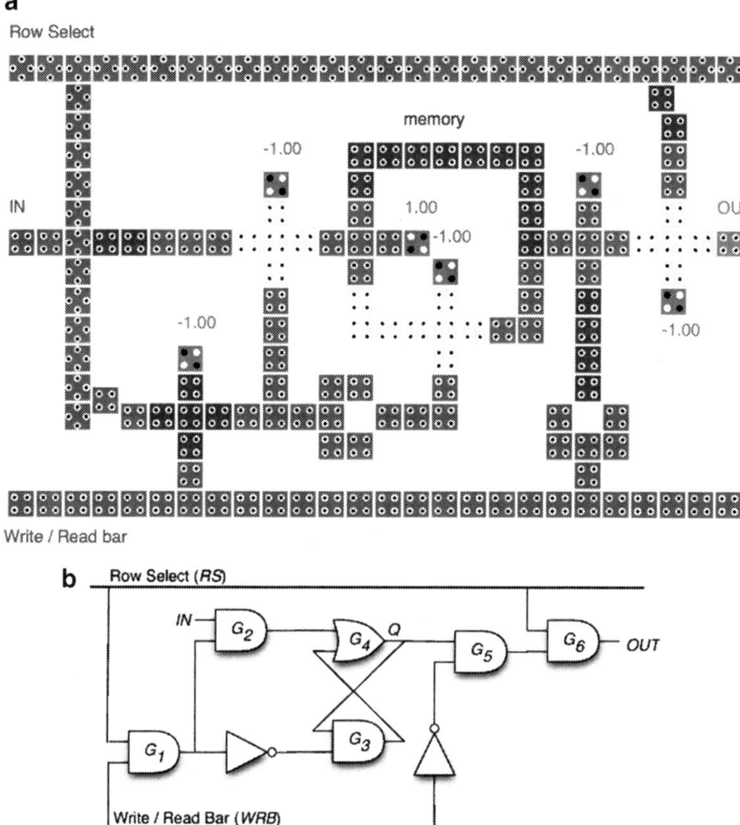

Fig. 8 (**a**) QCA memory cell and (**b**) equivalent logic circuit (Source: Ref. [6])

Since the fundamental QCA gates are a majority gate and INV, we need to map Boolean gates of a circuit onto these gates in order to implement them in QCA technology. However, directly mapping AND and OR gates to majority gates makes suboptimal use of such gates. Thus, researchers have developed techniques to synthesize majority logic circuits. MAjority Logic Synthesizer (MALS) [7] is a logic synthesis tool for QCA that synthesizes efficient majority logic circuits for arbitrary multi-output Boolean functions. The input is an algebraically-factored n-input, m-output Boolean circuit and the output is an optimized circuit consisting only of majority gates and inverters. In an algebraically-factored circuit that does not make use of within-circuit inverters, all the primary inputs and their complements are available. Thus, there is no need to use inverters inside such a circuit. Although MALS can generate a netlist solely comprising majority and INV gates, there are currently no back-end tools that can generate an optimized QCA layout from the netlist. This is an open research problem in QCA design.

The synthesis methodology implemented by MALS is shown in Fig. 9. The objective is to minimize the number of gates. The procedure begins by

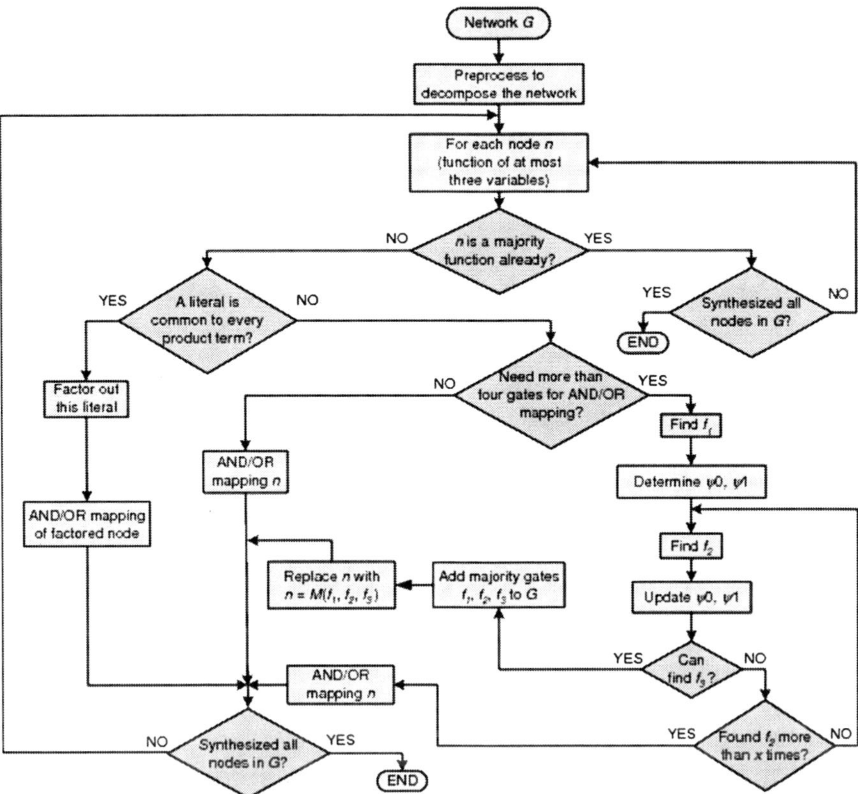

Fig. 9 Flow diagram providing an overview of the majority network synthesis methodology

preprocessing network G where it is decomposed such that no network node has more than three input variables (because a QCA majority gate has three inputs). Then, each node in the decomposed network is checked to determine if it is a majority function. If it is, the next node can be synthesized. Otherwise, we check to see if there exists a common literal (variable) in all the product terms of the node function. If a common literal exists, it is factored out. An AND/OR mapping is then applied to the factored node. In AND/OR mapping, a Boolean function is represented using AND and OR gates only. If no common literal exists, we check to see if this node can be implemented with four or fewer AND/OR gates. If so, we perform AND/OR mapping on this node. Otherwise, we map the node onto at most four majority gates using a Karnaugh-map (K-map) based method that is discussed later. The procedure terminates when all the nodes in the decomposed network have been synthesized.

Let us consider two examples to demonstrate parts of the majority logic synthesis methodology. The first example shows that the gate count can be reduced by factoring out the common literal in the product terms. Consider $n = x_1 x_2' + x_2' x_3$. If AND/OR mapping is applied, three majority gates are needed for n as $f_1 = x_1 x_2'$, $f_2 = x_2' x_3$, and $n = f_1 + f_2$. However, since literal x_2 appears in all the product terms of n, it can be factored out. Node n can, therefore, be expressed as $n = f_3 x_2$', where $f_3 = x_1 + x_3$, thus requiring only two majority gates. The next example demonstrates that when a node requires fewer than or equal to four majority gates by AND/OR mapping, there is no need to spend time during synthesis using the K-map based method. Consider $g = x_1 x_2 + x_2' x_3$. AND/OR mapping requires three majority gates, $f_4 = x_1 x_2$, $f_5 = x_2' x_3$, and $g = f_4 + f_5$. However, four majority gates will be required if we used the K-map based method. If such a scenario occurs, we choose the AND/OR mapping.

Network preprocessing. As has been mentioned earlier, the algebraically-factored network G is first decomposed, such that all the nodes in the network have at most three input variables. Majority network synthesis is then performed on each node in the decomposed network. Let the total number of nodes in the decomposed network be N. There is a result that says that all Boolean functions of three variables can be realized by at most four majority gates in two levels. Thus, the total number of nodes in the synthesized majority network Gm is N and $4N$ in the best and worst cases, respectively. Therefore, to minimize the gate count in the synthesized network, we must try to reduce N.

To reduce N, we can make use of the decomposition method in SIS (a logic synthesis tool from University of California, Berkeley) that is targeted at programmable gate arrays (PGAs). There are two main architectures in PGAs: lookup tables (LUTs) and multiplexer-based. The building block of a LUT architecture implements any function that has up to r inputs. If a function has at most r inputs, it is called r-feasible. Otherwise, it is called r-infeasible. A network is r-feasible if every node in the network is r-feasible. SIS provides a method to make any infeasible network feasible, and also minimizes the number of blocks used in the network. Therefore, if we set r to three, we can make any arbitrary network three-feasible and also minimize the number of nodes (i.e. N) in the process. Majority synthesis can then be performed on this optimized decomposed network.

K-map based majority synthesis. As mentioned earlier, K-map based majority synthesis is performed on a node if required. This method has been developed for three-input functions. First, we get the K-map for the logic function of node n, which is a function of at most three inputs (note that we have pre-processed the network to make it three-feasible). Next, we find an admissible pattern from Fig. 10 which gives the first majority function f_1. Then, we try to find the second admissible pattern based on the first admissible pattern and the original K-map of node n. This admissible pattern gives the second majority function f_2. Finally, based on the two previously found admissible patterns and the original K-map, we find the third admissible pattern. This admissible pattern gives the third majority function f_3. These three majority functions are chosen such that the original node can be represented as their majority function, i.e. $n = M(f_1, f_2, f_3) = f_1 f_2 + f_2 f_3 + f_1 f_3$.

Admissible patterns. An admissible pattern is a pattern of 1 cells in a K-map that can be realized by a majority gate. Figure 10 shows all the positive unate functions of three variables that can be realized by a majority gate. A function is positive unate if all its variables appear in uncomplemented form. As shown in the figure, a name has been assigned to each admissible pattern.

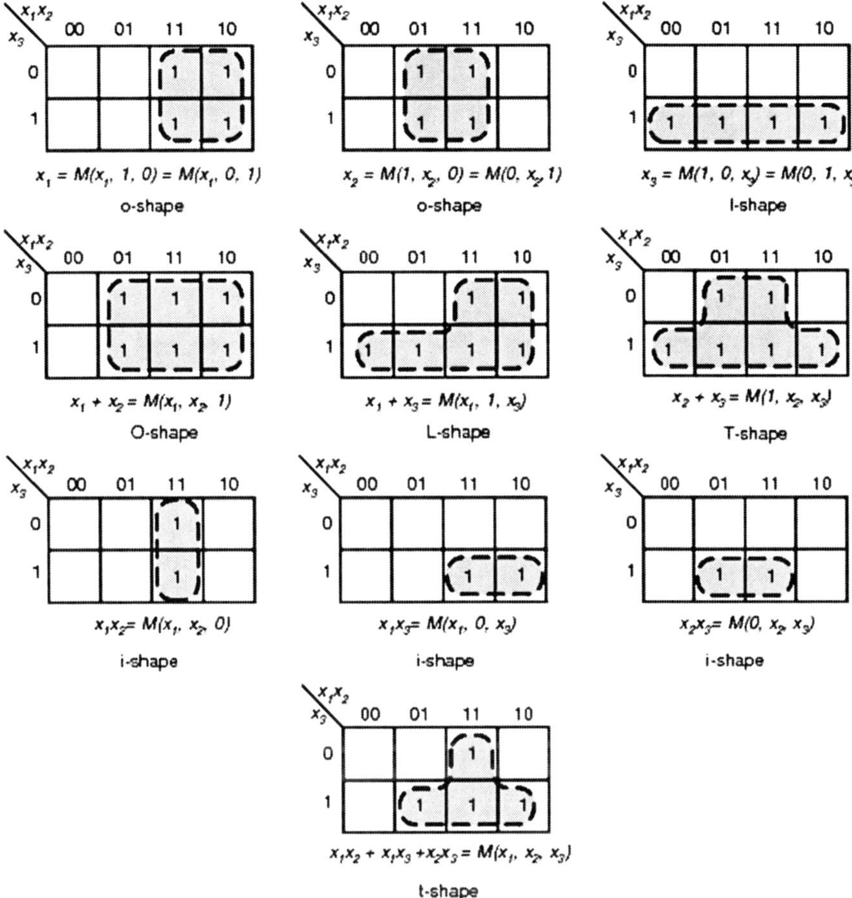

Fig. 10 All positive unate functions that can be realized by a three-input majority gate and their corresponding admissible patterns on the K-map

Finding f_1. To find f_1, we must find an admissible pattern that satisfies the following requirements:

1. It must contain all the minterms (terms where the function equals 1) in the set of minterms that must be included in the ON-set of a function. This set is denoted y1.
2. It must contain none of the maxterms (terms where the function equals 0) in the set of maxterms that must be included in the OFF-set of a function. This set is denoted y0.
3. It can contain minterms of the original node n. The more original minterms it covers, the better.
4. It can have some "makeup" minterms if such an admissible pattern can be found by adding minterms that are not minterms of the original node n. There may be

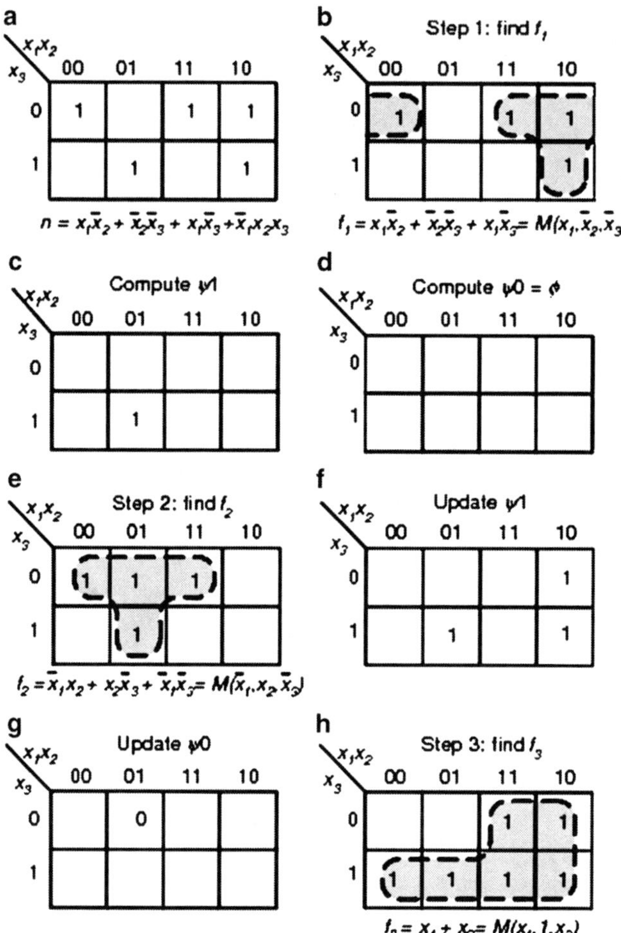

Fig. 11 Example to demonstrate K-map based majority synthesis method for $n = x_1x_2' + x_1 x_3' + x_2'x_3' + x_1'x_2 x_3$

many choices of admissible patterns that satisfy the above requirements. Therefore, the pattern that is chosen should be such that it covers the maximum number of minterms in the original node n and requires as few "make-up" minterms as possible.

As an example, consider the node $n = x_1 x_2' + x_1 x_3' + x_2'x_3' + x_1'x_2 x_3$, which must be mapped to a set of majority functions. Initially, f_1, $\psi 1$, and $\psi 0$ are all set to empty sets. The K-map of n is shown in Fig. 11a. It can be seen that there is a "t-shaped" admissible pattern in this K-map without the need for any make-up minterm. Therefore, f_1 is set to this admissible pattern, as shown in Fig. 11b.

Finding f_2. We now need to find f_2 and f_3 given that we have found f_1. We use the following rule for finding f_2 and f_3. An ON-set (OFF-set) minterm (maxterm) of n

must also be an ON-set (OFF-set) minterm (maxterm) of at least two of the three functions f_1, f_2, and f_3. This rule is reflected in the two sets $\psi 1$ and $\psi 0$. To find f_2, we set $\psi 1$ as follows: if a minterm of n is not a minterm of f_1, we add this minterm to $\psi 1$. Similarly, we set $\psi 0$ as follows: if a maxterm of n is not a maxterm of f_1, we add this maxterm to $\psi 0$.

To illustrate how f_2 is found, let us continue with our example from Fig. 11. Since f_1 has been determined, sets $\psi 1$ and $\psi 0$ are then computed. In this case, minterm $x_1`x_2x_3$ is a minterm of $n$ but a maxterm of $f_1$. Therefore, $\psi 1$ contains $x_1`x_2x_3$ only (see Fig. 11c). Set $\psi 0$ remains empty because all the maxterms of n are also maxterms of f_1 (see Fig. 11d). Given the restrictions imposed by $\psi 1$ and $\psi 0$, we search the admissible patterns in Fig. 10 that satisfy these constraints. One possible admissible pattern that meets the restrictions on f_2 is the "t-shaped" pattern shown in Fig. 11e.

Finding f_3. A process similar to finding f_2 is carried out for finding f_3 except that sets $\psi 1$ and $\psi 0$ are updated according to the following rule. If a minterm (maxterm) of node n is not a minterm (maxterm) of both f_1 and f_2, we add this minterm (maxterm) to $\psi 1$ ($\psi 0$). Given these new restrictions, we can now attempt to find f_3.

It turns out that f_3 cannot be guaranteed to be found based on the two previously chosen functions f_1 and f_2. Therefore, if we fail to find f_3 based on the current choices of f_1 and f_2, we backtrack to find a new f_2. We repeat this process y times. If we are still unable to find f_3, the AND/OR mapping is used to speed up the process. Since we are dealing with three-input functions, we set $y = 3$. This is because there are only three adjacent minterms (maxterms) for each minterm (maxterm). Therefore, there are at most three directions for expansion starting from a particular minterm or maxterm.

We continue with our running example to demonstrate how f_3 is found. Majority functions f_1 and f_2 have been obtained and sets $\psi 1$ and $\psi 0$ are then updated. Minterms $x_1`x_2x_3$, $x_1x_2`x_3`$, and $x_1x_2`x_3`$ belong to either $f_1$ or $f_2$ but not both. Therefore, they are in $\psi 1$. Maxterm $x_1`x_2x_3`$ belongs to $f_1$, but not $f_2$. Therefore, $\psi 0$ contains $x_1`x_2x_3`$ only. The updated $\psi 1$ and $\psi 0$ are shown in Fig. 11f, g, respectively. The last step is to find an admissible pattern for $f_3$. One possibility is the "L-shaped" admissible pattern shown in Fig. 11h. Therefore, the final decomposed majority functions are: $f_1 = x_1x_2` + x_2`x_3` + x_1x_3` = M(x_1, x_2`, x_3`), f_2 = x_1`x_2 + x_2 x_3` + x_1`x_3` = M(x_1`, x_2, x_3`)$, $f_3 = x_1 + x_3 = M(x_1, 1, x_3)$, and $n = f_1f_2 + f_2f_3 + f_1f_3 = M(f_1, f_2, f_3)$. This requires four majority gates to implement n. Note that a straightforward AND/OR mapping would have required eight majority gates to realize n.

3.3 Tile-Based QCA Design

In the previous subsections, we saw designs and methods for a gate-based methodology. An alternate approach to QCA logic design is to use a tile-based methodology [12]. The motivation behind this approach is that molecular QCA

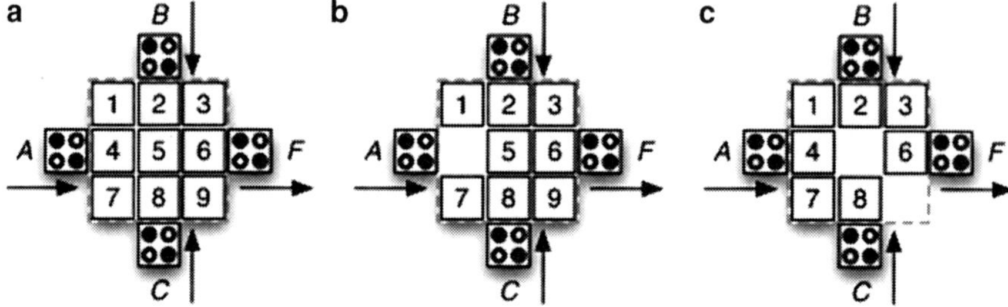

Fig. 12 Examples of (**a**) fully populated and (**b**), (**c**) non-fully populated QCA grids (Source: Ref. [12])

manufacturing techniques (to be discussed later) are well suited for modularization through a structured QCA design. Modularization in QCA can be achieved by using Manhattan-style interconnect. Although the area overhead is comparable to gate-based design, tile-based design is very simple, yet powerful, and holds significant promise. It is also very amenable to design automation.

In tile-based design, a tile is built using an $n \times n$ grid of QCA cells. Fully populated (FP) and non-fully populated (NFP) 3×3 grids are shown in Fig. 12. The basic idea is to use these grids and inputs/outputs to form tiles which are interconnected to form circuits. The final step is to assign clocking zones to the tiles and then, the layout can be generated. To prevent unwanted interactions, the tiles are isolated by physical separation or by using input/output cells. Through simulation, researchers have shown that tiles not only have the ability to implement a range of logic functions but are also highly defect-tolerant.

The layout phase in tile-based design is described next. Consider a two-dimensional square matrix A of dimension N; we can think of A as being a QCA layout prior to cell deposition on the substrate. Two types of tiles are used in the layout: *active* and *passive*. Active tiles implement a logic function of at least two variables. Passive tiles implement interconnects or the inverting function. Once we have a QCA circuit that has been synthesized and mapped onto a set of QCA gates (these tasks can be done by tools like MALS), we can think of generating A by iterating through the following sequence of operations:

1. Divide A into K^2 grids where $K = N/n$. Note that the size of a grid is $n \times n$.
2. Map each logic function onto a tile. If there is a horizontal or vertical adjacency among the tiles, spacing (an area with no cells) is inserted between the titles to prevent unwanted interactions.
3. Use passive tiles to route the signals. If unsuccessful, repeat Step 2 using different tile mappings.

A 3×3 grid is a good choice for the construction of tiles. This is because a QCA majority gate has three inputs. Furthermore, it has been shown that functions containing more than four literals in a minterm do not occur frequently. It should be obvious that a majority gate and INV can be implemented using the FP 3×3 grid of Fig. 12. An NFP grid is generated by selectively undepositing cells from an FP grid.

In our case, Fig. 13 shows five such configurations that we will consider. Depending upon which of the cells numbered 1–9 is undeposited, the logic function of the tile might change. While it is possible for multiple cells to be undeposited, we will only focus our attention on a single undeposited cell to keep the discussion simple.

The orthogonal tile in Fig. 13a has three inputs A–C and one output F. If the tile is fully populated, it functions as a majority gate. Thus, it is the fundamental building block of tile-based design. If one cell is undeposited, the logic function of the gate is determined by Table 1. These new functions are possible due to the interaction of the cells at the corners of the tile with the center cell (cell 5) of the majority gate.

The double and triple fanout tiles, shown in Fig. 13b, c, respectively, have a single input A, and two and three outputs F_1–F_3, respectively. All the outputs follow the value of the input and, hence, the tile forms a fanout point in an interconnect. If one cell is undeposited in the tiles, the tile operation is determined by Tables 2 and 3 for double and triple fanout tiles, respectively. Note that in both cases, the operation is either that of a wire or an inverter.

The baseline tile shown in Fig. 13d has two inputs A, B, and two outputs F_1, F_2 where $F_1 = A$ and $F_2 = B$. This tile accomplishes coplanar wire crossings. Table 4 shows the resulting logic function if one cell is undeposited. Finally, the fanin tile shown in Fig. 13e has two inputs A, B and one output F. Table 5 shows the output value depending upon which cell is undeposited.

To demonstrate tile-based design, let us analyze some examples. Figure 14 shows a one-bit full-adder with both the QCA layout and the circuit schematic. The deposited and undeposited cells are shown using black and white squares, respectively. Three baseline tiles for wire crossings, two double fanout tiles, one triple fanout tile, and three orthogonal tiles (highlighted in red) are utilized. These tiles have been connected using passive tiles which function as wires. Note that unlike the gate-based layout in Fig. 6a, inverters are not required since the orthogonal tiles can implement that internally by using undeposited cells (see Table 1). It should be mentioned that clocking zones are not shown in the figure. However, it is assumed that the output of the tile can be only used in a clocking zone to the right of the clocking zone in which the tile is located.

A four-bit parity checker is shown in Fig. 15. Three orthogonal and two double fanout tiles are used. In this case, two cells are undeposited in the orthogonal tile and one of the inputs is connected permanently to logic 1. This makes the tile function as an XOR gate. The fanout tiles are needed to propagate the logic 1 to the inputs of the orthogonal tiles.

4 Testing of QCA Circuits

Now that we have covered the basics in QCA logic design, we focus on testing of such circuits in the presence of defects. Due to the sub-lithographic pitches involved, it is expected that QCA circuits will exhibit a high number of

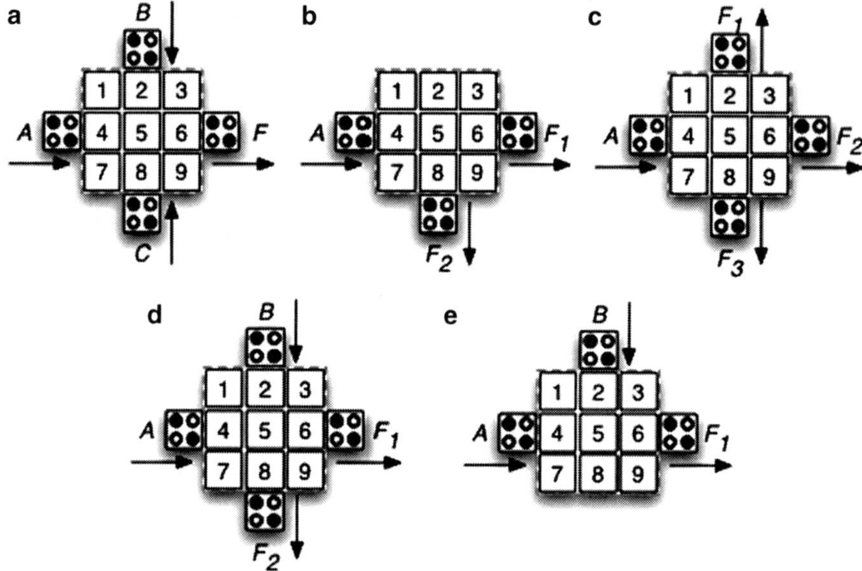

Fig. 13 Types of 3 × 3 FP tiles: (**a**) orthogonal tile, (**b**) double fanout tile, (**c**) triple fanout tile, (**d**) baseline tile, and (**e**) fanin tile (Source: Ref. [12])

Table 1 Undepositing a single cell in an orthogonal tile

Loc.	F	Loc.	F
None	$M(A, B, C)$	5	$M(A', B, C')$
1	$M(A, B, C)$	6	$M(A, B, C)$
2	$M(A', B, C)$	7	$M(A, B, C)$
3	$M(A, B, C)$	8	$M(A, B, C')$
4	$M(A, B, C)$	9	$M(A, B, C)$

Table 2 Undepositing a single cell in a double fanout tile

Loc.	F_1	F_2	Loc.	F_1	F_2
None	A	A	5	A'	A'
1	A	A	6	A'	A
2	A	A	7	A	A
3	A	A	8	A	A'
4	A	A	9	A	A

manufacturing defects. Thus, it is important to develop techniques that can assist in detecting these defects for proper QCA operation. Furthermore, the techniques must utilize the massive testing infrastructure that has been developed for CMOS circuits over the last few decades [8].

Table 3 Undepositing a single cell in the triple fanout tile

Loc.	F_1	F_2	F_3	Loc.	F_1	F_2	F_3
None	A	A	A	5	A'	A'	A'
1	A	A	A	6	A	A'	A
2	A'	A	A	7	A	A	A
3	A	A	A	8	A	A	A'
4	A'	A'	A'	9	A	A	A

Table 4 Undepositing a single cell in the baseline tile

Loc.	F_1	F_2	Loc.	F_1	F_2
None	A	B	5	B'	A'
1	A	B	6	B'	B
2	A	A	7	B	B
3	A	A	8	A	A'
4	B	B	9	A	B

Table 5 Undepositing a single cell in the fanout tile

Loc.	F	Loc.	F
None	A	5	A
1	A	6	B'
2	A	7	A
3	A	8	A
4	B	9	A

The types of defects that are likely to occur in the manufacturing of QCA devices have been investigated by researchers and are illustrated in Fig. 16 [10]. They can be summarized as follows:

1. In a *cell displacement defect*, the defective cell is displaced from its original position. For example, in Fig. 16b, the cell with input B is displaced to the north by Δnm from its original location (see Fig. 16a).
2. In a *cell misalignment defect*, the direction of the defective cell is not properly aligned. For example, in Fig. 16c, the cell with input B is misaligned to the east by Δnm from its original position.
3. In a *cell omission defect*, the defect cell is missing as compared to the defect-free case. For example, in Fig. 16d, the cell with input B is not present.

Since QCA circuits contain a lot of cells that are used for interconnect, cell displacement and omission defects on binary wires and inverter chains must also be considered. These defects are modeled using a dominant fault model in which the output of the dominated wire is determined by the logic value on the dominant wire. Many scenarios can occur in the presence of a bridging fault, as illustrated in Fig. 17. In the first scenario shown in Fig. 17b, the second cell is displaced to the

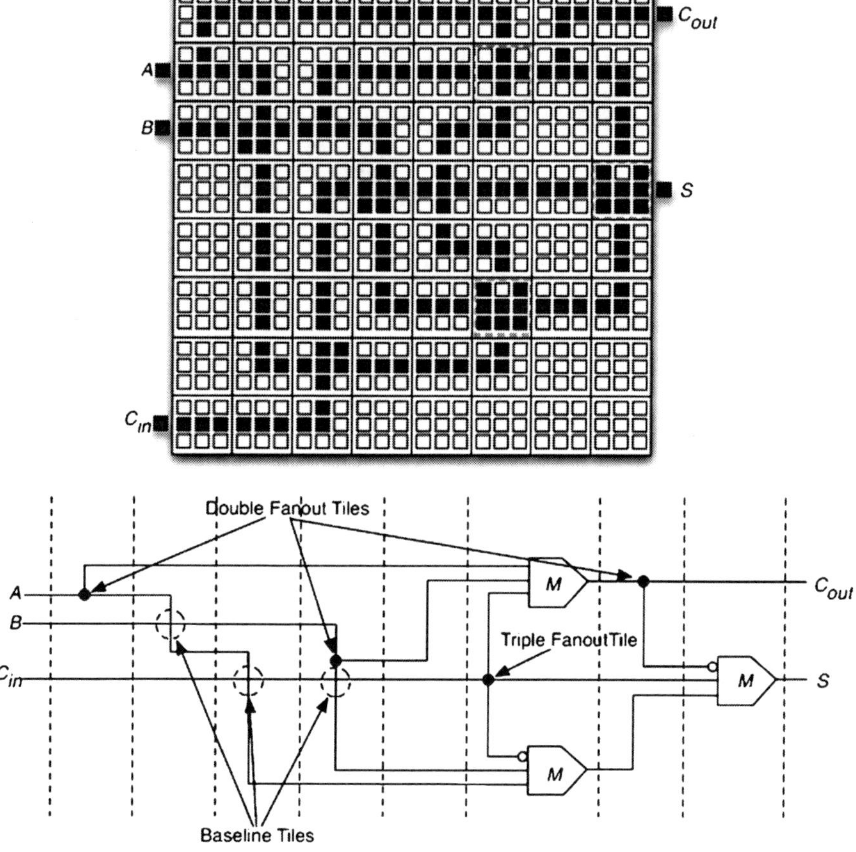

Fig. 14 A one-bit QCA full-adder implemented using tile-based design (Source: Ref. [12])

north from its original position by Δnm. In this case, the dominated wire O_2 will have a logic value equal to that of the dominant wire O_1. However, if the fourth cell is displaced, as shown in Fig. 17c, then O_2 will have a logic value equal to the complement of the logic value on O_1 (i.e. $O_2 = O_1{}'$). Finally, if multiple cells are displaced, as shown in Fig. 17d, then O_2 will also equal $O_1{}'$.

As an example, consider the majority circuit shown in Fig. 18a, which contains three majority gates (M_1–M_3), seven primary inputs (A–G), and one primary output (O). Because there are 10 lines in this circuit, there 20 single stuck-at faults (SSFs). A stuck-at fault is a fault where a wire is stuck permanently at logic 1 (i.e. stuck-at-one (SA1)) or logic 0 (i.e. stuck-at-zero (SA0)). However, if we target all SA0/SA1 faults at the primary inputs of this circuit, we can guarantee that any single SSF will be detected. This result is due to the *checkpoint* theorem – a well-known theorem in Boolean testing, which defines primary inputs and fanout branches as checkpoints that need to be tested.

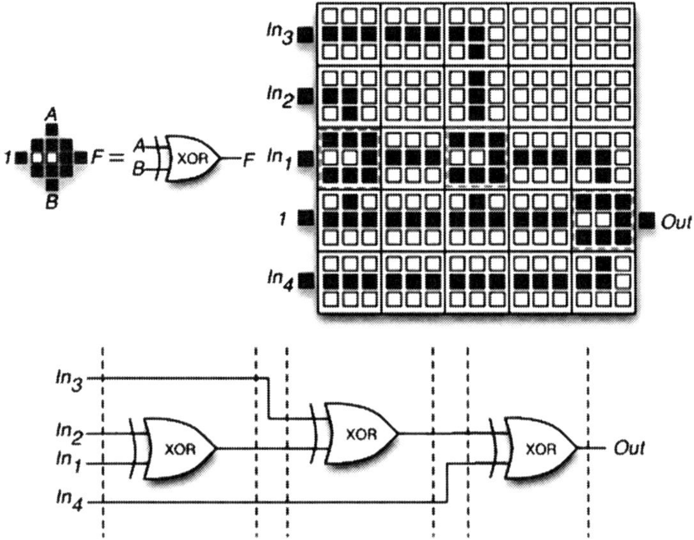

Fig. 15 A four-bit parity checker using tile-based design (Source: Ref. [12])

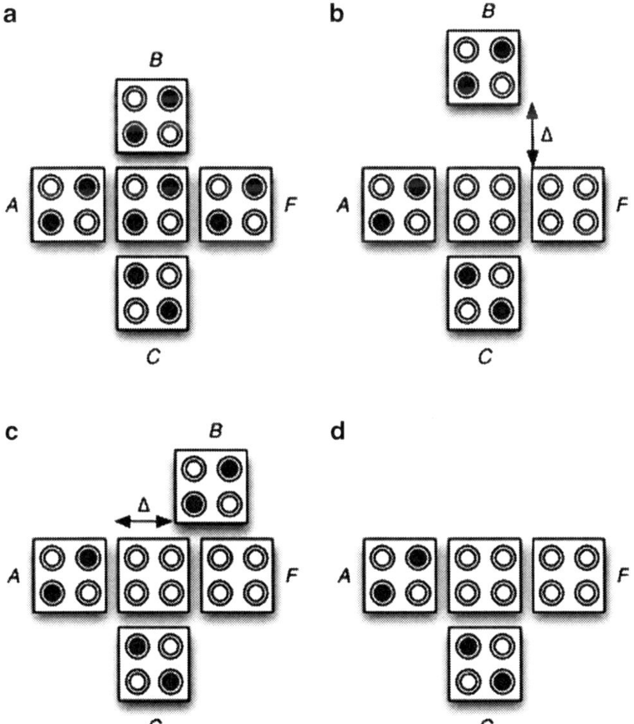

Fig. 16 (a) Defect-free majority gate, (b) displacement defect, (c) misalignment defect, and (d) omission defect

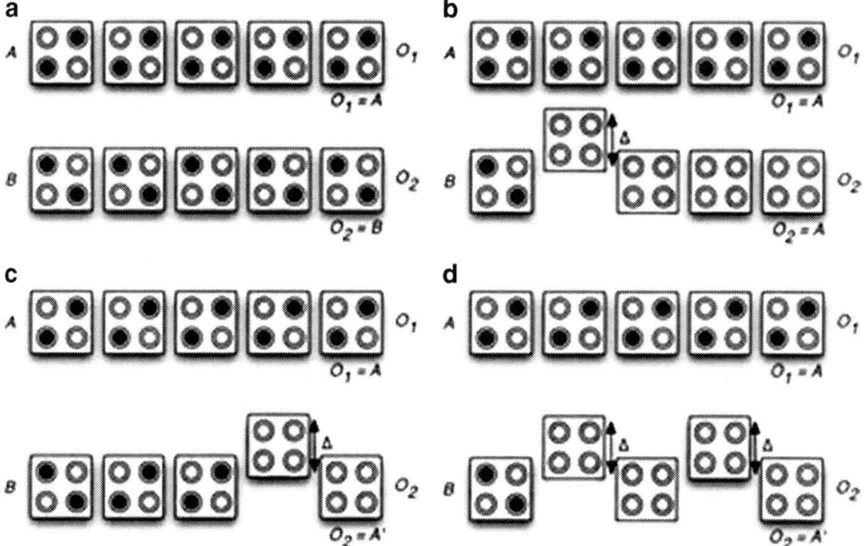

Fig. 17 (**a**) Fault-free binary wire and (**b**)–(**d**) binary wire with a bridging fault

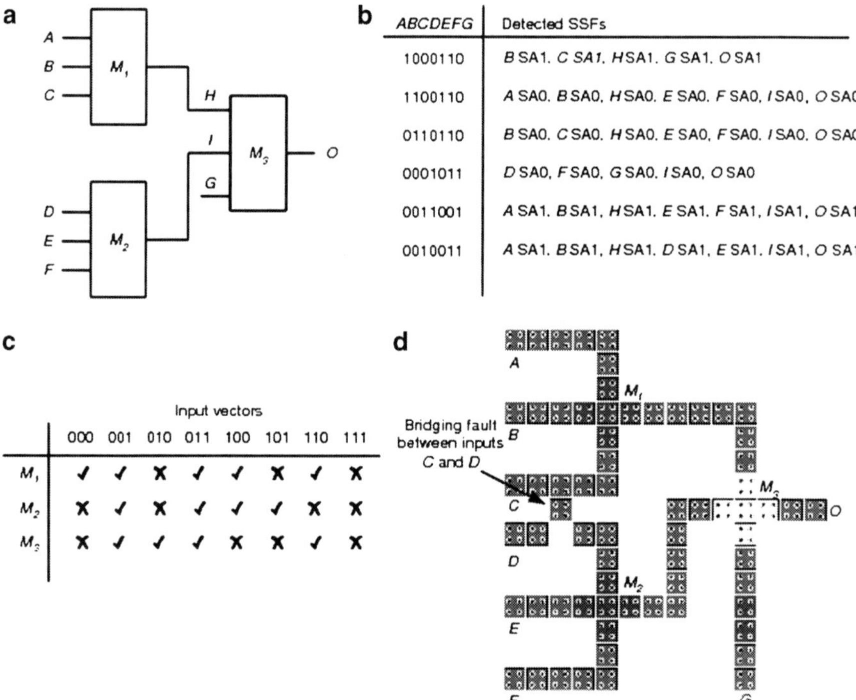

Fig. 18 (**a**) Example majority circuit, (**b**) minimal SSF test set, (**c**) input vectors received by each majority gate, and (**d**) QCA layout of the circuit with a bridging fault between inputs C and D

Because there are seven primary inputs, there are 14 SSFs that need to be targeted during test generation. Figure 18b shows a minimal single SSF test set for the circuit. For each test vector in Fig. 18b, all the SSFs detected by it are also shown.

Given the test set in Fig. 18b, c shows the input vectors that the individual majority gates of the circuit receive in the fault-free case. For example, note that M_2 receives input vectors 001, 011, 100, and 101. Now, consider the fault I SA0, which is detected by three test vectors, namely 1100110, 0110110, and 0001011. Given the application of vectors 1100110 and 0110110, M_3 will receive input vector 110 in the fault-free case. If vector 0001011 is applied, M_3 will receive input vector 011. In all these cases, the expected fault-free logic value at the output O is logic 1. However, in the presence of I SA0, output O becomes logic 0, thus detecting the fault.

Figure 18c shows that M_1 receives input vectors 000, 001, 011, 100, and 110. Even though these vectors from a complete SSF test set for a majority gate, they are not a complete test set for detecting all possible defects. In fact, five defects (three misalignments and two displacements on the QCA cells in M_1) cannot be detected by these inputs vectors. If we add vector 0100110 to our original test set, then M_1 will also receive input vector 010. Note that the effect of the last four of the preceding five vectors is also propagated from the output of M_1 to output O. This is now a complete test set for all defects in M_1. Gates M_2 and M_3 already receive inputs vectors, which form a complete defect test set and, hence, require no additional test vectors (the effect of these vectors is also propagated to output O).

Figure 18d shows a possible QCA layout for the circuit in Fig. 18a. Consider the QCA cell displaced from the binary wire of input D so that there exists a bridging fault between inputs C and D. In this case, C dominates D. Furthermore, it is unknown whether the bridging fault will result in $D = C$ or $D = C'$. To detect this defect, we need vectors to test for two of four possible conditions. Condition D SA0 with $C = 0$ or D SA1 with $C = 1$ must be tested and condition D SA1 with $C = 0$ or D SA0 with $C = 1$ must be tested. We will explain how we obtained these conditions later. The first and second conditions can be tested by vectors 0001011 and 0010011 that are already present in our test set. However, we currently have no vector that can test either of the latter two conditions. Therefore, additional test generation is required, and vectors 0000011 and 0011011 are derived as tests for the third and fourth conditions, respectively. Adding either of these two vectors is sufficient as we only need to satisfy either the third or fourth condition for fault detection. The bridging fault between C and D, when C is dominant and D is dominated, can now be completely tested for all defects.

SSF Test Set	100% Defect Coverage?	SSF Test Set	100% Defect Coverage?	SSF Test Set	100% Defect Coverage?
{001, 010, 011, 101}	✓	{001, 100, 101, 110}	✓	{010, 011, 100, 101}	✗ 3 Uncovered Defects
{001, 011, 100, 101}	✓	{010, 100, 101, 110}	✓	{001, 010, 011, 110}	✓
{001, 010, 101, 110}	✗ 1 Uncovered Defect	{001, 011, 100, 110}	✗ 5 Uncovered Defects	{010, 011, 100, 110}	✓

Fig. 19 Minimal SSF test sets for a majority gate. Some of these test sets, however, are not 100% defect test sets for a QCA majority gate

The preceding example illustrates that the SSF test set of a circuit cannot guarantee the detection of all defects in QCA majority gates. Therefore, test generation will be required to cover the defects not covered by the SSF test set. In addition, test generation will also be needed to cover bridging faults on QCA interconnects. For a majority gate, there are nine minimal SSF test sets that contain four test vectors each. Researchers applied these tests sets to a QCA majority gates and all the possible defects mentioned earlier were simulated. Figure 19 shows the results of this experiment. Of the nine minimal test sets, three tests were unable to detect all the defects.

Consider the majority gate in Fig. 20 that is embedded in a larger circuit. Suppose that after SSF test generation for the entire circuit, it is determined that M receives input vectors 010, 011, 100, and 101. According to Fig. 19, this is not a complete defect test set as three defects remain uncovered. We can make it complete by ensuring that M also receives either 001 or 110 as its input vector.

It is very important to test for defects on QCA interconnects because, as mentioned earlier, a significant number of cells are used as wires. As Fig. 17b–d shows, a displacement of the QCA cells on a wire will result in a bridging fault between the two wires. Using a tool called QCADesigner (to be discussed later), it can be verified that such defects can be modeled using the dominant bridging fault model. However, depending upon the displacement distance, the lower wire will assume either the upper wire's logic value or its complement.

Figure 20 shows the possible scenarios that can result if a bridging fault is present between two wires. The scenario that will occur depends on the displacement distance Δ of the defective cell or the number of cells that get displaced. It also shows the conditions that must be satisfied in order to detect the particular scenario. In the first scenario in Fig. 21a, the lower wire's logic value is equal to that of the upper wire, whereas in Fig. 21b, the lower wire's logic value is equal to the complement of that of the upper wire. If the first scenario occurs, then a vector is required to test for one of the two conditions shown in Fig. 21a. Similarly, if the second scenario occurs, a vector is required to test for either of the two conditions shown in Fig. 21b. In reality, it will not be known which scenario

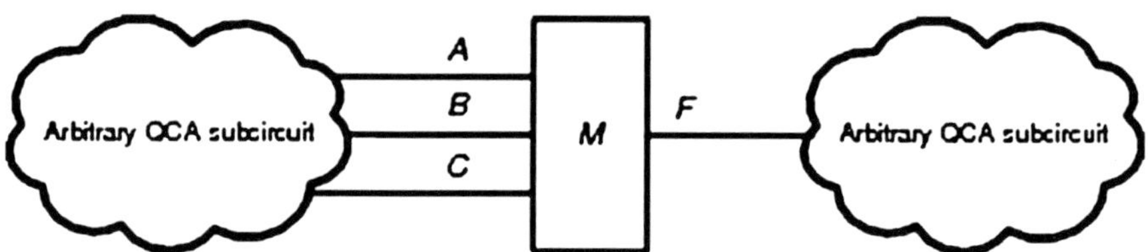

Fig. 20 Testing of defects in a QCA majority gate

a   Scenario 1: Δ is such that $B = A$

Fault-free $A\,B$	Faulty $A\,B$	Equivalent Condition
0 0	0 0	-
0 1	0 0	B SA0 with $A=0$
1 0	1 1	B SA1 with $A=1$
1 1	1 1	-

Note: Assumption is that B is dominated by A.

b   Scenario 2: Δ is such that $B = A'$

Fault-free $A\,B$	Faulty $A\,B$	Equivalent Condition
0 0	0 1	B SA1 with $A=0$
0 1	0 1	-
1 0	1 0	-
1 1	1 0	B SA0 with $A=1$

Note: Assumption is that B is dominated by A.

Fig. 21 Modeling of bridging faults in a QCA binary wire

occurred in the presence of the defect because Δ will not be known beforehand. Consequently, we need to have vectors that test for both scenarios to obtain a test that can completely detect this fault.

As an example, consider once again the majority gate in Fig. 20. Assume there is a bridging fault between inputs A and B and it is not known whether the defective cell is on wire A or wire B. In addition, Δ is unknown. To test for this fault, four conditions need to be satisfied (the first two result from A dominating B, and the last two result from B dominating A). The conditions are as follows:

$$B \text{ SA0 with } A = 0 \text{ or } B \text{ SA1 with } A = 1$$

$$B \text{ SA1 with } A = 0 \text{ or } B \text{ SA0 with } A = 1$$

$$A \text{ SA0 with } B = 0 \text{ or } A \text{ SA1 with } B = 1$$

$$A \text{ SA1 with } B = 0 \text{ or } A \text{ SA0 with } B = 1$$

If the test set contains vectors that can satisfy the preceding conditions, then this bridging fault can be detected. Otherwise, the fault is not completely testable.

Given a QCA circuit with n lines, there are at most $n(n-1)$ possible bridging faults involving two wires. This is based on the assumption that layout information is not available. Consequently, $2n(n-1)$ conditions will need to be satisfied in order to obtain a test set for all bridging faults. However, it should be noted that at least $n(n-1)$ (i.e. 50%) of these conditions will already be satisfied, given any complete SSF test set. This is because, given a pair of wires, when testing for a SA0/SA1 fault on one wire, the other line must have a value of 0 or 1. If the QCA layout of the circuit is available, then the designer can find out a priori which pairs of adjacent wires to test for bridging faults.

Because a majority network is also a threshold network, the testing methodologies developed for threshold networks that were presented in Chap. 12 are applicable. Although there are many similarities with testing of Boolean networks, traditional test generation approaches need to be augmented to take into account the different logic primitive (majority gate) as well as the inadequacy of the SSF model in the case of QCA.

5 CAD Tools for QCA Design

In this section, we briefly survey existing CAD tools for QCA design. Because QCA research is in its infancy, there are currently few tools available. With time, it is expected that this will change.

An important tool for layout and simulation for QCA circuits is QCADesigner [6] from the University of British Columbia. It allows the designer to rapidly explore different QCA layout topologies. Two simulation engines (coherence vector and bistable) are provided. However, simulation is slow as it solves quantum equations. MALS [7], as described earlier, is a tool for majority logic synthesis from Princeton University. It is integrated with SIS – a Boolean logic synthesis tool developed at University of California, Berkeley. MALS also has the ability to generate test vectors [8] based on the additional criteria described in the previous section. Finally, HDLQ provides Verilog models for basic QCA primitives that can be used to perform logic simulation of QCA circuits.

6 Fabrication Technology and Challenges

In this section, we discuss two of the several proposed fabrication technologies for physically implementing QCA. It is important to remember that these are currently in their experimental phase. While basic functionality of gates has been demonstrated, there remain several hurdles that must be overcome for each technology before it can challenge conventional CMOS.

6.1 *Metal-Dot QCA*

A scanning electron microscope (SEM) micrograph of a semiconductor-based QCA cell and its equivalent schematic diagram are shown in Fig. 22. The cell is fabricated using electron beam lithography (EBL) on an oxidized silicon wafer

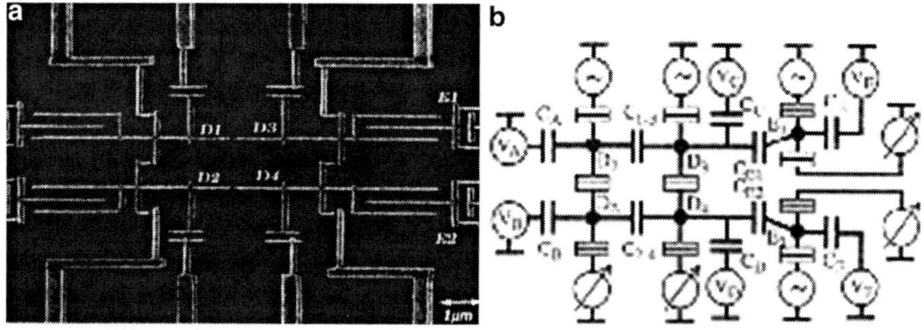

Fig. 22 (a) SEM micrograph and (b) schematic diagram of a four-dot metal QCA device (Source: Ref. [1])

and contains four aluminum islands (dots) connected with aluminum-oxide tunnel junctions. The area of the tunnel junction comprises the island capacitance which determines the island's charging energy. This indirectly controls the operating temperature of the device. The device in Fig. 22a has an area of 60×60 nm^2, and for characterization, is mounted on a surface at 10 mK ($-273.14°$C).

As shown in Fig. 22, the dots are positioned at locations D1–D4 and coupled in a circular fashion by tunnel junctions. The tunnel junction connects each dot in the cell with an electrode acting as a source or drain. E1 and E2 are single-electron transistor (SET) electrometers connected to D3 and D4, respectively, to sense the output.

The cell operation can be understood by biasing the cell so that an excess electron is on the verge of switching between D1 and D2, and another electron is on the verge of switching between D3 and D4. A differential voltage, V_1 and V_2 ($V_2 = -V_1$), is then applied while all other bias voltages are kept constant. As the differential voltage transitions from negative to positive, the electron in D1 moves to D2. This, in turn, forces the electron from D4 to move to D3. By using the electrometer signals E1 and E2, we can calculate the differential potential in the output half-cell, $V_{D3} - V_{D4}$, and plot it as a function of the input differential voltage. This is shown in the top panel in Fig. 23 where both the observed and theoretical results at 70 mK are demonstrated. Note that when $V_1/V_2 = -1/1$ mV, $V_{D3} - V_{D4}$ is positive, indicating the presence of an extra electron in D4 (remember, current is defined as positive electron flow using the conventional flow notation). At the other extreme, $V_1/V_2 = 1/-1$ mV, $V_{D3} - V_{D4}$ is negative indicating that V_{D3} is at a lower potential and, hence, contains the extra electron. The same idea is expressed in the second and third panel in Fig. 23 by calculating the charge on the dots [1].

Recall that for sequential QCA circuits, we need to use the six-dot clocked cell in Fig. 4. To understand the operation of this cell, we will first consider the operation of a half-cell or a QCA latch, as shown in Fig. 24. In the half cell, a modulated barrier is replaced with an extra dot, D_2, situated between two end dots, D_1 and D_3. Each half-cell is precharged with a single electron through a tunneling junction connected to ground. Each device dot is capacitively coupled to a corresponding gate and the differential input signal. The input signal may come from either external leads V_{in} or a predecessor latch, e.g. L_1 driving L_2 in Fig. 24. The clocking signal, V_{clk}, controls the voltage on D_2 which modulates the Coulomb barrier between D_1 and D_3. In this way, the clock can control the tunneling between the half-cell [2].

In the *null* state, V_{in} is zero and V_{clk} is such that an electron remains trapped at D_2. In the active state, a small V_{in} is first applied to the top and bottom gates. The magnitude is deliberately kept small to prevent the electron from tunneling out of D_2 into one of the end dots. If a negative clock signal with magnitude $V_{clk} > e/C_{clk}$ is applied, it is energetically unfavorable for the electron to remain at D_2. C_{clk} is the clock capacitance. In this case, there are two possibilities: (1) if no input bias (i.e. $V_{in} = 0$) is applied, the electron has equal likelihood of tunneling to either of the end dots, or (2) if an input bias is applied, the electron will transfer to the end dot to which a positive V_{in} is applied. That is, the electron will tunnel to D_1 if V_{in} is positive, and D_2 otherwise. For practical circuits, it should be clear that the first condition is undesirable.

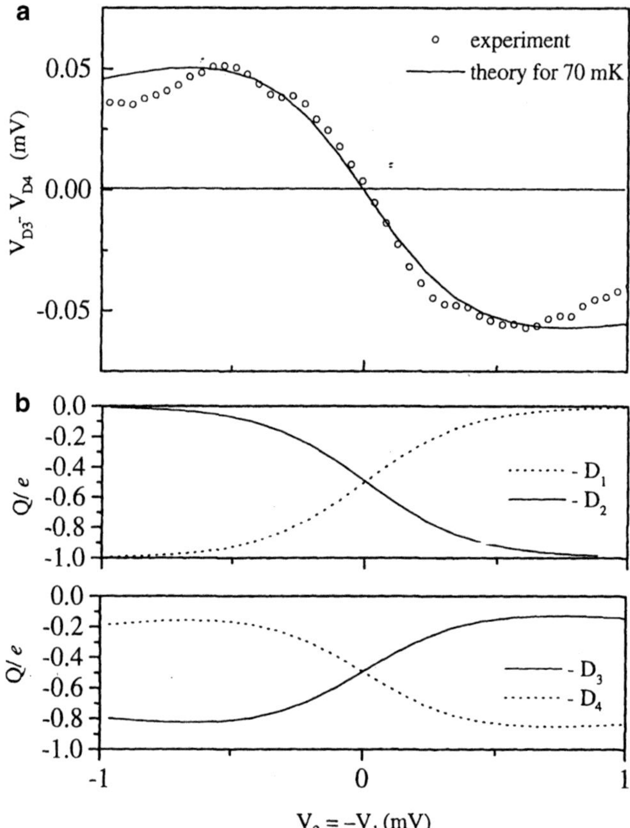

Fig. 23 Cell measurement data showing the differential potential on the right hand of the cell in Fig. 22b. The *bottom* two panels show the charge on the quantum dots (Source: Ref. [2])

The charge distribution between the dots reaches its minimal energy configuration as a result of the switching. In the locked state, the electron remains trapped at the location to which it was transferred in the active state. This is due to the presence of the clock bias which creates a Coulomb barrier to prevent the electron from switching. The ability of the cell to store an excess electron (hence, maintain "state") is the reason why it is called a QCA latch.

As a final and slightly more complex example, consider a QCA shift register. The SEM micrograph and schematic diagram for the fabricated device is shown in Fig. 25. The circuit comprises two QCA latches (L_1 and L_2) and two SET electrometers (E_1 and E_2) to measure the state of each latch. The latches are capacitively coupled to each other. Each latch contains three dots (D_1–D_3 for L_1, D_4–D_6 for L_2) separated by multiple tunneling junctions. A two-phase clock (CLK_1 and CLK_2) is also necessary.

The timing diagram for the shifter register's operation is shown in Fig. 26. Initially, V_{in} corresponding to logic 0 (1) is applied to V^+ and V^- at t_1 (t_7) (Fig. 26a). However,

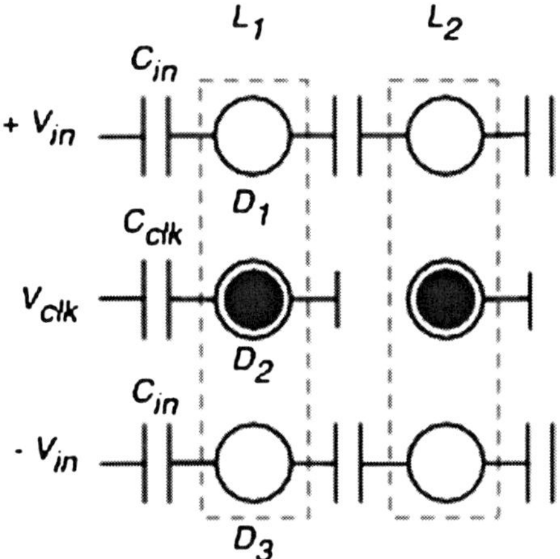

Fig. 24 Schematic diagram of a clocked QCA cell in the null state. The *dashed-line box* forms a QCA latch that is precharged with an extra electron. Two latches form a clocked QCA cell (Source: Ref. [2])

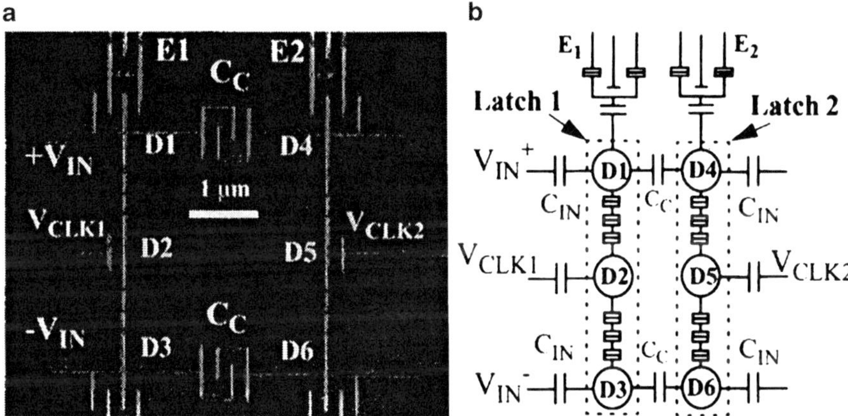

Fig. 25 (**a**) SEM micrograph and (**b**) schematic diagram of a two-stage QCA shift register (Source: Ref. [1])

L_1 remains in the null state because CLK_1 is low and does not go to logic 1 until t_2 (t_8) (Fig. 26b). At this time, an electron tunnels in L_1. Since V^+ is negative, the electron tunnels towards D_3 and, thus, V_{D1} is positive (Fig. 26c). After this, the input signal is removed at t_3 (t_9) and L_1's state no longer depends on the input. Next, CLK_2 is applied to L_2 at t_4 (t_{10}) (Fig. 26d) and an electron in L_2 tunnels in the direction determined by the state of L_1 (Fig. 26e). Since V_{D1} is positive (at t_4), the electron in L_2 tunnels towards D4 and, thus, V_{D4} is negative. At t_{10}, V_{D1} is negative and the electron in L_2 tunnels towards D_6, making V_{D4} positive. L_2 holds the bit after CLK_1 has been removed at t_5 (t_{11}), as long as CLK_2 is high (until t_6 (t_{12})). At this time (t_6), L_1 becomes inactive and is ready to receive the next bit. In this way, information continues to shift through the register every clock cycle [1].

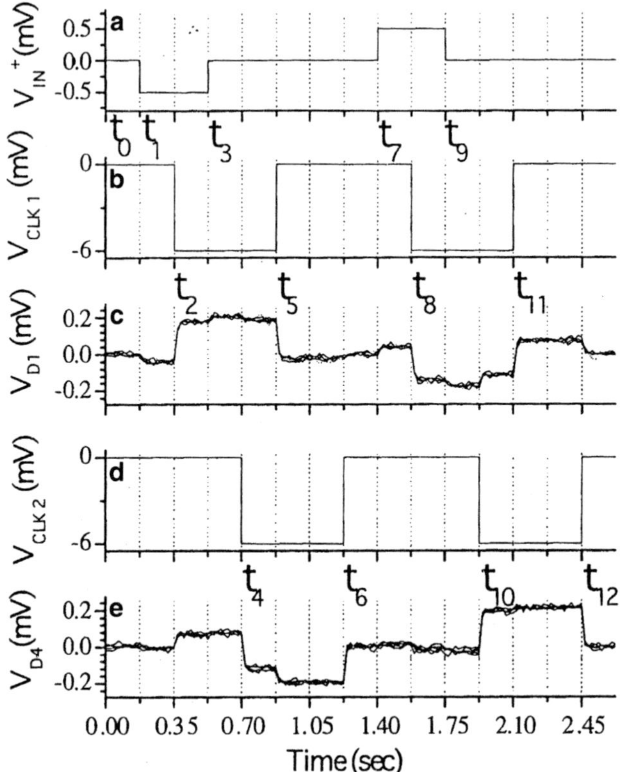

Fig. 26 Operation of a QCA shift register (Source: Ref. [1])

A semiconductor implementation for QCA is very advantageous because we understand the behavior of semiconductors very well and for which a significant number of tools, techniques, and infrastructure have been developed in the last five decades. Furthermore, it is easy to apply inputs and observe outputs. The examples that were presented used experimental cells with large dimensions (60 nm) operating at near absolute zero (70 mK) temperatures! However, they demonstrated proof-of-concept even though they stretched the idea of a quantum "dot." Researchers believe that it will be difficult for photolithography techniques to mass-produce QCA cells of small dimensions (few nanometers) that operate at room temperature.

6.2 Molelcular QCA

As an alternative to metal-dot QCA, molecular QCA provides several advantages. First, creating QCA cells from molecules allows us to realize the fundamental limits of miniaturization and achieve high density (up to 10^{13} devices/cm^2). Second, the small size of a molecule implies that Coulomb energies are much larger, making room temperature operation feasible. Third, QCA cells can be made homogeneous using chemical synthesis and self-assembly techniques. This is in stark contrast to photo-lithographic techniques, which introduce variations in device characteristics at the nanometer scale. Finally, improvements in switching speed by two orders of magnitude have been reported in the literature for molecular QCA over metal-dot QCA.

It is necessary to understand what desirable properties potential molecular QCA candidates must exhibit [1, 2]. They are as follows:

1. Parts of the molecule must be able to localize charge to form "dots."
2. The charge must be able to tunnel between dots to enable "switching."
3. The field from one molecule must be able to change the state of its neighboring molecule. Furthermore, the behavior should be highly non-linear with full state polarization (see Fig. 1b).
4. Controllability of molecule placement on a substrate is necessary to form circuits.
5. Clocked control is necessary while preserving the charge neutrality of the molecule and substrate.

Researchers believe that the first criterion can be accomplished by molecular redox (reduction–oxidation) centers with bridging ligands (a ligand that connects two or more molecules) between the redox centers providing the tunneling barriers for charge localization. The second requirement can be satisfied by chemical synthesis. The third condition is met by the choice of the molecule themselves. The fourth property could be accomplished with covalent bonds (i.e. the attraction-to-repulsion stability formed between atoms when sharing electrons). The final requirement can be met by having middle "null" dots within the molecule.

The first work done on fabricating molecular structures that exhibit QCA-like behavior was done at The University of Notre Dame [2]. Very recently, researchers at the University of Alberta have discovered that single silicon (Si) atoms on an ordinary silicon crystal serve as "quantum dots" [3]. The basic idea is demonstrated in Fig. 27 where scanning tunneling microscopy (STM) images of an H-terminated Si surface are shown. As shown in the inset, each Si atom at the surface shares in a surface-parallel Si–Si dimer bond (blue) and is capped by a single H atom (white). The diagonal bar-like features in Fig. 27 are rows of Si dimers. Figure 27a shows many Si atom dangling bonds (DB) that do not have an H atom. The low n-type dopant concentration of the substrate makes these DBs neutral, corresponding to one electron in the DB state. On the other hand, Fig. 27b shows DBs on a highly n-type doped H-terminated Si substrate. Due to high Fermi levels, each DB contains two electrons, making them negatively charged. Note that there is a dark halo around these DBs.

Figure 27c shows the interactions between two DBs as a result of their separation distance. If the distance between the DBs is large, they both appear as individual DBs, as shown at location "I." However, as the separation is decreased (location "II" and "III"), the DBs show an inversely proportional distance-dependent brightness. This is a signature of tunnel-coupled DBs! What essentially happens

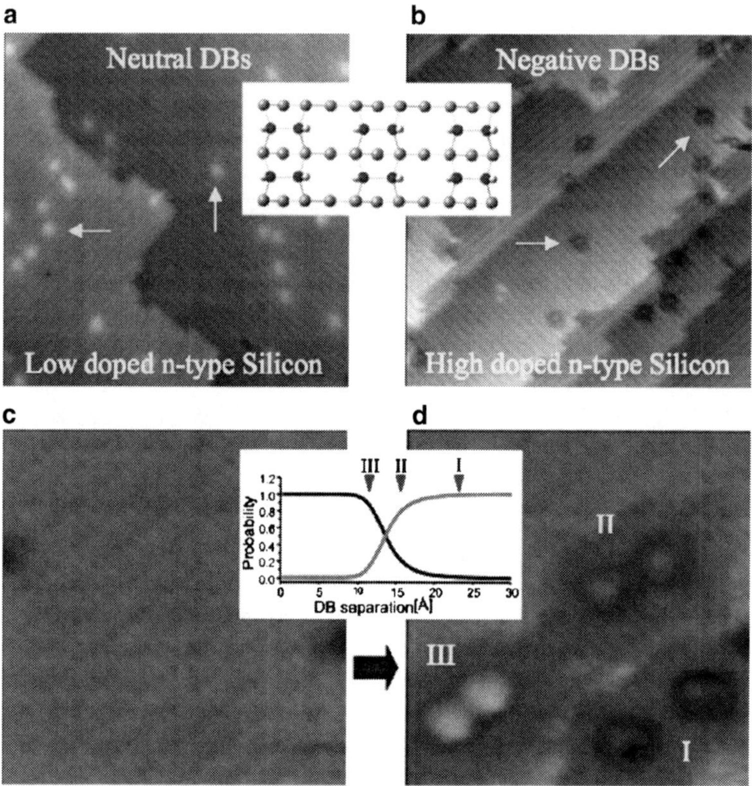

Fig. 27 STM images of (**a**) low *n*-type doped Si where DBs appear bright, (**b**) high *n*-type doped Si were DBs appear dark, (**c**) interactions between DB (non-coupled DBs at 2.32 nm (I), coupled DBs at 1.56 nm (II) and 1.15 nm (III)), and (**d**) one- and two-electron occupation probabilities corresponding to the DB pairs in (**c**) (Source: Ref. [3], © 2009 by the American Physical Society)

is that the occupation state with two electrons per DB becomes destabilized due to Coulombic repulsion. If one electron is removed from the DB pair and put into the bulk energy level, the pair becomes more stable. This, in turn, creates a partially empty state in the DB pair and provides a destination for a tunneling electron (i.e. tunnel-coupling). This reasoning is demonstrated in Fig. 27d where the probability of a DB containing one or two electrons is calculated as a function of the separation distance. It can be observed that if the distance is less than 15 Å, then the probability a DB contains a single electron is close to unity. If the distance is greater than 15 Å, the DB contains two electrons (and is not coupled: location "I" in Fig. 27c).

This concept can be extended to a four-DB system shown in Fig. 28. If isolated, the two extra electrons will be close to being equally shared between the four DBs. However, if two perturbing DBs are placed diagonally, they affect charge localization. As can be seen in Fig. 28, the two DBs closest to the perturbing DBs

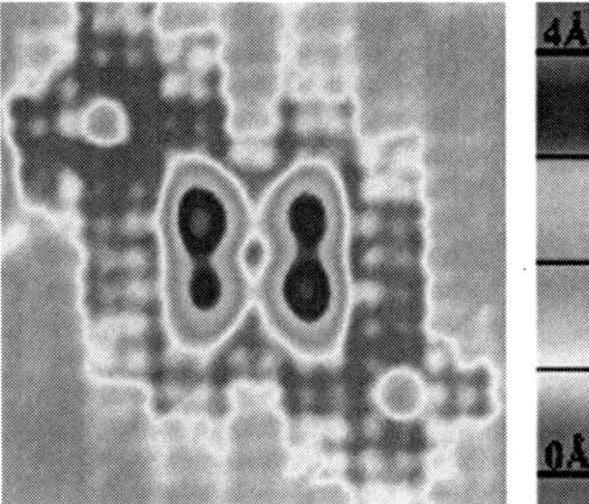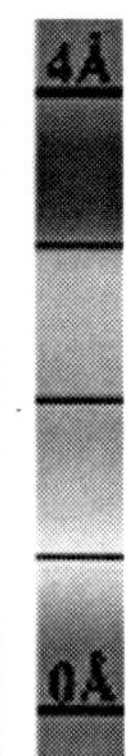

Fig. 28 STM image of a four-DB system under the influence of two perturbing DBs (Source: Ref. [3], © 2009 by the American Physical Society)

appear brighter than the other two DBs – indicating that less negative charge is localized there. This figure demonstrates the setting of a "state" at room temperature! Note that the structure is very similar to the QCA cell of Fig. 1a. Of course, the next logical step would be to demonstrate the dynamic setting of such states. The development of such gates is ongoing research.

While molecular QCA holds significant promise, there are some unique challenges that must be overcome. It remains to be seen if hundreds of, if not more, atoms can be precisely synthesized and oriented to make QCA computation feasible. Dynamic state setting of the molecules under the influence of a reference signal (i.e. clock) is a must if any worthwhile circuits are to be realized. Furthermore, another major hurdle that must be circumvented is the interfacing or integration of molecular QCA to the external (and possibly non-molecular QCA) environment. Research in the material science and device community will continue to attempt to seek solutions to these issues.

7 Future of QCA

QCA is a promising emerging technology that can complement or serve as an alternative to CMOS. With molecular implementation reaching high densities and allowing for room-temperature operation, it is necessary to develop design techniques and CAD tools for QCA. In this chapter, we have provided an introduction to the fundamentals of QCA, discussed logical design and testing of QCA circuits, and surveyed possible fabrication approaches.

While research in QCA will flourish, as with any emerging technology, QCA faces significant challenges for widespread adoption. There are a multitude of open research problems. Novel manufacturing techniques are needed so that large-scale QCA circuits can be developed reliably, yet economically. At sub-lithographic pitches, defects rates are expected to be very high. Thus, significant effort needs to be placed in investigating defect-free and fault-tolerant circuits and architectures. How QCA will interface or communicate with the external environment is an open and difficult problem that needs to be addressed. Finally, for widespread usage, powerful CAD tools need to be developed for front-end and back-end design for QCA. Currently, there are no automatic placement and routing tools for QCA. Essentially, an entire design automation flow needs to be developed. Nevertheless, it is an exciting time to be doing research in emerging nanoscale technologies and attempting to investigate solutions for the aforementioned problems.

8 Online Educational Links

Please visit the following links to learn more about QCA:

- QCA Java demos: http://www.virlab.virginia.edu/VL/contents.htm
- QCADesigner: http://ww.qcadesigner.ca
- MALS: http://nanohub.org/resources/3353
- HDLQ: http://www.ece.neu.edu/~mottavi/hdlq
- Atomic quantum dots: http://www.phys.ualberta.ca/~wolkow/news

Exercises

1. Prove that a majority gate and an inverter are functionally complete.

2. Show that the majority logic equations of a 1-bit full adder are equivalent to the Boolean logic equations.

3. Use QCADesigner to layout and simulate the following circuits:
   (a) 2-input XOR/XNOR
   (b) 4:1 decoder
   (c) 2:1 and 4:1 multiplexers

4. Design the following circuits using the QCA tile-based concept:
   (a) 4:1 decoder
   (b) 2:1 multiplexer

5. Generate a complete defect test set for a 4:1 multiplexer, assuming it is built from three 2:1 multiplexers. You do not need to consider bridging faults on the wires.

6. Demonstrate the K-map based majority synthesis method on $f = x_1 x_2 + x_1 x_3 + x_1 x_4$.

7. Demonstrate the K-map based majority synthesis method on $f = x_1x_2 + x_1{}^{'}x_3 + x_2{}^{'}x_3{}^{'}$.

8. Consider the QCA circuit in Fig. 18d and assume there is a bridging fault between inputs E and F. It is unknown which wire is dominant. Δ is also unknown. Derive test vectors to completely detect all defects on this pair of wires.

9. Consider the 4:1 multiplexer shown in Fig. 6c. Draw a QCA timing diagram to show the operation of this circuit.

10. Find three recent scholarly articles on any aspect of QCA design. Prepare a 15-min presentation for your class.

References

1. A. O. Orlov, R. Kummamuru, J. Timler, C. S. Lent, and G. L. Snider, *Mesoscopic Tunneling Devices*. Research Signpost, ch. Experimental studies of quantum-dot cellular automata devices, 2004.
2. G. L. Snider, A. O. Orlov, I. Amlani, G. H. Bernsetin, C. S. Lent, J. L. Merz, and W. Porod, "Experimental demonstration of quantum-dot cellular automata," *Semicond. Sci. Technol.*, vol. 13, no. 8A, pp. A130–A134, Aug. 1998.
3. M. B. Haider, J. L. Pitters, G. A. DLabio, L. Livadaru, J. Mutus, and R. A. Wolkow, "Controlled coupling and occupation of silicon atomic quantum dots at room temperature," *Phys. Rev. Lett.*, vol. 102, no. 4, pp. 0468051–4, Jan. 2009.
4. M. Liu and C. S. Lent, "Reliability and defect tolerance in metallic quantum-dot cellular automata," *J. Electron. Test. Theory & Appl.*, vol. 23, no. 2–3, pp. 211–218, June 2007.
5. P. D. Tougaw and C. S. Lent, "Logical devices implemented using quantum cellular automata," *J. Appl. Phys.*, vol. 75, no. 3, pp. 1818–1825, Feb. 1994.
6. K. Walus, T. J. Dysart, G. A. Jullien, and R. A. Budiman, "QCADesigner: A rapid design and simulation tool for quantum-dot cellular automata," *IEEE Trans. Nanotech.*, vol. 3, no. 1, pp. 26–31, Mar. 2004.
7. R. Zhang, P. Gupta, and N. K. Jha, "Majority and minority networks synthesis with application to QCA, SET, and TPL based nanotechnologies," *IEEE Trans. Computer-Aided Des.*, vol. 26, no. 7, pp. 1233–1245, July 2007.
8. P. Gupta, N. K. Jha, and L. Lingappan, "A test generation framework for quantum cellular automata (QCA) circuits," *IEEE Trans. VLSI Syst.*, vol. 15, no. 1, pp. 24–36, Jan. 2007.
9. R. Zhang, K. Walus, W. Wang, and G. A. Jullien, "A method of majority logic reduction for quantum cellular automata," *IEEE Trans. Nanotech.*, vol. 3, no. 4, pp. 443–450, Dec. 2004.
10. M. B. Tahoori, J. Huang, M. Momenzadeh, and F. Lombardi, "Testing of quantum cellular automata," *IEEE Trans. Nanotech.*, vol. 3, no. 4, pp. 432–442, Dec. 2004.
11. D. A. Antonelli, D. Z. Chen, T. J. Dysart, X. S. Hu, A. B. Kahng, P. M. Kogge, R. C. Murphy, and M. T. Niemier, "Quantum-dot cellular automata (QCA) circuit partitioning: problem modeling and solutions," in *Proc. Design Automation Conf.*, 2004, pp. 363–368.
12. J. Huang, M. Momenzadeh, L. Schiano, M. Ottavi, and F. Lombardi, "Tile-based QCA design using majority-like logic primitives," *J. Emerg. Technol. Comput. Syst.*, vol. 1, no. 3, pp. 163–185, Oct. 2005.
13. D. Berzon and T. J. Fountain, "A memory design in QCAs using the SQUARES formalism," in *Proc. Great Lakes Symp. on VLSI*, June 1999, pp. 166–169.

Index

CPSIA information can be obtained at www.ICGtesting.com
Printed in the USA
238335LV00005B/71/P